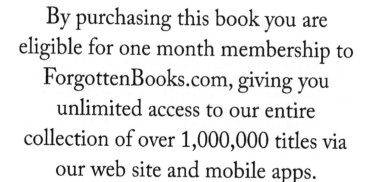

ISBN 978-0-365-41853-5
PIBN 10850415

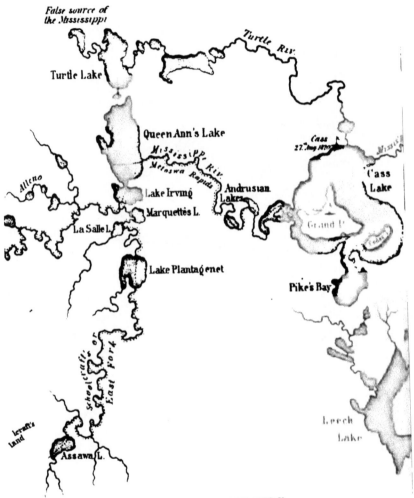

False source of the Mississippi

Turtle Riv.

Turtle Lake

Queen Ann's Lake

Cass 27.ᵗʰ Aug 1820

Cass Lake

Alleno

Mississippi Riv.

Metoswa Rapids

Lake Irving

Andrusian Lakes

La Salle L.

Marquettes L.

Grand L.

Lake Plantagenet

Pike's Bay

Schoolcraft or East Fork

leraft's land

Leech Lake

Assawa L.

SOURCES
of the
MISSISSIPPI RIVER,
Drawn to Illustrate
Schoolcraft's Discoveries.
By Capt. S Eastman U.S.A.

SUMMARY NARRATIVE

OF AN

EXPLORATORY EXPEDITION

TO THE

SOURCES OF THE MISSISSIPPI RIVER,

IN 1820:

RESUMED AND COMPLETED,

BY THE

DISCOVERY OF ITS ORIGIN IN ITASCA LAKE, IN 1832.

By Authority of the United States.

WITH APPENDICES,

COMPRISING THE

ORIGINAL REPORT ON THE COPPER MINES OF LAKE SUPERIOR, AND OBSERVATIONS ON THE GEOLOGY OF THE LAKE BASINS, AND THE SUMMIT OF THE MISSISSIPPI;

TOGETHER WITH

ALL THE OFFICIAL REPORTS AND SCIENTIFIC PAPERS OF BOTH EXPEDITIONS.

By HENRY R. SCHOOLCRAFT.

PHILADELPHIA:
LIPPINCOTT, GRAMBO, AND CO.
1855.

TO THE

HON. JOHN C. CALHOUN,

SECRETARY OF WAR.

SIR: Allow me to inscribe to you the following Journals, as an illustration of my several reports on the mineral geography of the regions visited by the recent Expedition under Governor Cass.

I beg you will consider it, not only as a proof of my anxiety to be serviceable in the station occupied, but also as a tribute of individual respect for those exertions which have been made, during your administration of the War Department, to develop the physical character and resources of all parts of our Western country; for the patronage it has extended to the cause of geographical science; for the protection it has afforded to a very extensive line of frontier settlements by stretching a cordon of military posts around them; and for the notice it has bestowed on one of the humblest cultivato natural science.

HENRY R. SCHOOLCR.

ALBANY, 1821.

riginal
ganiza-
rfectly
in iso-
he link
passed
Events
f our

PREFACE.

THE following pages embrace the substance of the narratives of two distinct expeditions for the discovery of the sources of the Mississippi River, under the authority of the United States. By connecting the incidents of discovery, and of the facts brought to light during a period of twelve years, unity is preserved in the prosecution of an object of considerable importance in the progress of our geography and natural history, at least, from the new impulse which they received after the treaty of Ghent.

Geographers deem that branch of a river as its true source which originates at the remotest distance from its mouth, and, agreeably to this definition, the combined narratives, to which attention is now called, show this celebrated stream to arise in Itasca Lake, the source of the Itasca River.

Owing to the time which has intervened since these expeditions were undertaken, a mere revision of the prior narrations, in the *journal form*, was deemed inexpedient. A concise summary has, therefore, been made, preserving whatever information it was thought important to be known or remembered, and omitting all matters not partaking of permanent interest.

To this summary, something has been added from the original manuscript journals in his possession. The domestic organization and social habits of the parties may thus be more perfectly understood. The sympathies which bind men together in isolated or trying scenes are sources of interest long after the link is severed, and the progress of science or discovery has passed beyond the particular points at which they then stood. Events pass with so much rapidity at present, in the diffusion of our population over regions where, but lately, the Indian was the only tenant, that we are in danger of having but a confused record of them, if not of losing it altogether. It is some abatement of this fear to know that there is always a portion of the community who take a pleasure in remembering individuals; who have either ventured their lives, or exerted their energies, to promote knowledge or advance discovery. It is in this manner that, however intent an age may be in the plans which engross it, the sober progress and attainments of the period are counted up. An important fact discovered in the physical geography or natural history of the country, if it be placed on record, remains a fact added to the permanent stores of information. A new plant, a crystal, an insect, or the humblest invertebrate object of the zoological chain, is as incontestable an addition to scientific knowledge, as the finding of remains to establish a new species of mastodon. They only differ in interest and importance.

It is not the province of every age to produce a Linnæus, a Buffon, or a Cuvier; but, such are the almost endless forms of vegetable and animal life and organization—from the infusoria upward—that not a year elapses which may not enlarge the boundaries of science. The record of discovery is perpetually accumulating, and filling the list of discoverers with humbler, yet worthy names. Whoever reads with care the scientific desiderata here offered will find matter of description or comment which has employed the pens of a Torrey, a Mitchell, a Cooper, a Lea, a Barnes, a Houghton, and a Nicollet.

It is from considerations of this nature, that the author has appended to this narrative the original observations, reports, and descriptions made by his companions or himself, while engaged in these exploratory journeys, together with the determinations made on such scientific objects as were referred to other competent hands. These investigations of the physical geography of the West, and the phenomena or resources of the country, constitute, indeed, by far the most important permanent acquisitions of the scrutiny devoted to them. They form the elements of classes of facts which will retain their value, to men of research, when the incidents of the explorations are forgotten, and its actors themselves have passed to their final account.

It would have been desirable that what has here been done

should have been done at an earlier period; but it may be sufficient to say that other objects engrossed the attention of the author
for no small part of the intervening period, and that he could not
earlier control the circumstances which the publication demanded.
After his permanent return from the West—where so many years
of his life passed—it was his first wish to accomplish a long-cher-
ished desire of visiting England and the Continent, in which
America, and its manners and institutions, might be contemplated
at a distance, and compared by ocular proofs. And, when he determined on the task of preparing this volume, and began to look
around for the companions of his travels, to avail himself of their
notes, he found most of them had descended to the tomb. For the
narrative parts, indeed, the manuscript journals, kept with great
fulness, were still preserved; but the materials for the other division
of the work were widely scattered. Some of them remained in the
archives of the public offices to which they were originally communicated. Other papers had been given to the pages of scientific
journals, and their reprint was inexpedient. The rich body of
topographical data, and the elaborately drawn map of this portion
of the United States, prepared by Captain Douglass, U. S. A.,
which would have been received with avidity at the time, had
been in a great measure superseded by subsequent discoveries.*
The only part of this officer's observations employed in this work,
are his determinations of the geographical positions. The latter
have been extended and perfected by the subsequent observations
of Mr. Nicollet. At every point, there have been difficulties to
overcome. He has been strenuous to award justice to his deceased
companions, to whose memory he is attached by the ties of sym-
pathy and former association. If more time has elapsed in preparing the work than was anticipated, it is owing to the nature
of it; and he can only say that still more time and attention would
be required to do justice to it.

* This remark is limited to the country south of about 46°. North of that point,
there are no explorations known to me, except those of Lieutenant James Allen,
who accompanied me above Cass Lake, in 1882, and those of J. N. Nicollet, in 1836,
which were reported by him to the Topographical Bureau, and by the latter transmitted to Congress.—Vide *Senate Doc.* No. 237, 1848. These observations relate to
the line of the Mississippi. Maj. Long's journey, in 1823, was *west* and *north* of
that river.

A word may be added respecting the period of these explorations. The year 1820 marked a time of much activity in geographical discovery in the United States. The treaty of Ghent, a few years before, had relieved the frontiers from a most sanguinary Indian war. This event enlarged the region for settlement, and created an intense desire for information respecting the new countries. Government had, indeed, at an earlier period, shown a disposition to aid and encourage discoveries. The feeling on this subject cannot be well understood, without allusion to the name of John Ledyard. This intrepid traveller had accompanied Captain Cook on his last voyage round the world. In 1786, he presented himself to Mr. Jefferson, the American minister at Paris, with a plan of extensive explorations. He proposed to set out from St. Petersburg, and, passing through Russia and Tartary to Behring's Straits, to traverse the north Pacific to Oregon, and thence cross the Rocky Mountains to the Missouri Valley.* Mr. Jefferson communicated the matter to the Russian plenipotentiary at Paris—and to the Baron Grimm, the confidential agent of the Empress Catherine—through whose influence he received the required passports. He proceeded on this adventure, and had reached within two hundred miles of Kamschatka, where he was arrested, and taken back, in a close carriage, to Moscow, and thence conducted to the frontiers of Poland. On reaching London, the African Association selected him to make explorations in the direction of the Niger. Reaching Egypt, he proceeded up the Nile to Cairo, where, having completed his preparations for entering the interior of Africa, he sickened and died, in the month of November, 1788.—*Life of Ledyard*, Sparks's *Amer. Biog.* vol. xvi.

The suggestion of Ledyard to explore Oregon became the germ of the voyages of Lewis and Clark. It appears that, in 1792, Mr. Jefferson proposed the subject to the American Philosophical Society at Philadelphia.* It is not known that its action resulted in anything practical. After Mr. Jefferson himself, however, came to the presidency, in 1801, he called the attention of Congress to the matter. Louisiana had been acquired, under his auspices, in 1803, which furnished a strong public reason for its exploration. To conduct it, he selected his private secretary and relative, Mer-

* Lewis and Clark.

riweather Lewis, of Virginia; Captain William Clark was named as his assistant. Both these gentlemen were commissioned in the army, and the expense thus placed on a public basis. Captain Lewis left the city of Washington, on this enterprise, on the 5th of July, 1803, and was joined by Captain Clark west of the Alleghanies. Having organized the expedition at St. Louis, they began the ascent of the Missouri River on the 14th of May, 1804. They wintered the first year at Fort Mandan, about 1,800 miles up the Missouri, in the country of the Mandans. Crossing the Rocky Mountains the next year, and descending the Columbia to the open shore of the Pacific, they retraced their general course to the waters of the Missouri, in 1806, and returned to St. Louis on the 23d of September of that year. (*Lewis and Clark*, vol. ii. p. 433.)

To explore the Missouri to its source, and leave the remote summits of the Mississippi untouched, would seem to have ill-accorded with Mr. Jefferson's conceptions. It does not appear, however, from published data, that he selected the person to perform the latter service, leaving it to the military commandant of the district. (*Life of Pike*, Sparks's *Amer. Biog.* vol. xv. pp. 220, 281.) General Wilkinson, who had been directed to occupy Louisiana, appears to have made the selection. He designated Lieutenant Zebulon Montgomery Pike. This officer left Bellefontaine, Missouri, on the 9th of August, 1805, with a total force of twenty men, at least four months too late in the season to reach even the central part of his destination, without an aid in the command, without a scientific observer of any description, and without even an interpreter to communicate with the Indians. That he should have accomplished what he did, is altogether owing to his activity, vigilance, and enterprise, his knowledge of hunting and forest life, and his well-established habits of mental and military discipline. Winter overtook him, on the 16th of October, in his ascent, when he was about one hundred and twenty miles (as now ascertained) above the Falls of St. Anthony.* Severe cold, snow, and ice, rendered it impossible to push his boats further. Devoting twelve days in erecting a blockhouse, and leaving his heavy stores and disabled men in charge of a non commissioned officer, he proceeded onwards, on snow shoes, with

* Estimated by him at 233 miles.

small hand-sledges, and, by great energy and perseverance, reached, at successive periods, Sandy Lake, Leach Lake, and Upper Red Cedar Lake, on the third great plateau at the sources of the Mississippi. On the opening of the river, he began his descent, and returned to his starting-point, at Bellefontaine, on the 30th of April, 1806, having been absent a little less than nine months. On his visiting the country above the point where the climate arrested his advance, the whole region was found to be clothed in a mantle of snow. On his journey, the deer, elk, buffalo, and wolf, were found on the prairies—the waters were inhabited by wild fowl; as he acted the part of hunter, and, to some extent, guide, these furnished abundant employ for his efficient sportsman-like propensities. Of its distinctive zoology, minerals, plants, and other physical desiderata, it was not in his power, had he been ever so well prepared, to make observations. Even for the topography, above the latitude of about 46°, he was dependent, essentially, on the information furnished by the factors of the Northwest British Fur Company, who, at that period, occupied the country.* This information was readily given, and enabled him, with general accuracy, to present the maps and descriptions which accompany his account of the region. He was, however, misled in placing the source of the river in Turtle Lake, and in the topography of the region south and west of that point.

Pike's account of his expedition did not issue from the press till 1810. The narrative of the expedition of Lewis and Clark was still longer delayed—owing to the melancholy death of Lewis—and was not given till 1814; a period of political commotion by no means favorable to literary matters. It was, however, at once hailed as a valuable and standard accession to

---

* The surrender of the lake country by Great Britian, in 1796, at the close of what is known as General Wayne's war, extended to Michilimackinac, the remotest British garrison. The region northwest of this post was occupied by numerous tribes of Indians, who continued to be supplied with goods by British traders till after the close of the war of 1812. In 1816, Congress passed an act confining the trade to American citizens. Under this state of affairs, the Northwest Company of Montreal sold out their trading-posts and fixtures, northwest of Michilimackinac, to Mr. John Jacob Astor, of New York, who, from an account of one of his active factors, invested about $800,000 per annum in merchandise adapted to the Indian habits.

geographical science. Public opinion had for years been called to this daring enterprise.

Such was the state of geographical discovery in the United States in 1816. The war with Great Britain had had an exhausting effect upon the resources and fiscal condition of the country. But, owing to the information gained by the operation of armies in the ample area west of the Alleghanies, it opened a new world for enterprise in that quarter. The treaty of 1814 with Great Britain, which affirmed the original boundaries of 1783, by terminating, at the same time, the war and the fallacious hopes of sovereignty set up for the Indian tribes, truly opened the Mississippi Valley to settlement.

All eyes were turned to the general climate of the West, and its capacities of growth and expansion. The universal ardor which then arose and was spread, of its fertility, extent, and resources, has, from that era, filled the public mind, and fixed the liveliest hopes of the extension of the Union.

The accession of Mr. Monroe to the presidency, 4th March, 1817, formed the opening of this new epoch of industrial empire and progress in the West. This period brought into the administration a man of great grasp of intellect and energy of character in Mr. Calhoun. By placing the army in a series of self-sustaining posts on the frontiers, in advance of the settlements, he gave them efficient protection against the still feverish tribes, who hovered—feeble and dejected from the results of the war, but in broken, discordant, and hostile masses—around the long and still dangerous line of the frontiers, from Florida to Detroit and the Falls of St. Anthony. He encouraged every means of acquiring true information of its geography and resources. In 1819, the military line was extended to Council Bluffs, on the Missouri, and to the Falls of St. Anthony, on the Mississippi. Major S. H. Long, of the Topographical Engineers, was directed to ascend the Missouri, for the purpose of exploring the region west to the Rocky Mountains. During the same year, he approved a plan for exploring the sources of the Mississippi, submitted by General Cass, who occupied the northwestern frontiers.

The author having then returned from the exploration of the Ozark Highlands, and the mine country of Missouri and Arkansas,* received from Mr. Calhoun the appointment of geologist and mineralogist on this expedition; and having, at a subsequent period, been selected, as the leader of the expedition of 1832, to resume and complete the discoveries under the same authority, commenced in 1820, it is to the journals and notes kept on these separate occasions, that he is indebted for the data of the narratives and for the body of information now submitted.

* *Vide* Scenes and Adventures in the Semi-Alpine Region of the Ozark Mountains of Missouri and Arkansas, with a View of the Lead-Mines of Missouri. New York, 1819. Philadelphia: Lippincott, Grambo, and Co. 1 vol. 8vo. pp. 256. 1853.

WASHINGTON, D. C., October 24, 1854.

CONTENTS.

EXPEDITION OF 1820.

CHAPTER I.

Departure—Considerations on visiting the northern summits early in the season—Cross the Highlands of the Hudson—Incidents of the journey from Albany to Buffalo—Visit Niagara Falls—Their grandeur the effect of magnitude—Embark on board the steamer Walk-in-the-Water—Passage up Lake Erie—Reach Detroit

CHAPTER II.

Preparations for the expedition—Constitution of the party—Mode of travel in canoes—Embarkation, and incidents of the journey across the Lake, and up the River St. Clair—Head winds encountered on Lake Huron—Point aux Barques—Cross Saganaw Bay—Delays in ascending the Huron coast—Its geology and natural history—Reach Michilimackinac

CHAPTER III.

Description of Michilimackinac—Prominent scenery—Geology—Arched Rock—Sugarloaf Rock—History—Statistics—Mineralogy—Skull Cave—Manners—Its fish, agriculture, moral wants—Ingenious manufactures of the Indians—Fur trade—Etymology of the word—Antique bones disclosed in the interior of the island

CHAPTER IV.

Proceed down the north shore of Lake Huron to the entrance of the Straits of St. Mary's—Character of the shores, and incidents—Ascend the river to Sault Ste. Marie—Hostilities encountered there—Intrepidity of General Cass .

CHAPTER V.

CHAPTER VI.

CHAPTER VII.

CHAPTER VIII.

CHAPTER IX.

CHAPTER X.

CHAPTER XI.

CHAPTER XII.

CHAPTER XIII.

CHAPTER XIV.

CHAPTER XV.

CHAPTER XVI.

1

EXPEDITION OF 1832.

CHAPTER XXI.

CHAPTER XXII.

CHAPTER XXIII.

CHAPTER XXIV.

CHAPTER XXV.

CHAPTER XXVI.

CHAPTER XXVII.

APPENDIX NO. 1.

APPENDIX NO. 2.

INTRODUCTION.

CHARLEVOIX informs us that the discovery of the Mississippi River is due to father Marquette, a Jesuit missionary, who manifested the most unwearied enterprise in exploring the north-western regions of New France; and after laying the foundation of Michilimackinac, proceeded, in company with Sieur Joliet, up the Fox River of Green Bay, and, crossing the portage into the Wisconsin, first entered the Mississippi in 1673.

Robert de la Salle, to whom the merit of this discovery is generally attributed, embarked at Rochelle, on his first voyage of discovery, July 14, 1678; reached Quebec in September following, and, proceeding up the St. Lawrence, laid the foundation of Fort Niagara, in the country of the Iroquois, late in the fall of that year. In the following year, he passes up the Niagara River; estimates the height of the falls at six hundred feet; and proceeding through Lakes Erie, St. Clair, and Huron, reaches Michilimackinac in August. He then visits the Sault de St. Marie, and returning to Michilimackinac, continues his voyage to the south, with a view of striking the Mississippi River; passes into the lake of the Illinois; touches at Green Bay; and enters the River St. Joseph's, of Lake Michigan, where he builds a fort, in the country of the Miamies.

In December of the same year, he crosses the portage between the St. Joseph's and the Illinois; descends the latter to the lake, and builds a fort in the midst of the tribes of the Illinois, which he calls Crevecœur. Here he makes a stand; sends persons out to explore the Mississippi, traffics with the Indians, among all of whom he finds abundance of Indian corn; and returns to Fort Frontenac, on Lake Ontario, in 1680. He revisits Fort Crevecœur

2

late in the autumn of the following year, and finally descends the
Illinois, to its junction with the Mississippi, and thence to the
embouchure of the latter in the Gulf of Mexico, where he arrives
on the 7th of April, 1683, and calculates the latitude between 23°
and 24° north.

The Spaniards had previously sought in vain for the mouth of
this stream, and bestowed upon it, in anticipation, the name of
Del Rio Ascondido. La Salle now returns to Quebec, by way of
the Lakes, and from thence to France, where he is well received
by the king, who grants him an outfit of four ships, and two
hundred men, to enable him to continue his discoveries, and found
a colony in the newly discovered territories. He leaves Rochelle
in July, 1684, reaches the Bay of St. Louis, which is fifty leagues
south of the Mississippi, in the Gulf of Mexico, in February follow-.
ing, where he builds a fort, founds a settlement, and is finally
assassinated by one of his own party. The exertions of this enterprising individual, and the account which was published of his
discoveries by the Chevalier Tonti, who had accompanied him in
all his perilous expeditions, had a greater effect, in the French
capital, in producing a correct estimate of the extent, productions,
and importance of the Canadas, than all that had been done by
preceding tourists; and this may be considered as the true era,
when the eyes of politicians and divines, merchants and specula-
tors, were first strongly turned towards the boundless forests, the
sublime rivers and lakes, the populous Indian tribes, and the profitable commerce of New France.

Father Louis Hennepin was a missionary of the Francisan order
of Catholics, who accompanied La Salle on his first voyage from
France; and after the building of Fort Crevecœur, on the Illinois,
was dispatched in company with three French voyageurs to explore the Mississippi River.

They departed from Fort Crevecœur on the 29th of February,
1680, and dropping down the Illinois to its junction with the
Mississippi, followed the latter an indeterminate distance towards
the Gulf, not believed to be great, where they left some memorial
of their visit, and immediately commenced their return. When
they had proceeded up the Mississippi a hundred and fifty
leagues above the confluence of the Illinois, they were taken
prisoners by some Indian tribes, and carried towards its sources

nineteen days' journey into the territories of the Naudowessies and Issati, where they were detained in captivity three or four months, and then suffered to return. The account which Hennepin published of his travels and discoveries, served to throw some new light upon the topography, and the Indian tribes of the Canadas; and modern geography is indebted to him for the names which he bestowed upon the Falls of St. Anthony and the River St. Francis.

In 1703, the Baron La Hontan, an unfrocked monk, published, in London, his voyages to North America, the result of a residence of six years in the Canadas. La Hontan served as an officer in the French army, and first went out to Quebec in 1688. During the succeeding four years he was chiefly stationed at Chambly, Fort Frontenac, Niagara, St. Joseph, at the foot of Lake Huron, and the Sault de St. Marie.

He arrives at Michilimackinac in 1688, and there first hears of the assassination of La Salle. In 1689 he visits Green Bay, and passes through the Fox and Wisconsin Rivers into the Mississippi. So far, his work appears to be the result of actual observation, and is entitled to respect; but what he relates of Long River appears wholly incredible, and can only be regarded as some flight of the imagination, intended to gratify the public taste for travels, during an age when it had been highly excited by the extravagant accounts which had been published respecting the wealth, population, and advantages of Peru, Mexico, the English and Dutch colonies, New France, the Illinois, and various other parts of the New World.

To convey some idea of this part of the Baron's work, it will be sufficient to observe that after travelling ten days above the mouth of the Wisconsin, he arrives at the mouth of a large stream, which he calls Long River, and which he ascends eighty-four days successively, during which he meets with numerous tribes of savages, as the Eskoros, Essenapes, Pinnokas, Mozemleeks, &c. He is attended a part of the way by five or six hundred, as an escort; sees at one time two thousand savages upon the shore; and states the population of the Essenapes at 20,000 souls; but this tribe is still inferior to the Mozemleeks in numbers, in arts, and in every other prerequisite for a great people. "The Mozemleek nation," he observes, "is numerous and puissant. The four slaves of that country informed me that, at the distance of 150 leagues from the place where I then was, their principal river empties itself into a salt lake of three hundred leagues in circumference, the mouth of which is about two leagues broad; that the lower part of that river is adorned with six noble cities, surrounded with stone, cemented with fat earth; that the houses of these cities have no roofs, but are open above like a platform; that, besides the above-mentioned cities, there are an hundred towns, great and small, round that sort of sea; that the people of that country make stuffs, copper axes, and several other manufactures, &c."

In 1721, P. De Charlevoix, the historian of New France, was commissioned by the French Government to make a tour of ob-servation through the Canadas, and in addition to his topographical and historical account of New France, published a journal of his voyage through the Lakes. He was one of the most learned divines of his age, and although strongly tinctured with the doctrines of fatality, and disposed to view everything relative to the Indian tribes with the over-zealous eye of a Catholic missionary, yet his works bear the impress of a strong and well-cultivated mind, and abound in philosophical reflections, enlarged views, and accurate deductions; and, notwithstanding the lapse of a century, he must still be regarded as the most polished and illustrious traveller of the region. He first landed at Quebec in the spring of 1721, and immediately proceeded up the St. Lawrence to Fort Frontenac and Niagara, where he corrects the error into which those who pre-ceded him had fallen, with respect to the height of the cataract. He proceeds through Lakes Erie, Huron, and Michigan, descends the Illinois and Mississippi to New Orleans, then recently settled, and embarks for France. The period of his visit was that, when the Mississippi Scheme was in the height of experiment, and excited the liveliest interest in the French metropolis; people were then engaged, in Louisiana, in exploring every part of the country, under the delusive hope of finding rich mines of gold and silver; and the remarks he makes upon the probability of a failure, were shortly justified by the event.

In 1760, Alexander Henry, Esq. visited the upper lakes, in the character of a trader, and devoted sixteen years to travelling over different parts of the north-western region of the Canadas and the United States. The result of his observations upon the topography, Indian tribes, and natural history of the country, was first published in 1809, and, as a volume of travels and adventures, is a valuable acquisition to our means of information. This work abounds in just and sensible reflections upon scenes, situations, and objects of the most interesting kind, and is written in a style of the most charming perspicuity and simplicity. He was the first English traveller of the region.

The date of Carver's travels over those regions is 1766. Carver, whose travels have been treated with too indiscriminate censure, was descended from an ancient and respectable English family in Connecticut, and had served as a captain in the provincial army, which was disbanded after the treaty of peace of Versailles, of 1763, and united to great personal courage a persevering and observing mind. By his bravery and admirable conduct among the powerful tribes of Sioux and Chippewas, he obtained a high standing among them; and, after being constituted a chief by the former, received from them a large grant of land, which was not, however, ratified by the British government. The fate of this enterprising traveller cannot but excite regret. After having escaped the massacre of Fort William Henry, on the banks of Lake George, in 1757, and the perils of a long journey through the American wilderness, he was spared to endure miseries in the heart of the British metropolis, which he had never encountered in the huts of the American savages, and perished of want in the city of London, the seat of literature and opulence!

Between the years 1769 and 1772, Samuel Hearne performed a journey from Prince of Wales's Fort, in Hudson's Bay, to the Coppermine River of the Arctic Ocean. McKenkie's voyages to the Frozen and Pacific Oceans were performed in 1789 and 1793. Pike ascended the Mississippi in 1805 and 1806.

Such is a brief outline of the progress of discovery in the north-western regions of the United States, by which our sources of information have been from time to time augmented, and additional light cast upon the interesting history of our Indian tribes—their numbers and condition, and other particulars connected with the regions they inhabit. Still, it cannot be denied that, amidst much sound and useful information, there has been mingled no inconsiderable proportion that is deceptive, hypothetical, or false; and, upon the whole, that the progress of infor-

mation has not kept pace with the increased importance which
that section of the Union has latterly assumed—with the great
improvements of society—and with the spirit and the enterprise
of the times. A new era has dawned in the moral history of our
country, and, no longer satisfied with mere geographical outlines
and boundaries, its physical productions, its antiquities, and the
numerous other traits which it presents for scientific research,
already attract the attention of a great proportion of the reading
community; and it is eagerly inquired of various sections of it—
whose trade, whose agriculture, and whose population have been
long known—what are its indigenous plants, its zoology, its
geology, its mineralogy, &c. Of no part of it, however, has the
paucity of information upon these, and upon other and more
familiar subjects, been so great, as of the extreme north-western
regions of the Union, of the great chain of lakes, and of the
sources of the Mississippi River, which have continued to be the
subject of dispute between geographical writers.

Impressed with the importance of these facts, Governor Cass,
of Michigan, projected, in the fall of 1819, an expedition for exploring the regions in question, and presented a memorial to the
Secretary of War upon the subject, in which he proposed leaving
Detroit the ensuing spring, in Indian canoes, as being best adapted
to the navigation of the shallow waters of the upper country, and
to the numerous portages which it is necessary to make from
stream to stream.

The specific objects of this journey were to obtain a more correct knowledge of the names, numbers, customs, history, condition, mode of subsistence, and dispositions of the Indian tribes;
to survey the topography of the country, and collect the materials for an accurate map; to locate the site and purchase the
ground for a garrison at the foot of Lake Superior; to investigate
the subject of the north-western copper mines, lead mines, and
gypsum quarries, and to purchase from the Indian tribes such
tracts as might be necessary to secure to the United States the
ultimate advantages to be derived from them. To accomplish
these objects, it was proposed to attach to the expedition a topographical engineer, an astronomer, a physician, and a mineralogist
and geologist, and some other scientific observers.

Mr. Calhoun not only approved of the proposed plan, but

determined to enable the governor to carry it into complete effect, by ordering an escort of soldiers, and enjoining it upon the commandants of the frontier garrisons, to furnish every aid that the exigencies of the party might require, either in men, boats, or supplies.

It is only necessary to add, that I was honored with the appointment of mineralogist and geologist to the expedition, in which capacity I kept the following journal. In presenting it to the public, it will not be deemed improper if I acknowledge the obligations which I have incurred in transcribing it, by availing myself of a free access to the valuable library of His Excellency De Witt Clinton, and of the taste and skill of Mr. Henry Inman, in drawing a number of the views which embellish the work.

HENRY B. SCHOOLCRAFT.

ALBANY, May 14, 1821.

PRELIMINARY DOCUMENTS.

PRELIMINARY DOCUMENTS.

I.

DETROIT, November 18, 1819.

SIR: The country upon the southern shore of Lake Superior, and upon the water communication between that Lake and the Mississippi, has been but little explored, and its natural features are imperfectly known. We have no correct topographical delineation of it, and the little information we possess relating to it has been derived from the reports of the Indian traders.

It has occurred to me that a tour through that country, with a view to examine the productions of its animal, vegetable, and mineral kingdoms, to explore its facilities for water communication, to delineate its natural objects, and to ascertain its present and future probable value, would not be uninteresting in itself, nor useless to the Government. Such an expedition would not be wholly unimportant in the public opinion, and would well accord with that zeal for inquiries of this nature which has recently marked the administration of the War Department.

But, however interesting such a tour might be in itself, or however important in its result, either in a political or geographical point of view, I should not have ventured to suggest the subject, nor to solicit your permission to carry it into effect, were it not, in other respects, intimately connected with the discharge of my official duties.

Mr. Woodbridge, the delegate from this Territory, at my request, takes charge of this letter, and he is so intimately acquainted with the subject, and every way so competent to enter into any explanations you may require, that I shall not be compelled to go as much into detail as, under other circumstances, might be necessary.

The route which I propose to take, is from here to Michilimackinac, and from thence, by the Straits of St. Mary's, to the river which contains the body of copper ore (specimens of which have been transmitted to the Government), and to the extremity of Lake Superior.

From that point, up the river which forms the water communication between that lake and the Mississippi, to the latter river, and, by the way of Prairie du Chien and Green Bay, to Lake Michigan.

The political objects which require attention upon this route are:—

1. A personal examination of the different Indian tribes who occupy the country; of their moral and social condition; of their feelings towards the United States; of their numerical strength; and of the various objects connected with them, of which humanity and sound policy require that the Government should possess an intimate knowledge. We are very little acquainted with these Indians, and I indulge the expectation that such a visit would be productive of beneficial effects.

The extract from the letter of Colonel Leavenworth, herewith inclosed, and the speech of the Winnebago Indians, transmitted to the War Department by Mr. Graham, from Rock Island, February 24, 1819, will show how much we have yet to learn respecting these tribes, which are comparatively near to us.

2. Another important object is, to procure the extinction of Indian titles to the land in the vicinity of the Straits of St. Mary's, Prairie du Chien, Green Bay, and upon the communication between the two latter places.

I will not trouble you with any observations respecting the necessity of procuring these cessions. They are the prominent points of the country—the avenues of communication by which alone it can be approached.

Two of them—Prairie du Chien and Green Bay—are occupied by a considerable population, and the Straits of St. Mary's by a few families. The undefined nature of their rights and duties, and the uncertain tenure by which they hold their lands, render it important that some step should be taken by the Government to relieve them. I think, too, that a cession of territory, with a view to immediate sale and settlement, would be highly important in the event of any difficulties with the Indians.

My experience at Indian treaties convinces me that reasonable cessions, upon proper terms, may at any time be procured. At the treaty recently concluded at Saginaw, the Indians were willing to cede the country in the vicinity of Michilimackinac, but I did not feel authorized to treat with them for it.

Upon this subject, I transmit extracts from the letters of Mr. Boyd and Colonel Bowyer, by which it will be seen that these gentlemen anticipate no difficulty in procuring these cessions.

3. Another important object is the examination of the body of copper in the vicinity of Lake Superior. As early as the year 1800, Mr. Tracy, then a senator from Connecticut, was dispatched to make a similar examination. He, however, proceeded no farther than Michilimackinac. Since then, several attempts have been made, which have proved abortive. The specimens of virgin copper which have been sent to the seat of Government have been procured by the Indians, or by the half-breeds, from a large mass, represented to weigh many tons, which has fallen from the brow of a hill.

I anticipate no difficulty in reaching the spot, and it may be highly important to the Government to divide this mass, and to transport it to the seaboard for naval purposes.

It is also important to examine the neighboring country, which is said to be rich in its mineral productions.

I should propose that the land in the vicinity of this river be purchased of the Indians. It could doubtless be done upon reasonable terms, and the United States could then cause a complete examination of it to be made.

Such a cession is not unimportant in another point of view. Some persons have already begun to indulge in speculations upon this subject. The place is remote, and the means of communicating with it are few. By timely presents to the Indians, illegal possessions might be gained, and much injury might be done, much time might elapse, and much difficulty be experienced, before such trespassers could be removed.

4. To ascertain the views of the Indians in the vicinity of Chicago, respecting the removal of the Six Nations to that district of country, an extract from the letter of Mr. Kenzie, sub-agent at Chicago, upon this subject, will show the situation in which this business stands.

5. To explain to the Indians the views of the Government
• respecting their intercourse with the British authorities at Malden,
and distinctly to announce to them that their visits must be dis-
continued.

It is probable that the annunciation of the new system which
you have directed to be pursued upon this subject, and the explanations connected with it, can be made with more effect by
me than by ordinary messengers.

6. To ascertain the state of the British fur trade within that
part of our jurisdiction. Our information upon this subject is
very limited, while its importance requires that it should be fully
known.

In addition to these objects, I think it very important to carry
the flag of the United States into those remote regions, where it
has never been borne by any person in a public station.

The means by which I propose to accomplish this tour are
simple and economical. All that will be required is an ordinary
birch canoe, and permission to employ a competent number of
Canadian boatmen. The whole expense will be confined within
narrow limits, and no appropriation will be necessary to defray
it. I only request permission to assign to this object a small
part of the sum apportioned for Indian expenditures at this place,
say from 1,000 to 1,500 dollars.

If, however, the Government should think that a small display
of force might be proper, an additional canoe, to be manned with
active soldiers, and commanded by an intelligent officer, would
not increase the expense, and would give greater effect to any
representations which might be made to the Indians.

An intelligent officer of engineers, to make a correct chart
for the information of the Government, would add to the value of
the expedition.

I am not competent to speculate upon the natural history of the
country through which we may pass. Should this object be deemed
important, I request that some person acquainted with zoology,
botany, and mineralogy may be sent to join me.

It is almost useless to add that I do not expect any compensation for my own services, except the ordinary allowance for nego-
tiating Indian treaties, should you think proper to direct any to
be held, and intrust the charge of them to me.

I request that you will communicate to me, as early as convenient, your determination upon this subject, as it will be necessary to prepare a canoe during the winter, to be ready to enter upon the tour as soon as the navigation of the Lakes is open, should you think proper to approve the plan.

Very respectfully, &c.

LEWIS CASS.

Hon. JOHN C. CALHOUN, *Secretary of War.*

II.

DEPARTMENT OF WAR, January 14, 1820.

SIR: I have received your letters of the 18th and 21st November last. The exploring tour you propose has the sanction of the Government, provided the expenditure can be made out of the sum allotted your superintendency for Indian affairs, adding thereto one thousand dollars for that special purpose.

The objects of this expedition are comprised under the five heads stated in your letter of the 18th of November, and which you will consider—with the exception of that part which relates to holding Indian treaties, upon which you will be fully instructed hereafter—as forming part of the instructions which may be given you by this Department.

Should your reconnoissance extend to the western extremity of Lake Superior, you will ascertain the practicability of a communication between the Bad, or Burntwood River, which empties into the Lake, and the Copper, or St. Croix, which empties into the Mississippi, and the facility they present for a communication with our posts on the St. Peter's.

The Montreal rivers will also claim your attention, with a view of establishing, through them, a communication between Green Bay and the west end of Lake Superior.

To aid you in the accomplishment of these important objects, some officers of Topographical Engineers will be ordered to join you. Perhaps Major Long, now here, will be directed to take that route to join the expedition which he commands up the Missouri. In that event, a person acquainted with zoology and botany will be selected to accompany him. Feeling, as I do, great interest in obtaining a correct topograpical, geographical, and military survey of our country, every encouragement, consistent with the means in my power, will be given by the Department. To this end, General Macomb will be ordered to afford you every facility you may require.

I have, &c.,

J. C. CALHOUN.

His Excellency, LEWIS CASS, Detroit, M. T.

III.

DEPARTMENT OF WAR, February 25, 1820.

SIR: Mr. Schoolcraft, a gentleman of science and observation, and particularly skilled in mineralogy, has applied to me to be permitted to accompany you on your exploring tour upon Lake Superior. I have directed him to report to you, for that duty, under the belief that he will be highly useful to you, as well as serviceable to the Government and the promotion of science.

You will furnish him with the necessary supplies and accommodation while employed, and every facility necessary to enable him to obtain a knowledge of the mineralogy of the country as far as practicable.

I have, &c.,

J. C. CALHOUN.

His Excellency, LEWIS CASS, Detroit.

IV.

DETROIT, March 10, 1820.

SIR: I have the honor to acknowledge the receipt of your letter of the 17th ult., inclosing a copy of a letter from Giles Sanford & Co.

Their statement with respect to the discovery of plaster of Paris upon one or more of the islands in the vicinity of Michilimackinac, to which the Indian title has not been extinguished, is correct. Specimens of this plaster have been brought here, and it is reported, by competent judges, to be of the best and purest kind. The quantity is stated to be inexhaustible, and, as vessels generally return empty, or nearly so, from the upper lakes, it could be transported to any part of Lake Erie at a trifling expense. ' I have great doubts, however, whether it would be proper for the Government to grant any permission to remove this plaster until the Indian title to the land is extinguished. The power of granting permission for that purpose is not given in the "act to regulate trade and intercourse with the Indian tribes, and to preserve peace on the frontiers," and appears, in fact, to be inconsistent with its general spirit and objects. To authorize these gentlemen to negotiate with the Indians for such a permission, is contrary to the settled policy which has always been pursued by the United States. I know of no case in which individuals have been or should be permitted to hold any councils with the Indians, except to procure the extinction of their title to lands, claimed under grants from one of the States. The application here must be to the tribe, because in all their land there is a community of interest, which cannot be severed or conveyed by the acts of individuals.

But, independent of precedent, there are strong objections to this course in principle. If private persons are authorized to open such negotiations for any object, the Government will find it very difficult to procure from the Indians any cession of land upon reasonable terms.

Were these islands the property of the United States, I think it would be very proper to permit the plaster upon them to be removed by every person making application for that purpose. The supply being inexhaustible, the agricultural interest would be greatly promoted by such a measure, and the dependence upon a foreign country for this important article would be removed.

I therefore take the liberty of recommending that a cession of these islands be procured by the United States from the Indians. I presume that this may be done without the payment of any annuity to them, and without any expense, except, perhaps, a few trifling presents. The plaster would then be at the disposal of Government, and its free distribution, under such regulations as might be adopted to prevent disputes between the adventurers, or a monopoly by any of them, would be equally proper and beneficial.

Very respectfully, sir,

I have the honor to be

Your most obedient servant,

LEWIS CASS.

Hon. JOHN C. CALHOUN, *Secretary of War.*

3

Extract of a letter from the Secretary of War to Governor Lewis Cass, dated

April 5, 1820.

Sir: I have received your letters of the 10th, 11th, and 17th ultimo. In relation to procuring cessions of land from the Indians, the Government has decided that it would be inexpedient to obtain any farther extinguishment of Indian title, except at the Sault de St. Marie, where it is the wish of the Department, that an inconsiderable cession, not exceeding ten miles square (unless strong reasons for a greater cession should present themselves from an actual inspection of the country), should be acquired upon the most reasonable terms, so as to comprehend the proposed military position there.

Herewith you will receive a plate of the country about the Sault de St. Marie, on which is indicated the military site intended to be occupied for defence. You will also procure the cession of the islands containing plaster, provided these islands are clearly within the boundary of the United States, and can be obtained without any considerable expense.

A commission, authorizing you to hold these treaties, will be forwarded to you in a few days.

As it is desirable to know by what title the people at Green Bay and Prairie du Chien hold their lands, and whether or not the Indian titles to those lands were extinguished by the French, at any period subsequent to their possession of the country (which is the impression of this Department), you will communicate such information as you possess, or may obtain, during your tour, on this subject.

In addition to Mr. Schoolcraft, Captain Douglass, of the engineer corps, has been ordered to join you, and Mr. Whitney (in whose behalf application has been made for that purpose) may accompany you, if you can accommodate him. Should he accompany you, he will be allowed the same compensation made to Mr. Schoolcraft, who will be allowed one dollar and fifty cents a day for the time actually employed.

VI.

(DIVISION ORDER.)

Major-General Macomb, commandant of the 5th military department, will, without delay, concentrate at Detroit the 5th regiment of Infantry, excepting the recruits otherwise directed by the general order herewith transmitted. As soon as the navigation of the Lakes will admit, he will cause the regiment to be transported to Fort Howard; from thence, by the way of the Fox and Wisconsin Rivers, to Prairie du Chien, and, after detaching a sufficient number of companies to garrison Forts Crawford and Armstrong, the remainder will proceed to the mouth of the River St. Peter's, where they will establish a post, at which the headquarters of the regiment will be located. The regiment, previous to its departure, will receive the necessary supplies of clothing, provisions, arms, and ammunition. Immediate application will be made to Brigadier-General Jesup, Quartermaster-General, for funds necessary to execute the movements required by this order.

By order of Major-General Brown.

(Signed)   JOHN E. WOOL,
Inspector- General.

VII.

(DEPARTMENT ORDER.)

The season having now arrived when the lakes may be navigated with safety, a detachment of the 5th regiment, to consist of Major Marston's and Captain Fowle's companies, under the command of Major Muhlenburg, will proceed to Green Bay. Surgeon's mate R. M. Byrne, of the 5th regiment, will accompany the detachment. The assistant deputy quartermaster-general will furnish the necessary transport, and will send by the same opportunity two hundred barrels of provisions, which he will draw from the contractor at this post. The provisions must be examined and inspected, and properly put up for transportation. Colonel Leavenworth will, without delay, prepare his regiment to move to the posts on the Mississippi, agreeably to the Division order of the 10th of February. The assistant deputy quarter-master-general will furnish the necessary transportation, to be ready by the first of May next. The Colonel will make requisition for such stores, ammunition, tools, and implements as may be required, and he be able to take with him on the expedition. Particular instructions will be given to the Colonel, explaining the objects of his expedition.

Mr. Melvin Dorr is appointed Inspector of Provisions, and he will inspect all provisions intended for the use of the army, before they are received and issued. Lieutenant Brooks, of the 3d regiment will forward, by the first detachment, such recruits as he has for the companies of the 3d regiment at Mackinac.

By order of MAJOR-GENERAL MACOMB.

(Signed)   CHESTER ROOT, *A. D. company, and*

Actg. Assist. Adjt.-General.

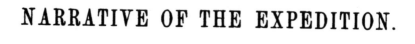

NARRATIVE OF THE EXPEDITION.

NARRATIVE OF THE EXPEDITION.

•

CHAPTER I.

Departure—Considerations on visiting the northern summits early in the season—Cross the Highlands of the Hudson—Incidents of the journey from Albany to Buffalo—Visit Niagara Falls—Their grandeur the effect of magnitude—Embark on board the steamer Walk-in-the-Water—Passage up Lake Erie—Reach Detroit.

THE determination to penetrate to the source of the Mississippi, during the summer months, made an early departure important. I had, while at Potosi, in Missouri, during the prior month of February, written to Hon. J. B. Thomas, U. S. S., Washington, to endeavor to secure an appointment to explore the mineralogy and natural features of the upper Mississippi River; and as soon as I had published my treatise on the mines and minerals of Missouri, I proceeded to Washington, and submitted to the proper officers of the Government, my account of the mineralogical wealth of the western domains, with a plan for the management of the public mines. Mr. Calhoun decidedly favored these views; but, foreseeing the necessity of congressional action on the subject, and the necessary delays of departmental references, said to me, that he had just received a memoir from Governor Cass, of Michigan, proposing an expedition to the source of the Mississippi, to leave Detroit early in the spring, and offered me the position of mineralogist and geologist on that service. This agreeing, as it did, with my prior views of exploring the public domains, I gladly accepted, and immediately returned to the city of New York to prepare for the journey.

. The year 1820 had commenced with severe weather, the Hudson being frozen hard, as high as West Point, on the 1st of January;

and there was a fall of snow between the 10th and 11th of Feb-
-ruary, which laid four feet deep in the streets of New York.
March opened with mildness, and every appearance denoted an
early spring, which led me to hasten my movement north. I left
New York on the 5th of March, in the citizens' post-coach, on
sleighs, for Albany, taking the route through Westchester, and
over the Highlands of Putnam and Dutchess; sleeping at Fishkill
and Kinderhook, the first and second nights, and reaching Albany
on the morning of the 7th, a distance of one hundred and
sixty miles. This distance we made in forty hours actual travelling, averaging four miles per hour, incidental stops included,
which is about the rate of travelling by the trekschuits of Holland,*
and by sledges over the frozen grounds of Russia.† In crossing
the Highlands, some one, in the change of the stage-sleighs,
pilfered a small box of choice minerals which I set store by; the
thief thinking, probably, from the weight and looks of the box,
which had been a banker's, that it was still filled with coin. We
crossed the Hudson from Greenbush, in a boat drawn through a
channel cut in the ice. Snow still laid in the streets of Albany,
and a cold north wind presaged a change of temperature. Next
day there was a hail-storm from the north-west, with rain and sleet,
and on the morning of the 9th, the hail lay six inches deep in
the streets. In the evening, proceeded by stage to the city of
Schenectady, a distance of sixteen miles, across the arenaceous
tract of the Pine Plains, by a turnpike, which forms the shorter
line of a triangle, made by the junction of the Mohawk with the
Hudson River. This tract is bounded southerly by the blue
summits of the Helderberg, a prominent spur of the Catskill
Mountain. At Schenectady, we experienced a night of severe
cold, and the next day, at an early hour, I took a seat in the stage-
sleigh for Utica, which we reached at seven in the evening. The
distance is ninety-six miles, which we passed in seventeen hours,
going an average rate of five miles per hour. The road lies up
the valley of the Mohawk, a name which recalls the history of
one of the most celebrated members of the Iroquois, a confederacy
of bold and indomitable tribes, who, at an early day, either pushed
their conquests or carried the terror of their arms from the St.
Lawrence to the Mississippi.

* Professor F. Hall.   † Clarke's Travels.

The winter was still unbroken, and the weather had assumed so unpropitious an aspect, since leaving New York, that there was no probability of the navigation of the lakes being open so as to embark at Buffalo before May. I proceeded seventeen miles west to my father's residence, in the village of Vernon, to await the development of milder weather. On the 10th of April, I resumed my journey, taking the western stage, which had left Utica at two o'clock in the morning. We lodged the first night at Skeneateles, at the foot of the beautiful and sylvan lake of the same name, and reached Geneva the next day, at one o'clock in the afternoon. The roads were now dry and dusty; indeed, the last traces of snow had been seen in sheltered positions, in passing through Oneida County, and every appearance in the Ontario country indicated a season ten days more advanced than the valley of the Mohawk. The field poplar put forth leaves on the 18th, and apricots were in bloom on the 22d.

At Geneva I remained until the 28th of April, when I again took my seat in the mail-stage, passing, in the course of the day, the lower margin of Canandaigua Lake, and through the attractive and tastefully laid-out village of the same name, and, after continuing the route through a most fertile country, with a constantly expanding vegetation, reached Avon, on the banks of the Genesee River. Here we slept. The next morning (the 29th), we crossed this noble stream, and, after a long and fatiguing day's staging, reached Buffalo in the evening. I was now at an estimated distance of two hundred and ten miles west of Utica, and three hundred and twenty-two from Albany. We had found the peach and apple-tree in blossom, and the vegetation generally in an advanced state, until reaching within eight or ten miles of Lake Erie, where the force of the winds, and the bodies of floating ice, evidently had the effect to retard vegetation. No vessel had yet ventured from the harbor, and although the steamer Walk-in-the-Water was advertised for the 1st of May, it was determined to delay her sailing until the 6th. This gave me time to visit Niagara*

---

* This is an Iroquois word, said to signify the thunder of waters. The word, as pronounced by the Senecas, is Oniágarah. For additional information on this subject, see *Notes on the Iroquois*, p. 453. The etymology of the word has not, however, been fully examined. It is clear the pronunciation of the word in Goldsmith's day was Niagára.

Falls, and some other places of historical interest in the neighbor-hood. This object I executed immediately, taking a horse and buggy, and keeping down the American shore. The distance is twenty-two miles, in which the Tonewanda River is crossed by a bridge. The day was clear and warm, with a light breeze blow-ing down the river. I stopped several times to listen for the sound of the Falls, but at the distance of fifteen, ten, eight, and even five miles, could not distinguish any; the course of the wind being, indeed, adverse to the transmission of sound, in that direction, until reaching within some two or three miles. There is nothing in the character of the country, in the approach from Buffalo, to apprise the visitor of the difference in its level and geological stratification, and thus prepare the mind to expect a cataract. It is different, I afterwards learned, in the approach from Lewiston, in which quite a mountain must first be ascended, when views are often had of the most striking parts of the gulf, which has been excavated by the passage of the Niagara River. It was not easy for me to erect standards of comparison for the eye to estimate heights. The ear is at first stunned by the inces-sant roar, and the eye bewildered by the general view. I spent two days at the place, and thus became familiarized with indi-vidual traits of the landscape. I found the abyss at the foot of the Falls to be the best spot for accomplishing that object. By far the greatest disproportion in the Falls exists between the height and great width of the falling sheet. The water is most thick and massy at the Horseshoe Fall, which gives one the most striking and vivid idea of creative power. In fitting positions in the gulf, with good incidences of light, the Falls look like a mighty torrent pouring down from the clouds. At the time of my visit, the wind drove immense fields of ice out of Lake Erie, with floating trees and other drift-wood, but I never saw any vestiges of these below the Falls. In front of the column of water falling on the American side, there stood an enormous pyramid of snow, or congealed spray.

What has been said by Goldsmith, and repeated by others, respecting the destructive influence of the Rapids above to ducks and water-fowl is imaginary—at least, as to the American sheet. So far from it, I saw the wild ducks swim down the Rapid, as if in pursuit of some article of food, and then rise and fly out at the brink, and repeat the descent, as if delighted with the gift of wings, which enabled them to sport over such frightful precipices without danger. I found among the debris in the abyss, pieces of hornstone, and crystals of calcareous spar, radiated quartz, sulphuret of zinc, and sulphate of lime. Its geology is best explained by observing that the river, in falling over the precipice of the Niagara ridge into the basin of Lake Ontario, leaps over horizontal strata of limestone, slate, and red sandstone. In this respect, nothing can be more simple and plain. It is magnitude alone that makes the cataract sublime.

On returning to Buffalo, I found the lake rapidly discharging its ice, which had been recently broken up by a storm of wind; and, while awaiting the motion of the steamer, I was joined by Captain D. B. Douglass, Professor of Engineering at West Point, who had been appointed topographer and astronomer of the expedition. We embarked on the 6th of May, at nine o'clock in the morning, in the steamer Walk-in-the-Water, an elegant and conveniently-planned vessel, with a low-pressure Fulton engine. This boat had been put upon the lake two years before, when it made a trip to Michilimackinac, and was, indeed, the initial boat in the history of steam navigation on the Lakes. We embarked at Black Rock, and it was necessary to use a tow-line, drawn by oxen on the shore, to enable the boat to ascend the Rapids. This Captain Rodgers, a gentlemanly man, facetiously termed his hornbreeze. The oxen were dismissed a short distance before reaching the mouth of Buffalo Creek, where we reached the level of Lake Erie, five hundred and sixty feet above the tide-waters of the Hudson River.* We were favored with clear weather, and, a part of the time, with a fair wind. The boat touched at Erie, at the mouth of Grand River, at Cleveland, and at Portland, in Sandusky Bay, on coming out of which we passed Cunningham Island, and the Put-in-Bay Islands, from a harbor in which Perry issued to achieve his memorable naval victory on the 10th of September, 1813. Passing through another group of islands, called the Three Sisters, we entered the mouth of the Detroit River late on the afternoon of the 8th, just as the light became dim and shadowy. The scale of these waters is magnificent.

* Report of the New York Canal Commissioners.

We had a glimpse of the town and fort of Malden, or Amherstburg, and of Boisblanc, and Gross Isle, which were the last objects distinctly seen in our ascent. The boat pushed on her way, under the guidance of good pilots, although the night was dark, and we reached our destination, and came to, at the city of Detroit, at twelve o'clock P. M., thus completing the passage in sixty-two hours.

The next morning, an official from the Executive of the Michigan Territory came on board with inquiries respecting Captain Douglass and myself, and we soon found ourselves in a circle where we were received with marked respect and attention. It was pleasing to behold that this respect arose, in a great degree, from the high interest which was manifested, in all classes, for the objects of the expedition, and the influence which its exploratory labors were expected to have on the development of the resources and prosperity of the country at large.

General Cass, who was to lead the expedition, received us cordially, and let us know that we were in season, as some days would still elapse before the preparations could be completed, and that the canoes in which we were to travel had not yet reached Detroit. We were also cordially welcomed by General Macomb, commanding the military district, Major John Biddle, commanding officer of the fort, and by the citizens generally. I was now, by the computations, about seven hundred and fifty miles from my starting-point at New York. We took up our lodgings at the old stone house occupied by Major Whipple, which, from its prominent position on the banks of the river, had sustained a random cannonade during the late war. We were here introduced to Dr. Alexander Wolcot, who filled the post of physician to the expedition, and to Lieutenant Eneas Mackey, United States artillery, commanding the escort, Major Robert A. Forsyth, private secretary of the Executive, and commissary of the expedition, and superintendent of embarkation; and to James D. Doty and Charles C. Trowbridge, Esqs., who occupied, respectively, the situations of official secretary and assistant topographer.

Detroit, the point to which I have now been conducted, is eligibly situated on the south bank of the straits of the same name, and enjoys the advantage of a regular plan and spacious streets, which have been introduced since the burning of the old

French town in 1805, not a building of which, within the walls, was saved. Its main street, Jefferson Avenue, is elevated about forty feet above the river. The town consists of about two hundred and fifty houses of all descriptions, public and private, and has a population of fourteen hundred and fifty,* exclusive of the garrison.

To the historian it is a point of great interest. It was the site of an Indian village called Teuchsagondie in 1620, the date of the landing of the Pilgrims at Plymouth. Quebec was founded in 1608; Albany in 1614. But no regular settlement or occupancy took place here, till the close of the seventeenth century. In June, 1687, the French took formal possession of the straits by erecting the arms of France. On the 24th of July, 1701, M. Cadillac established the first military post. Charlevoix, who landed here in 1721, found it the site of Fort Pontchartrain.

In 1763 the garrison, being then under British colors, sustained a notable siege from the confederate Indians under Pontiac. It remained under English rule till the close of the American Revolution, and was not finally surrendered to the United States until 1790, the year following Wayne's treaty at Greenville. Surrendered by Hull in 1812, it was reoccupied by General Harrison in October, 1813. It received a city charter 24th October, 1815. Indeed, the prominent civil and military events of which Detroit has been the theatre, confer on it a just celebrity, and it is gratifying to behold that to these events it adds the charm of a beautiful local site and fertile surrounding country. A cursory view of the map of the United States, will indicate its importance as a central military and commercial position. Situated on the great chain of lakes, connecting with the waters of the Ohio, Mississippi, St. Lawrence, Hudson, and Red River of the North, and communicating with the Atlantic at so many points, and with a harbor free of entrance at all times, its business capacities and means of expansion are very great. And when the natural channels of communication of the great lake chain shall be improved, it will afford a choice of markets between the most distant points of the Atlantic seaboard. It is thus destined to be to the regions of the north-west, what St. Louis is rapidly becoming to the south-

---

* The census of Detroit in 1850 gives it 21,019.

west, the seat of its commerce, the repository of its wealth, and the grand focus of its moral, political, and physical energies.*

* MICHIGAN. This Territory contained, at this period, a population of 8,896 inhabitants, principally Frenchmen, who were the descendants of the original settlers of the time of Louis XIV. In 1835, the population had so increased, chiefly by emigration from the older States, that the inhabitants applied for admission into the Union. The act of Congress admitting it was passed in 1836. In 1846, it had 212,267 souls. By the seventh national census, in 1850, it is shown to have a population of 397,654, entitling it to four representatives in Congress, with a large fraction. Its resources, its healthful climate, fertile soil, and very advantageous position on the great chain of navigable waters of the Upper Lakes, must insure a rapid development of its means and resources, and place the State, in a few years, in a high rank among the circle of American States.

CHAPTER II.

Preparations for the expedition—Constitution of the party—Mode of travel in canoes —Embarkation, and incidents of the journey across the Lake, and up the River St. Clair—Head winds encountered on Lake Huron—Point aux barques—Cross Saganaw Bay—Delays in ascending the Huron coast—Its geology and natural history—Reach Michilimackinac.

FROM the moment of our arrival at Detroit, we devoted ourselves, with intensity, to the preparation necessary for entering the wilderness. We were to travel, from this point, by a new mode of conveyance, namely, the Indian bark canoe, called a chimaun, a vehicle not less novel than curious. Constructed of large and thick sheets of the rind of the betula papyracea, or northern birch, which are cut in garment-like folds, and sewed together with the thin fibrous roots of the spruce, on a thin framework of cedar ribs, and having gunwales, with a sheathing of the same material, interposed between the bark and ribs. The seams are carefully gummed with the pitch of the pine. The largest of these canoes are thirty-six feet in length, and seven feet wide in the centre, tapering to a point each way. They carry a mast and sail, and are steered and propelled with light cedar paddles. They are at once light, so as to be readily carried over the portages, and so strong as to bear very considerable burdens. Those intended for us, were ordered from the Chippewas of Lake Huron, near Saganaw Bay. It was necessary to have mosquito-bars, portfolios, knapsacks, and various contrivances, and to make baggage of every sort assume the least possible bulk and space. The public armorer had orders to furnish me suitable hammers and other minerological apparatus for preparing and packing specimens. The expedition was quite an event in a remote town, and everybody seemed to take an interest in the preparation. A fortnight passed away in these preparations, and in awaiting the arrival of the canoes, respecting which there was some delay. It was the 24th of May before we were ready to embark. Besides the gentlemen mentioned as constituting the travelling party, ten Canadian *voyageurs* were taken to manage the canoes, ten United States soldiers to serve as an escort, and ten Ottowa, Chippewa, and Shawnee Indians to act as hunters, under the directions of James Riley, an Anglo-American, and Joseph Parks, a Shawnee captive (at present, head chief of the Shawnee nation), as interpreters. This canoe contained a chief called Kewaygooshkum, a sedate and respectable man, who, a year afterwards, played an important part at the treaty of Chicago.

The grand point of departure and leave-taking, was at Grose Point, at the foot of Lake St. Clair, a spot nine miles distant. For this point, horses and carriages, with the numerous friends of Gov. Cass, pushed forward at an early hour; and there was as much enthusiasm manifested, by all classes, as if a new world was about to be discovered. I had a strong wish to witness the mode of canoe travelling, and, declining an opportunity to join the cavalcade by land, took my seat beside Major Forsyth in the Governor's canoe. The Canadians immediately struck up one of their animating canoe songs, the military escort at the same moment displayed its flag and left the shore, and the auxiliary Indians, fired with the animation of the scene, handled their paddles briskly, and shot their canoe rapidly by us. A boat-race was the consequence. The Indians at first kept their advantage, but the firmer and more enduring nerves of the Canadians soon began to tell on our speed, and as we finally passed them, the Indians gracefully yielded the contest. We were two hours in going to Grose Point, with the wind slightly ahead.

The banks of the River Detroit present continuous settlements, in which the appearance of large old orchards and windmills, among farm-houses and smooth cultivated fields, reminds the visitor that the country has been long settled. And he will not be long in observing, by the peculiarity of architecture, dress, manners, and language, that the basis of the population is French. We found our land party had preceded us, and as the winds were adverse, we encamped in linen tents along the open shore. The next day the wind increased, blowing quite a gale down the Lake. I busied myself by making some meteorological and geological observations. The shores of Lake St. Clair are formed of a fertile alluvium, resting on drift. There are some heavy boulders of primitive rock resting on this, which denote a vast field of former drift action around the shores of these lakes.

The wind abated about eleven o'clock on the morning of the 26th, when the men commenced loading the canoes. It was twelve before we embarked. The mode of their embarkation, is peculiar. The canoes, when laden, are hauled out in deep water; the men then catch up the sitters on their backs, and deposit them in their respective seats; when this was done, they struck up one of their animated songs, and we glided over the smooth surface of the lake with rapidity, holding our course parallel with its shores, generally, until reaching a prominent point of land near Huron River.*

From Point Huron we crossed the lake, to reach the central mouth of the St. Clair River, thereby saving a tedious circuit; by the time we had half accomplished the transit, we encountered a head wind, which put the strength of the men severely to the test, and retarded our reaching the mouth of the river till dark. The River St. Clair has several mouths, which branch off above through a broad delta, creating large islands. These channels discharge a vast amount of argillaceous drift and mud, which has so far filled up the lake itself, that there is anchorage, I believe, in every part of it; and the principal ship channel is scooped, by the force of the current, out of a very compact blue clay—the geological residuum of ancient formations of clay-slates in the upper country.

The shores are often but a few inches *above*, and often a few inches *below* the surface, where they give origin to a growth of reeds, flags, and other aquatic plants, which remind the traveller of similar productions at the Balize of the Mississippi. In this nilotic region, myriads of water-fowls find a favorite resort. To us, however, these jets of alluvial formation, bearing high grass and rushes were as so many friendly arms stretched out to shelter us from the wind; but they were found to be so low and wet, that we were compelled to urge our way through them, in search of a

* Now called Clinton River, a change made by Act of Legislature, the frequent repetition of this name by the French having been found inconvenient in the lake geography. 1853.

4

dry encampment, till within two hours of midnight. This brought us to the upper end of Lawson's Island, where we arrived, wet, weary, and cold. We had advanced about twenty-five miles, having been ten hours, in a cramped posture, in our canoes. This initial day's journey was calculated to take away the poetry of travel from the amateurs of our party, and to let us all know, that there were toils in our way that required to be conquered.

We slept little this night, and waited for daylight and sunrise, as if the blessed luminary would have an animating effect upon our actual condition. We again embarked at seven o'clock in the morning. We now stowed away things with more handiness than at the first embarkation, and we began, ourselves, to feel a little more at home in this species of voyaging.

We had three canoes in our little squadron provided with masts and sails, and a small United States pennant to each, so that the brigade, when in motion, and led, as it usually was, by the chanting canoe-men, had a formidable and animated appearance.

The River St. Clair is a broad and noble stream, and impressed us as justifying the highest encomiums bestowed on it by Charlevoix, La Hontan, and other early French travellers. We ascended it thirty miles, which brought us to Fort Gratiot, at the foot of the rapid which marks the outlet of Lake Huron. In this distance, we passed, at separate places, nine vessels at anchor, being detained by head winds, and encountered several Chippewa and Ottowa canoes, each of which were generally occupied by a single family, with their females, blankets, guns, fishing apparatus, and dogs. They evinced the most friendly disposition.

In landing at Oak Point,* I observed a green snake (coluber æstivus) in the act of swallowing a frog, which he had succeeded in taking down, except the extremity of its hind legs. A blow was sufficient to relieve the frog, which still had sufficient animation to hop towards the river. The snake I made to pay the forfeit of his life.

At Fort Gratiot, we were received by Major Cummins, U. S. A., who occupied the post with sixty men. The expedition was received with a salute, which is due to the Governor of a Territory.

---

* Now the site of Algonac.

Two soldiers who were sickly, were here returned, and five able-bodied men received to supply their places, thus increasing the aggregate of the party to forty persons.*

The banks of the River St. Clair are wholly alluvial or diluvial. There is not a particle of rock in place. One idea presses itself prominently to notice, in reflecting on the formation of the country. It is the vast quantum of clay, mixed drift, and boulders, which have evidently been propelled, by ancient forces, down these straits, and afterwards arranged themselves according to affinities, or gravitation. At the precipitous banks between the inlet of Black River and Fort Gratiot, this action has been so clearly within the erratic block period of De la Buck, that it has imbedded prostrate forest-trees, and even freshwater shells, beneath the heavy stratum of sand, resting immediately upon the fundamental clay beds, upon which the city of Detroit, and indeed the alluvions of the entire straits rest.† We again encountered at this place, blocks of the primitive or crystalline boulders, which were first seen at Grosse Point. There are some traces of iron sand along the shore of this river, the only mineral body, indeed, which has thus rewarded my examinations.

We left our encampment, at Fort Gratiot, at eight o'clock next morning. A strong and deep rapid is immediately encountered, up which, however, vessels having a good wind find no difficulty in making their way. On surmounting this, we found ourselves on the level of Lake Huron. The lake here bursts upon the view in one of those magnificent landscapes which are peculiar to this region. Nature has everywhere operated on the grandest scale. Wide ocean expanses and long lines of shore spread before the eye, which gazes admiringly on the broad and often brilliant horizon, and then turns, for something to rest on, along the shore. Long ridges of gravel, sand, and boulders, meet it here. Beyond and above this storm-battered beach, are fringes of woods, or banks of clay. The monotony of travelling by un-varied scenes is relieved by an occasional song of the boatmen,

* To cover any arrangements of this kind, general orders had been issued by Gen. Macomb, to the commandants of the western posts.

† In the artesian borings for water, undertaken by Mr. Lucius Lyon, at Detroit, in 1833, these clay beds were found to be one hundred and fifteen feet deep.—Vide *Historical and Scientific Sketches of Michigan*, p. 177.

or an occasional landing—by changes of forest-trees—of the wind,
or flights of the gull, duck, plover, and other birds; but the traveller, is apt, before evening comes, to fancy himself very much in
the position of a piece of merchandise which is transported from
place to place. Glad were we when night approached, and the
order to encamp was heard. It was estimated we had advanced
thirty-five miles.

On passing along the Huron coast about fifteen miles, a bank
of dark clay is encountered, which has an elevation of thirty or
forty feet, and extends six or eight miles. We soon after came
to the White Rock—an enormous detached mass, or boulder of
transition,* or semi-crystalline limestone. It is a noted landmark
for *voyageurs* and travellers, and an equally celebrated place of
offerings by the Indians. I requested to be landed on it, and
detached some specimens. Geologically, it is a member of the
erratic block group, and we must look for its parent bed at a
more westerly point. There is no formation of limestone, in this
quarter, to which it can be referred. It bears marks of attrition,
which shows that it has been rubbed against other hard bodies;
and if transported down the lake on ice, it is necessary to consider these marks as pre-existing at the era of its removal.

On embarking in the morning, the wind was slightly ahead,
which continued during the forenoon, changing in the after-part
of the day, so that we were able to hoist sail. About four o'clock
the weather became cloudy and hazy, the wind increasing, at
the same time attended with thunder and lightning. A storm
was rapidly gathering, and the lake became so much agitated
that we immediately effected a landing, which was not done without some difficulty, on a shallow and dangerous shore, thickly
strewn with boulders. We pitched our tents on a small peninsula, or narrow neck of land, covered with beautiful forest-trees,
which was nearly separated from the main shore. Shortly after
our arrival a vessel hove in sight, and anchored on the same
dangerous lee shore. We were in momently expectation of her
being driven from her moorings, but were happily relieved, the
next morning, to observe that she had rode out the storm.

---

* This term has disappeared from the geological vocabulary under the researches
of Sir Roderick J. Murchison, Mr. Lyell, and other distinguished generalizers.

The lake was still too rough on the following day, and the wind too high, to permit our embarking. We made an excursion inland. The country proved low, undulatory, and swampy. The forest consisted of hemlock, birch, ash, oak, and maple, with several species of mosses, which gave it a cold, bleak character. The margin of the forest was skirted with the bulrush, briza canadensis, and other aquatic plants. The whole day passed, a night, and another day, with nothing but the loud sounding lake roar in our ears. A heavy bed of the erratic block formation commences at this point, and continues to Point aux Barques, the eastern cape of Saganaw Bay.

In one of these displaced masses—a boulder of mica slate, I discovered well-defined crystals of staurotide. This formed my second mineralogical acquisition.* There were, also, some striking water-worn masses of granitical and hornblende porphyry.

It was the 1st of June before we could leave the spot where we had been confined. We embarked at six o'clock, the lake being sufficiently pacific, though not yet settled. But after proceeding about a league, it again became agitated, and drove us ashore, where we lay without encamping. Kewaygushkum was requested to send some of his young men in quest of game. The soldiers and engagees also formed fishing parties, at a contiguous river; but about three o'clock in the afternoon all the parties returned completely unsuccessful. There was neither fish nor game to be had. At the same time the agitation of the lake ceased, the wind springing up from an opposite quarter, which enabled us to hoist sail. This put every one in a pleasant humor, and we proceeded along the coast till evening, and encamped on a small sandy bay, which puts into the land, immediately beyond the promontory of Point aux Barques—an estimated distance of twenty-five miles from our starting-point in the morning.

At the distance of a league before reaching this point, the first

* In passing along this coast in 1824, an Indian picked up, in shallow water, a small boulder imbedding a mass of native silver. Breaking off the most prominent mass, he still observed the metal forming veins in the rock, and brought both specimens to an officer of the British Indian department at Amherst (Lieut. Lewis S. Johnson), who presented them to me. This discovery is described in the *Annals of the New York Lyceum of Natural History*, vol. i. part 8, page 247.

stratum of rock, *in situ*, presents itself. It is a gray friable sand-stone, elevated from ten to twenty feet above the water, but attaining a greater height in the approach to this noted cape. This stratum of sandstone rock, which is of a perishable cha-racter, is exposed to receive the shock of the waves of Lake Huron for several hundred miles from the north and west. It exhibits the force and fury of the lake action by the numerous cavities which have been worn into it, at the water's edge, and by the sub-bays which have, in some localities, been formed in the line of dark opposing cliffs. It was in one of these sub-bays that we encamped, on a smooth sandy beach, which appears to have been a favorite encamping ground of the natives. But al-though we had met several canoes of Chippewas, on the route between Fort Gratiot and this point, none were found at the place of our encampment. Such of them as we approached, on the lake, were invariably in want of food, and received it with evident marks of gratification.

On going inland, back from our encampment, we found a suc-cession of arid ridges of sand, which had been evidently produced by the prostrated sandstone of the coast, which, after comminution by the waves, had been carried to this position by the winds. These ancient dunes and ridges were covered sparsely with pitch pines and aspen, and having their surfaces covered with the uva ursi, pyrola, and smaller shrub-growth common to arenaceous soils.

On the day following, we ascended along the eastern shores of Saganaw Bay, a distance of eighteen miles, which brought us to Point aux Chenes. At this place the guides pointed to a group of islands about midway of the bay, for which we steered. The calmness of the weather favored the traverse. We reached and landed on the largest of the group, called Shawangunk, by the Indians, probably from its southernmost position. I found it to consist of a dark, compact limestone, imbedding masses of chal-cedony and calcareous spar. I also picked up a detached mass of argillaceous oxide of iron, and some fragments of striped hornstone. Anxious to improve the favorable time for effecting the passage, we pushed on for the opposite western shore, which was safely reached. We then steered down the bay, skirting a low sandy shore some twenty miles or more, till entering the open lake, and reaching the River aux Sables. On entering this river, and after having pitched our camp, we were visited by a band of Chippewa Indians, with friendly salutations. It appeared that the arrival of the expedition had been anticipated by them, they having themselves constructed and furnished the canoes for it, and being well acquainted with the official position, at Detroit, of the leader of our party. The principal Chief, the Black Eagle, addressed a speech to Governor Cass, in which he appropriately recognized these relations, welcomed him to his village, and recommended the condition of his people to his notice. The calumet was then smoked in the usual style of Indian ceremony, the pipe-bearer beginning with persons of first rank, and handing it in the supposed order of grade, to the lowest member of the official family. The ceremony was ended by shaking of hands. All this was done with the ease and dignity of an oriental sheikh. We had anticipated savages, and savage manners, and armed ourselves to the teeth, pushing a point with an army official at Detroit, until we were each provided with a short rifle. But this first formal council with the sons of the forest, began to open our eyes to the true character of the Indian manners and diplomacy, in their intercourse with government officials.

The chiefs, after their departure, sent to our encampment a present of fresh sturgeon, a species which is caught abundantly in the aux Sables at this time, for which returns were made of such articles as were most acceptable to them. Being out of the Bay, we employed the following day making advances along the Huron coast, an estimated distance of forty-eight miles. In this distance, we passed Thunder Bay. Encamped on a low, calcareous shore, bearing cedar and spruce, which the Indians call Sho-she-ko-naw-be-ko-king, or Flat Rock Point. A few miles after leaving River aux Sables, the Highlands of Sables present themselves at a short distance back from the shore. This ridge, which is a landmark for mariners, runs from southeast to northwest, and is visible as far as Thunder Bay. The limestone, which is dark and of an earthy fracture, is very much broken up on the shore, and contains various species of organic remains. On crossing the Bay, we landed on an island covered with debris, where we observed one of those imitative, water-worn, primitive boulders, resembling altars, which are frequently set up by the Indians as the places of depositing some offering, or out of mere respect for some local god.

At six o'clock the next morning we were again in our canoes, assiduously moving along the Huron coast; but, after proceeding about a league, a storm of wind and rain suddenly arose, driving us from the lake. A few hours served to restore its calmness, but we had not gone over a couple of leagues when we were again compelled by the rising wind to take to the shore, where we were detained the rest of the day, listening to the capricious murmurs of the lake. This position was directly opposite Middle Island, a noted anchorage about six miles distant. All night the waves of the lake were heard. The morning broke without change. Lake Huron still evinced an angry aspect, threatening to renew the struggle of yesterday. It was concluded to send the canoes forward, relieved of our weight, and proceed ourselves on foot along the beach. Walking on this became difficult on those parts of it where the fossiliferous and shelly limestone had been broken up and heaped in small fragments. Among these, we recognized specimens of the cornu-ammonis, and the maderpore, with some other species. The cedars and brushy growth generally stood so thick, and grew so closely to this line of debris, that it was impracticable to take the woods. The toil, however, rewarded us with some specimens of the organic forms imbedded in the rock, while it enabled the topographers to secure the data for a very perfect map of the coast. At ten o'clock in the morning we reached the east cape of Presque Isle Bay, where the canoes came to take us across to the peninsula of that name. After completing this, the men landed the canoes and baggage on the peninsula side, and carried them across the narrow sandy neck of land; but, on reaching the open lake beyond it, the wind was found too strongly adverse to permit embarkation. The Canadians have the not inappropriate term of *degrade* for this species of detention; we were here foiled, indeed, in our high hopes of pushing ahead, and compelled to wait on the naked sands for many weary hours. While thus detained, the Indians brought in a brown rabbit,* a species of lake tortoise, and some pigeons, being their only fruits of success in hunting, except a single

* This is presumed to be a variety of the American Hare, and may be distinguished by the following characters: Body eighteen inches long; color of the hair grayish-brown on the back, grayish-white beneath. Neck and body rusty and cenerous. Legs pale rust color. Tail short, brown above, white beneath. Hind legs

grouse, or partridge, which had crowned their efforts since leaving Detroit. It must be borne in mind, however, that there has been very little opportunity for hunting, that we have had abundant supplies, and that our mode of travelling is such as to alarm all game within sound of our track. They have, indeed, brought reports at several points of seeing the footprints of the deer and black bear, but they have not had the leisure to pursue them.

At five o'clock, the wind abated so much as to permit embarkation, and our canoe-men hastened forward with the intention of travelling all night, but at eleven o'clock it freshened to such a degree, and at the same time became so intensely dark, that we were compelled to land and encamp. Neither the topography, mineralogy, or any branch of the physical geography of a country can be ascertained without minute examination; and this constitutes, indeed, the object of the investigations, which have been, thus far, so toilsomely pursued against adverse winds since the commencement of the expedition; but they have disclosed facts which reveal the true structure and physical history of this bleak, ungenial coast; this hope serves, every day, to give new impetus to the voyage.

Another day along the Huron coast. It was now the 6th of June. The *voyageurs* began now to manifest great anxiety to reach Michilimackinac, and had their canoes in the water at a very early hour. We all participated in this feeling, and saw with pleasure the long lines of sandy shores, strewed with boulders and pebbles, that were swiftly passed. We had traced about forty miles of the coast when we reached the foot of Bois Blanc Island, and pushed over the intervening arm of the lake to get its south or lee shore. This was a labor of hazard, as the wind was directly ahead, and drove the waves into the canoes. When accomplished, we had the shelter of this island for twelve miles, till reaching its southwest part. We then passed, due north, between it and Isle Ronde, which brought the wind again ahead. But the men had not kept this course long, when Michilimackinac, with its picturesque and imposing features, burst upon our view.

longest, and callous a short distance from the paws up. Ears tipped with black. Covering of the body rusty fur, beneath long coarse hair. Probable weight six pounds.

Nothing can present a more refreshing and inspiring landscape. From that moment the *voyageurs* appeared to disregard the wind. Striking into the water with bolder paddles, and opening one of their animating boat-songs, all thought of past toils was forgotten, and, urged forward with a new impetus, we entered the handsome little crescent-shaped harbor at four o'clock. The expedition was received with a salute from the fort, in command of Capt. B. K. Pierce, U. S. A.,* in compliment to the Governor of the Territory, and we landed amid the congratulations of the citizens, who pressed forward to welcome us.

Thus terminated the first part of our journey, after a tedious voyage of fourteen days, in which we had encountered a series of almost continued head-winds and foul weather. The distance by ship is usually estimated at three hundred miles; by following the indentations of the coast, and entering Saganaw Bay, we found it three hundred and sixty.† We found the Huron coast, to the line of which our observations were limited, bearing, in its vegetation, indubitable marks of its exposure to the northern winds. As a section of the lake geology, it is simple and instructive, exhibiting strata of sandstone and non-crystalline and fossiliferous limestone in horizontal positions, without the slightest disturbance in their dip or inclinations. Its mineralogy is scanty, being nearly confined, so far as observed, to some common silicious minerals, and traces of argillaceous and magnetic oxides of iron. The erratic block-stratum or drift, is remarkable, and prepares the mind for the still heavier accumulations of this kind which are perceived to be spread over the northern latitudes.‡

---

* Of this officer, who was a brother of Franklin Pierce, President of the United States, Gardner's *Army Dictionary* gives the following notice: Benjamin K. Pierce (N. H.), First Lieutenant Third Artillery, March, 1812; Adjutant, 1813; Captain, October, 1818; retained May 15, in artillery; in Fourth Artillery, May 21; Major ten years fa. service, Oct. 1, 1823; Major First Artillery, June 11, 1836 (Lieutenant-Colonel Eighth Infantry, July 7, 1838, declined); Brevet Lieutenant-Colonel "for distinguished service in affair at Fort Drane," Aug. 21, 1836 (Oct. 1836), in which he commanded: Colonel Regular Creek Mounted Volunteers, in Florida War, Oct. 1836; Lieutenant-Colonel First Artillery, March 19, 1842. Died April 1, 1850, at New York.

† Among the erratic block or drift stratum, I observed on the south Huron coast singularly striking, round fragments of white quartz, imbedding red fragments of coarse jasper; a rock, which I afterwards found in places on the south end of Sugar Island, in St. Mary's Straits, which lies directly north of the general position, and may serve as a proof of the course of the drift.

‡ *Vide* Geo. Report, Appendix.

CHAPTER III.

NOTHING can exceed the beauty of this island. It is a mass of
calcareous rock, rising from the bed of Lake Huron, and reaching
an elevation of more than three hundred feet above the water.
The waters around are purity itself. Some of its cliffs shoot up
perpendicularly, and tower in pinnacles like ruinous Gothic
steeples. It is cavernous in some places; and in these caverns,
the ancient Indians, like those of India, have placed their dead.
Portions of the beach are level, and adapted to landing from
boats and canoes. The harbor, at its south end, is a little gem.
Vessels anchor in it, and find good holding. The little old-
fashioned French town nestles around it in a very primitive
style. The fort frowns above it, like another Alhambra, its
white walls gleaming in the sun. The whole area of the island
is one labyrinth of curious little glens and valleys. Old green
fields appear, in some spots, which have been formerly cultivated by the Indians. In some of these there are circles of
gathered-up stones, as if the Druids themselves had dwelt here.
The soil, though rough, is fertile, being the comminuted materials
of broken-down limestones. The island was formerly covered
with a dense growth of rock-maples, oaks, ironwood, and other
hard-wood species, and there are still parts of this ancient forest
left, but all the southern limits of it exhibit a young growth.
There are walks and winding paths among its little hills, and precipices of the most romantic character. And whenever the visitor
gets on eminences overlooking the lake, he is transported with

sublime views of a most illimitable and magnificent water prospect. If the poetic muses are ever to have a new Parnassus in America, they should inevitably fix on Michilimackinac. Hygeia, too, should place her temple here, for it has one of the purest, driest, clearest, and most healthful atmospheres.

We remained encamped upon this lovely island six days, while awaiting the arrival of supplies and provisions for the journey, or their being prepared for transportation by hand over the northern portages. Meats, bread, Indian corn, and flour, had to be put in kegs, or stout linen bags.

The traders and old citizens said so much about the difficulties and toils of these northern portages that we did not know but what we, ourselves, were to be put in bags; but we escaped that process. This delay gave us the opportunity of more closely examining the island. It is about three and a half miles long, two in its greatest width, and nine in circumference. The site of Fort Holmes, the apex, is three hundred and twelve feet above the lake. The eastern margin consists of precipitous cliffs, which, in many places, overhang the water, and furnish a picturesque rocky-fringe, as it were, to the elevated plain. The whole rock formation is calcareous. It exhibits the effects of a powerful diluvial action at early periods, as well as the continued influence of elemental action, still at work. Large portions of the cliffs have been precipitated upon the beach, where the process of degradation has been carried on by the waves. A most striking instance of such precipitations is to be witnessed at the eastern cliff, called Robinson's Folly, which fell, by its own gravitation, within the period of tradition. The formation, at this point, formerly overhung the beach, commanding a fine view of the lake and islands in all directions, in consequence of which it was occupied with a summer-house, by the officers of the British garrison, after the abandonment of the old peninsular fort, about 1780.

The mineralogical features of the island are not without interest. I examined the large fragments of debris, which are still prominent, and which exhibit comparatively fresh fractures. The rock contains a portion of sparry matter, which is arranged in reticulæ, filled with white carbonate of lime, in such a state of loose disintegration that the weather soon converts it to the condition of agaric mineral. These reticulæ are commonly in the slate of calcspar, crystallized in minute crystals. The stratum on which this loose formation rests is compact and firm, and agrees in structure with the encrinal limestone of Drummond Island and the Manitouline chain. But the vesicular stratum, which may be one hundred and ten or twenty feet thick, has been deposited in such a condition that it has not had, in some localities, firmness enough permanently to sustain itself. The consequence is, that the table-land has caved in, and exhibits singular depressions, or grass-covered, cup-shaped cavities, which have no visible outlet for the rain-water that falls in them, unless it percolates through the shelly strata. Portions of it, subject to this structure, have been pressed off, during changing seasons, by frosts, and carried away by rains, creating that castellated appearance of pinnacles, which gives so much peculiarity to the rocky outlines of the island.

The ARCHED ROCK is an isolated mass of self-sustaining rock, on the eastern facade of cliffs; it offers one of those coincidences of geological degradation in which the firmer texture of the silicious and calcareous portions of it have, thus far, resisted decomposition. Its explanation, is, however, simple: The apex of this geological monument is on a level, or nearly so, with the Fort Holmes summit. While the diluvial action, of which the whole island gives striking proofs, carried away the rest of the reticulated or magnesian limestone, this singular point, having a firmer texture, resisted its power, and remains to tell the visitor who gazes at it, that waters have once held dominion over the highest part of the island.

Before dismissing the subject of the geological phenomena of this island, it may be observed that it is covered with the erratic block or drift stratum. Primitive or crystalline pebbles and boulders are found, but not plentifully, on the surface. They are observed, however, on the highest summit, and upon the lower plain; one of the best localities of these boulders, exists on the depressed ground, leading north, in the approach to Dousman's Farm, where there is a remarkable accumulation of blocks of granite and hornblende drift boulders. The principal drift of the island consists of smooth, small, calcareous pebbles, and, at deeper positions, angular fragments of limestone. Sandstone boulders are not rare. Over the plain leading from the fort north by

way of the Skull Rock, are spread extensive beds of finely comminuted calcareous gravel, the particles of which often not exceeding the size of a buck-shot, which makes one of the most
solid and compact natural macadamized roads of which it is
possible to conceive. Carriage wheels on it run as smoothly,
but far more solid, than they could over a plank floor. This
formation appears to be the diluvial residuum or ultimate wash,
which arranged itself agreeably to the laws of its own gravitation,
on the recession of the watery element, to which its comminution
is clearly due. It would be worth transportation, in boxes, for
gravelling ornamental garden-walks. The soil of the island is
highly charged with the calcareous element, and, however barren
in appearance, is favorable to vegetation. Potatoes have been
known to be raised in pure beds of small limestone pebbles,
where the seed potatoes had been merely covered in a slight way,
to shield them from the sun, until they had taken root.

The historical reminiscences connected with this island are of
an interesting character. It appears from concurrent testimony,
that the old town on the peninsula was settled about 1671,* which
was seven years before the building of Fort Niagara. In that
year, Father Marquette, a French missionary, prevailed on a party
of Hurons to locate themselves at that spot, and it was therefore
the first point of settlement made northwest of Fort Frontenac, on
Lake Ontario. It was probably first garrisoned by La Salle, in
1678, and continued to be the seat of the fur trade, and in many
respects, the metropolis of the extreme northwest, during the
whole period of French domination in the Canadas. After the
fall of Quebec, in 1759, it passed by treaty to the British government, but much against the wishes of the Indian tribes, who
retained a strong partiality for their early friends, the French.
Pontiac arose at this time, to dispute the English authority in the
northwest, and with confederates projected a series of bold
attacks upon the forts extending from the Ohio to this post.
Most of these were successful, but he was defeated at Detroit,
where he commanded in person, after a series of extraordinary

---

* Neither Fort Niagara nor Fort Ponchartrain (at the present site of Detroit) were
then in existence. The foundation of the former was laid by La Salle, in 1678; the
latter had not been erected when La Hontan passed through the country, in 1688.—
Herriot's Travels through Canada, p. 196.

movements. While he was pressing the siege of the garrison, he enjoined neutrality upon the French inhabitants, who were nevertheless called on to furnish cattle and corn for the subsistence of his warriors. It is remarked on good authority that, for these supplies, he issued evidences of debt. When General Bradstreet marched to the relief of the fort, with an army of three thousand men, the spirit and laconic temper of the warrior were at the same time evinced. He sent a deputation of chiefs to meet the herald of the British general, at Maumee, with the laconic and symbolic message: " I stand in the path."

The execution of the plan of attack on Old Fort Mackinac appears to have been intrusted to Minnawanna, a Chippewa chief, who, in addition to his own people, was aided by the Sacs. The Ottowas afterwards expressed displeasure in not having been admitted to a participation in the attack. The plan was ingeniously laid. The king's birthday, the 4th of June (1763), having arrived, the Chippewas and Sacs turned out to play, for a high wager, at ball. Many of the garrison, and the commanding officer himself, came out to witness the sport; and there was such a feeling of security that the gates of the fort were left open. To put the troops more off their guard, the ball had been thrown over the picket, and when once there, it was natural that it should be followed by the opposite parties, heated with the contest and eager for victory. But this artifice was the accomplishment of the plan. The war-whoop was immediately sounded, and an indiscriminate slaughter commenced. A few moments of intense anxiety ensued. They were passed by the officers eagerly listening for the roll of the drum. But they were passed in disappointment. There was no call of this kind to concentrate resistance. Panic and slaughter raged in their most fearful forms. None were spared who were deemed friendly to the English interest but such as were effectually secreted. Some of the soldiers who escaped the first onset, were incarcerated in a room, where they were sacrificed to glut the vengeance of a chief, who did not arrive till the principal work of slaughter had been accomplished.

This event sealed the fate of the old fort and the town on the peninsula. The British afterwards took possession of the island, which had served to give name to the peninsular fort. The town was gradually removed, by pulling down the buildings, and transporting the timber to the island, till there was not a building or fixture left; and the site is now as silent and deserted as if it had never been the scene of an active resident population.

The Island of Michilimackinac appears to have been occupied first as a military position by the British, about 1780, say some seven years after the massacre of the garrison of the old peninsular fort of the same name.

Wherever Michilimackinac is mentioned in the missionary letters or history of this period, it is the ancient fort, on the apex of the Michigan peninsula, that is alluded to.

The present town is pleasantly situated around a little bay that affords good clay anchorage and a protection from west and north winds. It has a very antique and foreign look, and most of the inhabitants are, indeed, of the Canadian type of the French. The French language is chiefly spoken. It consists of about one hundred and fifty houses and some four hundred and fifty permanent inhabitants.

It is the seat of justice for the most northerly county of Michigan. According to the observation of Lieut. Evelith, the island lies in north latitude 45° 54', which is only twenty-three minutes north of Montreal, as stated by Prof. Silliman.* It is in west longitude 7° 10' from Washington.

Col. Croghan's attempt to take the island, during the late war, was most unfortunate. He failed from a double spirit of dissension in his own forces, being at odds with the commanding officer of the fleet, and at sword's points with his second in command, Major Holmes. After entering the St. Mary's, and taking and burning the old post of St. Joseph's, where nobody resisted, instead of sailing direct to Mackinac, a marauding expedition was sent up this river to St. Mary's, and when the fleet and troops finally reached Mackinac, instead of landing at the town, under the panic of the inhabitants, it sailed about for several days. In the mean time the island filled with Indians from the surrounding shores.

Fort "Mackina" is eligibly situated on a cliff overlooking the town and harbor, and is garrisoned by a company of artillery. The ruin of Fort Holmes, formerly Fort George, occupies the apex of the island, and has been dismantled since the British evacuated it in 1815.*

* Tour from Hartford to Quebec, p. 341.

It happened that the British authorities on the island of St. Joseph, got intelligence of the declaration of war, in 1812, through Canada, before the American commander at Mackinac heard of it. Mustering their forces with such volunteers, militia, and Indians as could be hastily got together, they proceeded in boats to the back of the island, where they secretly landed at night with some artillery, and by daylight the next morning got the latter in place on the summit of Fort Holmes, which completely commanded the lower fort, when they sent a summons of surrender, which Captain Hanks, the American commanding officer, had no option but to obey.

Colonel Croghan, the hero of Sandusky, attempted to regain possession of it, in 1814, with a competent force, and after several demonstrations of his fleet about the island, by which time was lost and panic in the enemy allayed, he landed on the northern part of it, which is depressed, and his army marched through thick woods, most favorable for the operations of the Indians, to the open grounds of Dousman's Farm, where the army was met by Colonel McDouall, who was eligibly posted on an eminence with but few regular troops, but a heavy force of Indian auxiliaries and the village militia. Major Holmes, who gallantly led the attack, swinging his sword, was killed at a critical moment, and the troops retreated before Colonel Croghan could reach the field with a reinforcement. Thus ended this affair.

My attention was directed to the plaster stated to exist on the St. Martin Islands. These islands compose a small group lying about nine or ten miles north-northeast of Michilimackinac. Captain Knapp, of the revenue service, had been requested to take me to the spot with the revenue cutter under his command. I was accompanied by Captain Douglass, of the expedition, and by Lieutenant John Pierce, U. S. A., stationed at the fort.

The gypsum exists in a moist soil, not greatly elevated, during certain winds above the lake. Pits had been dug by persons visiting the locality for commercial purposes. It occurs in granular lumps of a gray color, as also in foliated and fibrous masses, white, gray, chestnut color, or sometimes red. No difficulty was encountered in procuring as many specimens as were required. This group of islands is noticeable, also, for the large boulder masses of hornblende and granite rock, which are found imbedded in, or lying

5

on the surface, along with fragments of breccia, quartz, &c. This drift is more abundant, on all the islands I have seen, as we approach the north shores of Lake Huron. Having completed the examination of these islands, we returned to the harbor after an agreeable excursion.

To observe the structure and character of the Island of Michilimackinac, I determined to walk entirely around it, following the beach at the foot of the cliffs. This, although a difficult task, from brush and debris, became a practicable one, except on the north and northwest borders, where there was, for limited spaces, no margin of debris, at which points it became necessary to wade in the water at the base of low precipitous rocks. In addition to the reticulated masses of limestone covered with calcspar from the fallen cliffs, the search disclosed small tabular pieces of minutely crystallized quartz and angular masses of a kind of striped hornstone, gray and lead colored, which had been liberated from similar positions in the cliffs. On passing the west margin of the island, I observed a bed of a species of light-blue clay, which is stated to part with its coloring matter in baking it, becoming white.

While the British possessed the island, they attempted to procure water by digging two wells at the site of Fort George (now Holmes), but were induced to relinquish the work without success, at the depth of about one hundred feet. Among the fragments of rock thrown out, are impressions of bivalve and univalve shells, with an impression resembling the head of a trilobite. These are generally in the condition of chalcedony, covered with very minute crystals of quartz. I also discovered a drift specimen of brown oxide of iron, on the north quarter. This sketch embraces all that is important in its mineralogical character.

This island appears to have been occupied by the Indians, from an early period. Human bones have been discovered at more than one point, in the cavernous structure of the island; but no place has been so much celebrated for disclosures of this kind, as the SKULL CAVE. This cave has a prominent entrance, shaded by a few trees, and appears to have been once devoted to the offices of a charnel-house by the Indians. It is not mentioned at all, however, by writers, till 1763, in the month of June of which year the fort of old Mackinac on the peninsula, was treacherously taken by the Sac and Chippewa Indians. An extensive and threatening confederation of the western Indians had then been matured, and a large body of armed warriors was then encamped around the walls of Detroit, under the leadership of Pontiac, who held the garrison in close siege day and night. The surrender of Canada to Great Britian, which had followed the victory of General Wolfe at Quebec, was distasteful to these Indians, and they attempted the mad project of driving back beyond the Alleghanies the English race ; making a simultaneous assault upon all the military posts' west of that great line of demarcation, and preaching and dealing out vengeance to all who had English blood in their veins. Alexander Henry, a native of Albany,* was one of those enterprising men who had pushed his fortunes West, with an adventure of merchandise, on the first exchange of posts, and he was singled out for destruction, as soon as the fort was taken. He had taken refuge in the house of a Frenchman named Longlade, where he was concealed in a garret by a Pawnee slave, and where he hid himself under a heap of birch-bark buckets, such as are employed in the Indian country, in the spring season, in carrying the sap of the sugar-maple. But this temporary reprieve from the Indian knife seemed only the prelude to a series of hairbreadth escapes, which impressed him as the direct interposition of Providence. At length, when the scenes of blood and intoxication began to abate a little, an old Indian friend of his, called Wawetum, who'had once pledged his friendship, but who had been absent during the massacre, sought him out, and having reclaimed him by presents, in a formal council, took him into his canoe and conducted the spared witness of these atrocities three leagues' across the waters of Lake Huron in safety to this island.

To this place they were accompanied by the actors in this tragedy to the number of three hundred and fifty fighting men,† and he would now, under the protection of Wawetum, have been safe from immediate peril, but that in a few days a prize of two canoes of merchandise in the hands of English traders was made, amongst which was a large quantity of liquor. Hereupon, Wawetum, foreseeing another carousal, and always fearful of his friend, requested him to go up with him to the mountain part of the island. Hav-

---

* *Vide* Henry's Travels, New York, 1809, 1 vol. 8vo. † Henry, p. 109.

ing ascended it, he led him to this cave, and recommended him to abide here in concealment until the debauch was over, when he promised to visit him.

Breaking some branches at its mouth for a bed, he then sought its recesses, and spreading his blanket around, laid down and slept till morning. Daylight revealed to him the fact that he had been reposing on dry human bones, and that the cave had anciently been devoted by the Indians as a sepulchre. On announcing this fact to his deliverer, two days afterward, when he came to seek him, Wawetum expressed his ignorance of it, and a party of the Indians, who came to examine it in consequence of the announcement, also concurred in declaring that they had no tradition on the subject. They conjectured that the bones were either due to the period when the sea covered the earth—which is a common belief with them—or to the period of the Huron occupancy of this island, after that tribe were defeated by the Iroquois, in the St. Lawrence valley.

So much for tradition.

This island has been long known as a prominent point in the fur trade. But of this I am not prepared to speak. It was selected by Mr. J. J. Astor, in 1816, as the central point of outfit for his clerks and agents in this region; and the warehouses erected for their accommodation constitute prominent features in its modern architecture. The capital annually invested in this business is understood to be about three hundred thousand dollars. This trade was deemed an object of the highest consequence from the first settlement of Canada, but it was not till 1766, agreeably to Sir Alexander Mackenzie, that it commenced from Michilimackinac.* The number of furred animals taken in a single year, the same author states to be one hundred and eighty-two thousand two hundred; of which number, the astonishing proportion of one hundred and six thousand were beavers.† Estimating each skin at but one pound, and the foreign market price at four dollars per pound, which are both much below the average at this era, this item of beaver alone would exceed by more than one-third the whole capital employed, taking the data before men-

---

* Mackenzie's Voyages, Hist. Fur Trade, vii.
† Mackenzie, xxiv.

tioned, and leave the seventy-six thousand smaller furred animals to be put on the profit side.   No wonder that acts of perfidy arose between rivals, such as the shooting of Mr. Waden at his own dinner-table, where he was entertaining an opponent or copartner in the trade; or the foul assassination of Owen Keveny on the Rainy Lakes.*   Indeed, the fur trade has for a long period been more productive, if we are to rely on statements, than the richest silver mines of Mexico or Peru.

Society at Michilimackinac consists of so many diverse elements, which impart their hue to it, that it is not easy for a passing traveller to form any just estimate of it.   The Indian, with his plumes, and gay and easy costume, always imparts an oriental air to it.   To this, the Canadian, gay, thoughtless, ever bent on the present, and caring nothing for to-morrow, adds another phase.   The trader, or interior clerk, who takes his outfit of goods to the Indians, and spends eleven months of the year in toil, and want, and petty traffic, appears to dissipate his means with a sailor-like improvidence in a few weeks, and then returns to his forest wanderings; and boiled corn, pork, and wild rice again supply his wants.   There is in these periodical resorts to the central quarters of the Fur Company, much to remind one of the old feudal manners, in which there is proud hospitality and a show of lordliness on the one side, and gay obsequiousness and cringing dependence on the other, at least till the annual bargains for the trade are closed.

We were informed that there is neither school, preaching, a physician (other than at the garrison), nor an attorney, in the place. There are, however, courts of law, a post-office, and a jail, and one or more justices of the peace.

There is a fish market every morning, where may be had the trout—two species—and the white fish, the former of which are caught with hooks in deep water, and the latter in gill nets. Occasionally, other species appear, but the trout and white fish, which is highly esteemed, are staples, and may be relied on in the shore market daily; whole canoe-loads of them are brought in.

The name of this island is said to signify a great turtle, to which it has a fancied resemblance, when viewed from a distance. Mike-

* Report of the Trials of De Reinhard, &c.   Montreal, 1818.

nok, and not Mackenok, is, however, the name for a tortoise. The term, as pronounced by the Indians, is Michinemockinokong, signifying place of the Great Michinamockinocks, or rock-spirits. Of this word, *Mich* is from *Michau* (adjective-animate), great. The term *mackinok*, in the Algonquin mythology, denotes in the singular, a species of spirits, called turtle spirits, or large fairies, who are thought to frequent its mysterious cliffs and glens. The plural of this word, which is an animate plural, is *ong*, which is the ordinary form of all nouns ending in the vowel *o*. When the French came to write this, they cast away the Indian local in .*ong*, changed the sound of *n* to *l*, and gave the force *mack* and *nack*, to *mök* and *nök*. The vowel *e*, after the first syllable, is merely a connective in the Indian, and which is represented in the French orthography in this word by *i*. The ordinary interpretation of great turtle is, therefore, not widely amiss; but in its true meaning, the term enters more deeply into the Indian mythology than is conjectured. The island was deemed, in a peculiar sense, the residence of spirits during all its earlier ages. Its cliffs, and dense and dark groves of maples, beech, and ironwood, cast fearful shadows; and it was landed on by them in fearfulness, and regarded far and near as the *Sacred Island*. Its apex is, indeed, the true Indian Olympus of the tribes, whose superstitions and mythology peopled it by gods, or monitos.

Since our arrival here, there has been a great number of Indians of the Chippewa and Ottowa tribes encamped near the town. The beach of the lake has been constantly lined with Indian wigwams and bark canoes. These tribes are generally well dressed in their own costume, which is light and artistic, and exhibit physiognomies with more regularity of features and mildness of expression than it is common to find among them. This is probably attributable to a greater intermixture of blood in this vicinity. They resort to the island, at this season, for the purpose of exchanging their furs, maple-sugar, mats, and small manufactures. Among the latter are various articles of ornament, made by the females, from the fine white deer skin, or yellow birch bark, embroidered with colored porcupine quills. The floor mats, made from rushes, are generally more or less figured. Mockasins, miniature sugar-boxes, called mo-cocks, shot-pouches, and a kind of pin and needle-holders, or housewives, are elaborately beaded. But nothing exceeds in value the largest merchantable mockocks of sugar, which are brought in for sale. They receive for this article six cents per pound, in merchandise, and the amount made in a season, by a single family, is sometimes fifteen hundred pounds. The Ottowas of L'Arbre Croche are estimated at one thousand souls, which, divided by five, would give two hundred families; and by admitting each family to manufacture but two hundred pounds per annum, would give a total of forty thousand pounds; and there are probably as many Chippewas within the basins of Lakes Huron and Michigan. This item alone shows the importance of the Indian trade, distinct from the question of furs.

During the time we remained on this island, the atmosphere denoted a mean temperature of 55° Fahrenheit. The changes are often sudden and great. The island is subject to be enveloped in fogs, which frequently rise rapidly. These fogs are sometimes so dense, as to obscure completely objects at but a short distance. I visited Round Island one day with Lieut. Mackay,* and we were both engaged in taking views of the fort and town of Michilimackinac,† when one of these dense fogs came on, and spread itself with such rapidity, that we were compelled to relinquish our designs unfinished, and it was not without difficulty that we could make our way across the narrow channel, and return to the island. This fact enabled me to realize what the old travellers of the region have affirmed on this topic.

We were received during our visit here in the most hospitable manner, as well as with official courtesy, by Capt. B. K. Pierce, the commanding officer, Major Puthuff, the Indian agent, and by the active and intelligent agents of Mr. John Jacob Astor, the great fiscal head of the Fur Trade in this quarter.

* Lieut. Eneas Mackay. This officer, after the return from this expedition, went through the regular grades of promotion in the army, and had at the period of his death, which took place in 1850, at St. Louis, Missouri, reached the brevet rank of colonel.

† For the view from this point, see Information respecting the History, Condition, and Prospects of the Indian Tribes of the United States, vol. iv. Plate 42.

CHAPTER IV.

Proceed down the north shore of Lake Huron to the entrance of the Straits of St. Mary's—Character of the shores, and incidents—Ascend the river to Sault Ste. Marie—Hostilities encountered there—Intrepidity of General Cass.

HAVING spent six days on the island, rambling about it, and making ourselves as well acquainted with its features and inhabitants as possible, we felt quite recruited and cheered up, after the tedious delays along the southern shores of Lake Huron. And we all felt the better prepared for plunging deeper into the northwestern forest. Before venturing into the stronghold of the Chippewas, whose territories extend around Lake Superior, it was deemed prudent to take along an additional military force as far as Sault de Ste. Marie. But five or six years had then passed since this large tribe had been arrayed in hostilities against the United States (in the war of 1814), and they were yet smarting under the wounds and losses which they had received at Brownstown and the River Thames, where they had lost some prominent men. Generals Brown and Macomb,* when making a reconnoissance, with their respective staffs, a couple of years before, had been fired on in visiting Gros Cape, at the foot of Lake Superior, and although no one was killed on that occasion, the circumstance was sufficient to indicate their feeling.

This additional force was placed under the command of Lieu-

* The following are the official data of this distinguished officer:—
Alexander Macomb, Jr., born April 8, 1782, Detroit, N. Y.; Cornet Cavalry, January 10, 1799; Second Lieutenant, February, 1801; retained, April, 1802, in Second Infantry; First Lieutenant of Engineers, October, 1802; Captain, June, 1805; Major of Engineers, February 23, 1808; Lieutenant-Colonel, July 23, 1810; Acting Adjutant-General of the Army, April 28, 1812; Colonel Third Artillery, July 6, 1812; Brigadier-General, January 24, 1814; Brevet Major-General, "for distinguished and gallant conduct in defeating the enemy at Plattsburg, September 11, 1814" (October 1, 1814); received the "thanks of Congress" of November 8,

·tenant John S. Pierce, U. S. A., a brother of the commanding officer,* and of Franklin Pierce, President of the United States. It consisted of twenty-two men, with a twelve-oared barge. The whole expedition, now numbering sixty-four persons, embarked at ten o'clock on the 15th, with a fair wind, for our first destination, at Detour, being the west cape of the Straits of St. Mary's. The distance is estimated at forty miles, along a very intricate, masked shore of islands, called Chenos. The breeze carried us at the rate of five miles per hour. The first traverse is an arm of the Lake, three leagues across, over which we passed swimmingly. This traverse is broken near its eastern terminus by Goose Island, the Nekuhmenis (literally Brant Island) of the Chippewas—a noted place of encampment for traders. We did not, however, touch at it. A couple of miles beyond this brought us to Outard Point, where the men rested a few moments on their oars and paddles. This point forms the commencement of those intricate channels which constitute the Chenos group. Our steersman gave them, however, a wide berth, and did not approach near the shore till it began to be time to look out for the mouth of the St. Mary's. After passing Point St. Vitel, a distance of about thirty miles, the guides led into a sandy bay, under the impression that we had reached the west cape of the St. Mary's; but in this we were deceived. While landing here a few moments, in a deep bay, the animal called Káug by the Chippewas (a porcupine), was discovered and killed by one of the men, called Baptiste, by a blow from a hatchet. Buffon gives two engravings of this animal, as found in Canada, under separate names; but it is apprehended that he has been misled by

1814, "for his gallantry and good conduct in defeating the enemy at Plattsburg, on the 11th of September, repelling with 1,500 men, aided by a body of militia and volunteers from New York and Vermont, a British veteran army, greatly superior in numbers," with the presentation of a *gold medal*, "emblematical of this triumph;" retained, April 8, 1815; retained, May 21, as Colonel and Principal Engineer, with Brevets Major-General and General-in-Chief of the Army, May 24, 1828; commanded the army of Florida 1836; died June 25, 1841, at his head-quarters, Washington City.—*Gardner's Army Dictionary.*

* John Sullivan Pierce (N. H., brother to Colonel Benjamin K. Pierce), Third Lieutenant Third Artillery, April 5, and Second Lieutenant, May, 1814; retained, May, 1815, in Artillery; First Lieutenant, April 1818; resigned February 1, 1828. —*Gardner's Army Dictionary.*

the same animal seen in its summer and winter dress. To the
Indian, this animal is valuable for its quills, which are dyed of
bright colors, to ornament their dresses, moccasons, shot-pouches,
and other choice fabrics of deer skin, or birch bark. This animal
has four claws on the fore paw, and five on the hinder ones. It
has small ears hid in the hair, and a bushy tail, with coarse black
and white hair. The specimen killed would weigh eight pounds.

Soon after coming out from this indentation of the lake, we
came in sight of Point Detour, on turning which, from E. to N.,
we found no longer use for sails. Mackenzie places this point
in north latitude 45° 54'.

The geology of this coast appears manifest. Secondary compact limestone appears in place, in low situations, on the reef of
Outard Island and Point, and in the approach to Point Detour.
A ridge of calcareous highlands appears on the mainland east of
Michilimackinac, stretching off towards Sault de Ste. Marie, in a
northeast direction. This ridge appears to belong to a low
mountain chain, of which the Island of Michilimackinac may be
deemed as one of the geological links. Just before turning, we
passed a very heavy angular block of limestone, much covered
with moss, which could not have been far removed, in the drift
era, from its parent bed. The largest angle of this stone, which
I have since examined, must be eight or ten feet. This block is
of the ortho-cerite stratum of Drummond Island. The shores
are heavily charged with various members of the boulder drift,
with a fringe beyond them of spruce and firs, giving one the idea
of a cold, exposed, and most unfavorable coast. Turning the
Point of Detour, we ascended the strait a few miles, and encamped
on its west shore, off Frying-pan Island, at a point directly opposite the British post of Drummond Island, which we could not
perceive, but the direction of which was clearly denoted by the
sound of the evening bugles.

The entrance into this strait forms a magnificent scene of waters
and islands, of which a map conveys but a faint conception. The
straits here appeared to be illimitable, we seemed to be in a
world of waters. It is stated to be thirty miles across to Point
Thessalon. The large group of the Manatouline Islands, stretching transversely through Lake Huron, terminates with the isle
Drummond—a name bestowed in compliment to the bold leader,

Col. Drummond, who led the night storming party, and was blown up on the bastion of Fort Erie, in 1818. This station was first occupied on the withdrawal of the British troops from Mackinac, in 1815. This day's trip gave us a favorable idea of canoe travelling. It also gave us an exalted idea of the gigantic system of these lake waters, and their connecting straits. We had never done gazing at the prospect before us, after turning the Detour, and did not retire from our camp fires early. The next morning we embarked at five o'clock, a light dreamy mist hanging over the waters. When this cleared away, we descried the ruined chimneys and buildings of St. Joseph, the abandoned British post burned by Col. Croghan, in 1814.* The day turned out a fine one, and we proceeded up the straits with pleasurable feelings, excited by the noble and novel views of scenery continually before us. Keeping the west side of a high limestone island called Isle a la Crosse, we then entered a sheet of water called Lac Vaseau, or Muddy Lake. We had proceeded northwardly perhaps twenty miles, when we encountered another of those large islands for which these straits are remarkable, called Nebeesh,† or Sailor's Encampment Island. Our guides held up on its western side, which soon brought us to the first rapids, and the commencement of St. Mary's River. A formation of sandstone is here observed in the bed of the stream. The waters are swift and shallow, and the men encountered quite a struggle in the ascent, and so much injured one of our canoes that it became necessary to unlade and mend it. In the mean time, the atmosphere put on a threatening aspect, with heavy peals of thunder, but no rain followed till we again re-embarked and proceeded five or six miles, when a shower fell. It did not, however, compel us to land, and by six o'clock in the afternoon, the sky again became clear. We had now ascended the strait and river so far, that it became certain we could reach our destination before night, and the men worked with the greater alacrity. At eight o'clock we had surmounted the second rapid, called the Little Rapid, Nebeetung of

* This fort was first erected by the British in 1795, the year before Michilimackinac was evacuated under Wayne's treaty with the Indians.
† From Nebee, water; hence Nebeesh, rapid water, or strong water, the name of the rapids which connect the straits with the River St. Mary's. This word is the *derogative* form of the Chippewa noun.

the Indians, where we encountered a swift current. We were now within two miles of our destination. The whole river is here embodied before the eye, and is a mile or three-fourths of a mile wide, and the two separate villages on the British and American shores began to reveal themselves to view, with the cataract of the Sault de Ste. Marie in the distance; and a beautiful forest of elms, oaks, and maples on either hand. We ascended with our flags flying, our little squadron being spread out in order, and the Canadian boatmen raising one of their enlivening songs. Long before reaching the place, a large throng of Indians had collected on the beach, who, as we put in towards the shore, fired a salute, and stood ready to greet us with their customary *bosho*.* We landed in front of the old Nolan house,† the ancient head-quarters of the Northwest Company; and immediately formed our encampment on the wide green, extending along the river. Day-light in this latitude is protracted, and although we had ascended a computed distance of forty-five miles, and had had the mishap to break a canoe in the Nebeesh, there was abundant light to fix our encampment properly. Lieut. Pierce encamped his men on our extreme right. Leaving an interval, Lieut. Mackay's escort came next, and our tents formed the northern line of his encampment, nearest to the Indians. The latter occupied a high plateau, in plain view, several hundred yards west, with an intervening gulley, and a plain, well-beat footpath. We had, in case of difficulty, thirty-four muskets, Pierce's command included, in addition to which, each of the savans, or Governor's mess, were armed with a short rifle. Our line may have looked offensively demonstrative to the Chippewas, who regarded it, from their ancient eminence, with unfriendly feelings. These particulars are given from the perilous position we were brought into next day.

Meantime, we passed a quiet night in our tents, where the deep sound of the Falls fell on the wakeful ear, interspersed with the distant monotonous thump of the Indian täwäegon. It required but little observation, in the morning, to explore the village of St. Mary's. It consisted of some fifteen or twenty buildings of all sorts, occupied by descendants of the original French settlers, all of whom drew their living from the fur trade. The principal

---

* From the French *bon jour*.   † The present site of Fort Brady.

buildings and outhouses were those of Mr. John Johnston, and the group formerly occupied by the Northwest Company. Most of the French habitations stood in the midst of picketed lots. There were about forty or fifty lodges, or two hundred Chippewas, fifty or sixty of whom were warriors. But, although this place was originally occupied as a missionary centre, by the Roman Catholic missionaries of New France, about the middle of the seventeenth century, no trace of the ancient church could be seen, unless it was in an old consecrated graveyard, which has continued to be used for interments. Mr. Johnston, the principal inhabitant, is a native of the County of Antrim, Ireland, where his connections are persons of rank. He is a polite, intelligent, and well-bred man, from a manifestly refined circle; who, soon after the close of the American Revolution, settled here, and married the daughter of a distinguished Indian chief.* Although now absent on a visit to Europe, his family received us with marked urbanity and hospitality, and invited the gentlemen composing the travelling family of Governor Cass to take all our meals with them. Everything at this mansion was done with ceremonious attention to the highest rules of English social life; Miss Jane, the eldest daughter, who had received her education in Ireland, presiding.

* INTER-EUROPEAN AMALGAMATION.—John Johnston was a native of the north of Ireland, where his family possessed an estate called "Craige," near the celebrated Giant's Causeway. He came to this country during the first Presidential term of Washington, and settled at St. Mary's, about 1798. He was a gentleman of taste, reading, refined feeling, and cultivated manners, which enabled him to direct the education of his children, an object to which he assiduously devoted himself; and his residence was long known as the seat of hospitality and refinement to all who visited the region. In 1814, his premises were visited, during his absence, by a part of the force who entered the St. Mary's, under Colonel Croghan, and his private property subjected to pillage, from a misapprehension, created by some evil-minded persons, that he was an agent of the Northwest Company. Genial, social, kind, and benevolent, his society was much sought, and he was sometimes imposed on by those who had been received into his employments and trusts (as in the reports which carried the Americans to his domicil in 1814). He died at St. Mary's, in 1828, leaving behind, among his papers, evidence that his leisure hours were sometimes lightened by literary employments. Mr. Johnston, by marrying the daughter of the ruling chief of this region, placed himself in the position of another Rolfe. Espousing, in Christian marriage, the daughter of Wabjeeg, he became the son-in-law of another Powhatan; thus establishing such a connection between the Hibernian and Chippewa races, as the former had done between the English and Powhetanic stocks.

The Sault (from the Latin *Saltus*, through the French) or Falls of St. Mary, is the head of navigation for vessels on the lakes, and has been, from early days, a thoroughfare for the Indian trade. It is equally renowned for its white fish, which are taken in the rapids with a scoop-net. The abundance and excellence of these fish has been the praise of all travellers from the earliest date, and it constitutes a ready means of subsistence for the Indians who congregate here.

The place was chiefly memorable in our tour, however, as the seat of the Chippewa power. To adjust the relations of the tribe with the United States, a council was convened with the chiefs on the day following our arrival. This council was assembled ·at the Governor's *marquée*, which was graced by the national ensign, and prepared for the interview with the usual presents. The chiefs, clothed in their best habiliments, and arrayed in feathers and British medals, seated themselves, with their usual dignity, in great order, and the business was opened with the usual ceremony of smoking the peace pipe. When this had been finished, and the interpreter* taken his position, he was directed to explain the views of the Government, in visiting the country, to remind them that their ancestors had formerly conceded the occupancy of the place to the French, to whose national rights and prerogatives the Americans had succeeded, and, by a few direct and well-timed historical and practical remarks, to secure their assent to its reoccupancy. The utmost attention was bestowed while this address was being made, and it was evident, from the glances of the hearers, that it was received with unfriendly feelings, and several chiefs spoke in reply. They were averse to the proposition, and first endeavored to evade it by pretending to know nothing of such former grants. This point being restated by the American commissioner, and pressed home strongly, was eventually dropped by them. Still, they continued to speak in an evasive and desultory manner, which had the effect of a negative. It was evident that there was a want of agreement, and some animated discussion arose among themselves. Two classes of persons appeared

---

* James Riley, a son of the late J. V. S. Riley, Esq., of Schenectady, N. Y., by a Saganaw woman; a man well versed in the language, customs, and local traditions of the Chippewas.

among the chiefs. Some appeared in favor of settling a boundary to the ancient precinct of French occupancy, provided it was not intended to be occupied by a garrison, saying, in the symbolic language of Indians, that they were afraid, in that case, their young men might kill the cattle of the garrison. Gov. Cass, understanding this, replied that, as to the establishment of a garrison, they need not give themselves any uneasiness—it was a settled point, and so sure as the sun that was then rising would set, so sure would there be an American garrison sent to that point, whether they renewed the grant or not. This decisive language had a sensible effect. High words followed between the chiefs. The head chief of the band, Shingabawossin, a tall, stately man, of prudent views, evidently sided with the moderates, and was evasive in his speech. A chief called Shingwauk, or the Little Pine, who had conducted the last war party from the village in 1814, was inclined to side with the hostiles. There was a chief present called Sassaba, a tall, martial-looking man, of the reigning family of chiefs of the Crane Totem, who had lost a brother in the battle of the Thames. He wore a scarlet uniform, with epaulets, and nourished a deep resentment against the United States. He stuck his war lance furiously in the ground before him, at the beginning of his harangue, and, assuming a savage wildness of air, appeared to produce a corresponding effect upon the other Indian speakers, and employed the strongest gesticulation. His address brought the deliberations to a close, after they had continued some hours, by a defiant tone; and, as he left the marquée, he kicked away the presents laid before the council. Great agitation ensued. The council was then summarily dissolved, the Indians went to their hill, and we to our tents.

It has been stated that the encampment of the Indians was situated on an eminence a few hundred yards west from our position on the shore, and separated from us by a small ravine. We had scarcely reached our tents, when it was announced that the Indians had raised the British flag in their camp. They felt their superiority in number, and did not disguise their insolence. Affairs had reached a crisis. A conflict seemed inevitable. Governor Cass instantly ordered the expedition under arms. He then called the interpreter, and proceeded with him, naked-handed and alone, to Sassaba's lodge at the hostile camp. Being armed with short rifles, we requested to be allowed to accompany him as a body-guard, but he decidedly refused this. On reaching the lodge of the hostile chief, before whose door the flag had been raised, he pulled it down with his own hands. He then entered the lodge, and addressing the chief calmly but firmly, told him that it was an indignity which they could not be permitted to offer; that the flag was the distinguishing symbol of nationality; that two flags of diverse kind could not wave in peace upon the same territory; that they were forbid the use of any but our own, and should they again attempt it, the United States would set a strong foot upon their rock and crush them. He then brought the captured flag with him to his tent.

In a few moments after his return from the Indian camp, that camp was cleared by the Indians of their women and children, who fled with precipitation in their canoes across the river. Thus prepared for battle, we momently expected to hear the war-whoop. I had myself examined and filled my shot-pouch, and stood ready, rifle in hand, with my companions, awaiting their attack. But we waited in vain. It was an hour of indecision among the Indians. They deliberated, doubtingly, and it soon became evident that the crisis had passed. Finding no hostile demonstration from the hill, Lieuts. Pierce and Mackay directed their respective commands to retire to their tents.

The intrepid act of Governor Cass had struck the Indians with amazement, while it betokened a knowledge of Indian character of which we never dreamed. This people possess a singular respect for bravery. The march of our force, on that occasion, would have been responded to, instantly, by eighty or a hundred Indian guns; but to behold an unarmed man walk boldly into their camp and seize the symbol of their power, betokened a cast of character which brought them to reflection. On one person in particular the act had a controlling effect. When it was told to the daughter of Wäbojeeg (Mrs. Johnston), she told the chief that their meditated scheme of resistance to the Americans was madness; the day for such resistance was passed; and this man, Cass, had the air of a great man, and could carry his flag through the country. The party were also under the hospitality of her roof. She counselled peace. To these words

Shingabowassin responded; he was seconded by Shingwäkonce, or the Little Pine. Of this effort we knew nothing at the moment, but the facts were afterwards learned. It was evident, before the day had passed, that a ·better state of feeling existed among the Indians. The chief Shingabowassin, under the friendly influences referred to, renewed the negotiations. Towards evening a council of the chiefs was convened in one of the buildings of this Pocahontean counsellor, and the treaty of the 16th June, 1820 (*vide* Ind. Treaties United States) signed. In this treaty every leading man united, except Sassaba. The Little Pine signed it, under one of his synonymous names, Lavoine Bart. By this treaty the Chippewas cede four miles square, reserving the right of a place to fish at the rapids, perpetually. The consideration for this cession, or acknowledgment of title, was promptly paid in merchandise.

The way being thus prepared for our entry into Lake Superior, it was decided to proceed the next day. Before leaving this point, it may be observed that the falls are produced by a stratum of red sandstone rock, which crosses the bed of the St. Mary's at this place. The last calcareous formation, seen in ascending the straits, is at Isle a la Crosse. As we proceed north, the erratic block stratum becomes heavier, and abraded masses of the granite, trap, sandstone, and hornblende series are confusedly piled together on the lake shores, and are abundant at the foot of these falls. In the central or middle channel, the waters leap from a moderate height, from stratum to stratum, at two or three points, producing the appearance, when seen from below, of a mass of tumbling waves. The French word *Sault* (pronounced *so*) accurately expresses this kind of pitching rapids or falls. The Indians call it Bawateeg, or Pawateeg, when speaking of the phenomenon, and Bawating or Pawating, when referring to the place. Paugwa is an expression denoting shallow water on rocks. The inflection *eeg* is an animate plural. *Ing* is the local terminal form of nouns. In the south or American channel, there is no positive leap of the water, but an intensely swift current, which is parted by violent jets, between rocks, still permitting canoes, skilfully guided, to descend, and empty boats to be drawn up. But these falls are a complete check to ship navigation. The descent of water has been stated by Colonel Gratiot, of the United States Engineers,

6

at twenty-two feet ten inches.*   They resemble a bank of rolling
foam, and with their drapery of trees on either shore, and the
mountains of Lake Superior in the distance, and the moving
canoes of fishing Indians in the foreground, present a most
animated and picturesque view.

To the Chippewas, who regard this spot as their ancient capital,
it is doubtless fraught with many associations, and they regard
with jealousy the advance of the Americans to this quarter.
This tribe, in the absence of any older traditions, are regarded as
the aboriginal inhabitants of the place. They are, by their language, Algonquins, and speak a pure dialect of it. They call
themselves Ojibwas. *Bwa*, in this language, denotes voice. Ojib-
wamong signifies Chippewa language, or voice. It is not manifest what the prefixed syllable denotes. They are a numerous
people, and spread over many degrees of latitude and longitude.
We have had them constantly around us, in some form, since
leaving Detroit, and they extend to the Great Winnipeg Lake
of Hudson's Bay. They appear, at the French era of discovery,
to have been confined almost exclusively to the north bank of
the St. Lawrence, below the influx of the Ottowa River, extending to Lake Nepising, and the geographical position seems to
have been the origin of the name Algonquin.

Whilst encamped here, we witnessed the descent down the
rapids of eleven barges and canoes laden with furs from the north.
This trade forms the engrossing topic, at this point, with all
classes. Hazardous as it is, the pursuit does not fail to attract
adventurers, who appear to be fascinated with the wild freedom
of life in the wilderness.

* ST. MARY'S CANAL.—Thirty-three years have produced an astonishing progress. A ship-canal is now (1853) in the process of being constructed at these
falls, by the State of Michigan, under a grant of public land for that purpose,
from Congress. It is to consist of two locks of equal lift, dividing the aggregate
fall.   This canal will add the basin of Lake Superior to the line of lake navigation. It will enable ships and steamers to enter the St. Louis River of Fond du
Lac, and to reach a point in latitude corresponding to Independence, on the Missouri. No other point of the lake chain reaches so far by some hundreds of
miles towards the Rocky Mountains; and this canal will eventually be the outlet to
the Atlantic cities of the copper and other mines of Lake Superior, and of the
agricultural and mineral products of all the higher States of the Upper Mississippi
and of the Missouri, and a part of Oregon and Washington on the Pacific.

CHAPTER V.

Embark at the head of the portage at St. Mary's.—Entrance into Lake Superior—Journey and incidents along its coasts—Great Sand Dunes—Pictured Rocks—Grand Island—Keweena peninsula and portage—Incidents thence to Ontonagon River.

HAVING accomplished the object of our visit, at this place, no time was lost in pushing our way into the basin of Lake Superior. The distance to it is computed to be fifteen miles above the Sault. It was nine o'clock of the morning following the day of the treaty, when the men began to take the canoes up the rapids, and transport the provisions and baggage. This occupied nearly the whole of the day. Taking leave of Lieutenant Pierce, who returned with his command, from this point, and our hospitable hostess, we proceeded to the head of the portage, long before the canoes and stores all arrived. To while away the time, while the men were thus employed, we tried our skill at rifle shooting. It was six o'clock in the evening before the work of transportation was finished, and the canoes loaded, when we embarked. The view from the head of the portage is imposing. The river spreads out like an arm of the sea. In the distance appear the mountains of Lake Superior.

We proceeded two leagues, and encamped at Point aux Pins, on the Canadian shore. At six o'clock the next morning we were again in our canoes, and crossed the strait, which is here several leagues wide, to the west, or Point Iroquois Cape. In this traverse we first beheld the entrance into Lake Superior. The scene is magnificent, and I could fully subscribe to the remark made by Carver, "that the entrance into Lake Superior affords one of the most pleasing prospects in the world." The morning was clear and pleasant, with a favoring breeze, but a tempest of wind and rain arose, with severe thunder, soon after we had

were observed, in some places, to be deposited over its vegetation
so as to arrest its growth. The largest trees were often half buried
and destroyed. Not less than nine miles of the coast, agreeably
to *voyageur* estimates, are thus characterized by dunes.

I found the sandstone formation of Cape Iroquois to reappear
at the western termination of these heights on the open shores of
the lake, where I noticed imbedded nodules of granular gypsum.
At this point, known to our men as La Pointe des Grandes Sables,
we pitched our tents, at nightfall, under a very threatening state
of the atmosphere. The winds soon blew furiously, followed by a
heavy rain-storm—and sharp thunder and lightning ensued. Our
line of tents stood on a gently rising beach, within fifty yards of
the margin of the lake, where they were prostrated during the
night by the violence of the waves. The rain still continued at
early daylight, the waves dashing in long swells upon the shore.
At sunrise the tempest abated, and by eight o'clock the atmosphere assumed a calm and delightful aspect. It was eleven
o'clock, however, before the waves sufficiently subsided to permit
embarkation. Indeed, a perfect calm now ensued. This calm
proved very favorable—as we discovered on proceeding three
leagues—to our passing the elevated coast of precipitous rock,
called Ishpäbecä,* and Pictured Rocks. This coast, which extends twelve miles, consists of a gray sandstone, forming a series
of perpendicular façades, which have been fretted, by the action
of the waves, into the rude architecture of pillared masses, and
open, cavernous arches. These caverns present their dark mouths
to observation as the voyager passes. At one spot a small stream
throws itself from the cliffs into the lake at one leap. In some
instances the cliffs assume a castellated appearance. At the spot
called the Doric Rock, near the commencement of these picturesque precipices, a vast entablature rests on two immense rude pil-
lars of the water-worn mass. At a point called Le Portail, the vast
wall of rock had been so completely excavated and undermined
by the lake, that a series of heavy strata of rock rested solely on
a single pillar standing in the lake. The day was fine as we passed
these geological ruins, and we sat silently gazing on the changing
panorama. At one or two points there are small streams which

* From *iupa*, high; *aubik*, a rock; and the substantive termination, *a*.

break the line of rock into quadrangles. A species of dark red clay overlies this formation, which has been carried by the rains over the face of the cliffs, where, uniting with the atmospheric sand and dust, it gives the whole line a pictorial appearance. We almost held our breath in passing the coast; and when, at night, we compared our observations around the camp-fire, there was no one who could recall such a scene of simple novelty and grandeur in any other part of the world; and all agreed that, if a storm should have arisen while we were passing, inevitable destruction must have been our lot. We came to Grand Island at a seasonable hour in the evening, and encamped on the margin of its deep and land-locked harbor. Our camp was soon filled with Chippewas from a neighboring village. They honored us in the evening by a dance. Among these dancers, we were impressed with the bearing of a young and graceful warrior, who was the survivor of a self-devoted war-party of thirteen men, who, having marched against their ancient enemies the Sioux, found themselves surrounded in the plain by superior numbers, and determined to sell their lives at the dearest rate. To this end, they dug holes in the earth, each of which thus becoming a fortification for its inmate, who dared their adversaries till overpowered by numbers. One person was selected to return with the news of this heroic sacrifice; this person had but recently returned, and it was from his lips that we heard the tragic story.

My mineralogical searches along the shores this day rewarded me with several water-worn fragments of agate, carnelian, zeolite, and prase, which gave me the first intimation of our approach to the trap and amygdaloidal strata, known to be so abundant in their mineral affluence in this quarter.

We left Grand Island the next morning at six o'clock, and passing through a group of sandstone islands, some of which had had their horizontality disturbed, we came to the mouth of Laughing-fish River, where a curious flux and reflux of water is maintained. From this place, a line of sandstone coast was passed, northwardly, till reaching its terminus on the bay of Chocolate River. This is a large and deep bay, which it would have required a day's travel to circumnavigate. To avoid this, the men held their way directly across it, steering N. 70° W., which, at the end of three leagues, brought us to Granite Point. Here we first struck the old crystalline rocks or primitive formation. This formation stretches from the north shores of the Gitche Sebeeng,* or Chocolate River, to Huron Bay, and gives the traveller a view of rough conical peaks. These characterize the coast for a couple of days' travel. They are noted for immense bodies of iron ore, which is chiefly in the condition of iron glance.† At Presque Isle, it assumes the form of a chromate of iron in connection with serpentine rock. We encamped on level ground on a sandstone formation, in the rear of Granite Point, and had an opportunity of observing the remarkable manner in which the horizontal sandstone rests upon and against the granitical, or, more truly, sienitic eminences. These sandstone strata lap on the shoulders of the primitive or crystalline rocks, preserving their horizontal aspect, and forming distinct cliffs along parts of the coast. This sandstone appears, from its texture and position, to be the "old red sandstone" of geologists.

The next morning (23d) we quitted our encampment at an early hour, in a haze, and urged our way, with some fluctuations of weather, an estimated distance of eleven leagues. This brought us, at four o'clock in the afternoon, to Huron River. Sitting in the canoe, in a confined position, makes one glad at every opportunity to stretch his limbs, and we embraced the occasion to bathe in the Huron. The shore consists of a sandy plain, where my attention was called to the Kinnikenik, a plant much used by the Indians for smoking. It is the *uva ursi*. I had seen it once before, on the expedition, at Point aux Barques.

We inspected here, with much attention, an Indian grave, as well from the care with which it was made, as the hieroglyphics cut on the head-posts. The grave was neatly covered with bark, bent over poles, and made roof-shaped. A pine stake was placed at the head. Between this and the head of the grave, there was placed a smooth tablet of cedar wood, with hieroglyphics. Mr. Riley, our interpreter, explained these. The figure of a bear denoted the chief or clan. This is the device called a Totem. Seven red strokes denoted his scalp honors in Indian heraldry,

* From *gitche*, great; *sebee*, a river; and the local terminal *ng*, signifying place.
† The extensive iron works of Carp River, which are now yielding such fine blooms, are seated on the verge of these mountains.

or that he had been seven times in battle. Other marks were not understood or interpreted. A paling of saplings inclosed the space.

On the following morning, our camp was astir at the customary early hour, when we proceeded to Point aux Beignes, a distance of six miles. Attaining this point, we entered Keweena Bay, coasting up its shores for an estimated distance of three leagues. We were then opposite the mouth of Portage River, but separated from it a distance of twelve miles. I was seated in Lieutenant Mackay's canoe. The whole squadron of five canoes unhesitatingly put out. The wind was adverse; before much progress had been made in crossing, three of our flotilla, after struggling against the billows, put back; but we followed the headmost one, which bore the Governor's flag, and, seizing hold of the paddles to relieve the men, we succeeded in gaining the river. The other canoes came up the next morning, at seven o'clock, when we all proceeded to cross the Portage Lake, and up an inlet, which soon exhibited a rank growth of aquatic plants, and terminated, after following a very narrow channel, in a quagmire. We had, in fact, reached the commencement of the Keweena Portage.

Before quitting this spot, it may be well to say, that the geology of the country had again changed. Portage Lake lies, in fact, in the direction of the great copper-bearing trap dyke. This dyke, estimating from the end of the peninsula, extends nearly southwest and northeast, probably seventy miles, with a breadth of ten miles. It is overlaid by rubble-stone and amygdaloid, which latter, by disintegration, yields the agates, carnelians, and other silicious, and some sparry crystalline minerals, for which the central shores of Lake Superior are remarkable. Nearly every part of this broad and extensive dyke which has been examined, yields veins, and masses of native copper, or copper ores.

The word was, when we had pushed our canoes into the quagmire, that each of the gentlemen of the party was to carry his own personal baggage across the portage. This was an awkward business for most of us. The distance was but two thousand yards, but little over a mile, across elevated open grounds. I strapped my trunk to my shoulders, and walked myself out of

breath in getting clear of the brushy part of the way, till reaching
the end of the first *pause*, or resting-place.   Here I met the
Governor (Cass), who facetiously said: "You see I am carrying
two pieces," alluding to his canoe slippers, which he held in his
hands.   "*A piece*," in the trade, is the back load of the *engagee*.

On reaching the termination of the second "pause," or rest, we
found ourselves on a very elevated part of the shore of Lake
Superior.   The view was limitless, the horizon only bounding
the prospect.   The waves rolled in long and furious swells from
the west.   To embark was impossible, if we had had our baggage all brought up, which was not the case.   The day was quite
spent before the transportation was completed.   This delay gave
us an opportunity to ramble about, and examine the shore.   In
a boulder of serpentine rock, I found an imbedded mass of native
copper, of two pounds' weight.   On breaking the stone, it proved
to be bound together by thin filaments of this metal.   Small
water-worn fragments of chalcedony, agate, carnelian, and other
species of the quartz family were found strewn along the beach,
together with fragments of zeolite.   Masses of the two former
minerals were also found imbedded in amygdaloid and trap-rock,
thus denoting the parent beds of rock.   In the zeal which these
little discoveries excited on the subject of mineralogy, the Chippewa, Ottowa, and Shawnee Indians attached to the expedition
participated, and as soon as they were made acquainted with the
objects sought, they became successful explorers.   They had
noticed my devotion to the topic, from the time of our passing
the Islands of Shawangunk, Michilimackinac, and Flat-rock Point,
in the basin of Lake Huron, where organic forms were chiselled
from the rock; and bestowed on me the name of Paguäbë-
kiegä.*

It turned out the next morning, that the whole of the baggage
and provisions had not been brought up, nor any of the canoes.
This work was early commenced by the men.   About half the
day was employed in the necessary toil.   When it was concluded,
the wind on the lake had become too high, blowing in an adverse
direction, to permit embarkation.   Nothing remained but to sub-

* The equivalent of geologist or mineralogist, from *pagua*, a tabular surface;
aubik, a rock; and *ëga*, the active voice of the verb to strike.

mit to the increased delay, during which we made ourselves as familiar with the neighboring parts of the lake shore as possible. During the time the expedition remained encamped at the portage, I made a short excursion up the peninsula northeastwardly, accompanied by Captain Douglass, Mr. Trowbridge, and some other persons. The results of this trip are sufficiently comprehended in what has already been stated respecting the geology and mineralogy of this prominent peninsula.

On the following morning (27th) the wind proved fair, and the day was one of the finest we had yet encountered on this fretful inland sea. We embarked at half-past four A. M., every heart feeling rejoiced to speed on our course. The prominent headlands, west of this point, are capped, as those on its southeastern border, with red sandstone. The wind proved full and adequate to bear us on, without endangering our safety, which enabled the steersmen to hold out boldly, from point to point. We had not proceeded far beyond the cliffs west of the portage, when the dim blue outlines of the Okaug or Porcupine Mountains* burst on our view.† Their prominent outline seemed to stretch on the line of the horizon directly across our track. The atmosphere was quite transparent, and they must have been seen at the distance of sixty miles. Captain Douglass thought, from the curve of the earth, that they could not be less than eighteen hundred feet in height. We successively passed the entrance of Little Salmon-Trout, Graverod, Misery, and Firesteel Rivers, at the latter of which a landing was made; when we again resumed our course, and entered the Ontonagon River, at half-past three in the afternoon. A large body of water enters the lake at the spot, but its mouth is filled up very much by sands. One of those curious refluxes is seen here, of which a prior instance has been noticed, in which its waters, having been impeded and dammed up by gales of wind, react, at their cessation, with unusual force. The name of the River Ontonagon‡ is, indeed, due to these refluxes, the prized dish of an Indian female having, agreeably to tradition, been carried out of the river into the lake.

Captain Douglass made observations for the latitude of the

* From *kaug*, a porcupine.

† For the view of this scene, see Information on the History, Condition, and Prospects of the Indian Tribes, vol. iv. Title iv.

‡ From the expression *nontonagon*, my dish; and *neen*, the pronoun *my*.

place, and determined it to be in north latitude 46° 52′ 2″. The
stationary distances of the route are given in the subjoined list,
in which it may be observed that they are probably exaggerated
about one-third by the voyagers and northwest traders, who
always pride themselves on going great distances; but they denote very well, in all cases, the *relative* distances.

*Stationary Distances between Michilimackinac and the River
Ontonagon.*

	Miles.	Total Miles.
From Michilimackinac to Detour	40	
Thence to Sault de St. Marie	45	85
Point aux Pins	6	91
Point Iroquois, at the entrance into Lake Superior	9	100
Taquamenon River	15	115
Shelldrake River	9	124
White-Fish Point	9	133
Two-Hearted River	24	157
Grande Marrais, and commencement of Grande Sables	21	178
La Point la Grande Sables	9	187
Pictured Rocks (La Portaille)	12	199
Doric Rock, and Miner's River	6	205
Grande Island	12	217
River aux Trains	9	226
Isle aux Trains	3	229
Laughing-Fish River	6	235
Chocolate River	15	250
Dead River (in Presque Isle Bay)	6	256
Granite Point	6	262
Garlic River	9	271
St. John's River, or Yellow Dog Run	15	286
Salmon-Trout, or Burnt River	12	298
Pine River	6	304
Huron River (Huron Islands lie off this River)	9	313
Point aux Beignes (east Cape of Keweena Bay)	6	319

	Miles.	Total Miles.
Mouth of Portage River	21	340
Head of Portage River (through Keweena Lake)	24	364
Lake Superior, at the head of the Portage	1	365
Little Salmon-Trout River . . .	9	374
Graverod's River (small, with flat rocks at its mouth)	6	380
Rivière au Misère	12	392
Firesteel River	18	410
Ontonagon, or Coppermine River . .	6	416

o

CHAPTER VI.

Chippewa village at the mouth of the Ontonagon—Organize an expedition to explore its mineralogy—Incidents of the trip—Rough nature of the country—Reach the copper rock—Misadventure—Kill a bear—Discoveries of copper—General remarks on the mineral affluence of the basin of Lake Superior.

A SMALL Chippewa village, under the chieftainship of Tshwee-tshweesh-ke-wa, or the Plover, and Kundekund, the Net Buoy, was found on the west bank of the river, near its mouth, the chiefs and warriors of which received us in the most friendly manner. If not originally a people of a serene and placid temperament, they have been so long in habits of intercourse with the white race that they are quite familiar with their manners and customs, and mode of doing business. They appeared to regard the Canadian-Frenchmen of our party as if they were of their own mode of thinking, and, indeed, almost identical with themselves.

The Ontonagon River had, from the outset, formed an object of examination, from the early and continued reports of copper on its borders. It was determined to lose no time in examining it. Guides were furnished to conduct a party up the river to the locality of the large mass of this metal, known from early days. This being one of the peculiar duties of my appointment, I felt the deepest interest in its success, and took with me the apparatus I had brought for cutting the rock and securing proper specimens.

The party consisted of Governor Cass, Dr. Wolcott, Captain Douglass, Lieutenant Mackay, J. D. Doty, Esq., and myself. We embarked in two canoes, with their complement of men and guides. It was six o'clock, when, leaving the balance of the expedition encamped at the mouth of the river, east shore, we took our departure, in high spirits, for the copper regions. A broad river with a deep and gentle current, with a serpentine channel, and heavily wooded banks with their dark-green foliage overhanging the water, rendered the first few miles of the trip delightful. At the distance of four miles, we reached a sturgeon-fishery, formed by extending a weir across the river. This weir consists of upright and horizontal stakes and poles, along the latter of which the Indians move and balance themselves, having in their hands an iron hook on a pole, with which the fish are caught. We stopped a few moments to look at the process, received some of the fish drawn up during our stay, which are evidently the *Acipenser oxyrinchus,* and went on a couple of miles higher, where we encamped on a sandbar. Here we were welcomed, during the sombre hours of the night, with a pertinacity we could have well dispensed with, by the mosquitos.

We resumed the ascent at four o'clock in the morning. The river is still characterized for some miles by rich alluvial banks, bearing a dense forest of elm, maple, and walnut, with a luxuriant growth of underbrush. But it was soon perceived that the highlands close in upon it and narrow its channel, which murmurs over dangerous beds of rocks and stones. Almost imperceptibly, we found ourselves in an alpine region of a very rugged character. The first rapid water encountered had been at the Indian wier, on the 27th. These rapids, though presenting slight obstacles, became more frequent at higher points. We had been in our canoes about three hours, the river having become narrower and more rapid, when the guides informed the party that we had ascended as far into the mountainous district as was practicable; that there was a series of bad rapids above; and that, by landing at this spot, the party could proceed, with guides, to the locality of the copper rock. Accordingly, arrangements were made to divide the party; Governor Cass placed at my service the number of men necessary to explore the country on foot, and carry the implements. Dr. Wolcott and Captain Douglass joined me. I took my departure with eight persons, including two Indian guides, in quest of the mineral region, over the highlands on the west bank of the river; while the Governor, Major Forsyth, and the other guides, remained with the canoes, which were lightened of half their burden, in hopes of their being able to ascend the stream quite to the Rock. Starting with my party with alacrity, this trip was found to be one of no ordinary toil.

Not only was the country exceedingly rough, carrying us up and down steep depressions, but the heat of the sun, together with the exercise, was oppressive, nor did our guides seem to move with a precision which betokened much familiarity with the region, if they did not feel, indeed, some compunction on leading whites to view their long superstitiously concealed mineral treasures. At one o'clock we came to an Indian path, leading directly to the place. The guides here sat down to await the party under Governor Cass, who were expected to join us at this spot. The thermometer at this hour stood at 90° in the shade of the forest. We had not been long seated when the other party made their appearance; but the Governor had been so much exhausted by clambering up the river hills, that he determined to return to his point of landing in the river. In this attempt he was guided by one of the Ontonagon Indians, named Wabiskipenais,* who missed his way, and wandered about he knew not whither. We leave him to thread his way back into the valley, with the Executive of the Territory, wearied and perplexed, at his heels, while the results of my excursion in search of the copper rock are detailed. After the reunion at the path, my mineralogical party proceeded some five or six miles, by estimation, farther, through a more favorable region, towards the object of search. On approaching the river, they passed some antique excavations in the forest, overgrown with saplings, which had the appearance of age, but not of a remote age. Coming to the brink of the river, we beheld the stream brawling over a rapid stony bed, at the depth of, perhaps, eighty or a hundred feet below. Towards this, its diluvial banks, charged with boulders and pebbles, sloped at a steep angle. At the foot, laid the large mass we were in search of, partly immersed in the water. Its position may be inferred from the following sketch :—

* From *wabiska*, white (transitive animate), and *penasee*, a bird.

Fig. 1.

The rock consists of a mass of native copper in a tabular boulder of serpentine. Its, face is almost purely metallic, and more splendent than appears to consist with its being purely metallic copper. There is no appearance of oxidation. Its size, roughly measured, is three feet four inches, by three feet eight inches, and about twelve or fourteen inches thick in the thickest part. The weight of copper, exclusive of the rock, is not readily estimated; it may be a ton, or a ton and a half. Old authors report it at more than double this weight. The quantity has been, however, much diminished by visitors, who have cut freely from it. I obtained adequate specimens, but found my chisels too highly tempered, and my hammer not heavy enough to separate large masses. Having made the necessary examinations, we took our way back up the elevated banks of the river, and across the forest about six miles, to the final place of debarkation of Gov. Cass and his party. But our fears were at once excited on learning that the Governor, with his guide, Wabishkepenais, had not reached the camp. It was already beginning to be dark, and the gloom of night, which is impressive in these solitudes, was fast closing around us. Guns were fired, to denote our position, and a light canoe was immediately manned, placed in charge of one of the gentlemen, and sent up the river in search. This canoe had not proceeded a mile, when the object of search was descried, with his companions, sitting on the banks of the river, with a real jaded air, with his Indian guide standing at no great distance. Wabishkepenais had been bewildered in his tracks, and

7

finally struck the river by the merest chance. The Governor, on reaching camp, looked as if he had been carried over steeps and through gloomy defiles, which had completely exhausted his strength, and he was not long in retiring to his tent, willing to leave such rough explorations for the present, at least, to other persons, or, if he ever resumed them, to do it with better guides. Poor Wabishkepenais looked chagrined and as woebegone himself as if he had encountered the bad influences of half the spirits of his Indian mythology ; for the fellow had really been lost in his own woods, and with a charge by whom he had felt honored, and employed his best skill to conduct. The camp-fires already threw their red glare among the trees as night spread her sable pall over us. The tents were pitched ; the canoes turned up on the shore to serve as a canopy for the men to sleep under. Indians and Canadians were soon engaged at their favorite pipes, and mingled their tones and hilarious conversation ; and we finally all slept the sounder for our eventful day's toils and misadventures. But deeply printed on our memory, and long to remain there, are the thrilling scenes of that day and that night.

At five o'clock the next morning, the entire camp was roused and in motion, when we began to descend the stream. We had descended about ten miles, when the Ontonagon Indians stopped the canoes to examine a bear-fall, on the east bank. It was a fine open forest, elevated some six or eight feet above the water. It was soon announced that a bear was entrapped. We all ascended the bank, and visited the locality. The structure had been so planned that the animal must needs creep lowly under a crib of logs to get at the bait, which he no sooner disturbed than a weight of logs fell on his prostrated legs. The animal sat up partially on his fore paws, when we advanced, the hinder being pressed heavily to the earth. One of the Indians soon fired a ball through his head, but it did not kill him, he still kept his upright position. Dr. Wolcott then requested permission to fire a shot, which was aimed at the heart, and took effect about that part, but did not kill him. One of the Indians then dispatched him with an axe. He was no sooner dead than one of the Indians, stepping up, addressed him by the name *Muk-wah*, shook him by the paw, with a smiling countenance, saying, in the Indian language, that he was sorry they had been under the necessity of killing him, and hoped the offence would be forgiven, as one of the shots fired had been from an American.*

This act of the Indian addressing the bear, will be better understood, when it is stated that their mythology tells them, that the spirit of the animal must be encountered in a future state, when the enchantment to which it is condemned in this life, will be taken off.

On passing down the river, an Indian had promised to disclose another mass of native copper, near the river, and we stopped at a spot indicated, to enable him to bring it. Whether he repented of his too free offer, agreeably to Indian superstition, or feared some calamity to follow the disclosure, or really encountered some difficulty in finding it, I know not, but it is certain that, after some time spent in the search, or affected search, he came back to the river without producing it.

Soon after this incident, we reached the mouth of the river, and found the party left encamped at that point, in charge of Mr. Trowbridge and Mr. Doty, well, nothing having occurred in our absence. The wind was, however, adverse to our embarkation, had it been immediately desired.

A council of the Ontonagon Indians was summoned, which met in the after part of the day; speeches were delivered, and replied to, and presents distributed. A silver medal was presented to Wabishkepenais.

Head winds continuing, we were farther detained at this spot the following day. While thus detained, an Ontonagon Indian brought in a mass of native copper, from the banks of this river, weighing eight or nine pounds. This mass was of a flattened, orbicular shape, and its surface coated with a green oxide. At a subsequent part of my acquaintance with this river, another mass of native copper (still deposited in my cabinet) was brought to me, from the east fork of the river, which weighed from forty to fifty pounds. This mass, of a columnar shape, originally embraced a piece of stone which the Indian finding it had detached. It was also coated with a dark green oxide of copper. Both of these masses appeared to have been volcanic. Neither of them had the slightest traces of gangue, or vein-matter, nor of attrition in

* Chemoquiman, from *gitches*, great, and *moquiman*, knife.

being removed from the parent beds. The following sketches depict the shapes of these masses.

Fig. 2.   Fig. 3.

With respect to the general question of the mineral character of this part of the country, and the probable value of its mineral and metallic deposits to the public domain, the entire class of facts, from which a judgment must be formed, are favorable.* Salts and oxides of copper are not only seen in various places in its stratification, but these indications of mineral wealth in this article are confirmed, by the subsequent discovery of masses of native copper, along the shore, and imbedded in its traps and amygdaloids. In addition to the opportunities of observation furnished by this expedition, subsequent public duties led me to perform seven separate trips along its shores, and each of these but served to accumulate the evidences of its extraordinary mineral wealth. Indications of the sulphurets, arseniates, and other ores of this metal are found in the older class of horizontal rocks; but it is to the trap-rocks alone that we must look for the veins of native metal. Some of these masses contain silver, in a state of combination. Traces of this metal, chiefly in the boulder form, are found in the metalliferous horizontal strata. Nor is there wanting evidence, that there are localities of virgin

---

* *Vide* Reports in the Appendix: 1. Report on the Copper Mines of Lake Superior, November 6, 1820. 2. Report on the Value of the Existing Evidences of Mineral Wealth in the Basin of Lake Superior to the Public Domain, October 1, 1822.

copper, which do not promise a considerable percentage of the metal. A mass of steatite, imbedding a heavy mass of pure native silver, which had been probably carried from the north-west, with the drift stratum, was found cast out quite into the Huron basin; and this rock, in its intimate associations with the serpentine formation of Lake Superior, should be closely scruti-nized. There is also a formation of slate and quartz in the primitive district, which is entitled to particular attention.

Inorganic masses are developed, throughout the globe, without regard to climate. Russia yields the precious metals in great profusion, and there are no laws governing the distribution of these metals, which forbid the expectation that they should be abundantly disclosed by the stratification of the basin of Lake Superior. With respect to the useful metals, particularly copper and iron, it is undeniably the richest and most extensive locality of these metals on the globe.*

* Geological Report, *vide* Appendix.

CHAPTER VII.

Proceed along the southern coast of Lake Superior from the Ontonagon, to Fond du Lac—Porcupine range of mountains—Streams that run from it, at parallel distances, into the lake—La Pointe—Group of the Federation Islands—River St. Louis—Physical geography of Lake Superior.

HEAD winds detained the expedition at the mouth of the Ontonagon, during the day and the day following that of our arrival from the copper rock. It was the first of July, at half-past four o'clock, A. M., when the state of the lake permitted us to embark. Steering west, we now had the prominent object of the Porcupine Mountains constantly in view. At the distance of fifteen miles, we passed the Pewabik Seebe, or Iron River. This stream, after ascending it a couple of miles, is a mere torrent, pouring from the Porcupine Mountains, over a very rough bed of grauwakke, which forbids all navigation. At the computed distance of five leagues beyond this stream, we passed the river called Pusábika, or Dented River, so called from standing rocks, which resemble broken human teeth. The Canadians, who, as previously remarked, appear to have had but a limited geographical vocabulary, called this Carp River, neglectful of the fact that they had already bestowed the name on a small river which flows into the bay south of Granite Point.* We were now at the foot of the Kaug range, which is one vast upheaval of trap-rock, and has lifted the chocolate-colored sandstone, at its base, into a vertical position. The Pusábika River originates in this high trap range, from which it is precipitated, at successive leaps, to the level of the lake, the nearest of which, a cascade of forty feet, is within three miles of the river's mouth.

Six miles further brought us to the Presque Isle River of the

---

* Now the seat of the Marquette Iron Works.

Canadians, for which I heard no Indian name. It also originates on this lofty trap range, and has worn its bed through frightful chasms in the grauwackke, through which it enters the lake. Within half a mile of its entrance, the river, hastening from its elevations, drops into a vast cauldron scooped in the grauwackke rock, whence it glides into the lake. Here are some picturesque and sublime views, worthy the pencil.

Two leagues beyond this river we reached and passed the entrance of Black River, another of the streams from the Kaug range. It is stated to be rapid, and to have its source south of the mountains, in a district sheltered from the lake winds, and suited to agriculture. Its borders bear at the same time indications of mineral wealth. Eight miles beyond this river, we encamped on the open shores of the lake, after travelling fifty miles. Having been doubled up in the canoe for all this distance, landing on terra firma, and being able to stretch one's legs, seemed quite a relief. "I will break a lance with you," quoth A to B, addressing Mr. Trowbridge, offering him at the same time a dried stalk, which had been cast up by the waves. We were, in fact, as much pleased to get ashore, after the day's confinement, as so many boys let loose from confinement in school. In strolling along the shore, I recognized the erismatolite, in the dark up-heaved sandstone at this locality.

We here observed a phenomenon, which is alluded to by Charlevoix as peculiar to this lake. Although it was calm, and had been so all day, save a light breeze for a couple of hours after leaving the Ontonagon, the waters near shore were in a perfect rage, heaving and lashing upon the rocks, in a manner which rendered it difficult to land. At the same time, scarce a breath of air was stirring, and the atmosphere was beautifully serene.

On passing thirteen miles, the next morning, we reached the mouth of the Montreal River, which is the last of the mountain streams of the Kaug range. It throws itself from a high precipice of the vertical sand-rock, within sight of the lake, creating quite a picturesque view.* (Vide *Information respecting the History, Customs, and Prospects of Indian Tribes*, vol. iv. plate 26.)

On landing here a few moments, at an early hour, the air being

---

* This river has subsequently been fixed on as the northwestern boundary of the State of Michigan, separating it from Wisconsin.

hazy, we knocked down some pigeons, which flew very low.*
This bird seems to be precisely the common pigeon of the Atlantic borders.   The Indians had constructed a fish-weir between
the lake and Montreal falls, where the lake sturgeon are caught.

After passing about a league beyond the Montreal, the voyager
reaches a curve in the lake shore, at which it bends to the north
and northwest.   This curve is observed to extend to the De Tour
of the great bay of Fond du Lac, a computed distance of the
voyageurs of thirty-six miles, which, as before indicated, is about
one-third overrated.   The immediate shore is a level plain of
sand, which continues to Point Chegoimegon, say eighteen miles.
About two-thirds of this distance, the Muskeego† River enters
through the sandy plain from the west.   This is a large stream,
consisting of two primary forks, one of which connects it with
Chippewa River, and the other with the River St. Croix of the
Mississippi.   The difficulties attending its ascent, from rapids and
portages, have led the French to call it Mauvaise, or Bad River.‡

---

* BIRDS OF LAKE SUPERIOR.—Of the species that frequent the vicinity of this
lake, the magpie is found to approach as far north as Lac du Flambeau, on the head
of the Montreal and Chippewa Rivers.   This bird is called by the Chippewas Wa-
bish Kagagee, a name derived from *Wabishkau*, white animate, and *Kaw-gaw-gee*, a
crow.   The three-toed woodpecker visits its forests.   The T. polyglottis has been
seen as far north as the Island of Michilimackinac.   In the spring of 1828, a species of grosbeak visited St. Mary's, of which I transmitted a specimen to the New
York Lyceum of Natural History, where it received the name of Evening Grosbeak.

† From *Muskeeg*, a swamp or bog, and o, the sign of the genitive.

‡ MUSKEEGO, or MAUVAIS RIVER.—In 1831, the United States government placed
under my charge an expedition into the Indian country which ascended this river,
with a view to penetrate through the intervening region to the Mississippi. Indian
canoes were employed, as being best adapted to its rapids and portages, which were
managed by *voyageurs*.   A detachment of infantry, under Lieut. R. Clary, was added.
The tribes in this secluded region were then meditating the outbreak which eventuated the next year in the Black Hawk War.   This expedition ascended the river
through a most embarrassing series of rapids and rafts, which often choked up its
channel for miles, into a long lake, on its summit, called K nogumaug.   From
the northwest end of this, it passed, from lake to lake, to the Namakagun fork of
the River St. Croix of the Mississippi, descended that stream to   w River, then retraced the Namakagun to a portage to Ottowa Lake, a source of Chippewa River, then
to a portage into Lac Chetac, the source of the Red Cedar, or Follavoine River, and
pursued the latter to the main channel of the Chippewa, and by the latter into the
Mississippi, which it enters at the foot of Lake Pepin; thence down the Mississippi to Prairie du Chien, and through the present area of the State of Wisconsin,
by the Wisconsin and Fox Rivers, to Green Bay; thence through Lakes Michigan
and Huron to Sault de Ste Marie.

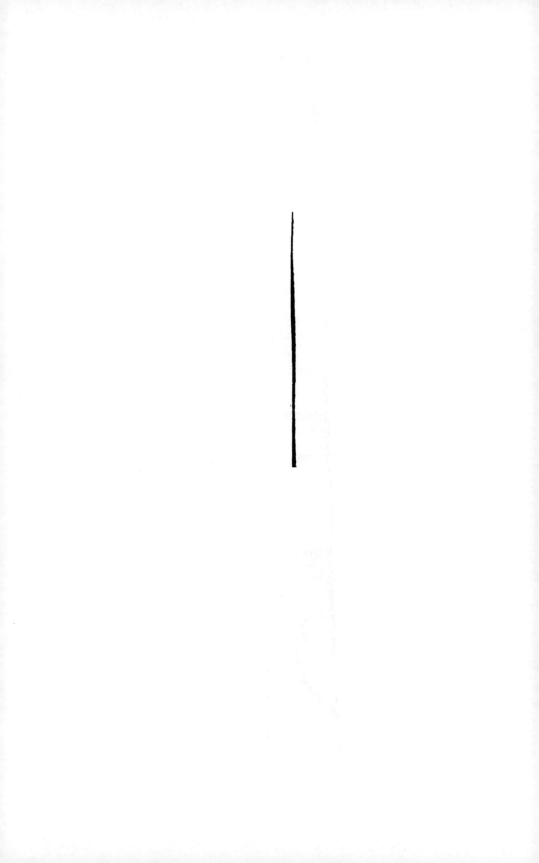

Passing this river, we continued along the sandy formation to its extreme termination, which separates the Bay of St. Charles by a strait from that remarkable group of islands, called the Twelve Apostles by Carwer. It is this sandy point, which is called La Pointe Chagoimegon* by the old French authors, a term now shortened to La Pointe. Instead of "twelve," there are, however, nearer thirty islands, agreeably to the subjoined sketch, by which it is seen that each State in the Union may stand sponsor for one of them, and they might be more appropriately called the *Federation Group*. Touching at the inner or largest of the group, we found it occupied by a Chippewa village, under a chief called Bezhike. There was a tenement occupied by a Mr. M. Cadotte, who has allied himself to the Chippewas. Hence we proceeded about eleven miles to the main shore, where we encamped at a rather late hour. I here found a recurrence of the granitic, sienitic, and hornblende rocks, in high orbicular hills, and improved the brief time of daylight to explore the vicinity. The evening proved lowering and dark, and this eventuated in rain, which continued all night, and until six o'clock the next morning. Embarking at this hour, we proceeded northwest about eight miles, to Raspberry River, and southwest to Sandy River. Here we were driven ashore by a threatening tempest, and before we had unladen the canoes, there fell one of the most copious and heavy showers of rain. The water seemed fairly to pour from the clouds. We had not pitched a tent, nor could the slightest shelter be found. There seemed but one option at our command, namely, that between sitting and standing. We chose the latter, and looked at each other, it may be, foolishly, while this rain tempest poured. When it was over, we were as completely wetted as if it had been our doom to lay at the bottom of the lake. When the rain ceased, the wind rose directly ahead, which confined us to that spot the rest of the day. The next day was the Fourth of July—a day consecrated in our remembrance, but which we could do no more than remember. The wind continued to blow adversely till about two o'clock, when we embarked, not without feeling the lake still laboring under the agitation into which it had been thrown. On travelling three miles, we turned the prominent point, called De Tour of

* From *Shaugwamegun*, low lands, and *ing*, a place.

Fond du Lac. At this point our course changed from northwest to south-southwest.

The sandstone formation here showed itself for the last time. The shore soon assumes a diluvial character, bordered with long lines of yellow sand and pebbles. In some places, heavy beds of pure iron sand were observed. The agitation which marked the lake soon subsided, under the change of wind, and our men seemed determined, by the diligence with which they worked, to make amends for our delay at Sandy River.

At eight o'clock in the evening we came to Cranberry River and encamped, having, by their estimation, come twenty-three miles. The evening was perfectly clear and calm, with a striking twilight, which was remarked all night. These lengthened twilights form a very observable feature as we proceed north. Mackenzie says that, in lat. 67° 47', on the 11th of July, 1789, he saw the sun above the horizon at twelve o'clock P. M.

The calmness and beauty of the night, and our chief's anxiety to press forward, made this a short night. Gen. Cass aroused the camp at a very early hour, so that at three o'clock we were again upon the lake, urging our way up the Fond du Lac Bay. The sun rose above the horizon at ten minutes before four o'clock. The morning was clear and brilliant. Not a cloud obscured the sky, and the waves of the lake spread out with the brightness of a mirror. At the distance of five leagues, we passed the mouth of the Wisakoda, or Broule River,* a stream which forms the connecting link with the Mississippi River, through the St. Croix. Three miles beyond this point we landed a short time, on the shore, where we observed a stratum of iron sand, pure and black, a foot in thickness.

* Wisacoda, or Broule River.—On returning down the Mississippi River, from the exploration of its sources, in 1832, I ascended the River St. Croix quite to its source in St. Croix Lake. A short portage, across a sandy summit, terminated at the head springs of the Wisacoda, which, from a very narrow and tortuous channel, is soon increased in volume by tributaries, and becomes a copious stream. Thus swelled in volume, it is dashed down an inclined plane, for nearly seventy miles, over which it roars and foams with the impetuosity of a torrent. It is not till within a few miles of Lake Superior that it becomes still and deep. The entire length of the river may be estimated at one hundred miles. It has two hundred and forty distinct rapids, at some of which the river sinks its level from eight to ten feet. It cannot fall, in this distance, less than 500. That it should ever have been used in the fur trade, is to be explained by the fact that it has much water.

At eleven o'clock, a northeast wind arose, which enabled the expedition to hoist sail. Land on the north shore had for some time been in sight, across the bay, and the line of coast soon closed in front, denoting that we had reached the head of the lake. At twelve o'clock, we entered the mouth of the River St. Louis, having been eighteen days in passing this lake, including the trip to the Ontonagon.

Before quitting Lake Superior, whose éntire length we have now traversed, one or two generic remarks may be made; and the first respects its aboriginal name. The Algonquins, who, in the Chippewa tribe, were found in possession of it, on the arrival of the French, early in the seventeenth century, applied the same radical word to it which they bestow on the sea, namely, Gum-ee (Collected water), or, as it is sometimes pronounced, Gom-ee, or Go-ma; with this difference, that the adjective big (gitchè) prefixed to this term for Lake Superior, is repeated when it is applied to the sea. The superlative is formed when it is meant to be very emphatic, in this language, by the repetition of the adjective; a principle, indeed, quite common to the Indian grammars generally. The word did not commend itself to French or English ears, so much as to lead to its adoption. By taking the syllable Al·from Algonquin, as a prefix, instead of gitchè, we have the more poetic combination of Algoma.

Geographers have estimated the depth of this lake at nine hundred feet. By the surveys of the engineers of the New York and Erie Canal, the surface of Lake Erie is shown to be five hundred and sixty feet above tide-water, which, agreeably to estimates kept on the present journey, lies fifty-two feet below the level of Lake Superior. These data would carry the bottom of the lake two hundred and eighty-eight feet below tide water. What is more certain is this, that it has been the theatre of ancient volcanic action, which has thrown its trap-rocks into high precipices around its northern shores and some of its islands, and lifted up vast ranges of sandstone rocks into a vertical position, as is seen at the base of the Porcupine Mountains. Its latest action appears to have been in its western portion, as is proved by the upheaval of the horizontal strata; and it may be inferred that its bed is very rough and unequal.

The western termination of the lake, in the great bay of Fond du Lac, denotes a double or masked shore, which appears to have been formed of pebbles and sands, driven up by the tempests, at the distance of a mile or two, outside of the original shore. The result is shown by an elongated piece of water, resembling a lake, which receives at the north, the River St. Louis, and the *Agoche*, or Lefthand River, at its south extremity.

About three miles above the mouth of the river, we landed at a Chippewa village. While exchanging the usual salutations with them, we noticed the children of an African, who had intermarried with this tribe. These children were the third in descent from Bongo, a freed man of a former British commanding officer at the Island of Michilimackinac. They possessed as black skins as the father, a fact which may be accounted for by observing, what I afterwards learned, that the marriages were, in the case of the grandfather and father, with the pure Indian, and not with Africano-Algonquin blood; so that there had been no direct advance in the genealogical line.

The St. Louis River discharges a large volume of water, and is destined hereafter to be a port of entry for the lake shipping, but at present it has shoals of sand at its mouth which would bar the entrance of large vessels. Proceeding up the river, we found it very serpentine, and abounding in aquatic plants, portions of it yielding the wild rice. At the computed distance of twenty-four miles, we reached the establishment of the American Fur Company. It was seven o'clock when we came to the place, where we encamped.

Lake Superior is called by the Chippewas a sea.

The superficial area of the lake has been computed by Mr. Darby at a little under nine hundred billions of feet, and its depth at nine hundred feet. By the latest surveys and estimate, the altitude of Lake Superior above tide water, is about six hundred and forty feet.* Allowing Mr. Darby's computation to be correct, this would sink its bed far below the surface of the Atlantic.

This lake has been the theatre of very extensive volcanic action. Vast dykes of trap traverse its northern shores. One of the principal of these has apparently extended across its bed, from northeast to southwest, to the long peninsula of Keweena, producing at the same time, the elevated range of the Okaug

* *Vide* Appendix.

Mountains. One of the most remarkable features of these dykes is the numerous and extensive veins of native copper which characterize them. Subsequent convulsions, and the demolition of these ancient dykes, by storms and tempests, have scattered along its shores abundant evidence of the metal and its ores and vein-stones, which have attracted notice from the earliest time. The geology of its southern coasts may be glanced at, and inferred, from the subjoined outlines.

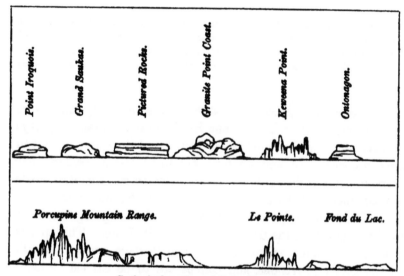

Geological outline of Lake Superior.

The teachings of topography, applied to commerce, are wonderful. A longitudinal line, dropped south, from this point, would cross the Mississippi at the foot of Lake Pepin, and pass through Jefferson city on the Missouri. When, therefore, a ship canal shall be made at St. Mary's Falls, vessels of large tonnage may sail from Oswego (by the Welland canal) and Buffalo, through a line of inter-oceanic seas, nearer to the foot of the Rocky Mountains, by several hundred miles, than by any other possible route. A railroad line from Fond du Lac west to the Columbia valley, would also form the shortest and most direct transit route from the Pacific to New York. Such a road would have the advantage of passing through a region favorable to agriculture, which cannot but develop abundant resources.

CHAPTER VIII.

Proceed up the St. Louis River, and around its falls and rapids to Sandy Lake in the valley of the Upper Mississippi—Grand Portage—Portage aux Coteaux—A sub-exploring party—Cross the great morass of Akeek Scepi to Sandy Lake—Indian mode of pictographic writing—Site of an Indian jonglery—Post of Sandy Lake.

WE had now reached above nine hundred and fifty miles from our starting-point at Detroit, and had been more than forty days in traversing the shores of Lakes Huron and Superior. July had already commenced, and no time was to be lost in reaching our extreme point of destination. Every exertion was therefore made to push ahead. By ten o'clock of the morning after our arrival at the Fond du Lac post, we embarked, and after going two miles reached the foot of the first rapids of the St. Louis. This spot is called the commencement of the Grand Portage—over this path all the goods, provisions, and canoes are to be carried by hand nine miles. During this distance, the St. Louis River, a stream of prime magnitude, bursts through the high trap range of what Bouchette calls the Cabotian Mountains, being a continuation of the upheavals of the north shore of Lake Superior, the river leaping and foaming, from crag to crag, in a manner which creates some of the most grand and picturesque views. We sometimes stood gazing at their precipices and falls, with admiration, and often heard their roar on our path, when we were miles away from them. Capt. Douglass estimated the river to fall one hundred and eight feet during the first nine miles; and from estimates furnished me by Dr. Wolcott, the aggregate fall from the mouth of the Savannè, to that point, is two hundred and twelve feet. We found the first part of the ascent of its banks very precipitous and difficult, particularly for the men who bore burdens, and what rendered the labor almost insupportable was the heat, which stood at 82°, in the shade, at noon. We made but five *pauses* the first day; and were three days on the portage. It rained the second day, which added much to the difficulty of our progress. We now found ourselves, at every step, advancing into a wild and rugged region. Everything around us wore the aspect of remoteness. Dark forests, swampy grounds, rocky precipices, and the distant roaring of the river, as it leapt from rock to rock, would have sufficiently impressed the mind with the presence of the wilderness, without heavy rains, miry paths, and the train of wild and picturesque Indians, who constituted a part of our carriers.

The rocks, at the foot of the portage, consisted of horizontal red sandstone. On reaching the head of it, we found argillite in a vertical position. I found the latter, in some places, pervaded by thin veins of quartz, and in one instance by grauwackke. At one spot there was a small vein of coarse graphite in the argillite. Large blocks of black crystallized hornblende rock lie along the shores, where we again reached the river, and are often seen on its bed, amid the swift-running water, but I did not observe this rock in place. Among the loose stones at the foot of the portage, I picked up a specimen of micaceous oxide of iron. Such are the gleams of its geology and mineralogy. The growth of the forest is pines, hemlock, spruce, birch, oak, and maple. In favorable situations, I observed the common red raspberry, ripe.

On embarking above the portage, the expedition occupied seven canoes, of a size most suitable for this species of navigation. Our Indian auxiliaries from Fond du Lac were here rewarded, and dismissed. On ascending six miles, we reached the Portage aux Coteaux, so called from the carrying path lying over a surface of vertical argillite. This rock, standing up in the bed, or on the banks of the stream, with a scanty overhanging foliage of cedar, gives a peculiarly wild and abrupt aspect to the scene; which is by no means lessened by the loud roaring of the waters. There is a fall and rapid at this portage, where the river, it may be estimated, sinks its level about fourteen feet.

We encamped at the head of this portage, where the water again permits the canoes to be put in. Thus far, we had found this stream a broad, flowing torrent, but owing to its rapids and rocks, anything but favorable to its navigation by boats, or canoes of heavy burden.   His excellency Gov. Cass, therefore, determined to relieve the river party, by detaching a sub-expedition across the country to Sandy Lake.   It was thought proper that I should accompany this party.   It consisted, besides, of Lieut. Mackay, with eight soldiers, and of Mr. Doty, Mr. Trowbridge, and Mr. Chase.   We were provided with an interpreter and two Chippewa guides, being sixteen persons in all.

Thus organized, we left the camp at the head of the portage, the following morning, at six o'clock.   Each one carried provisions for five days, a knife, a musquito bar, and a blanket or cloak.   There were a few guns taken, but generally this was thought to be an incumbrance, as we expected to see little game and to encounter a toilsome tramp.   The guides, taking their course by the sun, struck west into a close forest of pine, hemlock, and underbrush, which required energy to push through. On travelling a couple of miles, we fell into an Indian path leading in the required direction; but this path, after passing through two ponds, and some marshes, eventually lost itself in swamps. These marshes, after following through them, about four miles, were succeeded by an elevated dry sandy barren, with occasional clumps of pitch pine, and with a surface of shrubbery.   Walking over this dry tract was quite a relief.   We then entered a thick forest of young spruce and hemlock.   Two miles of this brought us to the banks of a small lake, with clear water, and a pebbly shore.   Having no canoe to cross it, our guides led us around its southern shores.   The fallen timber and brush rendered this a very difficult march.   To avoid these obstructions, as they approached the head of the lake, we eventually took its margin, occasionally leading into the water.   While passing these shores, I picked up some specimens of the water-worn agates, for which the diluvians in this quarter are remarkable.   We now fell into an old Indian path, which led to two small lakes, similar in size, to the former one, but with marshy borders, and reddish water. These small lakes were filled with pond lilies, rushes, and wild rice.   At the margin of the second lake, the path ceased, and the guides could not afterwards find it.   The path terminated abruptly at the second lake.   While searching about this, Chamees,*

* The pouncing hawk.

one of the Indian guides, found a large green tortoise, which he and his companion killed in a very ingenious and effectual way, by a blow from a hatchet on the neck, at the point where the shell or buckler terminates. After leaving this water, they appeared to be in doubt about the way; almost imperceptibly, we found ourselves in a great tamarak swamp. The bogs and moss served to cover up, almost completely, the fallen trees, and formed so elastic a carpet as to sink deep at every tread. Occasionally they broke through, letting the foot into the mire. This proved a very fatiguing tramp. To add to its toils, it rained at intervals all day. We were eleven hours in passing this swamp, and estimated, and probably over-estimated ourselves to have past twenty miles. We encamped at five o'clock near the shores of a third small lake, each one picking out for himself the most elevated spot possible, and the person who got a position most completely out of the water was the best man. It is fatigue, however, that makes sleep a welcome guest, and we awoke without any cause of complaint on that score.

The next morning, as we were about to depart, we observed near the camp-fire of our guides a pole leaning in the direction we were to go, with a birch-bark inscription inserted in a slit in the top of the pole. This was too curious an object not to excite marked attention, and we took it down to examine the hiero-glyphics, or symbols, which had been inscribed with charcoal on the birch scroll. We found the party minutely depicted by symbols. The figures of eight muskets denoted that there were eight soldiers in the party. The usual figure for a man, namely, a closed cross with a head, thus:—

and one hand holding a sword, told the tale that they were commanded by an officer. Mr. Doty was drawn with a book, they having understood that he was a lawyer. I was depicted with a hammer, to denote a mineralogist. Mr. Trowbridge and Mr. Chase, and the interpreter, were also depicted. Chamees and his companion were drawn by a camp-fire apart, and the figure of

8

the tortoise and a prairie-hen denoted the day's hunt. There were three hacks on the pole, which leaned to the N. W., denoting our course of travel. Having examined this unique memorial, it was carefully replaced in its former position, when we again set forward. It appeared we had rested in a sort of oasis in the swamp, for we soon entered into a section of a decidedly worse character than that we had passed the day before. The windfalls and decaying timber were more frequent—the bogs, if possible, more elastic—the spots dry enough to halt on, more infrequent, and the water more highly colored with infusions of decaying vegetable matter. We urged our way across this tract of morass for nine hours, during which we estimated our progress at fourteen miles, and encamped about four o'clock P. M., in a complete state of exhaustion. Even our Indian guides demanded a halt; and what had, indeed, added to our discouragements, was the uncertainty of their way, which they had manifested.

Our second night's repose in this swampy tract, was on ground just elevated above the water; the mosquitos were so pertinacious at this spot as to leave us but little rest. From information given by our guides, this wide tract of morass constitutes the sources of the Akeek Seebi, or Kettle River, which is one of the remotest sources of the Mille Lac, and, through that body of water, of Rum River. It is visited only by the Indians, at the proper season for trapping the beaver, marten, and muskrat. During our transit through it, we came to open spaces where the cranberry was abundant. In the same locality, we found the ripe fruit, green berries, and blossoms of this fruit.

It was five o'clock A. M. when we resumed our march through this toilsome tract, and we passed out of it, after pressing forward with our best might, during twelve hours. We had been observant of the perplexity of our guides, who had unwittingly, we thought, plunged us into this dreary and seemingly endless morass, and were rejoiced, on a sudden, to hear them raise loud shouts. They had reached a part of the country known to them, and took this mode to express their joy, and we soon found ourselves on the banks of. a small clear stream, called by them Bezhiki Seebi, or Buffalo Creek, a tributary to Sandy Lake. We had, at length, reached waters flowing into the Mississippi. On this stream we prepared to encamp, in high spirits, feeling, as those are apt to who have long labored at an object, a pleasure in some measure proportioned to the exertions made.

Any other people but the Indians would feel ill at ease in dreary regions like these. But these sons of the forest appear to carry all their socialities with them, even in the most forbidding solitudes. They are so familiarized with the notions of demons and spirits, that the wildest solitude is replete with objects of hope and fear. We had evidence of this, just before we encamped on the banks of the Bezhiki, when we came to a cleared spot, which had been occupied by what the Canadians, with much force, call a *jonglery*, or place of necromantic ceremonies of their priests or jossakeeds. There were left standing of this structure six or eight smooth posts of equal length, standing perpendicularly. These had been carefully peeled, and painted with a species of ochrey clay. The curtains of bark, extending between them, and isolating the powow, or operator, had been removed; but the precincts had the appearance of having been carefully cleared of brush, and the ground levelled, for the purposes of these sacred orgies, which exercise so much influence on Indian society.

We were awaked in our encampment, between four and five o'clock, the next morning, by a shower of rain. Jumping up, and taking our customary meal of jerked beef and biscuit, we now followed our guides, with alacrity, over a dry and uneven surface, towards Sandy Lake. We had now been three days in accomplishing the traverse over this broad and elevated, yet sphagnous summit, separating the valley of the St. Louis of Lake Superior from that of the Upper Mississippi. As we approached the basin of Sandy Lake, we passed over several sandy ridges, bearing the white and yellow pine; the surface and its depressions bearing the wild cherry, poplar, hazel, ledum latifolia, and other usual growth and shrubs of the latitude. On the dry sandy tracts the uva ursi, or kinnikinnik of the Indians, was noticed. In the mineral constitution of the ridges themselves, the geologist recognizes that wide-spreading drift-stratum, with boulders and pebbles of sienitic and hornblende, quartz, and sandstone rock, which is so prevalent in the region. As we approached the lake we ascended one of those sandy ridges which surround it, and dashing our way through the dense underbrush, were gratified

on gaining its apex to behold the sylvan shores and islands of
the lake, with the trading-post and flag, seen dimly in the distance.
The view is preserved in the following outlines, taken on the spot.

Sandy Lake, from an eminence north of the mouth of the West Creek of the
Portage of Savannah. 15th July, 1820.

I asked Chamees the Indian name of this lake. He replied,
Ka-metong-aug-e-maug. This is one of those compound terms,
in their languages, of which the particle *ka* is affirmative. Me-
tongaug, is the plural form of sandy lake. Maug is the plural
form of water, corresponding, by the usual grammatical duality
of meaning, to the plural form of the noun. The word might,
perhaps, be adopted in the form of Kametonga.

Having heard, on our passage through Lake Superior, that a
gun fired in the basin of Sandy Lake, could be heard at the fort,
that experiment was tried, while we sat down or sauntered about
to await the result. Having waited in vain, the shots were repeated. After the lapse of a long time, a boat, with two men,
was descried in the distance approaching. It proved to be occupied by two young clerks of the trading establishment, named
Ashmun and Fairbanks. They managed to embark the elite of
our party, in their small vessel, and, as we crossed the lake,
amused us with an account of the excitement our shots had
caused. Some Indian women affirmed to them that they had
heard warwhoops, and to make sure that a Sioux war party
were not upon them, they drove off their cattle to a place of
safety. In the actual position of affairs, the hunt being over for

the year, and the avails being sent to Michilimackinac (for this was the head-quarters of the factor whom we had met at Shelldrake River), the probabilities of its being a hunting party were less. We informed them that we were an advance party of an expedition sent out to explore the sources of the Mississippi River, under the personal order of his Excellency Governor Cass, who was urging his way up the St. Louis to the Savanna Portage, through which he intended to descend into Sandy Lake.

It was near sunset before we landed at the establishment. We found the trading fort a stockade of squared pine timber, thirteen feet high, and facing an area a hundred feet square, with bastions pierced for musketry at the southeast and northwest angles. There were three or four acres outside of one of the angles, picketed in, and devoted to the culture of potatoes. The stockade inclosed two ranges of buildings. This is the post visited by Lieut. Z. Pike, U. S. A., on snow-shoes, and with dog-trains, in the winter of 1806, when it was occupied by the British northwest trading company. As a deep mantle of snow covered the country, it did not permit minute observations on the topography or natural history; and there have been no explorations since. Pike's chief error was in placing the source of the Mississippi in Turtle Lake—a mistake which is due entirely, it is believed, to the imperfect or false maps furnished him by the chief traders of the time.

We were received with all the hospitality possible, in the actual state of things, and with every kindness; and for the first time, since leaving Detroit, we slept in a house. We were informed that we were now within two miles of the Mississippi River, into which the outlet of Sandy Lake emptied itself, and that we were five hundred miles above the Falls of St. Anthony. We had accomplished the transference of position from the head of the basin of Lake Superior, that is, from the foot of the falls of the St. Louis River, in seven days, by a route, too, certainly one of the worst imaginable, and there can be no temerity in supposing that it might be effected in light canoes in half that time.

CHAPTER IX.

Reunion of the expedition on the Savanna Portage—Elevation of this summit—
Descent to Sandy Lake—Council with the Chippewa tribe—Who are they?—
Traits of their history, language, and customs—Enter the Mississippi, with a
sub-exploring party, and proceed in search of its source—Physical characteristics
of the stream at this place—Character of the Canadian voyageur!

On rising on the next morning (14th July), our minds were
firmly set, at the earliest moment, to rejoin the main expedition,
which had been toiling its way up the St. Louis River to the
Savanna Portage. And as soon as we had dispatched our breakfast at the Post, we set out, accompanied by one of the trading
clerks, for that noted carrying place between the waters of the
St. Louis and Sandy Lake. We reached its northwestern terminus at about twelve o'clock, and were surprised to find Gov. Cass,
with some of his party, and a part of the baggage, already there;
and by five o'clock in the afternoon the last of the latter, together
with the canoes, arrived. And it was then, in the exhausted state
of the men, and at so late an hour, concluded to encamp, and
await the morning to commence the descent of the west Savannè
to the lake.

The expedition had, after we left them at the Portage aux
Coteaux on the 10th, and being thus relieved of our weight,
urged its way up the river, with labor, about fifty-six miles, to
the inlet of the east Savannè, having surmounted, in this distance,
rapids of the aggregate estimated height of two hundred and
twelve feet, which occupied two days. They then ascended the
Savannè twenty-four miles, rising eighteen feet. The portage,
from water to water, is six miles. It commences in a tamarak
swamp, from which the bog, in a dry season, has been burnt off,
leaving the path a mass of mire. Trees and sticks have, from
time to time, been laid in this to walk on, which it requires the

skill of a balancing master to keep. For the distance of three *pozes* [pauses] this is the condition of the path; afterwards, the footing becomes dry, and there are ascending sand ridges, which are easily crossed.

Dr. Wolcott, to whom I had handed my geological note-book, made the following observations. "We left the vertical strata of slate, about two miles above the Portage aux Coteaux. They were succeeded by rocks of hornblende, which continued the whole distance to the head of the Grand Rapid. These rocks were only to be observed in the bed of the river, and appeared to be much water-worn, and manifestly out of place. Soon after we left the Portage aux Coteaux, the hills receded from the river, and its banks for the rest of the way were generally low, often alluvial, and always covered with a thick growth of birch, elm, sugar-tree (acer saccharinum), and the whole tribe of pines, with an almost impenetrable thicket of underbrush.

"The appearances of this day (11th) have been similar to those of yesterday, except that the country bordering the river became entirely alluvial, and the poplar became the predominating growth, while the evergreen almost entirely disappeared. The rocks were seldom visible, except upon the rapids, and then only in the bed of the river, and were entirely composed of hornblende, all out of place, and exhibiting no signs of stratification, but evidently thrown confusedly together by the force of the current.

"The Savannè River is about twenty yards broad at its junction with the St. Louis, but soon narrows to about half the breadth, which it retains until it forks at the distance of about twelve miles from its mouth. Its whole course runs through a low marshy meadow, the timbered land occasionally reaching to the banks of the river, but generally keeping a distance of about twenty rods on either side. The meadow is, for the most part, covered with tufts of willow and other shrubs, common to marshes. The woods, which skirt it, are of the same kinds observed on the preceding days, except that a species of small oak frequently appears among it. The river becomes so narrow towards its head, that it is with great difficulty canoes can make their way through its·windings; and the portage commences a mile or two from its source, which is in a tamarak swamp."

The height of land between the east and west Savannè, Dr.

Wolcott estimates at about thirty feet. Adding to this elevation the estimates of Capt. Douglass, before mentioned, the entire elevation between the foot of the falls of the St. Louis and the apex of this summit is three hundred and sixty-eight feet.[*]

Having exchanged congratulations, and recited to each other the little personal incidents which had marked our respective tracks of entry into the country, we passed the night on the sources of this little stream; and the next morning, at five o'clock, began its descent. It is a mere brook, only deep enough, at this spot, to embark the canoes, and two men to manage them. At the distances of four, and of twelve miles, there are rapids, where half the loads are carried over portages. At the foot of the latter rapid, there is a tributary called Ox Creek, and from this point to the lake, a distance of six miles, the navigation is practicable with full loads. We entered the lake with pleasurable feelings, at the accomplishment of our transit over this summit, and after a passage of three miles over the calm and sylvan surface of the lake, the expedition reached and landed at the company's fort. It was now four o'clock in the afternoon of a most serene day, and the Indians, who were gathered on the shores, received us with a salute *a la mode de savage*, that is, with balls fired over our heads. Quarters were provided in the fort for such as did not prefer to lodge in tents. Understanding that there was to be a day's rest at this post, to reorganize the party, and hold intercourse with the Indians, each one prepared to make such use of his time as best subserved his purposes. Finding my baggage had been wetted and damaged on the portages in the ascent of the St. Louis, I separated the moulded and ruined from things still worth saving, and drying the latter in the sun, prepared them for further use.

On the day after our arrival (16th) a council of the Indians—the Chippewas—was convened. The principal chiefs were Kadewabedas,[†] or Broken Teeth, and Babisekundeba,[‡] or the Curly Head. This tribe, it appears, are conquerors in the country, having at

---

[*] For heights and distances, *vide* Appendix.

[†] From *ka*, an affirmative particle; *webeed*, teeth; and *eda*, a transitive objective inflection.

[‡] *Ba*, a repeating particle; *besaw*, fine, curly; and *kundib*, the human head.

an early, or ante-historical age, advanced from Lake Superior, driving back the Sioux. The war between these two tribes is known to have existed since the first entry of the French into the country—then a part of New France—early in the seventeenth century. Gov. Cass proposed to them to enter into a firm peace with the Sioux, and to send a delegation with him to St. Peter's, on his return from the sources of the Mississippi. To this they assented. Speeches were made by the Indians, which it is not my purpose to record, as they embraced nothing beyond the ordinary, every-day style of the native speakers.

It was determined to encamp the heavy part of the expedition at this place, and to organize a sub-expedition of two light canoes, well manned, to explore the sources of the Mississippi River. While these arrangements are in progress, it may be proper to state something more respecting the condition and history of the Chippewa nation. And first, they are Algonquins, having migrated, at ante-Cartierian* periods, from the vicinity of Lake Nippesing, on the Outawis summit. Anterior to this, their own traditions place them further eastward, and their language bears evidence that the stock from which they are sprung, occupied the Atlantic from the Chesapeake, extending through New England. The name Chippewa is derived from the term Ojibwa. The latter has been variously, but not satisfactorily derived. The particle *bwa*, in the language, signifies voice. They are a well-formed, active race of men, and have the reputation of being good hunters and warriors. They possess the ordinary black shining eyes, black straight hair, and general physiological traits of the Indian race; and do not differ, essentially, from the northern tribes in their manners and customs. Pike, who was the first American officer to visit them, in this region, estimates the whole number seated on the Upper Mississippi, and northwest of Lake Superior, in the year 1806, at eleven thousand one hundred and seventy-seven. This estimate includes the entire population, extending south to the St. Croix and Chippewa valleys, below St. Anthony's Falls. It is believed to be much too high, for which it can be plead in extenuation, that it was the rough estimate of foreign traders, who were interested in exalting their importance to the

---

* Cartier discovered the St. Lawrence in 1534.

United States. Certain it is, there are not more than half the numbers, in this region, at present. The number which he assigns to the Sandy Lake band is three hundred and forty-five.

The Chippewas of the Upper Mississippi are, in fact, the advanced band of the wide-spread Algonquin family, who, after spreading along the Atlantic from Virginia, as far as the Gulf of St. Lawrence, have followed up the great chain of lakes, to this region, leaving tribes of more or less variation of language on the way. There may have been a thousand years, or more, expended on this ethnological track, and the names by which they were, at various ages and places, known, are only important as being derivatives from a generic stock of languages whose radicals are readily recognized. Furthest removed, in the line of migration, appear the Mohicans, Lenno Lenawpees, Susquehannocks, and Powatans, and their congeners. The tribes of this continent appear, indeed, to have been impelled in circles, resembling the whirlwinds which have swept over its surface; and, so far as relates to the mental power which set them in motion, the comparison also holds good, for the effects of their migrations appear, everywhere, to have been war and destruction. One age appears to produce no wiser men than another. Having no mode of recording knowledge, experience dies with the generation who felt it, all except the doubtful and imprecise data of tradition; and this is little to be trusted, after a century or two. For the matter of exact history, they might as well trace themselves to the moon, as some of their mythological stories do, as to any other planet, or part of a planet. Of their language, the only certainly reliable thing in their history, a vocabulary is given in the Appendix. To the ear, it appears flowing and agreeable, and not of difficult utterance; and there is abundant reason, on beholding how readily they express themselves, for the plaudits which the early French writers bestowed on the Algonquin language.

We observed the custom of these Indians of placing their dead on scaffolds. The corpse is carefully wrapped in bark, and then elevated on a platform made by placing transverse pieces in forks of trees, or on posts, firmly set in the ground. This custom is said to have been borrowed by the Chippewas, of this quarter, from the Dacotahs or Sioux. When they bury in the ground, which is the general custom, a roof of bark is put over the deceased. This inclosure has an aperture cut in it at the head, through which a dish of food is set for the dead. Oblations of liquor are also sometimes made. This ancient custom of offering food and oblations to the dead, reminds the reader of similar customs among some of the barbarous tribes of the oriental world. We noticed also symbolic devices similar to those seen at Huron River or Lake Superior, inscribed on posts set at the head of Indian graves. It seems to be the prime object of these inscriptions to reveal the family name, or *totem*, as it is called, of the deceased, together with devices denoting the number of times he has been in battle, and the number of scalps he has taken. As this test of bravery is the prime object of an Indian's life, the greatest efforts are made to attain it.

A word may be said as to the climate and soil of this region, and their adaptation to the purposes of agriculture. By the tables of temperature annexed (*vide* Appendix), the mean solar heat, in the shade, during the time of our being in the country, is shown to be 67°. It is evident that it is the idle habits of the Indians, and no adverse circumstances of climate or soil, that prevent their raising crops for their subsistence.

Arrangements for a light party to ascend the Mississippi, and seek for its sources, having been made, we left Sandy Lake, in two canoes, at nine o'clock in the morning on the 17th. This party, in addition to his Excellency Gov. Cass, consisted of Dr. Alex. Wolcott, Capt. Douglass, Lieut. Mackay, Maj. Forsyth, and myself, with nineteen voyageurs and Indians, provisioned for twelve days. A voyage of about a mile across the western prolongation of the lake, brought us to its outlet—a wide winding stream, with a very perceptible current, and rich alluvial banks, bearing a forest. After pursuing it some mile and a half, we descended a small rapid, where the average descent of water in a short distance may be perhaps three feet; it appeared, however, to give the men no concern, for they urged their way down it, with full strength of paddle and song, and we soon found ourselves in the Mississippi. The first sight of this stream reminded me of one of its striking characteristics, at far lower points, namely, its rapidity. Its waters are slightly turbid, with a reddish tint. Its width, at this point, as denoted by admeasurements subsequently made,* is three hundred and thirty-one feet. Its banks are alluvial and of a fertile aspect, bearing a forest of oaks, maples, elms, ash, and pines, with a dense undergrowth of shrubbery. I observed a species of polyganum in the water's edge, and wherever we attempted to land it was miry and the borders wet and damp. We were now, from our notes, a hundred and forty-seven miles due west of the head of Lake Superior, by the curved lines of travelling, and probably one hundred in an air line; and had struck the channel of the Mississippi, not less, by the estimates, than two thousand five hundred miles above its mouth on the Gulf of Mexico. It could not, from the very vague accounts we could obtain from the traders, originate, at the utmost, more than three hundred miles higher, and our Canadian voyageurs turned up the stream, with that Troubadour air, or *gaite de cour*, keeping time with song and paddle, with which New France had at first been traversed by its Champlains, Marquetts, and Frontenacs. To conquer distance and labor, at the same time, with a song, has occurred to no other people, and if these men are not happy, in these voyages, they, at least, have the semblance of it, and are merry. To keep up this flow of spirits, and bravery of capacity in demolishing distances, they always overrate the per diem travel, which, as I have before observed, is put about one-third too high—that is to say, their league is about two miles. On we went, at this rapid rate, stopping every half hour to rest five minutes. During this brief rest, their big kettle of boiled corn and pork was occasionally brought forward, and dipped in, with great fervency of spoon; but, whether eating or working, they were always gay, and most completely relieved from any care of what might happen to-morrow. For the mess kettle was ever most amply supplied, and not according to the scanty pattern which these couriers de bois often encounter in the Indian trade on these summits, when they are sometimes reduced to dine on tripe de Roche and sup on buton de rose; but they bore in mind that their employer, namely, Uncle Sam, was a full-handed man, and they kept up a most commendable mental balance, by at once eating strong and working strong.

During the first twenty-seven miles, above the inlet of Sandy

* Expedition to Itasca Lake in 1832.

Lake, we passed six small rapids, at distances of three, four, three, one, five, and eleven miles, where the river sinks its level twenty-nine feet, in the estimated aggregate distance of seven hundred yards.* Above the latter, extending twenty miles, to the point of our encampment, there is no perceptible rapid. It was eight o'clock when we encamped, having been eleven hours in our canoes, without stretching our legs, and we had ascended forty-six miles.

* *Vide* Appendix—Elevations.

CHAPTER X.

OUR encampment was near the mouth of Swan River, a considerable stream, originating in Swan Lake, near the head of the St. Louis River of Lake Superior.

We had been pushing our way, daily, up to our arrival at Sandy Lake; but the word, from leaving that point, was, emphatically, push—and we can hardly be said to have taken proper time to eat or sleep. There was a shower of rain, during the night; it ceased at four o'clock, and we again embarked at five, in a cloudy and misty morning, and it continued cloudy all day. The current of the Mississippi continues to be strong; its velocity, during the ascent of this day, was computed by Capt. Douglass at two and a half miles per hour. We passed a rapid about six miles below Trout River, where there is a computed descent of three feet in a hundred and fifty yards. A few miles before reaching Trout River, we passed through a forest of dead pines, occupying ridges of sand, through which the river has cut its way. Four miles above the entrance of Trout River, we passed the mouth of a considerable stream, called by the Chippewas Mushkoda, or Prairie River, and encamped about five hundred yards above its mouth on a high sandy elevation. It was now eight o'clock P. M. We had ascended the river fifty-one miles, having been fifteen hours in our canoes, and we here first took our breakfast. This severity of fasting was, I think, quite unintentional, the mess-basket being in the other canoe, which kept ahead of us the entire day. We had this day observed specimens of the Unio and some other species of fresh-water shells along the shore. And of birds, besides the duck, plover, and loon, which frequent the water, we noticed the thrush, robin, blackbird, and crow. The comparative coolness of the day rendered the annoyance from mosquitos less severe than we had found them the preceding day. The night on this sandy and bleak elevation proved cool, with a heavy dew, which resulted in a dense fog in the morning. We found ice on the bottoms of the canoes, which are turned up at night, of the thickness of a knife-blade.

Our third day's ascent witnessed no diminution of the strength and alacrity with which our canoemen urged our way up the stream. We were off betimes, in a lowering and dense atmosphere, which obscured objects. After advancing some six miles, there are a series of small rapids, which are, taken together, called Ka-ka-bi-ka,* where I estimated the river to sink its level sixteen feet, in a short distance; at none of these is the navigation, however, impeded. The rock stratification appears too compact for sand-rock, and is obscured by contiguous boulders, which are indicative of the strong drift-formation, which has spread from the north and east over this region. Four miles after ascending the last of the Kakabika Rapids, we landed at the foot of the Pakagama Falls. Here the lading was immediately put ashore, the canoes landed, and the whole carried over an Indian portage path of two hundred and seventy-five yards. This delay afforded an opportunity to view the falls. The Mississippi, at this point, forces its way through a formation of quartzy rock, during which it sinks its level, as estimated, twenty feet, in a distance of about three hundred yards. There is no perceptible cascade or abrupt fall, but the river rushes with the utmost velocity down a highly inclined rocky bed towards the northeast. It forms a complete interruption to navigation, and must, hereafter, be the terminus of the navigation of that class of small steamboats which may be introduced above the Falls of St. Anthony. The general elevation of the geological stratum at the top of this fall must be

---

* From *ka*, a particle affirmative of an adverse quality, *aubik*, rock, and *ons*, a diminutive inflection.

but little under fourteen hundred feet above the Gulf of Mexico.*
This summit bears a growth of the yellow pine. I observed,
amongst the shrubs, the vaccinium dumosum. Immediately above
the falls is a small rocky island, bearing a growth of spruce and
cedars, being the first island noticed above Sandy Lake. This
island parts the channel into two, at the precise point of its precipitation. On coming to the head of these falls, we appear to
have reached a vast geological plateau, consisting of horizontal
deposits of clay and drift on the nucleus of granitical and metamorphic rocks, which underlie the sources of the Mississippi River.
The vast and irregular bodies of water called Leech Lake, Winnipek, and Cass Lakes, together with a thousand lesser lakes of a
mile or two in circumference, lie on this great diluvial summit.
These lakes spread east and west over a surface of not less than
two hundred miles; most of them are connected with channels of
communication forming a tortuous and intricate system of waters,
only well known to the Indians; and there seems the less wonder
that the absolute and most remote source of the Mississippi has
so long remained a matter of doubt.

By the time we had well seen the falls, and made some sketches
and notes, the indefatigable canoemen announced our baggage all
carried over the portage, and the canoes put into the water. Embarking, at this point, we found the river had lost its velocity; it
was often difficult to determine that it had any current at all.
We wound about, by a most tortuous channel, through savannas
where coarse species of grass, flags, reeds, and wild rice struggled
for the mastery. The whole country appeared to be one flat
surface, where the sameness of the objects, the heat of the weather,
and the excessively serpentine channel of the river, conspired to
render the way tedious. The banks of the river were but just
elevated above these illimitable fields of grass and aquatic plants.
In these banks the gulls had their nests, and as they were disturbed they uttered deafening screams. Water-fowl were intruded
upon at every turn, the blackbird and rail chattered over their
clusters of reeds and cat-tails; the falcon screamed on high, as he
quietly sailed above our heads, and the whole feathered creation
appeared to be decidedly intruded on by our unwonted advance

* Mr. Nicollet places the summit of the falls at 1,340 feet above the Gulf.

into the great watery plateau, to say nothing of the small and unimportant class of reptiles who inhabit the region.

Forty miles above the falls, the River Vermilion flows in through these savannas on the left hand; and three miles higher the Deer River is tributary on the right hand. We ascended six miles above the latter, and encamped in a dry prairie, on the same side, at a late hour. The men reported themselves to have travelled sixteen leagues, notwithstanding their detention on the Pakagama Portage. How far we had advanced, in a direct line, is very questionable. At one spot, we estimated ourselves to have passed, by the river's involutions, nine miles, but to have advanced directly but one mile. I noticed, on the meadow at this spot, a small and very delicious species of raspberry, the plant not rising higher than three or four inches. This species, of which I preserved both the roots and fruit, I referred to Dr. J. Torrey, of New York, who pronounced it the Rebus Nut-kanus of Moçino—a species found by this observer in the Oregon regions. It is now known to occur eastwardly, to upper Michigan. As night approached on these elevated prairies, we observed for the first time the fire-fly.

The next morning (20th) we were again in motion at half-past five o'clock. It had rained during the night, and the morning was cloudy, with a dense fog. At the distance of ten miles, we passed the Leech Lake River. This is a very considerable river, bringing in, apparently, one-third as much water as the main branch. It is, however, but fifty miles in length, and is merely the outlet of the large lake bearing that name. It was thought the current of the Mississippi denoted greater velocity above this point, while the water exhibited greater clearness. We had still the same savanna regions, with a serpentine channel to encounter. Through this the men urged their way for a distance of thirty-five miles, when Winnipek Lake displayed itself before us. The waters of this lake have a whitish, slightly turbid aspect, after the prevalence of storms, which appears to reveal its shallowness, with a probably whitish clay bottom. The Chippewa name of Winnebeegogish* is, indeed, derivative from this circumstance. This lake is stated to be ten miles in its greatest length. We

* From *weenud*, dirty, *beegog*, waters, and *ish*, a derogative inflection of nouns.

9

crossed it transversely in order to strike the inlet of the Mississippi, and encamped on the other side. In this transit we met a couple of Indian women in a canoe, who, being interrogated by the interpreter, stated that they came to observe whether the wild rice, which is quite an item of the Indian subsistence in this quarter, was matured enough to be tied into clusters for beating out. We estimated our advance this day, by the time denoted by the chronometer, at fifty-one miles.

We were again in our canoes the next morning at half-past four o'clock. In coasting along the north shores of Winnipek Lake, an object of limy whiteness attracted our attention, which turned out to be a small island composed of granitical and other boulders, which had served as the resting-place of birds, for which the region above the Pakagama Falls is so remarkable. On landing, a dead pelican was stretched on the surface. We had not before observed this species on the river, and named the island Shayta, from its Chippewa name. The buzzard, cormorant, brant, eagle, and raven had hitherto constituted the largest species. Along the shores of the river, the kingfisher and heron had been frequent objects. With respect to the cormorant, it was observed that the Indians classify it with the species of duck, their name for it, ka-ga-ge-sheeb, signifying, literally, crow-duck.

On again reaching the inlet of the Mississippi, its size and appearance corresponded so exactly to its character below the Winnipek, that it had evidently experienced but little or no change by passing through this lake. The same width and volume were observed which it had below this point; the same moderate velocity; the same borders of grassy savanna, and the same tendency to redouble its length, by its contortions, appeared. In some places, however, it approaches those extensive ridges of sandy formation, bearing pines, which traverse, or rather bound, these wide savannas. Through these channels the canoemen urged their course with their usual alacrity—now stopping a few moments to breathe, and then, striking their paddles again in the water with renewed vigor, and often starting off with one of their animated canoe-songs. From about eight o'clock in the morning till two in the afternoon we proceeded up the winding thread of this channel, when the appearance of a large body of water in the distance before us attracted attention. It was the first glimpse we had of the upper Red Cedar Lake. The Mississippi River here deploys itself in one of those large sheets of pellucid water which are so characteristic of its sources. On reaching the estuary at its entrance, a short halt was made. A large body of the most transparent water spread out before us. Its outlines, towards the south, were only bounded by the line of the horizon. In the distance appeared the traces of wooded islands. If Sandy Lake had, on emerging from the wilderness, impressed us with its rural beauty, this far transcended it in the variety and extent of outlines, and that oceanic amplitude of freshness, which so often inspires admiration in beholding the interior American lakes. It was determined to cross a part of the lake towards the northeast, in order to strike the site of an ancient Indian village at the mouth of Turtle River; and under the influences of a serene day, and one of their liveliest chants, the men pushed for that point, which was reached at three o'clock in the afternoon of the 21st July. The spot at which we landed was the verge of a green lawn, rising in a short distance to a handsome eminence, crowned with oaks and maples. One or two small log tenements stood on this slope occupied by two Canadians in the service of the American Fur Company. Several wigwams of bark and poles lifted their fragile conical forms on either side.

In one of these tenements, consisting of a small cabin of poles, sheathed with bark, we found an object of human misery which excited our sympathies. It was in the person of one of the Canadians, to whom reference has been made, of the name of Montruille. He had, in the often severe peregrinations of the fur trade in this quarter, been caught in a snow-storm during the last winter, and frozen both his feet in so severe a manner that they eventually sloughed off, and he could no longer stand upright or walk. He lay on the ground in a most pitiable state of dejection, with the stumps of his legs bound up with deer skins, with a gray, long-neglected beard, and an aspect of extreme despair. English he could not speak; and the French he uttered was but an abuse of the noble gift of language to call down denunciations on those who had deserted him, or left him thus to his fate. A rush mat lay under him. He had no covering. He was emaciated to the last degree, every bone in his body seemed visible through the skin. His cheeks were fallen in, and his eyes sunk in their sockets, but darting a look of despair. His Indian wife had deserted him. Food, of an inadequate quality, was occasionally thrown in to him. Such were the accounts we received. · Governor Cass directed groceries, ammunition, and presents of clothing to be made to him, to the latter of which, every member of the party added. He also engaged a person to convey him to Sandy Lake.

We examined the environs of the place with interest; the village occupies the north banks of Turtle River Valley. Turtle River, which cuts its way through this slope and plain, constitutes the direct line of intercourse for the Indian trade, through Turtle and Red Lakes, to the Red River Valley of Hudson's Bay. On inquiry, we learned that this river had constituted the ancient Indian line of communication by canoes and portages, from time immemorial, with that valley, the distance to the extreme plateau, or summit, being about sixty miles. On this summit, within a couple of miles of each other, lie Turtle and Red Lakes, the one having its discharge into the Gulf of Mexico and the other into Hudson's Bay. When Canada was settled by the French, this aboriginal route was adopted. The fur companies of Great Britain, on coming into possession of the country, after the fall of Quebec, 1759, followed the same route. The factors of these companies told Lieutenant Pike, in 1806, at Sandy Lake and Leech Lake, that the Turtle portage was the only practicable route of communication to the Red River, and that it was the true source of the Mississippi; and they furnished him manuscript maps of the country conformable to these views. The region has actually been in possession of the Americans only since 1806, adopting• the era of Pike's visit.

By inquiry from the Chippewa Indians at this village, sanctioned by the Canadian authorities, we are informed that the Mississippi falls into the south end of Cass Lake, at the distance of eight or ten miles; that it reaches that point from the west, by a series of sharp rapids stretching over an extent of about forty miles from a large lake;* and that this celebrated stream originates in Lac la Biche, about six days' journey from our present position, and has many small lakes, rapids, and falls. It is further asserted by the Indians, that the water in these remote streams, and upon these

* Called Andrusia. Expedition to Starca Lake in 1887.

rapids, is at all times shallow, but it is particularly so this season; and that it is not practicable to reach these remote sources of the river with boats, or large canoes of the size we have.

On submitting these facts to the gentlemen composing his party, Governor Cass asked each one to give his views, beginning with the youngest, and to express his opinion on the feasibility of further explorations. They concurred in opinion that, in the present low state of the water on these summits, considering the impossibility of ascending them with our present craft, and in the actual state of our provisions, such an attempt was impracticable. Thereon, he announced his decision to rejoin our party at Sandy Lake, and to pursue the exploration of the river down its channel to the Falls of St. Anthony, to the inlet of the Wisconsin and Fox Rivers, and to return into the great lake basins, and complete their circumnavigation.

Having reached the ultimate geographical point visited by the expedition, I thought it due to the energy and enlightened zeal of the gentleman who had led us, to mark the event by naming this body of water in my journal Cassina, or Cass Lake. There was the more reason for this in the nomenclature of the geography of the upper Mississippi, by observing that it embraces another Red Cedar Lake. The latitude of upper Red Cedar, or Cass Lake, is placed by Pike at 47° 42′ 40″.[*] Its distance above Sandy Lake, by the involutions of the river, is two hundred and seventy miles, and from Fond du Lac, at the head of Lake Superior, by the travelled route, four hundred and thirty miles. It is situated seventeen degrees north of the Gulf of Mexico, from which it is computed to be distant two thousand nine hundred and seventy-eight geographical miles. Estimating the distance to the actual origin of the river, as determined at a subsequent period, at one hundred and eighty-two miles above Cass Lake, the length of the Mississippi River is shown to be three thousand one hundred and sixty miles,[†] making a direct line over the earth's surface of more than half the distance from the arctic circle to the equator. It may also be observed of the Mississippi, that its sources lie in a region of snows and long-continued

[*] Nicollet, in the report of his exploration of 1836, places it in 47° 25′ 23″.

[†] *Vide* Expedition to Stasca Lake in 1832.

winter, while it enters the ocean under the latitude of perpetual
verdure; and at last, as if disdaining to terminate its career at the
ordinary point of embouchure of other large rivers, has protruded its banks into the Gulf of Mexico, more than a hundred
miles beyond any other part of the main.   To have visited both
the source and the mouth of the stream has fallen to the lot of but
few, and I believe there is no person living beside myself of
whom the remark can be made.   On the tenth of July, 1819, I
passed out of the mouth of the Mississippi in a brig bound for
New York, after descending it in a steamboat from St. Louis, but
little thinking I should soon visit its waters, yet, on the twenty-
first of July of the following year, I reached its sources in this
lake.

In deciding upon the physical character of the Mississippi
River, it may be advantageously considered under four natural
divisions, as indicated by permanent differences in its geological
and physical character—its vegetable productions, and its velocity
and general hydrographical character.   Originating in a region
of lakes upon the table-lands which throw their waters north
into Hudson's Bay, south into the Gulf of Mexico, and east into
the Gulf of St. Lawrence, it pursues its course south to the Falls
of Pakagama, a distance of two hundred and thirty miles, through
natural meadows or savannas covered with wild rice, rushes, reeds
and coarse grasses, and aquatic plants.   During the distance, it is
extremely devious in its course and width, often expanding into
lakes which connect themselves through a vast system of reticulated channels.   Leech Lake, Cass Lake, and Lake Andrusia
would themselves be regarded as small interior seas, were they
on any other part of the continent but that which develops Superior, Michigan, Huron, Erie, and Ontario.   Its velocity through
the upper plateau is but little, and it affords every facility for the
breeding of water fowl and the small furred quadrupeds, the
favorite reliance of a nomadic population.

At the Falls of Pakagama, the first rock stratum and the first
wooded island is seen.   Here the river has an aggregate fall of
twenty feet, and from this point to St. Anthony's Falls, a distance
of six hundred miles, it exhibits its second characteristic division.
The granitical and metamorphic rocks, which support the vast
plateaux and beds of draft of its sources, are only apparent above

this point, in boulders. The permanent strata are but barely concealed at several rapids below the Pakagama, but appear plainly below the influx of the De Corbeau, at Elk River, Little Falls, and near Sac River. And this system of rock is succeeded, before reaching the Falls of St. Anthony, by the horizonal white sand rock and its superior limestone series of the carboniferous formation.

Vegetation is developed as the river descends towards the south. A forest of maples, elm, oak, ash, and birch, is interspersed with spruce, birch, poplar, and pine above the Pakagama, and continues, in favorable positions, throughout this division. The black walnut is first seen below Sandy Lake, and the sycamore below the River De Corbeau. The river in this division has numerous well-wooded islands; its velocity is a striking feature; it abounds with rapids, none of which, however, oppose serious obstacles to its navigation. Agreeably to memoranda kept,* it has fifty-six distinct rapids, including the Little and Big Falls, in all of which the river has an aggregate estimated descent of two hundred and twenty-four feet, within a distance of fourteen thousand six hundred and forty yards, or about eight miles. The mean fall of the current, exclusive of these rapids, may be computed at nearly six inches per mile.

The course of the river below the Falls of Pakagama, is still serpentine, but strikingly less so than above, and its bends are not so short and abrupt. The general course of this river, till it reaches the rock formation of Pakagama, is from the west. Thence, to Sandy Lake inlet, it flows generally southeast; from this point to the inlet of the De Corbeau or Crow Wing, it is deflected to the southwest; thence almost due south, to the mouth of the Watab River; and thence again southeast to the Falls of St. Anthony. A geographical line dropped from the inlet of Sandy Lake, where the channel is first deflected to the southwest, ·to St. Anthony's Falls, or the mouth of the St. Peter's,† forms a vast bow-shaped area of prairie and forest lands of high agricultural capabilities, whose future products must be carried to a market through the Fond du Lac of Lake Superior. These

---

* *Vide* Appendix.
† Now called Minnesota River.

prairies and grove lands, which cannot square less than two by four hundred miles, constitute the ancient area of the Issati,* and are now the resort of great herds of the buffalo, elk, and deer; and it is a region known as the predatory border, or battle-ground of the Chippewas and Dacotas.

* *Vide* Hennepin.

CHAPTER XI.

Physical traits of the Mississippi—The elevation of its sources—Its velocity and mean descent—Etymology of the name Mississippi—Descent of the river to Sandy Lake, and thence to the Falls of St. Anthony—Recross the great Bitobi Savanna—Pakagama formation—Description of the voyage from Sandy Lake to Pine River—Brief notices of the natural history.

THE third geographical division in which it is proposed to consider the Mississippi, begins at the Falls of St. Anthony. Within half a day's march, before reaching this point from its sources, the primitive and crystallized, and the altered and basaltic rocks are succeeded by the great limestone and sandstone horizontal series of the carboniferous, magnesian, and metalliferous rocks, which constitute by themselves so extraordinary a body of geological phenomena. Entering on the level of the white sandstone stratum, which is fundamental in this column, about the inlet of Rum River, the Mississippi urges its way over a gently inclining bed of this rock, to the brink of this cataract, where it drops perpendicularly about sixteen feet; but the whole descent of its level from the head to the foot of the portage path, cannot be less than double that height.

The river, at this point, enters a valley which is defined by rocky cliffs, which attain various elevations from one to three hundred feet, presenting a succession of picturesque or sublime views. In some places these cliffs present a precipitous and abrupt façade, washed by the current. In far the greatest number of cases, the eminence has lost its sharp angles through the effects of frosts, rains, and elemental action, leaving a slope of debris at the foot. As the river descends, it increases in volume and in the extent of its alluvions. These form, in an especial manner, its characteristic features from St. Anthony's Falls to the junction of the Missouri, a distance of not less than eight hundred miles. The principal tributaries which it receives in this distance, are, on the right, the St. Peter's, Upper and Lower Iowa, Turkey River, Desmoines, and Salt Rivers; and, on the left, the St. Croix, Chippewa, Wisconsin, Rock River, and the Illinois. One hundred miles below St. Anthony, it expands for a distance of twenty-four miles into the sylvan sheet of Lake Pepin, at the foot of which it receives the large volume of the Chippewa River, which originates on the sandy tracts at the sources of the Wisconsin, Montreal, and Ontonagon; and it is from this point that its continually widening channel exhibits those innumerable and changing sand-bars, which so embarrass the navigation. But in all this distance, it is only at the Desmoines and Rock River rapids that any permanent serious impediment is found in its navigation, with the larger craft.

The fourth change in the physical aspect of this river, is at the junction of the Missouri, and this is an almost total and complete one; for this river brings down such a vast and turbid flood of commingled earths and floating matter, that it characterizes this stream to its entrance into the Gulf of Mexico. If its length of channel, velocity, and other leading phenomena had been accurately known at an early day, it should also have carried its name from this point to the ocean. Down to this point, the Mississippi, at its summer phases, carries the character of a comparatively clear stream. But the Missouri, which, from its great length and remote latitude, has a summer freshet, flows in with a flood so turbid and opaque, that it immediately communicates its qualities and hue to the milder Mississippi. At certain seasons, the struggle between the clear and turbid waters of the two streams can be seen, at opposite sides of the river, at the distance of twenty or thirty miles. Entire trees, sometimes ninety feet long, with their giant arms, are swept down the current; and it is not unusual, at its highest flood, to observe large, spongy masses of a species of pseudo pumice carried into its channel, from some of its higher western tributaries.

To such a moving, overpowering liquid mass, there are still, below the Missouri, rocky banks, and occasionally isolated cliffs, to stand up and resist its sweep; but its alluvions become wider and deeper opposite to these rocky barriers. Its bends stretch over greater distances, and its channel grows deeper at every accession of a tributary. The chief of these, after passing the

Missouri, are from the Rocky Mountains and Ozark slopes, the St. Francis, White, Arkansas, and Red Rivers; and from the other bank the Kaskaskia, the Ohio, Wolf, and Yazoo. It is estimated to flow twelve hundred miles below the Missouri. Its width is about one mile opposite St. Louis. It is narrower but more than twice the depth at New Orleans, and yet narrower, because more divided, at its embouchure at the Balize, where a bar prevents ships drawing over eighteen feet of water from entering.

No attempt has heretofore been made to determine the elevation of that part of the American continent which gives rise to the Mississippi River. From the observations made on the expedition, the elevation is confessedly less than would *à priori*. be supposed. If it is not, like the Nile, cradled among mountains, whose very altitude and position are unknown, there is enough of the unknown about its origin to wish for more information. Originating on a vast continental plateau, or water-shed, the superabundance of its waters are drained off by the three greatest rivers of North America, namely, the St. Lawrence, the Nelson's rivers of Hudson's Bay, and the Mississippi. Yet the apex of this height of land is moderate, although its distance from the sea at either point is immense. From the best data at command, I have endeavored to come at the probable altitude of this plateau, availing myself at the same time of the judgment of the several members of the expedition. Taking the elevation of Lake Erie above tide-water, as instrumentally determined, in the New York surveys, as a basis, we find Lake Superior lying at an altitude of six hundred and forty-one feet above the Atlantic. From thence, through the valley of the St. Louis, and across the Savanna summit, to the Mississippi, at the confluence of the Sandy Lake River, estimates noted on the route, indicate an aggregate rise of four hundred and ninety feet. The ascent of the river, from this point to Cass Lake, is estimated to be one hundred and sixty-two feet; giving this lake an aggregate elevation of thirteen hundred and ninety-three feet above the Atlantic. Barometrical admeasurements made in 1836, by Mr. Nicollet, in the service of the United States Topographical Bureau, place the elevation of this lake at fourteen hundred and two feet above the Gulf of Mexico,* being just twelve feet above these early estimates. The

* Senate Document No. 237, 26 Con. 2d Session, A.D. 1843.

same authority estimates its length from the Balize, at twenty-seven hundred and fifty miles. Its velocity below Cass Lake may be estimated to result from a mean descent of a fraction over five inches per mile.

The name of the Mississippi River is derived from the Algonquin language, through the medium of the French. The term appears first in the early missionary letters from the west end of Lake Superior about 1660. Sippi, agreeably to the early French annotation of the word, signifies a river. The prefixed word Missi is an adjective denoting all, and, when applied to various waters, means the collected or assembled mass of them. The compound term is then, properly speaking, an adverb. Thus, Missi-gago, means all things; Missi-gago-gidjetod, He who has made all things—the Creator. It is a superlative expression, of which great river simply would be a most lean, impracticable, and inadequate expression. It is only symbolically that it can be called the father of American rivers, unless such sense occurs in the other Indian tongues.

Finding it impracticable to proceed higher in the search of the remote sources of the river at this time, a return from this point was determined on. The vicinity had been carefully scanned for its drift specimens, and fresh-water conchology. Wishing to carry along some further memorial of the visit, members of the party cut walking-canes in the adjoining thickets, and tied them carefully together; and at five o'clock in the afternoon (21st July) we embarked on our descent. An hour's voyage over the surface of this wide lake, with its refreshing views of northern scenery, brought us to the point where the Mississippi issues from it. Never did men ply their paddles with greater animation; and having the descent now in their favor, they proceeded eighteen miles before they sought for a spot to encamp. Twilight still served, with almost the clearness of daylight, while we spread our tents on a handsome eminence on the right-hand shore. Daylight had not yet dawned the next morning, when we resumed the descent. It was eight o'clock A.M. when we reached the border of Lake Winnipek. This name, by the way, is derived from a term heretofore given, which, having the Chippewa inflection of nouns in *ish*, graphically describes that peculiarity of its waters created by the disturbance of a clay bottom.

The winds were high and adverse, which caused the canoemen to toil two hours in crossing. After reaching the river again, we passed its sedgy borders, to, and through Rush Lake, or the Little Winnipek; then by the inlet of Leech Lake River, and through the contortions of its channel, to within a few miles of the spot of our encampment at Deer River, on the 20th.

The great savannas, through which the Mississippi winds itself above the Pakagama, are called collectively, the Gatchi Betobeeg, Great Morasses, or bog meadows.

While descending the river, we encountered nine canoes filled with Chippewa Indians and their families. They were freighted with heavy rolls of birch-bark, such as their canoes are made from; together with bundles of rushes designed for mats. The annoyance suffered from mosquitos on this great plateau, was almost past endurance. We embarked again at a quarter past four, and reached the Falls of Pakagama at five o'clock. Just forty minutes were spent in making the portage. The rock at this spot is quartzite. The day was cloudy, with some rain. As night approached an animal, judged to be the wolverine, was seen swiming across the stream. The efforts of the men to over-take it were unavailing; it nimbly eluded pursuit, and dashed away into the thickets. In some queries sent to me by the New York Lyceum, this animal is alluded to as a species of the glutton. The Indians said there was no animal in their country deserving this name; the only animal they knew deserving of it, was the horse; which was eating all the time. We en-camped on an abrupt sandy bank, where, however, sleep was impossible. Between the humidity of the atmosphere and the denseness of the foliage around us, the insect world seemed to have been wakened into unusual activity. Besides, we en-camped so late, and were so jaded by a long day's travel, that the mosquito-nets were neglected. To get up and stand be-fore a camp-fire at midnight and switch off the mosquitos, re-quires as much philosophy as to write a book; and at any rate, ours completely failed. We were again in our canoes (24th), at an early hour. Daylight apprised us of the clearing up of the atmosphere, and brought us one of the most delightful days. Animated by these circumstances, we descended the stream with rapidity. Soon after midday, we entered and ascended the short channel of the Sandy Lake River, and, by two o'clock in the afternoon, we rejoined our camp at the Fur Company's Fort, having been three days in descending a distance which had consumed four and a half in the ascent.

We were received with joy and acclamation by the Sandy Lake party, and felicitated ourselves on the accomplishment of what had all along appeared as the most arduous part of our route. Nor had we indeed, overrated its difficulties; the incessant motion of travelling depriving us of mature opportunities of observation, and also rest at night, the stings of the mosquitos whenever we attempted to land, and the cravings of an often unsatisfied appetite, had made this visit one of peculiar privation and fatigue. Without such an effort, however, it is doubtful whether the principal objects of the expedition could have been accomplished. Nothing untoward had happened at the camp, no difficulty had occurred with the Indians, and all the party were in good health. Having left my thermometer with Mr. Doty, during my absence, the observations made by him are denoted in the appendix.

The following day was fixed on for our departure for the Falls of St. Anthony. The distance to these falls is generally put by the traders at from five to six hundred miles. These estimates denote, however, rather the difficulties and time employed by days' journeys in the trade than any other measurements.* Pike states the latitude some thirteen minutes too far north. It is found to be 46° 47' 10". It appears from Lieut. Pike (*Expt.* p. 60), that the stockade at this place was erected in 1794. Its elevation above the Gulf of Mexico is 1,253 feet. The soil of the environs yields excellent potatoes, and such culinary vegetables as have been tried. The mean temperature of July is denoted to be 78°. The post is one of importance in the fur trade. It yields the deer, moose, bear, beaver, otter, martin, muskrat, and some other species, whose skins or pelts are valuable.

It was twelve o'clock on the morning of the 25th, before we were ready to embark. Our flotilla now consisted of three canoes, of the kind called *Canoe-allege* in the trade, and a barge occupied

---

* Nicollet, in his report to the Top. Bureau, in 1836, states the direct distance from St. Peter's to Sandy Lake, at but 334 miles.

by the military. To this array, the chief Babesakundiba, or the Curly Head, added a canoe filled with Chippewa delegates, who accompanied him on a mission of peace to the Sioux. This chief is the same individual who met Lieut. Pike in this quarter, in 1806, and he appears to be a man of much energy and decision of character. His reputation also gives him the character of great skill, policy, and bravery in conducting the war against the Sioux. Indian wars are not conducted as with us, by opposing armies. It is altogether a guerilla affair. War parties are raised, marched, fight, and disperse in a few days. The war is carried on altogether by stealth and stratagem. Each one furnishes himself with food and weapons. In such a warfare, there is great scope for individual exploits and daring. In these wars the Curly Head had greatly distinguished himself, and he was, therefore, an ambassador of no mean power. In every view, the mission assumed an interesting character; and we kept an eye on the chief's movements, on our journey down the river, chiefly that we might notice the caution which is observed by the Indians in entering an enemy's country.

After entering the Mississippi, below Sandy Lake, the stream presents very much the character it has above. It was below this point that we first observed the juglans nigra in the forest. Its banks are diluvial or alluvial formations, elevated from six to ten feet. The elm, maple, and pine are common. There are some small grassy islands, with tufts of willows, and driftwood lodged. No rock strata appear. The river winds its way through vast diluvial beds, exhibiting at its rapids granitical, quartz, and trappose boulders. It appears to glide wholly over the primitive or crystalline rocks, which rise in some places through the soil, or show themselves at rapids. The expedition descended the stream twenty-eight miles, and encamped on a sandy elevation on the west shore, near Alder River, which seemed to promise an exemption from the annoyance of insects; but in this we were mistaken. In the hurry of a late encampment, it had been omitted to pitch the tents. The first ill effect of this was felt on being awakened at night by rain. A humid atmosphere is ever the signal for awakening hordes of insects, and the mosquitos became so troublesome that it was impossible to sleep at all after the shower. We got up and whiled away the time as best we could around the camp-fire.

We embarked a few minutes before 5 A. M., the morning being lowering and overcast, which eventuated in rain within an hour. The atmosphere resumed its serenity, and the sun shone out at noon. The river, as on the preceding day, has its course between alluvial and diluvial banks, sweeping its way over the smooth orbicular beds of the granitical age. The influx of rivers, the occurrence of islands, which bear witness of their entire submersion during the freshets, and the succession of bends, points, and rapids—these changes, with notices of the wild fowl, forest birds, and sometimes a quadruped, or a mass of boulders, absorbed my notices, which it seems unimportant, at this time, to refer to. No fixed stratification of rocks was encountered this day.

We encamped at about eight o'clock, on the east bank, on an open eminence, just below the rapids which mark the confluence of Pine River, having been in our canoes, with very brief and infrequent landings, fifteen hours. At the points of landing, I observed the rosa parviflora, and ipomea nil. As night approached, we heard the monotonous notes of the caprimulgus virginianus. We had also observed during the day, the bald eagle, king-fisher, turdus polyglottis, teal, plover, robin, and pigeon. The nimble sciuris vulgaris was also observed on shore. Boulders of sienite, hornblende rock, silicious slate, sandstone, and quartz, served as so many monuments to testify that heavy oceanic currents had heretofore disrupted the northern stratification, and poured down over these long and gradual geological slopes.

High and open as our position was on this eminence, our old friends the mosquitos did not forget us. Even the Indians could not endure their continued attacks. A fine fellow of our original auxiliaries, called Iaba Waddik, or the Buck, took this occasion to give us a specimen of his English, exclaiming, as he came to the camp-fire, "Tia!* no sneep!" putting the usual interchangeable *n* of the tribe for the *l* in the noun.

* An exclamation.

CHAPTER XII.

Description of the descent from Pine River—Pine tracts—Confluence of the Crow-wing River—Enter a sylvan region—prairies and groves, occupied by deer, elk, and buffalo—Sport of buffalo hunting—Reach elevations of sienitic and meta-morphic rocks—Discover a pictographic inscription of the Sioux, by which they denote a desire for peace—Pass the Osaukes, St. Francis's, Cornielle, and Rum Rivers—St. Anthony's Falls—Etymology of the name—Geographical considerations.

THE night dew was heavy on this elevation, and a dense fog prevailed at the hour of our embarkation (5 o'clock A. M., on the 27th). The pine lands come in with the valley of Pine River, a large and important stream tributary from the west, which has a connection with Leech Lake. These lands characterize both banks of the Mississippi to the entrance of the River De Corbeau. We were seven hours, with a strong current, in passing through this tract. It is to be observed that ancient fires have been permitted to run through these forests, destroying immense quantities of the timber. It was twelve o'clock, A. M., when we came opposite to the entrance of the great Crow-wing River.* This stream, which has a large island in its mouth, is a prime tributary with a large, full-flowing current, and must bring in one-third of the entire volume of water to this point.† Such is the effect of this current on the opposite shore, that, at the distance of a couple

---

* CROW-WING RIVER.—In returning from Itasca Lake, in 1882, I passed from Leech Lake by a series of old Indian portages into Lake Ka-ge-no-ge-maug, or Long Water Lake, which is its source; and from thence descended it to its entrance into the Mississippi.—Vide *Exp. to Itasca Lake.* N. Y., Harpers, 1834: vol. i. 8vo. with maps.

† The Indian name of this river is Kagiwegwon, or Raven's-wing, or Quill, which is accurately translated by the term *Aile de Corbeau,* but it is improperly called Crow-wing. The Chippewa term for crow is *andaig,* and the French, *cornielle*—terms which are appropriately applied to another stream, nearer St. Anthony's Falls.

of leagues below, at a spot called *Prairie Percié* by the French, it appears to have forced its way headlong, till, meeting obstructions from the primary rocks, it was again deflected south. At this point, the whole face of the country has an exceedingly sylvan aspect. It is made up of far-stretching plains, covered with grass and wild flowers, interspersed with groves of oak, maple, and other species. The elevation of these beautiful plains, above the river, is not less than twenty to thirty feet, placing them above the reach of high waters. We were now passing below the latitude of 46°. Everything indicated a climate favorable to the vegetable kingdom. While passing in the valley, through the fine bends which the river makes, through these plains, we came to a hunting-camp of probably one hundred and fifty Indians. They were Chippewas, who, on landing at their camp, saluted us in the Indian fashion, and were happy to exchange some dried buffalo meat and pemmican, for corn and flour. Some miles below we observed several buffalo, on the eastern shore, on the sub-plains below the open bluffs. Alarmed by our approach, these animals set out, with a clumsy, shambling trot, for the upper plains. Clumsy as their gait seemed, they got over the ground with speed. Our whole force was immediately landed, a little below, and we eagerly climbed the banks, to engage in the sport of hunting them. Quite a large drove of this animal was seen on the prairie. Our best marksmen, and the Indians, immediately divided themselves, to approach on different sides the herd. Cautiously approaching, they fired; the effect was to alarm and divide them. Most of the herd pushed directly to the spot on the banks of the river, where the non-combatants of the party stood; and there arose a general firing, and *mêlée* of men and buffaloes, which made it quite doubtful, for awhile, who stood in greatest danger of being hit by the bullets, the men or animals. I am certain the bullets whizzed about the position I occupied on the top of the alluvial cliffs. None of the herd were, however, slain at that time; but at our encampment, a short distance below, the flesh of both the buffalo and elk was profusely brought in by the Indians. It is stated that this animal lifts both the feet on one side, at the same time; but this remark, I presume, arises from a mode of throwing its feet forward, which is decidedly different from other quadrupeds.

On descending the river two miles, the next morning, we found ourselves opposite the mouth of Elk River, a stream coming in from the west. This point has been determined to be but four minutes north of latitude 46° [*Sen. Doc.* 237]. A short distance below the river, we passed, on the west shore, the Painted Rock, an isolated or boulder mass, having Indian devices, which we had no opportunity of examining. We were now passing down a channel of manifestly increased velocity, and at the distance of a couple of miles more, found ourselves hurried through the west channel of the Little Falls. At this point the primitive or basis stratification over which we had been so long gliding, crosses the river, rising up and dividing it, by an abrupt rocky island, into two channels. The breadth of the stream is much compressed, and the velocity of its current increased. By what propriety of language it is called "falls" did not, however, appear; perhaps there are seasons when the descent assumes a greater degree of disturbance and velocity. To us, it appeared to be about ten feet in a hundred and fifty yards. Here, then, in N. lat. 46°, the Mississippi is first visibly crossed by the primary series of rocks.

Being now in the region of buffalo, it was decided to land in the course of the day, for the purpose of entering into the chase. An occasion for this was presented soon after passing the Little Falls, by observing one of these animals on shore. On landing, and reaching the elevation of the prairies, two herds of them were discovered at a distance. An attack on them was immediately planned, for which the tall grass and gentle inequalities of surface, appeared favorable. The fire proved unsuccessful, but served to distract the herds, giving scope for individual marksmanship and hunter activity, during which, innumerable shots were fired, and three animals killed. While this scene was passing, I had a good stand for witnessing the sport, some of the herd passing by very near, as with the blindness of fury. The bison is certainly an animal as clumsy as the ox, or domestic cow; but, unlike these, it is of a uniform dun color, and ever without being spotted, or mottled. Its horns are nearly straight, short, very black, and set wide apart. The male is formidable in look, and ferocious when wounded. Its ordinary weight is eight hundred to a thousand pounds.

It may be said, in reference to this animal being found in this region, that it is a kind of neutral ground, between the Chippewas and Sioux, neither of which tribes permanently occupy the country between the mouth of the Raven's-wing and Rum Rivers.*

Having spent several hours in the chase, we again embarked, and proceeded down the river until three o'clock in the afternoon. On the left bank of the river two prominent elevations of the granitical series, rising through the prairie soil, attracted my attention. Immediately below this locality, a high and level prairie stretches on the west shore, which had a striking appearance from its being crowned with the poles and fixtures of a large, recently abandoned Sioux encampment. At this spot the expedition landed and encamped. The quick glances of Babasikundiba and his party of delegates immediately discovered a pole, at the site of the chief's lodge, bearing a birch bark scroll, or letter, inscribed with Indian hieroglyphics, or devices. It turned out that this spot was the northern terminus of a Sioux peace embassage, dispatched from St. Peter's shortly previous, under the direction of Col. H. Leavenworth, U. S. A., the newly-arrived commanding officer at that post. The message was eagerly received and read by the Chippewa delegates. By it they were informed that the Sioux also desired a termination of hostilities. The scroll was executed by tracing lines, with the point of a knife, or some sharp instrument. The pictographic devices thus drawn denoted the exact number of the party, their chiefs, and the authority under which these crude negotiations were commenced.

Of this mode of communicating ideas among the Algonquin tribes, we have before given details in crossing the boggy plateau of Akik Sepi, between the St. Louis River and Sandy Lake. The present instance of it is commented on in an interesting communication of the era, in the appendix, from the pen of Gov. Cass. It was now no longer doubtful that the Chippewa mission would be successful, and the satisfaction it produced was evident in the countenances and expressions of Babasikundiba and his colleagues.

I took a canoe and crossed the Mississippi, to inspect the geology of the opposite shore. On reaching the summit of the

* The Chippewas affirm that this was the last time the buffalo crossed the Mississippi eastwardly. It did not appear, in the same region, in 1821.

rock formations rising through the prairies, which had attracted my notice from the river, I found them to consist of sienite, which was almost exclusively made up of a trinary compound of white quartz, hornblende, and feldspar—the two former species predominating. The feldspar exhibited its splendent black crystals in fine relief in the massy quartz. This formation extended a mile or more. What excited marked attention, in surveying these rocks, was their smoothly rubbed surfaces, which seemed as if they must have been produced by equally hard and heavy masses of rock, driven over them from the north. I registered this locality, in my Geological Journal, as the Peace Rock, in allusion to the purport of the Indian mission, evidences of which were found at the opposite encampment.*

During our night's encampment at this spot we heard the howling of a pack of wolves, on the opposite bank—a sure indication, hunters say, that there are deer, or objects of prey in the vicinity. There are two species of wolves on the plains of the Mississippi—the canis lupus, and the animal called coyote by the Spanish. The latter is smaller, of a dingy yellow color, and bears the generic name of prairie wolf. I have also seen a black wolf on the prairies of Missouri and Arkansas, three feet nine inches long, with coarse, bristly, bear-like hair. As daylight approached, our ears were saluted with the hollow cry of the strix nictea, a species which is asserted to be found, sometimes, as far south as the Falls of St. Anthony.

On embarking, at an early hour, we found the humidity of the night atmosphere to be such, that articles left exposed to it were completely saturated. Yet, the temperature stood at 50° at half-past four o'clock, the moment of our embarkation. On descending six miles we passed the mouth of the Osakis, or Sac River, a

---

* In the treaty of Indian boundaries of Prairie du Chien, of 1825, this mission of the Sioux became a point of reference by the Sioux chiefs Wabishaw, Petite Corbeau, and Wanita, as denoting the limit of their excursions north. The Chippewas, on the contrary, by the mouths of Babasekundabi, Kadawabeda, and the Broken Arm of Sandy Lake, contended for Sac River as the line. I discussed this subject, having Indian maps, at length, with the chiefs and Mr. Taliaferro, the Sioux agent, of St. Peter's. An intermediate stream, the Watab River, was eventually fixed on, as the separating boundary between these two warlike tribes.—*Indian Treaties;* Washington, D. C. 1837. Vol. i. 8vo. p. 870.

considerable tributary from the west, which opens a line of communication with the Red River valley.

About ten o'clock we encountered a series of rapids extending some eight hundred or a thousand yards, in the course of which the river has a probable aggregate fall of sixteen feet. These rapids bear the malappropriate title of the Big Falls. Following these, were a series called Prairie Rapids. At half-past four we passed the entrance of the River St. Francis, a considerable stream on the left bank. At this spot, Hennepin terminated his voyage in 1681, and Carver in 1766. There is an island at the point of confluence. At six o'clock we passed the entrance on the west shore of the stream called *Cornielle*, by the French, which is the true interpretation of the Sioux name *Karishon*, and the Chippewa term *Andaig*, which mean the crow, and not the raven. We encamped five miles below, on the east bank, having been thirteen hours in our canoes, with a generally strong current. My mineralogical gleanings, during the day, had given some specimens of the interesting varieties of the quartz family, for which the geological drift is noted, and a single piece of agatized wood. The geological floor on which the river runs, has been indicated.

At five o'clock the following morning (30th) we resumed the descent, and at the distance of two leagues reached the entrance of the Missisagiegon, or Rum River. It is Carver, I believe, who first gives us this name, for a stream which the Indians describe as a river flowing from a lake of lakes—a term, by the way, which the French, with their usual adherence to Indian etymology, have called *Mille Lacs*. The term *missi*, in this word, does not signify great, but a collected mass, or all kinds, and sometimes everywhere—the allusion being to water. *Sa-gi-e-gon* is a lake, and when the prefixed term *missi*, is put to it, nothing could more graphically describe the large body of water, interspersed with islands, which give a confused aspect, from which the river issues. The Dacotas call this lake *Mini Wakan*, meaning Spirit-water, which is probably the origin of the name of Rum River.

About thirteen miles below Rum River, and when within half a mile of the Falls,* I observed calcareous rocks in horizontal

---

* It is recently asserted that this change in the stratification occurs about a mile above the Falls. [*Sen. Doc.* p. 237.] By the same authority it is shown that the aggregate fall of the Mississippi from the mouth of Sandy Lake River to the Falls of St. Anthony is 897 feet.

beds, on the left bank of the river. It was now evident we had passed out of the primitive range of deposits, and had entered that of the great sedimentary horizontal and semicrystalline or silurian system of the Mississippi Valley; and descending with a strong current, we came, rather suddenly, it appeared, to the Falls of St. Anthony, where the river drops, by a cascade, into a rock-bordered valley. Surprise and admiration were the first emotions on getting out of our canoes and gazing on this superlative scene; and we were not a little struck with the idea that the Sioux had named the Falls from manifestly similar impressions, calling it Rara, from the Dacota verb *irara*, to laugh. By another authority, the word is written *Ha Ha*, or *Dhaha*, the letters *h* in the word representing a strong guttural sound resembling the old Arabic r.* (S. R. Riggs's *Dakota Dict. and Gram.*) Nothing can exceed the sylvan beauty of the country which is here thrown before the eye; and we should not feel surprised that the Aboriginal mind has fallen on very nearly identical sounds with the English, to express its impressions. A not very dissimilar principle has been observed by the Chippewas, who have a uniform termination of their names in *ish*, which signifies the very same quality which we express by ish in whitish, blackish, saltish—meaning a lesser, or defective quality of the noun.

The popular name of these Falls, it is known, is due to Father Louis Hennepin, a missionary who accompanied La Salle to the Illinois, in 1679, and was carried captive into the country of the Issati, a Dacota tribe, in 1680. Lt. Pike states the portage to be two hundred and sixty poles. By the time we had taken a good view of the position, and made a few sketches, the men had completed carrying over our baggage and canoes. It was now one o'clock, when we embarked to proceed to the newly-established military encampment, a few miles below. It was a noticeable feature, in our descent of the river above the Falls, that Babasi-kundiba had always kept behind the flotilla of canoes; but the moment we advanced below the Falls, he shot ahead with his delegates, each one being dressed out in his best manner. His canoe had its little flag displayed—the Indian drum was soon heard sending its measured thumps and murmurs of vocal accom-

---

* Both words are derived from the verb *to laugh.*

paniment over the water, and ever and anon guns were fired. All this was done that the enemy might be apprised of the approach of the delegation in the boldest and most open manner. It was eight or nine miles to the post, near the influx of the St. Peter's, and long before we reached Col. Leavenworth's camp, which occupied a high bluff, the attention of the Sioux was arrested by their advance, and it was inferable from the friendly answering shouts which they gave, that the mission was received with joy. Although we had known nothing of the movement which produced the pictographic letter found on a pole at the Petite Roche, above Sac River, it was, in fact, regarded by the Dacotas as an answer to that letter. And the Chippewa chief, and his followers were received with a salute by the Sioux, by whom they were taken by the hand, individually, as they landed.

Col. Leavenworth, the commanding officer, received the expedition in the most cordial manner, and assigned quarters for the members. Gov. Cass was received with a salute due to his rank. We learn that the post was established last fall. Orders for this purpose were issued, as will be seen by reference to the *Preliminary Documents*, p. 35, early in the spring. The troops destined for this purpose, were placed under the orders of Col. Leavenworth, who had distinguished himself as the commander of the ninth and twenty-second regiments, in the war of 1812. They left Detroit in the spring (1819), and proceeding by the way of Green Bay and Prairie du Chien, where garrisons were left, they ascended to the mouth of the St. Peter's, in season to erect cantonments before winter. The site chosen, being on the alluvial grounds, proved unhealthy, in consequence of which the cantonment was removed, in the spring of 1820, to an eminence and spring on the west bank of the Mississippi, about a mile from the former position.

CHAPTER XIII.

Position of the military post established at the mouth of the St. Peter's—Beauty, salubrity, and fertility of the country—Pictographic letter—Indian treaty—The appearance of the offer of frankincense in the burning of tobacco—Opwagonite—native pigments—Salt; native copper—The pouched or prairie rat—Minnesota squirrel—Etymology of the Indian name of St. Peter's River—Antiquities—Sketch of the Dacota—Descent of the Mississippi to Little Crow's village—Feast of green corn.

In favor of the soil and climate, and of the salubrity of the position, the officers speak in terms of the highest admiration. The garrison has directed its attention to both horticulture and agriculture. About ninety acres of the choicest bottom land along the St. Peter's Valley, and the adjacent prairies, have been planted with Indian corn and potatoes, cereal grains, and esculents, inclusive of a hospital, a regimental, and private gardens. At the mess-table of Col. Leavenworth, and in our camp, we were presented with green corn in the ear, peas, beans, cucumbers, beets, radishes, and lettuce. The earliest garden peas were eaten here on the 15th of June, and the first green corn on the 20th July. Much of the corn is already too hard for the table, and some of the ears can be selected which are ripe enough for seed corn. Wheat, on the prairie lands, is found to be entirely ripe, and melons in the military gardens nearly so. These are the best practical commentaries on the soil and climate.*

The distance of the St. Peter's from the Gulf of Mexico is estimated to be about two thousand two hundred miles. Its position above St. Louis is estimated at nine hundred miles. Its elevation above the Gulf is but 744 feet. The precise latitude of this point

---

* This is now (1854) the central area of Minnesota Territory—a territory in a rapid process of the development of the population and resources of a State.

is 44° 52' 46".* The atmosphere is represented as serene and transparent during the summer and spring seasons, and free from the humidity which is so objectionable a trait of our eastern latitudes. The mean temperature is 45°.† Its geology and mineralogy will be noticed in my official reports. It will be sufficient here to say that the stratification, at and below St. Anthony's Falls, consists wholly of formations of sandstones and limestones, horizontally deposited, whose relative positions and ages are chiefly inferable from the evidences of organic life, in the shape of petrifactions, which they embrace. The lowest of this series of rocks is a white sandstone, consisting of transparent, loosely-cohering grains, special allusion to which is made by Carver, in his travels in 1766, and which may be received as testimony, were there no other, that this too much discredited author had actually visited this region.

I have mentioned the interest excited by our Chippewas finding the bark letter, or pictographic memorial at the deserted Sioux encampment above Sac River. It turned out, as we were informed, that this Aboriginal missive was a reply to a similar proposition transmitted from Sandy Lake, by the Chippewas. The very person, indeed, who inscribed the Chippewa bark message, was one of the ten persons who had accompanied us from that lake. Gov. Cass, on learning this fact, requested him to draw a duplicate of it on a roll of bark. He executed this task immediately. We thus had before us the proposition in this symbolic character, which is called *ke ke win* by the Chippewas, and its answer. By this mode of communication two nations of the most diverse language found no difficulty in understanding each other.‡

On the second day after our arrival, the Indians consummated their intentions, as signified by the bark letter, and the Sandy-Lake delegation assembled with the Sioux at the old quarters of the military, now occupied as an Indian agency, and smoked the pipe of peace. There were present at this pacification, besides the chiefs Shacopee and Babasikundiba, and minor chieftains, His Excellency Gov. Cass, Col. Leavenworth, and sundry officers

---

* Ex. Doc., No. 237.　　　　　　　　　　　　† Army Register.

‡ *Vide* Appendix, for a letter from Gen. Cass to the Secretary of War on this curious topic.

of the garrison and the expedition. The ceremonies were conducted under the auspices of the U. S. Indian Agent, Mr. Taliaferro. Every attention was given to make these ceremonies impressive, by a compliance with the Aboriginal customs on these occasions, and it is hoped not without leaving permanent effects on their minds.

The pipe employed by the native diplomatists, in these negotiations, is invested with a symbolic and sacred character, as if the fumes of the weed were offered, in the nature of frankincense, to the Deity. The genuflections with which it is presented, more than the words expressed, countenance this idea. The bowl of the pipe used on this occasion consisted of the well-known red pipe-stone, called opwagonite,[*] so long known in Indian history as being brought from the *Coteau des Prairies*. It is furnished with a wooden stem two or three feet long, and two and a half inches broad, shaved down thin so as to resemble a spatula. It is then painted with certain blue or green clays, and ornamented with braids of richly dyed porcupine quills, or the holcus fragrans, and the tuft feathers of the male duck or red-headed woodpecker. These state pipes are usually presented by the speakers as memorials of the speeches, and laid aside by the officials having charge of Indian affairs. Col. Leavenworth presented us with some of these carefully ornamented diplomatic testimonials.

I obtained from the Sioux some very carefully moulded pyramidal-shaped pieces of the blue and green clays from the valley of the St. Peter's, which they employ in painting their pipe-stems and persons. The coloring matter of these appears to be carbonate of copper. It is brought from the Blue Earth River. I also obtained from the Indians very small and carefully tied leathern bags of the red oxide of iron, which they obtain in the state of a dry, powdery mass, on the prairies near the Big Stone. The Indians brought me, from the same region, crystals of salt, scraped up from the margin of certain waters on the prairies, of a dark cast, mixed with impurities. The tendency of these crystals to assume a cubic form was quite distinct. The most interesting development, in the mineralogical way, consisted of small lumps of native copper, which I obtained on an eminence on the banks

---

[*] Schoolcraft's View of the Lead Mines of Missouri. Scenes and Adventures in the Ozark Mountains, the Catlinite of Dr. Jackson.

of the Mississippi, directly opposite the influx of the St. Peter's. They occupy, geologically, a diluvial position, being at the bottom of the prairie-drift stratum, and immediately above the superior limestone.

In the luxurious kitchen gardens of Camp Leavenworth, great depredations have been made by a small quadruped of a burrowing character, called gopher. By patient watching, gun in hand, one of these was killed, and its skin preserved and prepared. The animal is ten inches long to the termination of the tail, with a body very much the size and color of a large wharf-rat. It has five prominent claws, and two broad cutting teeth, but its most striking peculiarity is a duplicature of the cheek, which permits it to carry earth to the mouth of its burrow. It has been called the pouched rat. Sir Francis Drake found a similar animal in his visit to the Gulf of California, in 1587. The distribution of this species, of which this seems to be the northern limit, is very wide through Atlantic America, and it is known to be destructive to vegetation throughout Alabama, Georgia, and the Carolinas. I had, two years ago, been led to notice its ravages in Missouri and Arkansas. But the animal called gopher, in the southern country, is a burrowing tortoise, and the name is improperly applied to this species, which is the *Pseudostoma pinetorum.*

A peculiar species of squirrel was observed in this vicinity, which is also found to be a destructive visitor to the military gardens. In appearance, this species resembles the common striped squirrel, but it has a more elongated body, and shorter legs. The body has six black stripes, with the same number of intervening lines of spots, on a reddish-brown skin. This Minnesota squirrel has, since the return of the Expedition, been named, by the late Dr. Samuel L. Mitchell, *sciurus tredeceum.*

The River St. Peter's is called, by the Dacotas, *Watepa Minnesota.* The prefixed term *watepa*, is their word for river; *minni* is the name for water. The term *sota* has been variously explained. The Canadian French, who have proved themselves most apt translators of Indian phrases, render it by the word *brouille*, or *blear;* or, if we regard this as derivative from the verb *brouiller, mixed,* or *mottled*—a condition of the waters of this river, whenever the Mississippi is in flood, and consequently at a higher elevation when it rushes into the mouth of the St. Peter's, producing that addled aspect of the water, to which the Dacotas, it is believed, apply the term *sota*.

The scenery around St. Peter's is of the most sylvan and delightful character. About six miles west of the cantonment there are several beautiful lakes, in the prairies. The largest of these is about four miles in circumference, and is called Calhoun Lake, in compliment to the Secretary of War. Its waters are stored with bass and other varieties of fish. There are several pure springs of sparkling water, issuing from the picturesque cliffs which face the Mississippi at this place. I visited one about a mile from the cantonment, which deposits a yellow sulphurous flocculent mass along its course. On the prairies is found the *holcus fragrans*, which is braided by the Indian females, and employed in some instances to decorate their deer-skin clothing. This aromatic grass retains its scent in the dried state. Along the waters of the St. Peter's is found the *acer negundo*, the inner bark of which, mixed with the common nettle, is employed by the natives in the state of a strong decoction, as a cure for the *lues venerea*.

Mr. Carver having described certain antiquities near the foot of Lake Pepin, in 1766, inquiries were made after objects of this kind in the vicinity. I was informed that traces of such remains existed in the valley of the St. Peter's, but can say nothing concerning them from actual inspection.*

Of the Dacotas, or Sioux, for which St. Peter's forms the central point, some anecdotes have been related which denote that they are, on certain occasions, actuated by exalted motives. It is related that the chief Little Crow, going out to the confines of the Chippewa Territory, to examine his beaver-traps, discovered an individual of that tribe in the act of taking a beaver from the trap. As he was himself unperceived, the tribes being at war, and the offence an extreme one, a summary punishment would have been justified by Indian law. But the Sioux chief decided differently: "Take no alarm," said he, approaching the offender: "I come to present you the trap, of which I see you stand in need. Take my

---

* The last known platform mound in the spread of the mound-builders north, is at Prairie du Chien. The monuments, supposed to be mounds, in the St. Peter's region, are found by Mr. Owen to be geological elevations. The remains on Blue Earth River are attributed to a fort or inclosure built by Le Seur, in his search for copper on that stream, in 1700. Other remains, in the St. Peter's valley, appear to be old trading-houses, fallen in.

gun, also, as I see you have none of your own, and return to the
land of your countrymen; and linger not here, lest some of my
young men should discover your footsteps."

A still more striking and characteristic incident is related of a
chief called the Red Thunder. Col. Wm. Dixon, a Scotchman of
family, who made his influence felt in the late war of 1812 as a
leader of the Sioux and a merchant among them, married the
sister of this notable chief. So daring were the acts of Red Thunder, that he had put the Chippewa nation in awe of him. At
length, however, after a long series of the bravest acts, he was
taken prisoner, with a favorite dog, and condemned to expiate his
offences at the stake. It was a time of want by his captors. One
day he said to them: "Why do you not feed my dog?" They
replied, "feed him yourself." "Then," he said, "give me a knife."
This being thrown to him, he cut a piece of flesh from one of his
large and fleshy thighs, and threw it to the dog. Admiration of
this act ran through the Indian camp. They immediately released
him, and bestowed on him the highest attentions and honors.

The Dacota or Sioux nation constitute one of the families of
America who speak a peculiar language. Lieut. Pike, who visited
them in 1806, estimated their numerical strength at twenty-one
thousand six hundred and seventy-five; of which number he computed three thousand eight hundred to be warriors. They consist of six or seven independent tribes, or sub-tribes, bearing
different names, who occupy most of the country between the
Mississippi and Missouri, between N. latitude 43° and 46°. The
Mendawekantoñs are located on the Mississippi, below the Falls of
St. Anthony and the mouth of the St. Peter's. The Sessitoñs and
Yanktoñs occupy the upper waters of the St. Peter's. The Titoñs
only extend west of the Missouri. The several tribes regard
themselves as a confederacy, which is the signification of the term
Dacota. They do not acknowledge the name of Sioux as an
Indian word. We first hear of them from the early French mis-
sionaries, who visited the head of Lake Superior about the middle
of the 17th century, under the name of *Nadowasie*.* They speak

---

* This is an Algonquin expression, signifying enemy. It is derived from *Nodowa*,
an Iroquois, or a Dacota; the word was originally applied to a serpent. The termination in *sie* is from *awasie*, an animal or creature. This term is the root, it is
apprehended, of the French sobriquet *Sioux*.

a language which prevails over an immense area, which is now occupied by the prairie tribes towards the west and southwest, from whence, it is inferred, they came. They appear, at a former time, to have reached and dwelt at the sources of the Mississippi, and to have approached, if not reached, the west end of Lake Superior; for it is from these positions that the oldest traditions represent them to have been driven by the Chippewas. Lieut. Pike thinks they are, undeniably, descendants of Tartars. If so, I feel inclined to think that they must have made the circuit of the Mexican provinces before reaching the Mississippi Valley, for the track of their migration is traced towards the south certainly as far as the country of the Kansas and Osages; while they preserve some striking traits and characteristics which appear to be referable to those intertropical regions.

Having passed the better part of three days in the vicinity of St. Peter's, adding to our collections and portfolios, we left it on the second of August, and proceeded down the river to the village of La Petite Corbeau, or the Little Raven, situated on the east bank not far above the mouth of the St. Croix. The river, in this distance flows between lofty cliffs of the white sandstone and neutral-colored limestones, which are first conspicuously displayed at the Falls of St. Anthony. Springs of water, not infrequently, issue from these cliffs. We landed at one of these, flowing in through a gorge at the distance of four miles below St. Peter's, on the east bank, for the purpose of visiting a remarkable cave, from the mouth of which a small stream issues. The cave is seated wholly within the beautiful white crumbling sandstone rock. It is, in fact, the loose character of the rock which permits the superincumbent waters of the plains above to permeate through it, that has originated the cave. The stream consisted of the purest filtrated water, which is daily carrying away the loosened grains of sand into the Mississippi, and thus enlarging the boundaries of the cavern.* We had been erroneously informed that this was Carver's Cave, and looked in vain for this traveller's name on its walls.† The atmosphere in this

---

* St. Paul's, the present capital of Minnesota (1854), is situated on the high grounds, a few miles below this cave.

† Carver's Cave is four miles lower down, on the same side of the river, agreeably to subsequent observation. It is now obstructed by fallen rock and debris.

cave was found to be seven degrees higher than the water. We noticed nothing in the form of bones or antiquities.

The village of Petite Corbeau consists of twelve large lodges, which are said to give shelter to two hundred souls. They plant corn, and cultivate vines and pumpkins. They sallied from their lodges on seeing us approach, and, gathering along the margin of the river, fired a *feu de joie* on our landing. The chief was among the first to greet us. He is a man below the common size, but brawny and well proportioned, and, although above fifty years of age, retains the look and vigor of forty. He invited us to his lodge—a spacious building about sixty feet by thirty, substantially constructed of logs and bark. Being seated, he addressed himself to His Excellency Gov. Cass. He said that he was glad to see him in his village. That, in his extensive journey, he must have suffered many hardships. He must also have noticed much of the Indian mode of life, and of the face of the country, which would enable him to see things in their proper light. He was glad that he had not, like others who had lately visited the country, passed by his village without calling. He referred, particularly, to the military force sent to establish a garrison at St. Peter's, the year before, who had passed up on the other side of the river. He acquiesced in the treaty that had been recently concluded with the Chippewas. He referred to a recent attack of a party of Fox Indians on their people, on the head waters of the St. Peter's. He said it was dastardly, and that, if that *little* tribe should continue their attacks, they would at length drive him into anger, and compel him to do a thing he did not wish.

While this speech was being interpreted, the Indian women were employed in bringing basketsful of ears of Indian corn from the fields, which they emptied in a pile. This pile, when it had reached a formidable height, was offered as a present to the Expedition. It was, indeed, the beginning of the season of green corn, with them, and we were soon apprised, by the sound of music from another lodge, that the festival of the green-corn dance was going forward. Being admitted to see the ceremonies, the first thing which attracted notice was two large iron kettles suspended over a fire, filled with green-corn cut from the cob. The Indians, both men and women, were seated in a large circle around them; they were engaged in singing a measured chant in the Indian manner, accompanied by the Sioux cancega or drum and rattles; the utmost solemnity was depicted on every countenance. When the music paused, there were certain gesticulations made, as if a mysterious power were invoked. In the course of these ceremonies, a young man and his sister, joining hands, came forward to be received into the green corn society, of whom questions were asked by the presiding official. At the conclusion of these, the voice of each member was taken as to their admission, which was unanimous. At the termination of the ceremonies, an elderly man came forward and ladled out the contents of the kettles into separate wooden dishes for each head of a family present. As these dishes were received, the persons retired from the lodge by a backward movement, still keeping their faces directed to the kettles, till they had passed out.

11

CHAPTER XIV.

Descent of the river from the site of Little Crow's Village to Prairie du Chien.—Incidents of the voyage, and notices of the scenery and natural history.

THE next morning we embarked at 5 o'clock. On descending the river six miles, we passed the mouth of the St. Croix.* This stream heads on high lands, which form a rim of hills around the southern and western shores of Lake Superior, where it is connected with the River Misacoda, or Broulè of Fond du Lac. The Namakagon, its southern branch of it, is connected with the Maskigo,† or Mauvais River of La Pointe, Lake Superior. Immediately above its point of entrance into the Mississippi the St. Croix expands into a beautiful lake, which is some twelve miles long, and about two in width. The borders of the Mississippi about this point assume an increased height, and more imposing aspect. In many places, as the voyager descends from this spot to Lake Pepin, he observes the calcareous cliffs to terminate in pyramids; the crest of the hills frequently resemble the crumbling ruins of antique towers. At 12 o'clock we came to the vicinity of an isolated calcareous cliff, called La Grange, which may be regarded as one of those monuments resulting from geological denudation, which constitute a striking feature in the St. Peter's region. The top of this cliff affords a fine view of the scenery of the Mississippi for a long distance above and below it. It has been found to be three hundred and twenty-two feet above the river.‡

* This river was explored by me in 1832. Vide *Schoolcraft's Expedition to Itasca Lake.* 1 vol. 8vo. p. 307—1834: N. Y., Harpers.

† In 1831, this river was ascended by me with a public expedition, dispatched into the Indian country to quell the disturbances which eventuated the next year in the Sauk war. Vide *Schoolcraft's Thirty Years in the Indian Country.* Lippincott, Grambo, & Co., Philad.: 1 vol. p. 703, 1851.

‡ Doc. 287.

This spot is noted as being near the site of Tarangamani, or the Red Wing's Village. This chief is one of the notable men of his tribe. He has been long celebrated as a man skilled as a native magician. The village consists of four large, elongated, and of several small lodges. Tarangamani is now considered the first chief of his nation. He is noted for his wisdom and sagacity. He bears the marks of being sixty years of age. His grand-daughter married Col. Crawford, a man of commercial activity about Prairie du Chien and Michilimackinac, during the late war of 1812, who has left descendants in the lake country. We observed, at this village, several buffalo skins undergoing the Indian process of dressing. The hair having been removed, they were stretched on the ground, where they were subjected to a process analogous to tanning by being covered with a decoction of oak bark.

In ascending the hill of La Grange, we first encountered the rattlesnake, two of which we killed. This is the highest northern point at which we have observed this species on the Mississippi. I observed on this elevation small detached masses of radiated quartz, cinnamon-colored and white, together with an ore of iron crystallized in cubes. Having cursorily examined the environs, the expedition again embarked. It was 1 o'clock when we entered Lake Pepin. This admired lake is a mere expansion of the Mississippi, having a length of twenty-four miles by a varying width of from two to four miles. During this distance there is not the least current during calm weather. The prospects, in passing through this expanse of water, are of the most picturesque kind. Its immediate shores are circumscribed with a broad beach of gravel, in which may be found rolled pieces of the chalcedonies, agates, and other species of the quartz family, which are characteristic of the drift-stratum of the upper borders of the Mississippi. On the eastern shore, at a short distance from the margin, there is a lofty range of limestone cliffs. On the west, the eye rests on an elevated formation of prairie, nearly destitute of trees. From this plain several conical hills ascend, which have the appearance, but only the appearance, of artificial construction. The lake is quite transparent, and yields several species of fish. The most remarkable of these is the *acipenser spatularia*, of which we obtained a specimen. It is also remarkable for its numerous varieties, and the large size of its fresh-water shells. I procured several species of *unio*, which, from their size and character, attracted my attention, particularly to the subject of this branch of American conchology. Several of these, from the duplicates of my cabinet, have attracted the attention of conchologists.* Lake Pepin receives a river from the west called the Ocano, or more properly *Au Canot;* its mouth having been, in former times, a noted place for concealing canoes during the winter season.† At a point, on the east shore, about half-way down the lake, where a small stream enters, we were informed there existed the remains of an old French fort, or factory; but we did not land to examine them.

In passing through this lake the interpreters pointed to a high precipice in the cliffs on the east shore, which Indian tradition assigns as the locality of a tragical love tale, of which a Dacota girl was the heroine. To avoid the dilemma of being compelled to accept a husband of repulsive character, and to sacrifice her affections for another person, she precipitated herself down this precipice. The tale has been so differently told to travellers visiting the region, that nothing but the simple tradition appears worth recording. Olaita and Winona, have been mentioned as the name of the Dacota Sappho.

At 6 o'clock in the evening we encamped on a gravelly beach on the east shore of the lake, the weather threatening a storm. Rain commenced at 8 o'clock, and continued at intervals, with severe thunder and most vivid flashes of lightning during the night. At 5 o'clock the next morning (4th), the expedition was again in motion. The rain had ceased, but the morning remained cloudy. The scenery on the borders of the lake continued to be impressive. The precipices on the east shore shot up into spiral points; yet the orbicular elevations are covered with grass and shrubbery. These high grass-crowned elevations, without forest, terminate near the influx of the Chippewa River in a remarkable isolated elevation, called *Mont La Garde,* from the fact that it is,

---

* Silliman's Journal of Science, 1828 ; also, Trans. Am. Phil. Soc.

† Travellers who are disposed to regard La Hontan's fiction of his purported discoveries on *Rivier la Longue,* as entitled to notice, have suggested *this* river as the locality intended. Nicollet, otherwise reliable, has gone so far as to call it La Hontan River.

and long has been, a noted look-out station for Chippewa war parties, who descend this stream, against the Sioux. It commands an extensive view of Lake Pepin. This lake was thought to be two miles wide opposite our last night's encampment; it narrows to probably less than half a mile at its mouth. The west shore along this portion of the lake consists of singularly striking, picturesque, level, and elevated prairie lands.

Carver, in 1768, places his remains of ancient circumvallations in this vicinity, but "some miles below Lake Pepin."* This was a period when no attention had been directed to the subject of antiquities in the United States, and his mind appears to have been impressed strongly by what he saw. As opportunities did not allow me to land, nor was the precise spot, indeed, known to any of our guides or men, reference can only be made to the observations of a man who is known to have been the first American traveller that has called attention to our western antiquities. Mr. H. V. Hart, long a resident of this region, verbally assures me that he has visited these works.†

Chippewa River, just referred to, comes into the Mississippi on its left bank, within half a mile of the foot of Lake Pepin. It is a tributary of prime volume, draining the Chippewa territories lying around the south and west shores of Lake Superior. Originating on the sandy tracts extending over the elevated central plains of the Wisconsin, it brings a large deposit of sand into the Mississippi, the navigation of which is visibly more embarrassed below this point with sand-bars, willow, and cotton-wood islands.

At four o'clock in the afternoon we reached and landed at Wabashaw's village. It is eligibly seated on the west shore, and consists of four of the large elongated Sioux lodges before mentioned, containing a population of about sixty souls. The usual intercourse and speeches of congratulation by the Indians, and acknowledgment of the American authorities were made, and we again embarked, after a detention of forty minutes. A few miles below Wabashaw's village, we came to a high rocky or mountain island, called *La montaigne qui trompe dans l'eau*, a term which is shortened by western phraseology into TROMPLEDO mount-

---

* Carver's Travels, p. 80.

† Mr. G. W. Featherstonehaugh, in his *Geological Reconnoissance*, in 1884, landed at the location of these antiquarian remains, and is disposed to recognize their authenticity.

ain. This is a very remarkable feature in the geography of the Upper Mississippi. The rock is calcareous; it is, in fact, the only fast or rocky island we have encountered below the little islet at the head of the Packagama Falls. It is not only striking from its lofty elevation, but is several miles in circumference; standing in the bed of the river and parting its channel into two, it appears to be the first bold geological monument which has effectually resisted its course.

We had passed this island but a short distance, and the approaches of evening began to be manifest, when a large gray wolf sprang into the river to cross it. The greatest animation at once arose in our flotilla; the canoemen bending themselves to their paddles, the auxiliary Indians of our party shouting, and the whole party assuming an unwonted excitement. A shot was soon fired from one of our rifles, but either the distance was too great, or the aim incorrect. The wolf was fully apprised of his peril, put forth all his strength, outstripped his pursuers, reached the shore, and nimbly leaped into the woods.

We encamped on the west shore, a few miles below the island at seven o'clock, having been twelve hours in our canoes. The confinement of the position nobody can appreciate who has not tried it, and I hastened to stretch my legs, by ascending the river cliffs in our rear, to have a glimpse of its geology and scenery. The view westwardly was one of groves and prairies of most inviting agricultural promise. In front, the island mountain rises to an elevation which appears to have been the original geological level of the stratification before the Mississippi cut its way through it.

At the rapids of Black River, which enters opposite our encampment, a saw-mill, we were informed, had been erected by an inhabitant of Prairie du Chien. Thus the empire of the arts has begun to make its way into these regions, and proclaims the advance of a heavy civilization into a valley which has heretofore only resounded to the savage war-whoop. Or, if a higher grade of society and arts has ever before existed in it, as some of our tumuli and antiquities would lead us to infer, the light of history has failed to reach us on the subject.[*]

* *American Antiquities.* As the tumuli and earthworks of the Mississippi Valley are more closely scrutinized, they do not appear to denote a higher degree of

At the spot of our encampment, as soon as the shades of night closed in, we were visited by hordes of ephemera. The candles lighted in our tents became the points of attraction for these evanescent creations. They soon, however, began to feel the influence of the sinking of the thermometer, and the air was imperceptibly cleared of them in an hour or two. By the hour of three o'clock the next morning (5th) the expedition was again in motion descending the river. It halted for breakfast at Painted Rock, on the west shore. While this matter was being accomplished, I found an abundant locality of unios in a curve of the shore which produced an eddy. Fine specimens of U. purpureus, elongatus, and orbiculatus were obtained. With the increased spirit and animation which the whole party felt on the prospect of our arrival at Prairie du Chien, we proceeded unremittingly on our descent, and reached that place at six o'clock in the evening.

Prairie du Chien does not derive its name from the dog, but from a noted family of Fox Indians bearing this name, who anciently dwelt here. The old town is said to have been about a mile below the present settlement, which was commenced by Mr. Dubuque and his associates, in 1788. The prairie is most eligibly situated along the margin of the stream, above whose floods it is elevated. It consists of a heavy stratum of diluvial pebbles and boulders, which is picturesquely bounded by lofty cliffs of the silurian* limestones, and their accompanying column of stratification. The village has the old and shabby look of all the antique French towns on the Mississippi, and in the great lake basins; the

civilization than may be assigned to the ancestors of the present races of Indians, prior to the epoch of the introduction of European arts into America. Certainly there is nothing in our earthworks and mounds, to compare with the Toltec and Aztec type of arts at the opening of the 16th century; while the possession by our tribes of the zea maize, a tropical plant, and other facts indicative of a southern migration, appear to denote a residence in warmer latitudes. The distribution of the Mexican teocalli and pyramid is also plainly traceable from the south. Neither the platform nor the solid conical mound has been traced higher north than Prairie du Chien; nor have the earthworks (adopting Carver's notices) reached higher than Lake Pepin. There are no mounds or earthworks at the sources of the Mississippi nor in all British America to the shores of the Arctic Seas. We cannot bring arts or civilization from that quarter.

* This term, unknown to geology at the period, has been subsequently introduced by Sir Roderic Murchison.

dwellings being constructed of logs and barks, and the court-yards picketed in, as if they were intended for defence. It is called Kipisagee by the Chippewas and Algonquin tribes gene-rally, meaning the place of the jet or outflow of the (Wisconsin) River. It is, in popular parlance, estimated to be 300 miles below St. Peter's, and 600 above St. Louis.* Its latitude is 43° 3' 6". It is the seat of justice for Crawford County, having been so named in honor of W. H. Crawford, Secretary of the Treasury of the U. S. It is, together with all the region west of Lake Michigan, attached to the territory of Michigan. There is a large and fertile island in the Mississippi, opposite the place.

We found the garrison to consist of a single company of infantry, under the command of Capt. J. Fowle, Jun.,† who re-ceived us courteously, and offered the salute due to the rank of His Excellency, Gov. Cass. The fort is a square stockade, with bastions at two angles. There was found on this part of the prairie, when it came to be occupied with a garrison by the Amer-icans, in 1819, an ancient platform-mound, in an exactly square form, the shape and outlines of which were preserved with exact-itude by the prairie sod. This earthwork, the probable evidence of a condition of ancient society, arts, and events of a race who are now reduced so low, was, with good taste, preserved by the military, when they erected this stockade. One of the officers built a dwelling-house upon it, thus converting it, to the use, and probably the only use, to which it was originally devoted. No measurements have been preserved of its original condition ; but judging from present appearances, it must have squared seventy-five feet, and have had an elevation of eight feet.

* These distances are reduced by *Ex. Doc.* 237, respectively to 260 and 542 miles.
† This officer entered the army in 1812, serving with reputation. He rose, through various grades of the service, to the rank of Lieut. Col. of the 6th infantry. He lost his life on the 25th April, 1838, by the explosion of the steamer Moselle, on the Ohio River.

CHAPTER XV.

Mr. Schoolcraft makes a visit to the lead mines of Dubuque—Incidents of the trip —Description of the mines—The title of occupancy, and the mode of the mines being worked by the Fox tribe of Indians—Who are the Foxes?

I SOLICITED permission of Gov. Cass to visit the lead mines of Dubuque, which are situated on the west bank of the Mississippi, at the computed distance of twenty-five leagues below Prairie du Chien. Furnished with a light canoe, manned by eight voyageurs, including a guide, I left the prairie at half-past eleven A. M. (6th). Passed the entrance of the Wisconsin, on the left bank, at the distance of a league.* Opposite this point is the high elevation which Pike, in 1806, recommended to be occupied with a military work. The suggestion has not, however, been adopted; military men, probably, thinking that, however eligible the site might be for a work where civilized nations were likely to come into contact, a simpler style of defensive works would serve the purpose of keeping the Indian tribes in check. I proceeded nine leagues below, and encamped at the site of a Fox village,† located on the east bank, a mile below the entrance of Turkey River from the west. The village, consisting of twelve lodges, was now temporarily deserted, the Indians being probably absent on a hunt; but, if so, it was remarkable that not a soul or living thing was left behind, not even a dog. My guide, indeed, informed me that the cause of the desertion was the fears entertained of an attack from the Sioux, in retaliation for the massacre lately perpetrated by them on the heads of the St. Peter's, which was alluded to in the speech of the Little Crow, while we were at his village (*ante*, p. 160).

---

* It was at this spot, one hundred and thirty-seven years ago, that Marquette and M. Joliet, coming from the lakes, discovered the Mississippi.

† Now the site of Cassville, Grant County, Wisconsin. It is a post town, pleasantly situated, with a population of 200.

It was seven o'clock P. M. when I landed here, and having some hours of daylight, I walked back from the river to look at the village, and its fields, and to examine the geological structure of the adjacent cliffs. In their gardens I observed squashes, beans, and pumpkins, but the fields of corn, the principal article of cultivation, had been nearly all destroyed, probably by wild animals. I found an extensive field of water and musk melons, situated in an opening in a grove, detached from the other fields and gardens. None of the fruit was perfectly ripe, although it had been found so at Prairie du Chien; some of it had been bitten by wild animals.* The cliffs consisted of the same horizontal strata of sandstones and neutral colored limestone, prevailing at higher positions in this valley. Returning to the river beach, I perceived the same pebble drift which characterizes higher latitudes. This seems the only difference in its structure or form, namely, that the pieces of quartz pebble, limestone, and other fragments brought down, become smaller and smaller, as they are carried down.

There were frequent thunders, and a rain-storm, during the night, which, with a slight intermission, characterized the morning until noon. I embarked at half-past three A. M. (7th), and landed at the Fox village of the Kettle chief, at the site of Dubuque's house,† at ten o'clock; a moderate rain having continued all the way. It ceased an hour after my arrival.

* Fondness for melons, and annual vine fruits of the garden, is a striking trait of the Indians. Some curious facts on this head are published in the statistics.— *Indian Information*, vol. iii. p. 624, 1853, Philadelphia, Lippincott & Co.

† This is now (1854) the site of the city of Dubuque, State of Iowa, which is reputed to be the oldest settlement in that State. This city is eligibly situated on a broad plateau, between limestone cliffs. The soil rests on a rock foundation, which renders it incapable of being undermined by the Mississippi. Its streets are broad and laid out at right angles. It has several Protestant churches, a Catholic cathedral, a public land office, two banks, four printing offices, and by the last census contains a population of 7,500, the county of which it is the seat of justice, has 10,840. Two railroads have their terminal points at this place. At the time of my visit, in 1820, the house which had been built by Mr. Dubuque, had been burnt down; and there was not a dwelling superior to the Indian wigwam within the present limits of Iowa. The State of Iowa was admitted into the Union in 1837. By the 7th U. S. census, the population of this State, in 1850, is shown to be 192,214. The number of square miles is 50,914. No Western State is believed to contain a less proportionate quantity of land unsuited to the plough, and its population and resources must have a rapid development.

The Kettle chief's village is situated fifteen miles below the entrance of the Little Makokety River, consisting of nineteen lodges, built in two rows, pretty compact, and having a population of two hundred and fifty-souls. There is a large island in the Mississippi, directly opposite this village, which is occupied by traders. I first landed there to get an interpreter of the Fox language, and obtain some necessary information respecting the location of the mines, and the best means of accomplishing my object. Meantime the rain had ceased. I then proceeded across the Mississippi to the Kettle chief's lodge, to solicit his permission to visit the mines, and obtain Indian guides. I succeeded in getting Mr. Gates, as interpreter; and was accompanied by Dr. S. Muir, a trader, who politely offered to go with me. On entering the lodge of Aquoqua, the chief, I found him suffering under a severe attack of bilious fever. As I approached him, he sat upon his pallet, being unable to stand, and bid me welcome; but soon became exhausted by the labor of conversation, and was obliged to resume his former position. He appeared to be a man of eighty years of age, had a venerable look, but was reduced to the last stage of physical debility. Yet he retained his faculties of sight and hearing unimpaired, together with his mental powers. He spoke to me of his death with calm resignation, as a thing to be desired. On stating the object of my visit, some objections were made by the chiefs who surrounded him, and they required further time to consider the proposition. In the mean time, I learned from another source, that since the death of Dubuque, to whom the Indians had formerly granted the privilege of working the mines, they had manifested great jealousy of the whites, were afraid they would encroach on their rights, denied all former grants, and did not make it a practice even to allow strangers to view their diggings. Apprehending some difficulties of this kind, I had provided myself with some presents, and concluding this to be the time, because of the reluctance manifested, directed one of my voyageurs to bring in a present of tobacco and whiskey; and in a few moments I received their assent, and two guides were furnished. One of these was a minor chief, called Scabass, or the Yelling Wolf; the other, Wa-ba-say-ah, or the White Fox-skin. They led me up the cliff, where I understood the Indian woman, Peosta, first found lead ore; after reaching the level of the river bluffs, we pursued a path over undulating hills, exhibiting a half prairie, and quite picturesque rural aspect.   On reaching the diggings, the most striking part of them, but not all of them, exhibited excavations such as the Indians only do not seem persevering enough in labor to have made.

The district of country called Dubuque's Mines, embraces an area of about twenty-one square leagues, commencing at the mouth of the Little Maquaquity River, sixty miles below Prairie du Chien, and extending along the west bank of the Mississippi River, seven leagues in front by three in depth.   The principal mines are situated on a tract of one square league, beginning immediately at the Fox village of Aquoqua, or the Kettle chief, and extending westwardly.   This is the seat of the mining operations carried on by Dubuque, as well as of what are called the Indian Diggings.

Geologically it is the same formation that characterizes the mines of Missouri; but there are some peculiarities.   The ore found is the common sulphuret of lead, with a broad foliated, or lamellated structure, and high metallic lustre.   It occurs massive and disseminated, in a red loam, resting on a horizontal limestone rock.   Sometimes small veins of the ore are seen in the rock, but it has been generally explored in the soil. It generally occurs in narrow beds, which have a fixed direction; these beds extend three or four hundred feet, when they cease, or are traced into crevices in the rock.   At this stage, the pursuit of ore, at most of the diggings, has been abandoned, frequently with small veins of the metal in view.   No matrix, so far as I observed, is found with the ore which is dug out of the soil, unless we may consider such an ochery oxide of iron, with which it is slightly incrusted.   Occasionally, pieces of calcareous spar are thrown out with the earth in digging after ore.   I picked up from one of these heaps of earth a specimen of transparent crystallized sulphate of barytes; but this mineral appears to be rare. There appears to be none of the radiated quartz, or white opaque heavy spar, which are so abundantly found at the Missouri mines.*

The ore at these mines is now exclusively dug by the Indian

---

* *Vide* my View of the Lead Mines of Missouri, &c., New York, 1819.

women. Old and superannuated men also partake in the mining labor, but the warriors and men hold themselves above it. In this labor, the persons who engage in it employ the hoe, shovel, pick-axe, and crow-bar. These implements are supplied by the traders at the island, who are the purchasers of the crude ore. With these implements they dig trenches, till they are arrested by the solid rock. There are no shafts, even of the simplest kind, and the windlass and bucket are unknown to them—far more so the use of gunpowder in the mining operations. Their mode of going down into the deepest pits, and coming up from them, is by digging an inclined way, which permits the women to keep an erect position in walking.* I descended into one of these inclined excavations, which had probably been carried down forty feet, at the perpendicular angle.

When a quantity of ore has been got out, it is carried in baskets to the banks of the Mississippi, by the females, who are ferried over to the island. They receive at the rate of two dollars for a hundred and twenty pounds, payable in goods. At the profit at which these are usually sold, it may be presumed to cost the traders at the rate of seventy-five cents or a dollar, cash value, per hundred weight. The traders smelt the ore on the island, in furnaces of the same construction which I have described, and given plates of, in my treatise on the mines.† They observe that it yields the same per centum of metallic lead. Formerly, the Indians were in the habit of smelting the ore themselves on log-heaps, by which an unusual proportion of it was converted into lead-ashes and lost. They are now induced to search about the sites of these old fires to collect these lead-ashes, which consist, for the most part, of desulphuretted ore, for which they receive a dollar per bushel.

There are three mines in addition to those above mentioned, situated upon the Upper Mississippi, which are worked by the Indians. They are located at Sinsinaway, at Rivière au Fevre, and at the Little Makokety. 1. Sinsinaway mines. They are situated fifteen miles below Aquoqua's Village, on the east shore of the Mississippi, at the junction of the Sinsinaway River. 2.

* This is believed to be an oriental mode of excavation, which appears to have been practised in digging wells.

† New York, 1819.

Mine au Fevre. Situated on the River au Fevre, which enters
the Mississippi on its east banks, twenty-one miles below Dubuque's mines. The lead ore is found ten miles from its mouth.
At this locality, the ore is accompanied by the sulphate of barytes, and is sometimes crystallized in cubes or octohedrons.* 8.
Mine of the Makokety, or Maquoqueti. This small river enters
the Mississippi fifteen miles above Dubuque's mines. The mineral character and value of the country has been but little explored.

The history of the mines of Dubuque is brief and simple. In
1780, a discovery of lead ore was made by the wife of Peosta, a
Fox Indian of Aquoqua's Village. This gave the hint for explorations, which resulted in extensive discoveries. The lands
were formally granted by the Indians to Julien Dubuque, at a
council held at Prairie du Chien in 1788, by virtue of which he
permanently settled on them, erected buildings and furnaces, and
continued to work them until 1810. In 1796, he received a confirmation of his grant from Carondelet, the governor of Louisiana,
in which they are called "the mines of Spain." By a stone monument which stands on a hill near the mines, Dubuque died on the
24th March, 1810, aged forty-five years and six months. After his
death, the Indians burnt down his house and fences—he leaving,
I believe, no family†—and erased every vestige of civilized life;
and they have since revoked, or at least denied the grant, and
appear to set a very high value on the mines. Dubuque's claim
was assigned to his creditors, by whom it was presented to the
commissioners for deciding on land titles, in 1806. By a majority
of the board it was determined to be valid, in which condition it
was reported to Washington for final action. At this stage of

* The city of Galena has subsequently been built on this river, at the distance
of six miles from the Mississippi. The river is, indeed, thus far, an arm of the
Mississippi, which permits steamboats freely to enter, converting the place into a
commercial depot for a vast surrounding country. Not less than 40,000,000 pounds
of lead were shipped from this place in 1852, valued at one million six hundred
thousand dollars. It is the terminus of the Chicago and Galena Railroad, connecting it by a line of 180 miles with the lakes. It contains a bank, three newspaper
offices, and several churches of various denominations, and has, by the census of
1850, a population of 6,004.

† There is believed to be no instance, in America, where the Indians have disan-
nulled grants or privileges to persons settling among them, and leaving families
founded on the Indian element.

.the investigation, Mr. Gallatin, who was then Secretary of the Treasury, made a report on the subject, clearly stating the facts, and coming to the conclusion that it was not a perfect title, stating that no patent had ever been issued for it, at New Orleans, the seat of the Spanish authority, from which transcripts of the records of all grants had been transmitted to the Treasury.*

On the arrival of Lieut. Pike at Mr. Dubuque's on the 1st of September, 1805, he endeavored to obtain information necessary to judge of the value and extent and the nature of the grant of the mines; but he was not able to visit them. To the inquiries which he addressed to Mr. Dubuque on the subject, the latter replied in writing that a copy of the grant was filed at the proper office in St. Louis, which would show its date, together with the date of its confirmation by the Spanish authority, and the extent of the grant to him. He states the mine to be twenty-seven or twenty-eight leagues long, and from one to three leagues broad. He represents the per centum of metal to be yielded from the ore to be seventy-five, and the quantity smelted per annum at from 20,000 to 40,000 pounds. He stated that the whole product was cast into pig lead, and that there were no other metals at the mines but copper, of the value of which he could not judge.

Having examined the mines with as much minuteness as the time allowed me would permit, and obtained specimens of its ores and minerals, I returned to the banks of the Mississippi, before the daylight departed, and, immediately embarking, went up the river two leagues and encamped on an island.

It may be proper to add to this narrative of my mineralogical visit to these mines, a few words respecting the Fox Indians, by whom the country is owned. The first we hear of these people is from early missionaries of New France, who call them, in a list drawn up for the government in 1736, "Gens du Sang," and Miskaukis. The latter I found to be the name they apply to themselves. We get nothing, however, by it. It means Red-earths, being a compound from *misk-wau*, red, and *auki*, earth. They are a branch of the great Algonquin family. The French, who formed a bad opinion of them, as their history opened, bestowed

---

* For the facts in this case, see *Collection of Land Laws of the United States*, printed at Washington, 1817.

on them the name of Renouard, from which we derive their long-standing popular name. Their traditions attribute their origin to eastern portions of America. Mr. Gates, who acted as my interpreter, and is well acquainted with their language and customs, informs me that their traditions refer to their residence on the north banks of the St. Lawrence, near the ancient Cataraqui. They appear to have been a very erratic, spirited, warlike, and treacherous tribe; dwelling but a short time at a spot, and pushing westward, as their affairs led them, till they finally reached the Mississippi, which they must have crossed after 1766, for Carver found them living in villages on the Wisconsin. At Saginaw, they appear to have formed a fast alliance with the Saucs, a tribe to whom they are closely allied by language and history. They figure in the history of Indian events about old Michilimackinac, where they played pranks under the not very definite title of Muscodainsug, but are first conspicuously noted while they dwelt on the river bearing their name, which falls into Green Bay, Wisconsin.* The Chippewas, with whom they have strong affinity of language, call them Otagami, and ever deemed them a sanguinary and unreliable tribe. The French defeated them in a sanguinary battle at Butte de Mort, and by this defeat drove them from Fox River.

Their present numbers cannot be accurately given. I was informed that the village I visited contained two hundred and fifty souls. They have a large village at Rock Island, where the Foxes and Saucs live together, which consists of sixty lodges, and numbers three hundred souls. One-half of these may be Saucs. They have another village at the mouth of Turkey River; altogether, they may muster from 460 to 500 souls. Yet, they are at war with most of the tribes around them, except the Iowas, Saucs, and Kickapoos. They are engaged in a deadly, and apparently successful war against the Sioux tribes. They recently killed nine men of that nation, on the Terre Blue River; and a party of twenty men are now absent, in the same direction, under a half-breed named Morgan. They are on bad terms with the Osages and Pawnees of the Missouri, and not on the best terms with their neighbors the Winnebagoes.

* This name was first applied to a territory in 1836.

I again embarked at four o'clock A. M. (8th). My men were stout fellows, and worked with hearty will, and it was thought possible to reach the Prairie during the day, by hard and late pushing. We passed Turkey River at two o'clock, and they boldly plied their paddles, sometimes animating their labors with a song; but the Mississippi proved too stout for us; and some time after nightfall we put ashore on an island, before reaching the Wisconsin. In ascending the river this day, observed the pelican, which exhibited itself in a flock, standing on a low sandy spot of an island. This bird has a clumsy and unwieldy look, from the duplicate membrane attached to its lower mandible, which is constructed so as when inflated to give it a bag-like appearance. A short sleep served to restore the men, and we were again in our canoes the next morning (9th) before I could certainly tell the time by my watch. Daylight had not yet broke when we passed the influx of the Wisconsin, and we reached the Prairie under a full chorus, and landed at six o'clock.

12

CHAPTER XVI.

The expedition proceeds from Prairie du Chein up the Wisconsin Valley—Incidents of the ascent—Etymology of the name—The low state of its waters favorable to the observation of its fresh-water conchology—Cross the Wisconsin summit, and descend the Fox River to Winnebago Lake.

WE were now at the foot of the Wisconsin Valley—at the point, in fact, where Marquette and Joliet, coming from the forests and lakes of New France, had discovered the great River of the West, in 1673. Marquette, led by his rubrics, named it the River "Conception," but, in his journal, he freely employs the aboriginal term of Mississippi, which was in use by the whole body of the Algonquin tribes. While awaiting, at Prairie du Chein, the preparations for ascending the Wisconsin, the locality was found a very remarkable one for its large unios, and some other species of fresh-water shells. Some specimens of the unio crassus, found on the shores of the island in the Mississippi, opposite the village, were of thrice the size of any noticed in America or Europe, and put conchologists in doubt whether the species should not be named *giganteus.** I had, in coming down the Mississippi, procured some fine and large specimens of the unio purpureus of Mr. Say, at the Painted Rock, with some other species; and the discovery of such large species of the crassus served to direct new attention to the subject.

Our sympathies were excited, at this place, by observing an object of human deformity in the person of an Indian, who, to remedy the want of the power of locomotion, had adjusted his legs in a large wooden bowl. By rocking this on the ground, he supplied, in a manner, the lost locomotive power. This man of the bowl possessed his faculties of mind unimpaired, spoke seve-

* American Journal of Science, vol. vi. p. 119.

ral Indian languages, besides the Canadian French, and appeared cheerful and intelligent. An excursion into the adjacent country, to view some caves, and a reported mineral locality made by Mr. Trowbridge, during my descent to the mines of Dubuque, brought me some concretions of carbonate of lime, but the Indian guides either faltered to make the promised discoveries, through their superstitions, or really failed in the effort to find the object. By tracing the shores of the Mississippi, I found the rolled and hard agates and other quartz species, which characterize the pebble-drift of its sources, still present in the down-flowing shore-drift.

The aboriginal name of this place is Kipesági, an Algonquin word, which is applied to the mouth or outflow of the Wisconsin River. It appears to be based on the verb *kipa*, to be thick or turbid, and *sauge*, outflow—the river at its floods, being but little else than a moving mass of sand and water.

It was the 9th (Aug.) at half-past ten in the morning before the expedition left the Prairie to ascend the Wisconsin, the mouth of which we reached after descending the Mississippi three miles. This is an impressive scene—the bold cliffs of the west bank of the Mississippi, with Pike's-hill rising in front on the west, while those of the Wisconsin Valley stand at but little less elevation on the north and south. At this season of the year the water is clear and placid, and mingles itself in its mighty recipient without disturbance. But it is easy to conceive, what the Indians affirm, that in its floods it is a strong and turbid mass of moving waters, against which nothing can stand. This character of the stream is believed, indeed, to be the origin of the Indian name of Wisconsin. Miskawägumi, means a strong or mixed water, or liquid. By adding to this word *totoshabo* (milk), the meaning is coagulated or turning milk; it is often used to mean brandy, which is then called strong water; by adding *iscodawabo*, the meaning is fire-water. Marquette, in 1673, spells the name of the river indifferently Meshkousing, and Mishkousing. Of this term, the inflection *ing*, is simply a local form, the letter *s* being thrown in for euphony. This word appears to be a derivation from the term *mushkowa*, strong water. By admitting the transmutation of *m* to *w*, the initial syllable *mis* is changed to *wis*, and the interpretation is then river (or place) of strong waters. The term of *kipesagi*, applied to its mouth, is but another characteristic feature of it—the one laying stress on its *turbidity in flood*, and the other on its *strength of current*. These are certainly the two leading traits of the Wisconsin, which rushes with a great average velocity over an inclined plane, without falls, for a great distance. It originates in a remarkable summit of sandy plains, which send out to the west the Chippewa River of Lake Pepin, to the north the Montreal and Ontonagon of Lake Superior, and to the east the Menomonee of Green Bay, while the Wisconsin becomes its southern off-drain, till it finally turns west at the Portage, and flows into the Mississippi.

We ascended, the first day, eighteen miles; the next, thirty-six; the third day, thirty-four miles; the fourth, forty; the fifth, thirty-eight, and the sixth, sixteen, which brought us to the Fox and Wisconsin Portage, a spot renowned from the earliest French days of western discovery. For here, on the waters separating the Mississippi from the great lakes, there had, at successive intervals, been pitched the tents of Marquette, La Hontan, Carver, and other explorers, who have, in their published journals, left traces of their footsteps. La Salle, who excelled them all in energy of character, proceeded to the Mississippi from Lake Michigan, down the Illinois.

Our estimates made the distance from the Mississippi to this point one hundred and eighty-two miles. It is a wide, and (at this season) shallow stream, with transparent waters, running over a bed of yellow sand, checkered with numerous small islands, and long spits of sandbars. There is not a fall in this distance, and it must be navigable with large craft during the periodical freshets. It receives the Blue, Pine, and other tributaries in this distance. Its valley presents a geological section, on a large scale, of the series of lead-bearing rocks extending in regular succession from the fundamental sandstone to the topmost lime-stones. The water being shallow and warm, we often waded from bar to bar, and found the scene a fruitful one for its fresh-water conchology. The Indians frequently amused me by accounts of the lead mines and mineral productions of its borders; but I followed them in this search only to be convinced that they were without sincerity in these representations, and had no higher objects on this head, than, by assuming a conciliatory manner, to secure temporary advantages while the expedition was passing through their country. The valley belongs to the Winnebagoes, whom we frequently met, and received a friendly reception from. We also encountered Menomonies, who occupy the lower part of the adjacent Fox River Valley, but rove widely west and north over the countries of the tribes they are at peace with.

The Wisconsin Valley was formerly inhabited by the Sacs and Foxes, who raised large quantities of corn and beans on its fertile shores. They were driven by the French, in alliance with the Chippewas and Menomonies. It is now possessed exclusively by the Winnebagoes, a savage and bloodthirsty tribe, who came, according to tradition, many years ago from the south, and are thought to be related to some of the Mexican tribes. Their language is cognate with the great Sioux or Dakota stock west of the Mississippi, who likewise date their origin south. To those accustomed to hear the softer tones of the Chippewa and Algonquin, it sounds harsh and guttural. Their name for themselves is Hochungara; the French call them *Puants.*

In passing up this valley, an almost never-failing object of interest was furnished by the univalve shells found along its banks, and by the variety in size, shape, and color which they exhibited. Of these, the late Mr. Barnes has described, from my duplicates, the U. plicatus, U. verrucosus, U. ventricosus, U. planus, U. obliqua, and U. gracilis.* We frequently observed the scolipax minor, the plover, the A. alcyon, a small yellow bird, and C. vociferus, along its sandy shores; and, in other positions, the brant, the grouse, the A. sponsa, and the summer duck, and F. melodia. A range of hills extends from the Mississippi, on each shore, to within twenty miles of the Portage, where it ceases, on the south side, but continues on the north—receding, however, a considerable distance. This section is called the Highlands of the Wisconsin. The stratification is exclusively sandstone and limestone, in the usual order of the metalliferous series of the West, and lying in horizontal positions.

There are two kinds of rattlesnake in the Valley of the Wisconsin. The larger, or barred crotalis, is confined to the hills, and attains a large size. I killed one of this species at the mouth

* American Journal of Science, vol. vi. p. 120, &c.

of a small cave on the summit of a cliff to which I ascended, which measured four feet in length, and had nine rattles. Its great thickness attracted notice. Attaching a twig to its neck, I drew it down into the valley as a present to our Indians, knowing that they regard the reptile in a peculiar manner. They found it a female, having eleven young, who had taken shelter in their maternal abdominal covering. The Ottowas carefully took off the skin, and brought it with them. The second kind of this reptile is called prairie rattlesnake, is confined to the plains, and does not exceed fifteen or twenty inches in length.

The Indians had reported localities of lead, copper, and silver at various places, but always failed, as we ascended, to reveal anything of more value than detached pieces of sulphuret of iron, or brown iron-stone. When we reached the portage, a Winnebago, who had been the chief person in making these reports, came with great ceremony to present a specimen of his reported silver. On taking off the envelop it turned out to be a small mass of light-colored glistening folia of mica. We had found the horizontal rocks along the stream thus far, but the primitive shows itself, within a mile north of the portage, in orbicular masses in situ, coming through the prairies.

Having reached the summit, we proceeded across it to the banks of Fox River, where we encamped. It consists of a level plain. The distance is a mile and a half. It required, however, some time to have our baggage and canoes transported, which was done by a Frenchman residing at this summit. Such is the slight difference in the level of the two rivers, that Indian canoes are pushed through the marshy ridges when the rivers are swelled by freshets. It was half-past three o'clock of the 15th, the day following our arrival, before the transportation and loading of our canoes was completed. It was then necessary to push our canoes through fields of rushes and other aquatic plants, through which the river winds. This was a slow mode of progress, and we spent the remainder of the day in passing fifteen miles, which brought us to the FORKS, so called, where the northern unites with the southern branch of the river. At this spot we encamped. Next day we estimated our descent at sixty-three miles, having found the navigation less intricate and obstructed from the aquatic growth. In this distance we passed, at thirty miles below the fork, a piece of clear water of nine miles extent, called Buffalo Lake; and at the distance of twelve miles lower, another lake of some twelve miles in extent, called Puckaway Lake. Down to this point, the Fox River has scarcely a perceptible current. We found we had not only, in parting from the Wisconsin to the Fox, exchanged an open, swift, and strong flowing current, for a very quiet and still one, winding through areas of wild rice and the whole family of water plants; but had intruded into a region of water-fowl and birds of every plumage, who, as they rioted upon their cherished zizania aquatica, made the air resound with their screams. The blackbird appeared to be lord of these fields. We had also intruded upon a favorite region of the water-snake, who, coiled up on his bed of plants at every bend of the stream, slid off with spiteful glance into the stream. In passing these places of habitation, which the Chippewas call *wauzh*, we perceptibly smelt an unpleasant odor arising from it.

The next day we descended the river seventy miles. There is a perceptible current below Puckaway Lake. The river increases in width and depth, and offers no impediment whatever to its navigation. Fox River runs, indeed, from the portage to Winnebago Lake on a summit, over which it winds among sylvan hills, covered with grass and prairie-flowers, interspersed with groves of oak, elm, ash, and hickory, and dotted at intervals with lakes of refreshing transparent water. The height of this summit, above the Mississippi and the lakes, must be several hundred feet (stated at 284), which permits the stream to flow with liveliness, insuring, when it comes to be settled,* the erection of hydraulic works; and it would be difficult to point to a region possessing in its soil, climate, and natural resources, a more favorable character for an agricultural population. It has a diversified surface, without mountains; a fine dry atmosphere; an admirable

---

* WISCONSIN. This region was separated from Michigan, and formed into a separate territory in 1836; and admitted as a State in 1848. By the census of 1850, it has a population of 305,891, divided into 88,517 families, occupying 82,962 dwellings, and cultivating 1,045,499 acres of land. There are 48 organised counties, and 834 churches of all denominations, giving one church to every 1,250 inhabitants. It has three representatives in the popular branch of Congress. It was 16 years after my visit, before it had a distinct legal existence—it increased to become a State in twelve years; and, according to our ordinary rate of increase, will contain one million of inhabitants in 1890.

drainage east, west, north, and south, and a ready access to the great oceanic marts through the Great Lake and the Mississippi.

We passed, this day, several encampments and villages of Winnebagoes and Menomonies—tribes, who, with the erratic habits of the Tartars, or Bedouins, once spread their tents in the Fox and Wisconsin valleys, but have now (1853) relinquished them to the European race; and it does not, at this distance of time, seem important to denote the particular spots where they once boiled their kettles of corn, or thumped their magic drums. God have mercy on them in their wild wanderings! We also passed the entrance of Wolf River, a fine bold stream on the left; and soon below it the handsome elevation of La Butte de Morts, or the Hillock of the Dead. This eminence was covered by the frail lodges of the Winnebagoes. The spot is memorable in Indian history, for a signal defeat of the Foxes, by the French and their Indian allies in the seventeenth century, after which, this tribe was finally expelled from the Fox valley. Our night's encampment (17th) was below this spot. The night air was remarkably cold, and put an end to our further annoyance from mosquitos. We embarked at five o'clock the next morning during a dense fog, which was in due time dissipated by the rising sun. We had been five hours in our canoes, under the full force of paddles, when we entered Winnebago Lake. This is a most beautiful and sylvan expanse of water some twenty-four miles long by ten in width, surrounded by picturesque prairie and sloping plains. It has a stream at Fond du Lac, its southern extremity,* which is connected by a short portage with the principal source of Rock River of the Mississippi.

The Fox River, after having displayed itself in the lake, leaves it, at its northern extremity, flowing by a succession of rapids and falls over horizontal limestones to the head of Green Bay.

---

* This spot is now the site of the flourishing town of Fond du Lac, which was laid out in 1845. It had a population of 2,014 in 1850, including two newspaper offices, two banking houses, one iron foundry, a car factory, twelve drygoods stores, and sixty other stores. It is situated 72 miles N. N. W. from Milwaukie, and 90 N. E. from Madison, the capital of the State of Wisconsin. It is the shire town of a county containing a population of 14,510, with 17 churches, and 2,844 pupils attending public schools, and 85 attending academies. It has a plank road to Lake Michigan, and will soon be connected by a railroad with Chicago. It is by such means that the American wilderness is conquered.

There is a Winnebago village, under Hoo Tshoop, or Four Legs, at the point of outlet, where we landed, and as the first rapid begins at that point, creating a delay, I took the occasion to examine its geology more closely, by procuring fresh fractures of the masses of rock in the vicinity. This process, it appeared, was narrowly watched by the Indians, who wondered what such a scrutiny should mean. The French, said the chief to one of our interpreters, formerly held possession of this country; and, afterwards, came the British. They contented themselves with common things, and never disturbed these rocks, which have been laying here forever. But the moment the Americans get possession of the country, they must come and knock off pieces of the rock, and look at them. It is marvellous!

A brilliant mass of native copper, weighing ten or twelve pounds, was found by an Indian, some years ago, on the shores of this lake. The moment he espied it, his imagination was fired, and he fancied he beheld the form of a beautiful female, standing in the water. Glittering in radiancy, she held out in her hand a lump of gold. He paddled his canoe towards her, furtively and slow, but, as he advanced, a transformation gradually ensued. Her eyes lost their brilliancy, her face the glow of life and health, her arms disappeared; and when he reached the spot, the object had changed into a stone monument of the human form, with the tail of a fish. Amazed, he sat awhile in silence; then, lighting his pipe, he offered it the incense of tobacco, and addressed it, as the guardian angel of his country. Lifting the miraculous image gently into his canoe, he took his seat, with his face in an opposite direction, and paddled towards shore, on reaching which, and turning round to the object of his regard, he discovered, in its place, nothing but a lump of shining virgin copper.

Such are the imaginative efforts of this race, who look to the eyes of civilization as if they had themselves faces of stone, and hearts of adamant.

CHAPTER XVII.

Descent of the Fox River from Winnebago Lake to Green Bay—Incidents—
Etymology, conchology, mineralogy—Falls of the Konomic and Kakala—Population and antiquity of the settlement of Green Bay—Appearances of a tide, not
sustained.

A RAPID commences at the precise point where Fox River
issues from Winnebago Lake. This rapid, down which canoes
descend with half loads, extends a mile and a half, when the river
assumes its usual navigable form, presenting a noble volume.
Nine miles below this, a ledge of the semi-crystalline limestone
rock crosses the entire channel, lifting itself five feet above the bed
of the stream. Over this the Fox River throws itself by an abrupt
cascade. Down this shelf of rock, the canoes, previously lightened
of their burden, are lifted by the men. It was sometime after
dark when we reached and encamped on the north shore, at the
foot of this cascade, which bears the name of Konamik. The
syllable *kon*, in this word, appears to me to be the same as *con*
in Wisconsin, and is, apparently, a derivative from a term for
strong water, which has, in this case, the meaning of cascade or
fall. The word *amik*, its terminal, means a beaver. We thus
have the probable original meaning in beaver-water, or, by im-
plication, beaver cascade. There is a rapid below this fall. I
judged the water must sink its level, in this vicinity, about fifteen
feet. On examining the character of the limestone, I discovered
crystals of calcareous spar occupying small cavities. At other
localities, at lower points, there were found crystals of black sulphuret of zinc, and yellow sulphuret of iron. The rock appears
to be of the same age as the lead-bearing limestone of the West;
it is also overlaid by the red marly clay, and I should judge it to
contain deposits of sulphuret of lead.

The next morning, we resumed our descent of the Fox River

NARRATIVE OF THE EXPEDITION.

with difficulty. It was now the 19th of August, and the waters had reached their lowest summer stage. The entire distance of twelve miles from the Konamik to the Kakala fall may be deemed to be, at this season, a continuous rapid. Our barge was abandoned on the rapids. While the men toiled in these rapids to get down their canoes, it was found rather a privilege to walk, for it gave a more ample opportunity to examine the mineral structure and productions of the country.

It was high noon when we reached the rapids of the Kákala. This is a formidable rapid, at which the river rushes with furious velocity down a rocky bed, which it seems impossible boats or canoes should ever safely descend. It demands a portage to be made, under all circumstances, the water sweeping round a curve or bow, of which the portage path is the string. This is the apparent meaning of the term, in the Indian tongue; but it is disguised by early orthography, in which the letter *l* has taken the place of *n*, and the syllable *in* of *au*. The term *kakina* is the ancient French form of the Indian transitive-adjective *all*, inclusive, entirely. There is another root for the term in *kakiwa*, which is the ordinary term for a portage, or walk across a point of land, which is rendered local by the usual inflection, *o-nong*.

We found the portage path to be a well-beaten wagon road across a level fertile plain, which appeared to have been in cultivation from the earliest Indian period. Probably it had been a locality for the tribes, where they raised their favorite maize, long before the French first reached the waters of Green Bay. Evidence of such antiquity in the plain of Kákala appeared in an ancient cemetery of a circular shape, situated on one side of the road, on a comparatively large surface, which had reached the height of some eight or ten feet, by the mere accumulation of graves. This has all the appearance of a sepulchral mound, in the slow process of construction; for, on viewing it, I found a recent grave. We passed, on this plain, a Winnebago village of ten or twelve lodges, embracing two hundred souls. The portage is continued just one mile. Embarking again, at this point, we proceeded down the river, and encamped eight miles below this point, having, with every exertion, made but twenty miles this day.

The interest which had been excited by the conchology of the

Mississippi and Wisconsin valleys, was renewed in the descent of the Fox River, particularly in the section of it below Winnebago Lake. Shrunk to its lowest summer level, its shores disclosed almost innumerable species of unios, many of which had been manifestly dragged to the shores and opened by the musk-rat, thus serving to give hints for finding the living species. Among these, the U. obliqua, U. cornutus, U. ellipticus, U. carinatus, U. Alatus, U. prælongus, and U. parvus, were conspicuous; the latter of which, it is remarked by Mr. Barnes, is the smallest and most beautiful of all the genus yet discovered in America.* . In the duplicates, from this part of the Fox River, transmitted to Mr. Isaac Lea, of Philadelphia, he found a species with green-rayed beaks, on a yellow surface and iridescent nacre, having a peculiar structure, which he did me the honor to name after me.† The description of Mr. Lea is as follows: " Unio Schoolcraftensis. Shell subrotund, somewhat angular at posterior dorsal margin, nearly equilateral, compressed, slightly tuberculate posteriorly to umbonical slope. Substance of the shell rather thick ; beaks elevated; ligament short; epidermis smooth yellow, with several broad green rays ; teeth elevated, and cleft in the left valve, single, and rising from a pit in the right; lateral teeth elevated, straight, and lamellar; anterior cicatrices distinct, posterior cicatrices confluent; dorsal cicatrices within the cavity of the shell on the base of the cardinal tooth ; cavity of the beaks angular and deep; nacre pearly white and iridescent. Diameter ·7, length 1·1, breadth 1·3 inches."

The next morning (20th), a heavy fog in the Fox Valley detained us in our encampment till 7 o'clock. Six miles brought us to another rapid, called the Little Kakala, which, however, opposes no obstacle to the descent of canoes. At this spot, which is the apparent western terminus of the Bay settlement, we found a party of U. S. soldiers, from Fort Howard, engaged in digging the foundations for a saw-mill. Our appearance must have been somewhat rusty at this time, from our deficiences in the tonsorial and sempstrescal way, for these sons of Mars did not recognize their superior officers in Capt. Douglass and Lt. Mackay; glibly

* Amer. Journ. Science, vol. vi. pp. 120, 259, &c.
† Transactions of the American Philosophical Society, vol. v. p. 37 ; plate 8, fig. 9.

saying, in a jolly way, as they handed them a drink of water: "After me, sir, is manners;" and drinking off the first cup. At this rapid I got out of my canoe, wishing to see the geological formation more fully, and walked quite to the Rapide du Pere, where Fox River finds its level in the broad, elongated, and lake-like tongue of water, extending up from the head of Green Bay. On reaching this point, the scene of the settlement first burst on our view, with its farm-houses and cultivated fields stretching, for five miles, along both banks of the river; disclosing the flagstaff of the distant fort, and the bannered masts of vessels, all of which brought vividly to mind our approach to the civilized world. If the Canadian boat-song was ever exhilarating and appropriate, it was peculiarly so on the present occasion; and when our *voyageurs* burst out, in full chorus, with the ancient ditty, beginning,

> "*La fille du Roi son vout chasseau,*
> *Avec son grande fusee d'largent,*"

they waked up a responsive feeling, not alone in the breasts of the French *habitans*, lining the shores of the river, but in our own breasts. On reaching the fort, the salute due to the governor of a territory was paid, in honor of our leader, Governor Cass; and in exchanging congratulations with the officers and citizens, we began first to feel, in reality, that, after passing among many savage tribes, our scalps were still safely on our heads. I found, at the fort, letters from my friends, and was thus reminded that warm sympathies had been alive for our fate. Weary regions had now been past, and privations endured, of which we thought little, at the time; the flag of the Union had been carried among barbarous tribes, who hardly knew there was such a power as the United States, or, if they knew, despised it; and some information had been gathered, which it was hoped would enlarge the boundaries of science, and would at the same time send a thrill of satisfaction, and impart a feeling of security, along the whole line of the advanced and extended western settlements. If Berkeley, in the dark days of the Commonwealth of England, could turn to the West, with exultation, as the hope of the nation, it must be admitted that it is by some out-door means, like this, that the way for the car of "empire" must be prepared.

We found the fort, which bears the name of Howard, in charge of Capt. W. Wistler, during the absence of Col. Joseph L. Smith. Its strength consists of three hundred men, together with about the same number of infantry at Camp Smith, at Rock or Dupere Rapid, a few miles above, who are engaged in quarrying stone for a permanent fortification at that point. On visiting this quarry, I found it to consist of a bluish-gray limestone, semi-crystalline in its structure, containing small disseminated masses' of sulphuret of zinc, calc-spar, and iron pyrites, and corresponding, in every respect, with the beds of this rock observed along the upper parts of the Fox and Wisconsin valleys.

Fort Howard is seated on a handsome fertile plain, on the north banks of the Fox, near its mouth. It consists of a stockade of timber, thirty feet high, inclosing barracks, which face three sides of a quadrangle. This forms a fine parade. There are block-houses, mounting guns, at the angles, and quarters for the sur-geon and quartermaster, separately constructed. The whole is whitewashed, and presents a neat military appearance. The gardens of the military denote the most fruitful soil and genial climate. Data observed by the surgeon, indicate the site to be unexcelled for its salubrity, such a disease as fever, of any kind, never having visited it, in either an endemic or epidemic form.

The name of Green Bay is associated with our earliest ideas of French history in America. When La Salle visited the country in the 17th century, it had been many years known to the French, and was esteemed one of the prime posts for trading with the Indians. The chief tribes who were located here, and in the vi-cinity, making this their central point of trade, were the *Puants*, i. e. Winnebagoes, Malomonies, or Folle Avoins, known to us as Menomonies, Sacs, and Foxes, called also Sakis, Outagami, and Renouards, and it was also the seat of trade for the equivo-cal tribe of the Mascoutins. The present inhabitants are, with few exceptions, descendants of the original French, who inter-married with Indian women), and who still speak the French and Indian languages. They are indolent, gay, and illiterate. I was told there were five hundred inhabitants, and about sixty princi-pal dwellings, beside temporary structures. There are seventy inhabitants enrolled as militia-men, and the settlement has civil courts, being the seat of justice from Brown County, Michigan, so called in honor of Major-General Jacob Brown, U. S. A.   The place is surrounded by the woodlands and forests, and seems destined to be an important lake-port.* The Algonquin name for this place is Boatchweekwaid, a term which describes an eccentric or abrupt bay, or inlet. Nothing could more truly depict its singular position; it is, in fact, a kind of cul-de-sac—a duplicature of Lake Michigan, with the coast-shore of which it lies parallel for about ninety miles.

The singular configuration of this bay appears to be the chief cause of the appearances of a tide at the point where it is entered by Fox River. This phenomenon was early noticed by the French. La Hontan mentions it in 1689. Charlevoix remarks on it in 1721, and suggests its probable cause, which is, in his opinion, explained by the fact that Lakes Michigan and Huron, alternately empty themselves into each other through the Straits of Michilimackinac. The effects of such a flux and reflux, under the power of the winds, would appear to place Green Bay in the position of a siphon, on the west of Lake Michigan, and go far to account for the singular fluctuations of the current at the mouth of the Fox River. On reaching this spot of the rising and falling of the lake waters, Governor Cass caused observations to be made, which he greatly extended at a subsequent period.† These give no countenance to the theory of regular tides, but denote the changes in the level of the waters to be eccentrically irregular, and dependent, so far as the observations extend, altogether on the condition of the winds and currents of the lakes.

Something analogous to this is perceived in the Baltic, which has no regular tides, and therefore experiences no difference of height, except when the wind blows violently. "At such times," says Pennant,‡ "there is a current in and out of the Baltic, according to the points they blow from, which forces the water

* GREEN BAY. This town has just (1854) been incorporated as a city, the anticipations respecting it having been slow in being realised. It has now an estimated population of 8,000, with several churches in a healthy and flourishing state, two printing presses, a post-office, collectorship, and thriving agricultural and commercial advantages, which will be fully realised when the internal improvements in process of construction through the Fox and Wisconsin valleys are finished. Its extreme salubrity has, it seems, been disregarded by emigrants.

† American Journal of Science, vol. xvii.

‡ Arctic Geology.

through the sound, with the velocity of two or three Danish miles in the hour. When the wind blows violently from the German Sea, the water rises in several Baltic harbors, and gives those in the western tract a temporary saltness; otherwise, the Baltic loses that other property of a sea, by reason of the want of tide, and the quantity of vast rivers it receives, which sweeten it so much as to render it, in many places, fit for domestic use."

CHAPTER XVIII.

The expedition traces the west shores of Lake Michigan southerly to Chicago—
Outline of the journey along this coast—Sites of Manitoowoc, Sheboigan, Milwaukie, Racine, and Chicago, being the present chief towns and cities of Wisconsin and Illinois on the west shores of that Lake—Final reorganization of
the party and departure from Chicago.

Two days spent in preparations to reorganize the expedition,
enabled it to continue its explorations. For the purpose of
tracing the western and northern shores of Green Bay, and the
northern shores of Lake Michigan, a sub-expedition was fitted
out, under Mr. Trowbridge, our sub-topographer, who was accompanied by Mr. J. D. Doty, Mr. Alex. R. Chase, and James
Riley, the Chippewa interpreter. The auxiliary Indians, who
had, thus far, attended us in a separate canoe, were rewarded for
their services, furnished with provisions to reach their homes,
and dismissed. The escort of soldiers under Lieut. Mackay,
U. S. A., were returned to their respective companies at Fort
Howard and Camp Smith. The Chippewa chief, *Iaba Wawash-
kash*, or the Buck, who belonged to Michilimackinac, went with
Mr. Trowbridge, together with Jo Parks, the intelligent Shawnee
captive, and assimilated Shawnee of Waughpekennota,* Ohio.
The Ottowa chief, Kewaygooshkum, of Grand River, took the
rest of the party in a separate canoe to their destination. Our
collections in natural history were shipped in the schooner De-

---

* WAUGHPEKENNOTA. This place was then the residence of the Shawnee tribe,
under the Prophet Elksattawa, of war memory, the celebrated brother of Tecumseh,
who, seeing the intrusive tread of the Americans, headed, in 1827, the first exploring party of the tribe to the west of the Mississippi, where they finally settled. After
living twenty-seven years at this spot, they found themselves within the newly-
erected territory of Kansas, and sold their surplus lands to the U. States by a treaty
concluded at Washington in May, 1854, the said Parks being at this time first chief
of the Shawnee tribe.

catur, Capt. Burnham (Perry's boatswain in the memorable naval battle of Lake Erie, Sept. 11, 1813), to Michilimackinac, together with the extra baggage.

Thus relieved in numbers and canoe-hamper, we were reduced to two canoes; the travelling family of Gov. Cass now consisted of Capt. Douglass, Dr. Wolcott, Maj. Forsyth, Lieut. Mackay, and myself. Leaving Fort Howard at two o'clock P. M., we parted with Mr. Trowbridge and his party at the mouth of Fox River, at half-past two, and taking the other, or east side of the bay, proceeded along its shores about twenty-five miles, and encamped on the coast called Red Banks. This is a term translated from the Winnebago name, which is renowned in their traditions as the earliest spot which they can recollect. They dwelt here when the French first reached Green Bay in their discoveries in the seventeenth century. Here, then, is a test of the value and continuity of Indian tradition, so far as this tribe is concerned, for admitting, what is doubtful, that the French reached this point so early as 1650, the period of recognized Winnebago history, as proved by geography, reaches but 170 years prior to the above date.

In a short time after entering the bay, we were overtaken by Kewaygooshkum and his party, who travelled and encamped with us. In the course of the evening he pointed out a rocky island, at three or four miles distance, containing a large cavern, which has been used by the Indians from early times as a repository for the dead. The chief, as he pointed to it, as if absorbed in a spirit of ancestral reverence, seemed to say :—

> "It hath a charm the stranger knoweth not,
> It is the [sepulchre] of mine ancestry ;
> There is an inspiration in its shade,
> The echoes of its walls are eloquent,
> The words they speak are of the glorious dead ;
> Its tenants are not human—they are more !
> The stones have voices, and the walls do live ;
> It is the home of memories dearly honored
> By many a trace of long departed glory."

The appearance of ancient cultivation of this coast is such as to give semblance to the Winnebago tradition of its having been their former residence. The lands are fertile, alluvion, bearing a secondary growth of trees, mingled with older species of the acer saccharinum, elm, and oak.

The next day, after traversing this coast twenty miles further, we reached and passed up Sturgeon Bay, to a portage path leading to Lake Michigan. This path begins in low grounds, where several of the swamp species of plants occur. On reaching the open shores of Lake Michigan, the wind was found strongly ahead, and we were compelled to encamp. At this spot we found several species of madreperes, and some other organic forms, among the shore debris. The next day the wind abated, and, agreeably to the estimate of Capt. Douglass, we advanced along the shore, southwardly, forty-six miles. The day following, we made forty miles, and reached the River Manitowakie,* and encamped on the lake shore, five miles south of it.

In passing along the lake shore this day (25th), we observed it to be strewed abundantly with the carcasses of dead pigeons. This bird, we were told, is often overcome by the fatigue of long flights, or storms, in crossing the lake, and entire flocks drowned. This causes the shores to be visited by great numbers of hawks, eagles, and other birds of prey. The Indians only make use of those carcasses of pigeons, as food, when they are first cast on shore.

The next day the expedition passed the mouth of the Sheboigan River, a stream originating not remotely from the banks of Winnebago Lake, with which, as the name indicates, there is a portage or passage through.† Pushing forward with every force

* From *Manito*, a spirit, *suk*, a standing or hollow tree that is under a mysterious influence, and the generic inflection *is*, which is applied to vital or animate nouns. A town, at present, exists at the spot called Manitoowoo. It is the shire town of a county of the same name in Wisconsin; it has a good harbor, and by the census of 1850 contains four churches, twelve stores, two steam mills, two shipyards, a newspaper, post-office, and 2,500 inhabitants. We found the site inhabited by a village Monomonees of six lodges.

† *Shebiau*, is to look critically; *shebiabunjegun*, a spy-glass or instrument to look through. Sheboigan appears to have its termination from the word *gan*, a lake, and the combination denotes a river, or water pass from lake to lake. This place is now (1854) a town and county site of Wisconsin. The county was organized in 1839, and by the last census has seven churches, two newspapers, 624 pupils at schools, and a population of 8,379. The town of this name contains 2,000 inhabitants. It is 62 miles N. from Milwaukie, and 110 N. E. from Madison, the State capital. It has a plank road of 40 miles to Fond du Lac, and is noted for its lumber trade.

during the day, we reached the mouth of the Milwaukie River, and encamped on the beach some time after dark. This is a large and important river, and is connected by an Indian portage with the Rock River of the Mississippi. The next morning adverse winds confined us to this spot, where we remained a considerable part of the day, which enabled us to explore the locality. We found it to be the site of a Pottawattomie village. There were two American families located at that place, engaged in the Indian trade.

The name of Milwaukie,* exhibits an instance of which there are many others, in which the French have substituted the sound of the letter *l* in place of *n*, in Indian words. *Min*, in the Algonquin languages signifies *good*. *Waukie*, is a derivative from *auki*, earth or land, the fertility of the soil, along the banks of that stream, being the characteristic trait which is described in the Indian compound.

When the wind lulled so as to permit embarkation, we proceeded on our course. At the computed distance of five miles, we observed a bed of light-colored tertiary clay, possessing a compactness, tenacity, and feel, which denote its utility in the arts. This bed, after a break of many miles in the shores, reappears in thicker and more massive layers, at eight or ten miles distance. The waves dashing against this elevated bank of clay,† have liberated balls and crystallized masses of sulphuret of iron.

Some of the more recently exposed masses of this mineral are of a bright brass color. The tendency of their crystallization is to restore octahedral and cubical forms. We advanced along this shore about thirty-five miles, encamping on an eligible part of the beach before dark. I found, in examining the mineralogy of the coast, masses of detached limestone, containing fissures filled with asphaltum. On breaking these masses, and laying open the fissures, the substance assumed the form of naphtha. We observed among the plants along this portion of coast, the tradescantia vir-

---

* Milwaukie is the principal city of the State of Wisconsin. It lies in latitude 43° 3′ 45″ North. It is ninety miles north of Chicago and seventy-five east from Madison. It contains thirty churches, five public high schools, two academies, five orphan asylums, and other benevolent institutions, seven daily and seven weekly newspapers, four banks, and, by the census of 1850, 20,161 inhabitants.

† An admired kind of cream-colored bricks are manufactured from portions of the clay found near Milwaukie.

ginica, and T. liatris, and squarrosa scariosa.* By scrutinizing the
wave-moved pebble-drift along shore, it is evident that inferior
positions, in the geological basin of Lake Michigan, contain slaty,
or bituminous coal, masses of which were developed.

The next day's journey, 28th, carried us forty miles, in which
distance, the most noticeable fact in the topography of the coast,
was the entrance of the Racine, or Root River;† its eligible shores
being occupied by some Pottawattomie lodges. Having reached
within ten or twelve miles of Chicago, and being anxious to make
that point, we were in motion at a very early hour on the morning of the 29th, and reached the village at five o'clock A. M.
We found four or five families living here, the principal of which
were those of Mr. John Kinzie, Dr. A. Wolcott, J. B. Bobian, and
Mr. J. Crafts, the latter living a short distance up the river. The
Pottawattomies, to whom this site is the capital of their trade, appeared to be lords of the soil, and truly are entitled to the epithet,
if laziness, and an utter inappreciation of the value of time, be a
test of lordliness. Dr. Wolcott, being the U. S. Agent for this
tribe, found himself at home here, and constitutes no further, a
member of the expedition. Gov. Cass determined to return to
Detroit from this point, on horseback, across the peninsula of
Michigan, accompanied by Lt. Mackay, U. S. A., Maj. Forsyth,
his private secretary, and the necessary number of men and pack
horses to prepare their night encampments. This left Capt.
Douglass and myself to continue the survey of the Lakes, and
after reaching Michilimackinac and rejoining the party of Mr.
Trowbridge, to return to Detroit from that point.

The preparation for these ends occupied a couple of days, which
gave us an opportunity to scan the vicinity. We found the post
(Fort Dearborn) under the command of Capt. Bradley, with a
force of one hundred and sixty men. The river is ample and
deep for a few miles, but is utterly choked up by the lake sands,
through which, behind a masked margin, it oozes its way for a

---

* Dr. J. Torrey, *Am. Journ. Science*, vol. 4, p. 56.

† RACINE.—This is now the second city in size in the State of Wisconsin. By
the census of 1850, its population is 5,110. It has a harbor which admits vessels
drawing twelve feet water; it has fourteen churches, a high school, college, bank,
several newspapers, three ship-yards, and exhibits more than two millions of imports and exports. The settlement was commenced in 1835.

mile or two, till it percolates through the sands into the lake. Its banks consist of a black arenaceous fertile soil, which is stated to produce abundantly, in its season, the wild species of cepa, or leek. This circumstance has led the natives to name it the place of the wild leek. Such is the origin of the term Chicago,* which is a derivative, by elision and French annotation, from the word *Chikaug-ong*. *Kaug*, is the Algonquin name for the hystrix, or porcupine. It takes the prefix *Chi*, when applied to the mustela putorius. The particle *Chi*, is the common prefix of nouns to denote greatness in any natural object, but it is also employed, as here, to mean increase, or excess, as acridness, or pungency, in quality. The penultimate *ong*, denotes locality. The putorius is so named from this plant, and not, as has been thought, the plant from it. I took the sketch, which is reproduced in the fourth vol. of my *Ethnological Researches*, Plate xxvii., from a standpoint on the flat of sand which stretched in front of the place. This view embraces every house in the village, with the fort; and if the reproduction of the artist in vol. iv. may be subjected to any criticism, it is, perhaps, that the stockade bears too great a proportion to the scene, while the precipice observed in the shore line of sand, is wholly wanting in the original.

The country around Chicago is the most fertile and beautiful that can be imagined. It consists of an intermixture of woods and prairies, diversified with gentle slopes, sometimes attaining the elevation of hills, and it is irrigated with a number of clear streams and rivers, which throw their waters partly into Lake Michigan, and partly into the Mississippi River. As a farming country, it presents the greatest facilities for raising stock and

* CHICAGO is the largest city of the State of Illinois, excelling all others in its commercial and business capacities, and public and moral influences. Standing on the borders of the great western prairies, it is the great city of the plains, and its growth cannot be limited, or can scarcely be estimated. It began to be built about 1831, eleven years after this visit. It was incorporated as a city in 1836, with 4,853 inhabitants. In 1850, it had 29,963, and it is now estimated to exceed 60,000. This city lies in lat. 41° 52′ 20″. It is connected by lakes, canals, and railroads, with the most distant regions. Its imports and exports the last year, were twenty millions. Like all the cities and towns of America, its political and moral influence, are seen to keep an exact pace with its sound religious influences; the number of churches and newspapers, having a certain fixed relation. More than any other city of the West, its position destines it to be another Nineveh.

grains, and it is one of the most favored parts of the Mississippi Valley; the climate has a delightful serenity, and it must, as soon as the Indian title is extinguished,* become one of the most attractive fields for the emigrant. To the ordinary advantages of an agricultural market town, it must add that of being a depot for the commerce between the northern and southern sections of the' Union, and a great thoroughfare for strangers, merchants, and travellers.

The Milwaukie clays to which I have adverted, do not extend thus far, although the argillaceous deposits found, appear to be destitute of the oxide of iron, for the bricks produced from them burn white. There is a locality of bituminous coal on Fox River, about forty miles south. Near the junction of the Desplaines River with the Kankakee, there exists in the semi-crystalline or sedimentary limestone, a remarkable fossil-tree.†

* This was done in 1821; having been, myself, secretary to the Commissioners, Gov. Cass and Hon. Sol. Sibley, who were appointed to treat with the Indians. Vide *Indian Treaties*, p. 297.

† FOSSIL FLORA OF THE WEST.—Of this gigantic specimen of the geological flora of the newer rocks of the Mississippi Valley, I published a memoir in 1822, founded on a personal examination of the phenomena. Albany, E. and E. Hosford, 24 pp. 8vo. This paper (*Vide* Appendix) was prepared for the American Geological Society, at New Haven. See *American Journ. Science*, vol. 4, p. 285; See also, vol. 5, p. 23, for appreciating testimony of the value of geological science (then coming into notice), from Ex-Presidents John Adams, Thomas Jefferson, and James Madison, to whom copies of it were transmitted.

CHAPTER XIX.

It was now the last day of August. Having partaken of the
hospitalities of Mr. Kinzie, and of Captains Bradley and Green, of
Fort Dearborn, during our stay at Chicago, and completed the
reorganization of our parties, we separated on the last day of the
month, at two o'clock P. M.; Gov. Cass and his party, on horseback, taking the old Indian trail to Detroit, and Capt. Douglass
and myself being left, with two canoes, to complete the circumnavigation of the lakes. We did not delay our departure over
thirty minutes, but bidding adieu to Dr. Wolcott, whose manners,
judgment, and intelligence had commanded our respect during
the journey, embarked with two canoes; our steersmen immediately hoisted their square sails, and, favored by a good breeze,
we proceeded twenty miles along the southern curve, at the head
of Lake Michigan, and encamped.

Within two miles of Chicago, we passed, on the open shores
of the lake, the scene of the massacre of Chicago, of the 15th of
August, 1812, being the day after the surrender of Detroit by
Gen. Hull. Gloom hung, at that eventful period, over every
part of our western borders. Michilimackinac had already been
carried by surprise; and the ill-advised order to evacuate Chicago, was deemed by the Indians an admission that the Americans
were to be driven from the country. The Pottawattomies determined to show the power of their hostility on this occasion.
Capt. Heald, the commanding officer, having received Gen. Hull's
order to abandon the post, and having an escort of thirty friendly
Miamis, from Fort Wayne, under Captain Wells, had quitted the

fort at nine o'clock in the morning, with fifty-four regulars, a subaltern, physician, twelve militia, and the necessary baggage wagons for the provisions and ammunition, which contained eighteen soldiers, women and children. They had not proceeded more than a mile and a half along the shore of the lake, when an ambuscade of Indians was discovered behind the sand-hills which encompass the flat sandy shore. The horrid yell, which rose on the discovery being made, was accompanied by a general and deadly fire from them. Several men fell at the first fire, but Capt. Heald formed his men, and effected a charge up the bank, which dispersed his assailants. It was only, however, to find the enemy return by a flank movement, in which their numbers gave them the victory. In a few moments, out of his effective force of sixty-six men, but sixteen survived. With these, he succeeded in drawing off to a position in the prairie, where he was not followed by the Indians. On a negotiation, opened by a chief called Mukudapenais, he surrendered, under promise of security for their lives. This promise was afterwards violated, with the exception of himself and three or four men. Among the slain was Ensign Ronan; Dr. Voorhis, and Capt. Wells. The latter had his heart cut out, and his body received other shocking indignities. The saddest part of the tragedy was the attack on the women and children who occupied the baggage wagons, and were all slain. Several of the women fought with swords. During the action, a sergeant of infantry ran his bayonet through the heart of an Indian who had lifted his tomahawk to strike him; not being able to withdraw the instrument, it served to hold up the Indian, who actually tomahawked him in this position, and both fell dead together.* The Miamis remained neuter in this massacre. Mr. Kinzie, of Chicago, of whose hospitalities we had partaken, was a witness of this transaction, and furnished the principal facts of this narrative.

The morning (Sept. 1) opened with a perfect gale, and we were *degradé*, to use a Canadian term, all day; the waves dashed against the shore with a violence that made it impossible to take the lake with canoes, and would have rendered it perilous even to a large

* Gouverneur Morris recites a similar incident at the battle of Oriskany, in 1777. —*Coll. New York Hist. Soc.*

vessel. This violence continued, with no perceptible diminution, during the day. As a mode of relief from the tedium of delay, a short excursion was made into the prairie. I found a few species of the unio, in a partially choked up branch of the Konamek. Capt. Douglass improved the time by taking observations for the latitude, and we footed around ten miles of the extreme southern head of the lake. It is edged with sand-hills, bearing pines. A few dead valves of the fresh-water musole were found on the shore.

On the following day the wind lulled, when we proceeded fifty-four miles, passing in the distance the remains of the schooner Hercules, which went ashore in a gale, in November, 1816, and all on board perished; her mast, pump, spars, and the graves of the passengers, among which, was that of Lieut. W. S. Eveleth, U. S. A., were pointed out to us. We landed a few moments at the entrance of the River du Chemin,* where the trail to Detroit leaves the lake shore. The distance to that city is estimated at three hundred miles. Ten miles beyond this spot we passed the little River Galien, where, at this time, the town and harbor of New Buffalo, of Michigan, is situated, and we encamped on the shore twelve miles beyond it.

We had been travelling on a slightly curved line from Chicago to the spot, in the latitude of 41° 52′ 20″, and had now reached a point where the course tends more directly to the northeast and north. By the best accounts, the length of Lake Michigan, lying directly from south to north, is four hundred miles. There is no other lake in America, north or south, which traverses so many degrees of latitude, and we had reason to expect its flora and fauna to denote some striking changes. We had passed down its west, or Wisconsin shore, from Sturgeon Bay, finding it to present a clear margin of forest, with many good harbors, and a fertile, gently undulating surface. But we were now to encounter another cast of scenery. It is manifest, from a survey of the eastern shore of this lake, that the prevalent winds are from the west and northwest, for they have cast up vast sand dunes along the coast, which give it an arid appearance. These dunes are,

* Michigan City, of the State of Indiana, is located near this spot. This city has its harbor communicating with Lake Michigan through this creek. It has a news-paper, branch bank, railroad, and (in 1858) 2,858 inhabitants.

however, but a hem on the fertile prairie lands, not extending more than half a mile or more, and thus masking the fertile lands. Water, in the shape of lagoons, is often accumulated behind these sand-banks, and the force of the winds is such as to choke and sometimes entirely shut up the mouth of its rivers. We had found this hem of sand-hills extending around the southern shore of the lake from the vicinity of Chicago, and soon found that it gave an appearance of sterility to the country that it by no means merited. On reaching the mouth of St. Joseph's River (3d), a full exemplification of this striking effect of the lake action was exhibited. This is one of the largest rivers of the peninsula, running for more than a hundred and twenty miles through a succession of rich plains and prairies; yet its mouth, which carries a large volume of water into the lake, is rendered difficult of entrance to vessels, and its lake-borders are loaded with drifts of shifting sand.

The next day's journey carried us fifty miles; and, on proceeding ten miles further on the 4th, we reached the mouth of the Kalamazoo.* Before reaching this river, I discovered on the beach a body of detached orbicular masses of the calcareous marl called septaria—the ludus' helmontii of the old mineralogists. On breaking some of these masses, they disclosed small crystalline seams of sulphuret of zinc. The Kalamazoo irrigates a fine tract of the most fertile and beautiful prairies of Michigan, which, at the date of the revision of this journal, is studded with flourishing towns and villages.

Fifteen miles further progress towards the north, brought us to the mouth of Grand River—the Washtenong of the Indians— which is, I believe the largest and longest stream of the Michigan peninsula. It is the boundary between the hunting-grounds of the Pottowattomies (who have thus far claimed jurisdiction from Chicago) and the Ottowas. The latter live in large numbers at its rapids and on its various tributaries.† The next stream of

---

* KALAMAZOO. This word is the contraction of an Indian phrase descriptive of the stones seen through the water in its bed, which, from a refractive power in the current, resembles an otter swimming under water. Hence the original term, Negikanamazoo. This term has its root forms in *negik*, an otter, the verb *kaaa*, to hide, and *ozoo*, a quadruped's tail. The letter *l* is the mere transposition of *l* in native words passing from the Indian to the Indo-French language.

† OTTOWAS. So late as 1841, the number of the tribe, reported to the Superintendent of Indian Affairs for Michigan, was 1,391, which was divided into 13 villages,

note we encountered was the Maskigon, twelve miles north of Grand River, where we encamped, having travelled, during the day, fifty-four miles. The view of this scene was impressive from its bleakness, the dunes of sand being more at the mercy of the winds. I found here a large, branching specimen of the club-fungus, attached to a dead specimen of the populus tremuloides, which had been completely penetrated by these drifting sands, so as to present quite the appearance, and no little part of the hardness and consistency, of a fossil. The following figure of this transformation from a fungus to a semi-stony body, presents a perfect outline of it as sketched in its original position.

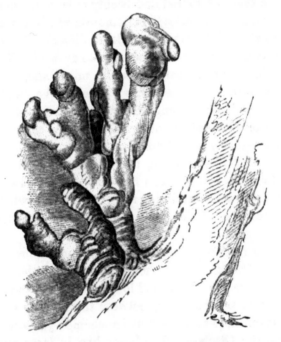

On the day of our departure from the Maskigon, we enjoyed fine weather and favorable winds, and proceeded, from the data of Captain Douglass, seventy miles, and encamped a few miles beyond the Sandy River. In this line of coast, we passed, successively, the White, Pentwater, and Marquette. Of these, the latter, both from its size and its historical associations, is by far the most

scattered over its whole valley.—*Schoolcraft's Report on Indian Affairs*, Detroit, A. S. Bagg, 1840.

important; for it was at this spot, after having spent years of devotion in the cause of missions in New France—in the course of which he discovered the Mississippi River—that this zealous servant of God laid down in his tent, after a hard day's travel, and surrendered up his life. The event occurred on the 8th of May, 1675, but two years after his grand discovery. Marquette was a native of Laon, in Picardy, where his family was of distinguished rank. The precise moment of his death was not witnessed, his men having retired to leave him to his devotions, but returning, in a short time, found him lifeless. They carried his body to the mission of old Michilimackinac, of which he was the founder, where it was interred.*

It rained the next morning (6th), by which we lost two hours, and we had some unfavorable winds, but, by dint of hard pushing, we made forty-five miles, and slept at Gravelly Point. In this line we passed successively, at distances of seventeen and thirty miles, the rivers Manistic and *Becsie*, which is the Canadian phrase for the anas canadensis. Clouds and murky weather still hovered around us on the next morning, but we left our encampment at an early hour. Thirteen miles brought us to the Omicomico, or Plate River, nine miles beyond which found us in front of a remarkable and very elevated sand dune, called the Sleeping Bear— a fanciful term, derived from the Indian, through the French *l'ours qui dormis*. Opposite this feature in the coast geology, lie the two large wooded islands called the Minitos—well-known objects to all mariners who venture into the vast unsheltered basin of the southern body of Lake Michigan. Thirty miles beyond this sandy elevation, brought us to the southern cape of Grand Traverse Bay, where we encamped, having advanced fifty-two miles. This was the first place where we had noticed rocks in situ, since passing the little Konamic River, near Chicago. It proved to be limestone, of the same apparent era of the calcareous rock

* PLACE OF INTERMENT OF MARQUETTE. It is known that the mission of Michilimackinac fell on the downfall of the Jesuits. When the post of Michilimackinac was removed from the peninsula to the island, about 1780, the bones of the missionary were transferred to the old Catholic burial-ground, in the village on the island. There they remained till a land or property question arose to agitate the church, and, when the crisis happened, the whole graveyard was disturbed, and his bones, with others, were transferred to the Indian village of La Crosse, which is in the vicinity of L'Arbre Croche, Michigan.

which we had observed at Sturgeon Bay and the contiguous west shore of Lake Michigan. The line of lake coast included in this remark is three hundred and twenty miles; during all which distance the coast seems, but only seems, to be the sport of the fierce gales and storms, for there is reason to believe that the formations of drift clay, sand, and gravel rest, at various depths, on a stratification of solid, permanent rock. To us, however, it proved a barren field for the collection of both geological and mineralogical specimens. There were gleaned some rolled specimens of organic remains, of no further use than to denote the occurrence of these in some part of a vast basin. There was a specimen of gypsum from Grand River. The few patches of iron sand I had noticed, were hardly worthy of record after the heavy beds of this mineral which we had passed in Lake Superior. The same remark may be made of the few rolled fragments of calcedonies, and other varieties of the quartz family, gleaned up along its shores, for neither of these constitute a reliable locality.

Of the floræ and fauna we had been observant, but the sandy character of the mere coast line greatly narrowed the former, in which Captain Douglass found but little to preserve, beyond the

Petrified leaf of the *Fagus Ferruginea.*

parnassia caroliniana and scottia cerna.* The fury of the waves renders it a region wholly unfitted to the whole tribe of fresh-water shells. A petrifaction of the fagus ferruginia, brought from a spring on the banks of the St. Joseph's River by Gov. Cass, on his home route, on horseback, presented the petrifying process in one of its most perfect forms (*vide* p. 206). Surfeited with a species of scenery in which the naked sand dunes were often painful to the eye, from their ophthalmic influence, and of geological pros-trations which seemed to lay the coast in ruins, we were glad to reach the solid rock formations, supporting, as they did, a soil favorable to green forests.

A partial eclipse of the sun had been calculated for the 5th of September (1820), to commence at seven o'clock, twenty minutes; but, though we were on the lake, and anxious to note it, the weather proved to be too much overcast, and no effects of it were observed. This eclipse was observed, according to the predictions, at Philadelphia.

The morning of the 8th proved calm, which permitted us to cross the mouth of Grand Traverse Bay. This piece of water is nine miles across, with an unexplored depth, and has some 800 Chippewas living on its borders. Six miles north of this point, we reached and crossed Little Traverse Bay, which is occupied by Ottawas. These two tribes are close confederates, speak dia-lects of the same language which is readily understood by both, and live on the most friendly terms. The Ottawas on the head of Little Traverse Bay, and on the adjoining coast of Lake Michi-gan—which, from its principal village, bears the names of Village of the Cross, and of Waganukizzie,† or L'Arbre Croche—are, to a great extent, cultivators of the soil, and have adopted the use of hats, and the French *capot*, having laid aside paints and feathers. They raise large quantities of Indian corn for the Mackinac mar-ket, and manufacture, in the season, from the sap of the acer sac-charinum, considerable quantities of maple sugar, which is put up, in somewhat elongated bark boxes, called muckucks, in which it is carried to the same market. We found them, wherever they were encountered, a people of friendly manners and comity.

* Dr. John Torrey, *Am. Journ. Science*, vol. iv.

† From *Waganuk*, a crooked or croched tree, and *izzie*, an animate termination, denoting existence or being, carrying the idea of its being charmed or enchanted.

We were now drawing toward the foot of Lake Michigan, at the point where this inland sea is connected, through the Straits of Michilimackinac, with Lake Huron. A cluster of islands, called the Beaver Islands, had been in sight on our left hand, since passing the coast of the Sleeping Bear, which are noted as affording good anchorage ground to vessels navigating the lake. It is twenty-five miles from the site of the old French mission, near L'Arbre Croche, to the end of point Wagoshance,* which is the southeast cape of the Straits of Michilimackinac, and nine miles from thence to the Island. Along the bleak coast of this storm-beaten, horizontal limestone rock, with a thin covering of drift, we diligently passed. Night overtook us as we came through the straits, hugging their eastern shore, and we encamped on a little circular open bay, long after it became pitchy dark. We had traversed a coast line of fifty-seven miles, and were glad, after a refreshing cup of tea and our usual meal, to retire to our pallets.

The next morning revealed our position. We were at the ancient site of old Michilimackinac—a spot celebrated in the early missionary annals and history of New France. This was, indeed, one of the first points settled by the French after Cadaracqui, being a missionary and trading station before the foundation of Fort Niagara, in 1678; for La Salle, after determining on the latter, proceeded, the same fall, up the lakes to this point, which he installed with a military element. The mission of St. Ignace had before been attempted on the north shore of the straits, but it was finally removed here by the advice of Marquette. On gazing at the straits, they were found to be agitated by a perfect gale. This gave time for examining the vicinity. It was found a deserted plain, overspread with sand, in many parts, with the ruins of former occupancy piercing through these sandy drifts, which gave it an air of perfect desolation. By far the most conspicuous among these ruins, was the stone foundation of the ancient fort, and the excavations of the exterior buildings, which had evidently composed a part of the military or missionary plan. Not a house, not a cultivated field, not a fence was to be seen. The remains of broken pottery, and pieces of black bottles, irri-

* Little Fox Point. This word comes from *Wagoush*, a fox, and the denominative inflection a *aims* or *ants*.

descent from age, served impressively to show that men had once eaten and drank here. It was in 1763, in the outbreak of the Pontiac war, that this fort, then recently surrendered to the English, was captured, by a *coup-de-main*, by the Indians. The English, probably doubting its safety, during the American Revolution, removed the garrison to the island, which had, indeed, furnished the name of Michilimackinac before; for the Indians had, *ab initio*, called the old post Peekwutinong, or Headland-place, applying the other name exclusively, as at this day, to the Gibraltar-like island which rises up, with its picturesque cliffs, from the very depths of Lake Huron. The sketch of this scene of desolation, with the Island in view, is given in the second volume of my *Ethnological Researches*, Plate LIII.

After pacing the plain of this ancient point of French settlement in every point, we returned to our tent about eleven o'clock A. M., and deemed it practicable to attempt the crossing to the island in a light canoe, for, although the gale was little if any abated, the wind blew fair. I concurred in the opinion of Captain Douglass that this might be done, and very readily assented to try it, leaving the men in the baggage canoe to effect the passage when the wind fell. It cannot be asserted that this passage was without hazard; for my own part, I had too much trust in my nature to fear it, and, if we were ever wafted on "the wings of the wind," it was on this occasion; our boatmen, volunteers for the occasion, reefing the sails to two feet, and we owed our success mainly to their good management. On rounding the Ottowa point, which is the south cape of the little harbor of 'Mackinac, our friends who had parted from us at Green Bay were among the first to greet us. By the union of these two parties, the circumnavigation of Lake Michigan had been completely made. The rate of travel along the line traversed by them was computed at forty-five miles per day. They had been eight days on the route. The coast line traversed by Captain Douglass and myself, since quitting Chicago, is four hundred and thirty-nine miles, giving a mean of forty-three miles per diem, of which one entire day was lost by head winds.

14

CHAPTER XX.

THE coast line traversed by the party detached from Green
Bay on the 22d of August, under Mr. Trowbridge, extended from
the north shore of Fox River to the entrance of the Monominee
River, and thence around the Little and Great Bay de Nocquet,
to the northwestern cape of the entrance of Green Bay. From
the latter point, the northern shore of Lake Michigan was traced
by the Manistic, and the other smaller rivers of that coast, to the
northern cape of the Straits of Michilimackinac, and through
these to Point St. Ignace and the Island of Michilimackinac. The
line of survey, agreeably to their reckoning, embraced two hundred and eighty miles, thus closing the topographical survey of
the entire coast line of the basin of Lake Michigan, and placing
in the hands of Captain Douglass the notes and materials for a
perfect map of the lake.*

Mr. Trowbridge, whom I had requested to note the features of
its geology and mineralogy, presented me with labelled specimens
of the succession of strata which he had collected on the route.
These denoted the continuance of the calcareous, horizontal series

---

* It is to be regretted that Capt. Douglass, who, immediately on the conclusion
of this expedition, was appointed to an important and arduous professorship in the
U. S. Military Academy of West Point, could not command the leisure to complete
and publish his map and topographical memoir of this part of the U. S. So long
as there was a hope of this, my report of its geology, &c., and other data intended
for the joint PUBLIC WORK, were withheld. But in revising this narrative, at this
time, they are submitted in the Appendix. Prof. Douglass, of whose useful and
meritorious life, I regret that I have no account to offer, died as one of the Faculty
of Geneva College, October 21, 1849.

of formations of the Fox Valley, and of the islands of Green Bay, quite around those northern waters to the closing up of the surveys at Point St. Ignace and Michilimackinac. Nor do the primitive rocks disclose themselves on any part of that line of coast. Of this collection, Mr. Trowbridge well observes, in his report to me, the most interesting will probably be the organic remains. These were procured on the northeast side of Little Nocquet Bay, where areas of limestone appear. They consist of duplicates of the pectinite. Three layers of this, the magnesian limestone, show themselves at this place, of which the intermediate bed is of a dull blue color and compact structure, and is composed in a great measure of the remains of this species. It is comparatively soft when first taken up, but hardens by exposure. About ten miles north of this point, the upper calcareous, or surface rock, embraces nodules of hornstone. Specimens of a semi-crystalline limestone, labelled "marble," were also brought from a cliff, composed of this rock, on the lake shore, about thirty to forty miles southwest from Michilimackinac. Mr. Doty also brought some specimens of sulphate of lime, cal. spar, and some of the common rolled members of the quartz-drift stratum.

Michilimackinac is a name associated with our earliest ideas of history in the upper lakes. How so formidable a polysyllabic term came to be adopted by usage, it may be difficult to tell, till we are informed that the inhabitants, in speaking the word, clip off the first three syllables, leaving the last three to carry the whole meaning. The full term is, however, perpetuated by legal enactment, this part of Michigan having been organized into a separate county some time, I believe, during the administration of Gen. Hull. The military gentlemen call the fort on the cliff, "Mackinā," the townspeople pronounce it Mackinaw; but if a man be hauled up on a magistrate's writ, it is in name of the sovereignty of Michilimackinac. Thus law and etymology grow strong together.

Commerce, we observe, is beginning to show itself here, but by the few vessels we have met, while traversing these broad and stormy seas, and their little tonnage, it seems as if they were stealthily making their way into regions of doubtful profit at least. The fur trade employs most of these, either in bringing up supplies, or carying away its avails. La Salle, when, in 1679, he built the first vessel on the lakes, and sent it up to traffic in furs, was greatly in advance of his age; but he could hardly have anticipated that his countrymen should have adhered so long to the tedious and dangerous mode of making these long voyages in the bark canoe. It is memorable in the history of the region, that last year (1819) witnessed the first arrival of a steamer at Michilimackinac. It bore the characteristic name of Walk-in-the-water,* the name of a Wyandot chief of some local celebrity in Detroit, during the last war.

The astonishment produced upon the Indian mind by the arrival of this steamer has been described to us as very great; but, from a fuller acquaintance with the Indian character, we do not think him prone to this emotion. He gazes on new objects with imperturbability, and soon explains what he does not understand by what he does. Perceiving heat to be the primary cause of the motion, without knowing how that motion is generated, he calls the steamboat Ishcoda Nabequon, *i. e.* fire-vessel, and remains profoundly ignorant of the motive power of steam. The story of the vessel's being drawn by great fishes from the sea, is simply one of those fictions which white loungers about the Indian posts fabricate to supply the wants of travellers in search of the picturesque.

The winds seem to be unloosed from their mythologic bags, on the upper lakes, with the autumnal equinox; and we found them ready for their labors early in September; but it was not till the 18th of that month, after a detention of two days, that we found it practicable for canoes to leave the island. Mustering now a flotilla of three canoes, we embarked at three o'clock P. M., with a wind from the east, being moderately adverse, but soon got under the shelter of the island of Boisblanc; we passed along its inner shore about ten miles, till reaching Point aux Pins—so named from the prevalence here of the pinus resinosa. At this point, the wind, stretching openly through this passage from the east, compelled us to land and encamp. The next day, we were confined to the spot by adverse winds. While thus detained, Captain Douglass, under shelter of the island, returned to

* So called from the water insect, called *Miera* by the Wyandots, one of the invertebrata which slips over the surface of water without apparently wetting its feet.—Vide *Ethnological Researches*, vol. ii. p. 226.

Mackinac, in a light canoe, doubly manned, for something he had left. When he returned, the wind had so far abated that we embarked, and crossed the separating channel, of about four miles, to the peninsula, and encamped near the River Cheboigan.* This was a tedious beginning of our voyage to Detroit; the first day had carried us only *ten* miles, the second but *four*.

We were now to retraverse the shores of the Huron, along which we had encountered such delays in our outward passage, and the men applied themselves to the task with that impulse which all partake of when returning from a long journey. Winds we could not control, but every moment of calm was improved. Paddle and song were plied by them late and early. A violent rain-storm happened during the night, but it ceased at daybreak, when we embarked and traversed a coast line of forty-four miles, encamping at Presque Isle. Rain fell copiously during the night, and the unsettled and changing state of the atmosphere kept us in perpetual agitation during the day. Notwithstanding these changes, we embarked at five o'clock in the morning (16th), and, by dint of perseverance, made thirty miles. We slept on the west cape of Thunder Bay. Next morning, we landed a few moments on the Idol Island, in Thunder Bay, and, continuing along the sandy shore of the *au sauble*, or Iosco coast, entered Saganaw Bay, and encamped, on its west shore, at Sandy Point. Indians of the Chippewa language were encountered at this spot, whose manners and habits appeared to be quite modified by long contact with the white race.

The morning of the 18th (Sept.) proved fair, which enabled us to cross the bay, taking the island of Shawangunk in our course, where we stopped an hour, and re-examined its calcedonies and other minerals. We then proceeded across to Oak Point, on its eastern shore, and, coasting down to, and around, the precipitous cliffs of Point aux Barques, encamped in one of its deeply-indented coves, having made, during the day, forty-two miles.

---

* CHEBOIGAN. This is a noted river of the extreme of the peninsula of Michigan, which has just been made the centre of a new land district by Congress. It affords a harbor for shipping, and communicates with Little Travers Bay on Lake Michigan. A canal across a short route, of easy excavation, would avoid the whole dangerous route through the Straits of Michilimackinac, converting the end of the peninsula into an island, and save ninety miles of dangerous travel.

The formation of this noted promontory consists of an ash-colored, not very closely-compacted sandstone, through original crevices in which the waves have scooped out entrances like vast corridors. In one of these, which has a sandy beach at its terminus, we encamped. He who has travelled along the shores of the lakes, and encamped on their borders, having his ears, while on his couch, close to the formation of sand, is early and very exactly apprised of the varying state of the wind. The deep-sounding roar of the waves, like the deep diapason of a hundred organs, plays over a gamut, whose rising or falling scale tells him, immediately, whether he can put his frail canoe before the wind, or must remain prisoner on the sand, in the sheltering nook where night overtakes him. These notes, sounded between two long lines of cavernous rocks, told us, long before daybreak, of a strong head wind that fixed us to the spot for the day. I amused myself by gathering some small species of the unio and the anadonta. Captain Douglass busied himself with astronomical observations. We all sallied out, during the day, over the sandy ridges of modern drift, in which the pinus resinosa had firmly imbedded its roots, and into sphagnous depressions beyond, where we had, in the June previous, found the sarracenia purpurea, which is the cococo mukazin, or oral's moccasin of the Indians. Here we found, as at more westerly points on the lake, the humble juniperus prostrata, and, in more favorable spots, the ribes lacustre.*

It was stated to us at Michilimackinac, that Lake Huron had fallen one foot during the last year. It was also added that the decrease in the lake waters had been noticed for many years, and that there were, in fact, periodical depressions and refluxes at periods of seven and fourteen years. A little reflection will, however, render it manifest that, in a region of country so extensive and thinly populated, observations must be vaguely made, and that many circumstances may operate to produce deception with respect to the permanent diminution or rise of water, as the prevalence of winds, the quantity of rain and snow which influences these basins, and the periodical distribution of solar heat. It has already been remarked, while at the mouth of Fox River,

* Am. Journ. Science, vol. iv. 1822.

that a fluctuation, resembling a tide, has been improperly thought to exist there, and, indeed, similar phenomena appear to influence the Baltic. Philosophers have not been wanting, who have attributed similar appearances to the ocean itself. "It has been asserted," observed Cuvier, "that the sea is subject to a continual diminution of its level, and proofs of this are said to have been observed in some parts of the shores of the Baltic. Whatever may have been the cause of these appearances, we certainly know that nothing of the kind has been observed upon our coast, and, consequently, that there has been no general lowering of the waters of the ocean. The most ancient seaports still have their quays and other erections, at the same height above the level of the sea, as at their first construction. Certain general movements have been supposed in the sea, from east to west, or in other directions; but nowhere has any person been able to ascertain their effects with the least degree of precision."*

On the next day (20th) the wind abated, so as to permit us, at six o'clock A. M., to issue from our place of detention; but we soon found the equilibrium of the atmosphere had been too much disturbed to rely on it. At seven o'clock, and again at nine o'clock, we were driven ashore; but as soon as it slackened we were again upon the lake; it finally settled to a light head wind, against which we urged our way diligently, until eight o'clock in the evening. The point where we encamped was upon that long line of deposit of the erratic block, or boulder stratum, of which the White Rock is one of the largest known pieces. At four o'clock the next morning, we were again in motion, dancing up and down on the blue waves; but after proceeding six miles the wind drove us from the lake, and we again encamped on the boulder stratum, where we passed the entire day. Nothing is more characteristic of the upper lake geology, than the frequency and abundance of these boulders. The causes which have removed them, at old periods, from their parent bed, were doubtless oceanic; for the area embraced is too extensive to admit of merely local action; but we know of no concentration of oceanic currents, of sufficient force, to bear up these heavy masses, over such

---

* Theory of the Earth. Modern geologists attribute these changes to the rising or sinking of the earth from volcanic forces.

extensive surfaces, without the supporting media of ice-floes. The boulders and pebbles are often driven as the moraines before glacial bodies, and there are not wanting portions of rock surface, in the west, which are deeply grooved or scratched by the pressing boulders. The crystallized peaks of the Little Rocks, above St. Anthony's Falls, have been completely polished by them.— *Vide* p. 149.

The next morning (22d) we were released from our position on this bleak drift-coast, although the wind was still moderately ahead, and after toiling twelve hours adown the closing shores of the lake, we reached its foot, and entered the River St. Clair. Halting a few moments at Fort Gratiot, we found it under the command of Lieut. James Watson Webb, who was, however, absent at the moment. Two miles below, at the mouth of Black River, we met this officer, who had just returned from an excursion up the Black River, where he had laid in a supply of fine watermelons, with which he liberally supplied us. From this spot, we descended the river seven miles, to Elk Island, on which we encamped at twilight, having made fifty-seven miles during the day. Glad to find ourselves out of the reach of the lake winds, and of Eolus, and all his hosts, against which we may be said to have fought our way from Michilimackinac, and animated with the prospect of soon terminating our voyage, we surrounded our evening board with unwonted spirits and glee. Supper being dispatched, with many a joke, and terminated with a song in full chorus, and the men having carefully repaired our canoes, it was determined to employ the night in descending the placid river, and at nine o'clock P. M. all was ready and we again embarked. Never did men more fully appreciate the melody of the Irish bard :—

> " Sweetly as tolls the evening chime,
> Our voices keep tune and our oars keep time."

At half-past three the next morning, we found ourselves at the entrance to Lake St. Clair, thirty miles from our evening repast. Owing to the dense fog and darkness, it was now necessary to await daylight, before attempting to cross. Daylight, which had been impatiently waited for, brought with it our old lake enemy, head winds, which made the most experienced men deem the passage impracticable. Counselled, however, rather by impatience than anything else, it was resolved on. Rain soon commenced, which appeared the signal for increased turbulence; but by dint of hard pushing in the men, with some help from our own hands, we succeeded in weathering Point Huron, the first point of shelter. The right hand shore then became a continued covert, and we successively saw point after point lessen in the distance. It was noon when we reached Grosse Point, the original place of our general embarkation on commencing the expedition; the rest of the voyage ran like a dream "when one awaketh," and we landed at the City of Detroit at half-past three o'clock P. M.

Gov. Cass, and his equestrian party from Chicago, had preceded us thirteen days, as will be perceived from the following article from the weekly press of that city, of September 15, 1820, which embraces a comprehensive notice of the expedition; its route, the objects it accomplished, and the effects it may be expected to have on the leading interests and interior policy of the country, as well as the drawing forth of its resources.

EXPLORING EXPEDITION.

FROM THE DETROIT GAZETTE.

Last Friday evening, Governor Cass arrived here from Chicago, accompanied by Lieutenant M'Kay and Mr. R. A. Forsyth,* both of whom belonged to the expedition—all in good health.

* Major Robert A. Forsyth was a native of the Detroit Country, of Canadian descent, and born a few years after its transfer to the United States. At the time of the expedition, he was the Secretary of Governor Cass, and was admirably qualified to take a part in it, by his energy and perseverance, his indomitable courage, and his physical power and activity. Some of these traits of character were developed at an early age. He was but yet a lad at the time of the surrender of Detroit, and was so much excited by that untoward event, that he insulted the British officers in the fort by his reproaches, and so irritated them that one of them threatened to pin him to the floor with a bayonet. During the war upon the frontier, he was actively employed, and on more than one occasion distinguished himself by his conduct and courage. He was with Major Holmes at the battle near the Long Woods, and behaved with great gallantry. In 1814, he was sent with Chandruai, a half-breed Pottowatamie, and with a small party of Indians, to invite the various Indian tribes to come to Greenville, at the treaties about to be held by Generals Harrison and Cass, with a view to detach the North-Western Indians from British influence. On the route, they met a superior party of Indians, led by an officer of the British Indian Department, who attempted to take them prisoners.. They resisted, and, by their

We understand that the objects of the expedition have been successfully accomplished. The party has traversed 4,000 miles of this frontier since the last of May. Their route was from this place to Michilimackinac, and to the Sault of St. Mary's, where a treaty was concluded with the Chippewas for the cession of a tract of land, with a view to the establishment of a military post. They thence coasted the southern shore of Lake Superior to the Fond du Lac; ascended the St. Louis River to one of its sources, and descended a small tributary stream of Sandy Lake to the Mississippi. They then ascended this latter river to the Upper Red Cedar Lake, which may be considered as the principal source of the Mississippi, and which is the reservoir where the small streams forming that river unite. From this lake they descended between thirteen and fourteen hundred miles to Prairie du Chien, passing by the post of St. Peter's on the route. They then navigated the Ouisconsin to the portage, entered the Fox River, and descended it to Green Bay. Then the party separated, in order to obtain a topographical sketch of Lake Michigan. Some of them coasted the northern shore to Michilimackinac, and the others took the route by Chicago. From this point they will traverse the eastern shore of the lake to Michilimackinac, and may be expected here in the course of a week. Governor Cass returned from Chicago by land. A correct topographical delineation of this extensive frontier may now be expected from the accurate observations of Captain Douglass, who is fully competent to perform the task. We have heretofore remained in ignorance upon this subject, and very little has been added to the stock of geographical knowledge since the French possessed the country. We understand that all the existing maps are found to be very

prompt and almost desperate courage, drove off the British party. Forsyth distinguished himself in the contest, in which the British leader of the party was killed. Soon after the war, he was appointed Private Secretary to Governor Cass, and continued in that capacity for fifteen years, till the latter was transferred to the War Department. He accompanied the General in all his expeditions into the Indian country, and rendered himself invariably useful, having a peculiar talent to control the rough men who took part in these dangerous excursions. He was ultimately appointed a paymaster in the army, in which capacity he served in Mexico, where he acquired the seeds of the disorder which proved fatal to him in 1849. He will be long recollected and regretted by those who knew him, for the shining qualities of head and heart which endeared him to all his acquaintances.

erroneous. The character, numbers, situation, and feelings of the Indians in those remote regions have been fully explored, and we trust that much valuable information upon these subjects will be communicated to the Government and to the public. We learn that the Indians are peaceable, but that the effect of the immense distribution of presents to them by the British authorities, at Malden and at Drummond's Island, has been evident upon their wishes and feelings through the whole route. Upon the establishment of our posts, and the judicious distribution of our small military force, must we rely, and not upon the disposition of the Indians. The important points of the country are now almost all occupied by our troops, and these points have been selected with great judgment. It is thought by the party, that the erection of a military work at the Saut is essential to our security in that quarter. It is the key of Lake Superior, and the Indians in its vicinity are more disaffected than any others upon the route. Their daily intercourse with Drummond's Island, leaves us no reason to doubt what are the means by which their feelings are excited and continued. The importance of this site, in a military point of view, has not escaped the observation of Mr. Calhoun, and it was for this purpose that a treaty was directed to be held. The report which he made to the House of Representatives, in January last, contains his views upon the subject.

We cannot but hope that no reduction will be made in the ranks of the army. It is by physical force alone, and by a proper display of it, that we must expect to keep within reasonable bounds, the ardent, restless, and discontented savages, by whom this whole country is filled and surrounded. Few persons living at a distance are aware of the means which are used, and too successfully used, by the British agents, to imbitter the minds of the Indians, and preserve such an influence over them as will insure their co-operation in the event of any future difficulties. A post at the Fond du Lac will, before long, be necessary, and it is now proper that one should be established at the portage between the Fox and Ouisconsin Rivers.

Mr. Schoolcraft has examined the geological structure of the country, and has explored, as far as practicable, its mineralogical treasures. We are happy to learn that this department could not have been confided to one more able or zealous to effect the objects connected with it. Extensive collections, illustrating the natural history of the country, have been made, and will add to the common stock of American science.

↓ We understand that copper, iron, and lead are very abundant through the whole country, and that the great mass of copper upon the Outanagon River has been fully examined. Upon this, as well as upon other subjects, we hope we shall, in a few days, be able to communicate more detailed information.

DISCOVERY .

OF THE

ACTUAL SOURCE OF THE MISSISSIPPI RIVER

IN

ITASCA LAKE,

BY AN EXPEDITION, AUTHORIZED BY THE·WAR DEPARTMENT OF
THE UNITED STATES, IN 1832.

BY HENRY R. SCHOOLCRAFT,

UNITED STATES SUPERINTENDENT OF INDIAN AFFAIRS FOR MICHIGAN, ETC.

CHAPTER XXI.

TWELVE years elapse between the closing of the prior, and the opening of the present narrative. In the month of August, 1880, instructions were received by Mr. Schoolcraft to proceed into the Upper Mississippi valley, to endeavor to terminate the renewed hostilities existing between the Chippewa and Sioux tribes. These directions did not come to hand at the remote post of Sault de Ste. Marie, at the outlet of Lake Superior, in season to permit the object to be executed that year. On reporting the fact that the tribes would be dispersed to their hunting-grounds before the scene could be reached, and that severe weather would close the streams with ice before the expedition could possibly return, the plan was deferred till the next year. Renewed instructions were issued in the month of April, 1881, and an expedition organized at St. Mary's to carry them into immediate effect.

These instructions did not require the broad table-lands on which the river originates to be visited, though the journey connected itself with preliminary questions; nor was it found practicable to extend the geographical examinations, in the Mississippi Valley, beyond about latitude 44°.

The force designed for this expedition consisted of twenty-seven men, including a botanist and geologist, and a small military party under Lieut. Robert E. Clary, U. S. A. Entering Lake Superior, in the month of June, with a bright pure atmosphere and serene weather, the party enjoyed a succession of those clear transporting vistas of rock and water scenery, which render this picturesque basin by far the most magnificent, varied, and affluent in its prospect in America. It is in this basin only,

of all the series of North American lakes which stretch west
from the St. Lawrence, that peaks and high mural walls of volcanic formation, pierce through, or lift up, the horizontal series
of the silurian system; and that, in the lake region, the latter is
found in singular juxtaposition, by means of these upheavals,
with the senites, sienitic granites, and metamorphic rocks composing the globe's nucleus, or primary out-pushed stony coats
of these latitudes.

I had passed through this varied and wonder-creating scene
of coast views and long-stretching vistas in 1820, when geology, in
America, at least, was in its infancy, as a member of the organic
government expedition into this quarter of the Union, as detailed
in the preceding pages. I had, in 1826, revisited the whole coast
from Point Iroquois to Fond du Lac, in the exercise of official
duties, connected with the Indian tribes; besides making sectional expeditions into the regions of the Gargontwa and Mishe-
pecotin, and of the Takwymenon sand-rock, interior, and coast
lines. But the beauty of the prospects presented in 1831, the
serenity of the weather, and the opportunity which it gave of
revisiting scenes which had before flitted by, as the fragments of
a gorgeous dream, gave to this visit a charm which no length of
time can obliterate. And these attractions were enhanced by
association with the agreeable men who accompanied me; of
whom it may be said that they represented the place of strings
in a melodious harp, whose concurrence was at all times necessary to produce harmony. The sainted and scene-loving Woolsey*—the self-poised and amiable Houghton, just broke loose
from the initial struggles of life to luxuriate on the geological
smiles of the face of nature in this scene—ah! where are they?
Death has laid his cold hand on them, to open their eyes on
other, and to us inscrutable scenes.

Passing through this lake, the expedition met the brigade of
boats of the late Mr. Wm. Aitken, from the Upper Mississippi
waters, with the annual returns of furs from that region. He
represented the urgent necessity of an official visit to that section
of the country, where the Indians were in turmoil; but stated, at
the same time, that the waters were too low in the streams at the

* Vide Letters on Lake Superior, in *Southern Literary Messenger*, 1886.

sources of the Mississippi to render explorations practicable. He also represented it impracticable, this season, to enter the Mississippi by the way of the *Broulé*, or Misakoda River. This information .was confirmed on reaching Chegoimegon, at the remarkable group of the Confederation Islands (*ante*, p. 105). Returning eight miles on my track, I entered the Muskigo, or Mauvais River, and ascended this stream by all its bad rafts, rapids, and portages, to the upper waters of the River St. Croix of the Mississippi. Crossing the intermediate table-lands, with their intricate system of lakes and portages to *Lac Courteroille*, or Ottawa Lake, I entered one of the main sources of Chippewa River, and descended this prime tributary stream to its entrance into the Mississippi, at the foot of Lake Pepin. From the latter point I descended to Prairie du Chien, and to Galena in Illinois. Dispatching the men and canoes from this place back to ascend the Wisconsin River, and meet me at the portage of Fort Winne-bago, I crossed the lead-mine country by land, by the way of the Pekatolica, Blue Mound, and Four Lakes, to the source of the Fox River, and rejoining my canoes here, descended this stream to Green Bay, and returned to my starting-point by the way of Michilimackinac and the Straits of St. Mary. Two months and twelve days were employed on the journey, during which a line of forests and Indian trails had been passed, of two thousand three hundred miles.

The Indians had been met, and counselled with at various points, at which presents and provisions were distributed, and the peace policy of the Government enforced. A Chippewa war party, under Ninaba, had been arrested on its march against the Sioux in descending the Red Cedar fork of the Chippewa River. Information was obtained that nine tribes or bands had united in their sympathies for the restless Sauks and Foxes, who broke out in hostility to the United States the following spring. Messages, with pipes and belts, and in one case notice, with a tomahawk smeared with vermilion, to symbolize war, had passed between these tribes.*

The information was communicated to the Government, with a

---

* An outline of the expedition of 1831 is found in Schoolcraft's "Thirty Years on the American Frontiers." Lippincott & Co. Phila. 1850.

suggestion that an expedition should be organized for visiting remoter regions the next year, and forwarding, at the same time, detailed estimates of the expenditures essential to its efficiency. These suggestions were approved by the Secretary of War on the 3d of May, 1832, and instructions forwarded to me for organizing an expedition to carry the reconnoissance and scrutiny to the tribes on the sources of the Mississippi. A small escort of U. S. infantry was ordered to accompany me, under Lieut. James Allen, U. S. A., who, being a graduate of the West Point Military Academy, undertook the departments of topography and trigonometry. I secured the services of Dr. Houghton, as physician and surgeon, and acting botanist and geologist—positions which he had occupied on the prior expedition of 1831. The American Board of Commissioners for Foreign Missions were invited to send an agent to observe the wants and condition of the Indian tribes in these remote latitudes; who directed the Rev. Wm. T. Boutwell to join me at St. Mary's. I charged myself especially with inquiring into the Indian history and languages, statistics, and general ethnography.

The expedition left the Sault de Ste. Marie on the 7th of June, taking the route through Lake Superior to Fond du Lac and the St. Louis River, and the Savanna Summit to Sandy Lake, which lies 500 miles above St. Anthony's Falls of the Upper Mississippi. The width of the Mississippi at the outlet of Sandy Lake, by a line stretched across, was found to be 331 feet. At my camp here, a general council was summoned of the lower tribes, who were notified to assemble at the mouth of the River Des Corbeau on the 20th of July; and a boat with presents and supplies was sent down the Mississippi to await the return of the expedition through that river. Lightened thus of baggage, and having fixed a point of time within which to finish the explorations above, I proceeded up the main channel of the river to, and across the Pakagama Falls, and its wide plateau of savannas, and through the Little and Great Winnipek Lakes, to the Upper Red Cedar, or Cass Lake, which we entered on the 10th of July. This is a fine lake of transparent water, about eighteen miles in length, with several large bays and islands as denoted in the accompanying sketch, which give it an irregular shape. The largest island, called *Grande Isle* by the French, which is the *Gitchiminis* of the Indians, and the *Colcaspi** of my initial narrative of 1832. This lake was the terminus of the respective explorations of Lieutenant Zebulon Pike, U. S. A., in 1806, and Governor Lewis Cass in 1820. The points at which they approached it were not, however, the same. Pike visited it in a dog train, on the snow, in the month of January, across the land, from the Northwest Company's trading post at Leech Lake. He visited an out-station of that company on Grand Island. Cass landed in July, after tracing its channel from Sandy Lake to the entrance of Turtle River, the line of communication to Turtle Lake, which was long the reputed source of the river. This has been called by a modern traveller in the region Lake Julia, that he might call it the *Julian* source of the Mississippi.†

I found the Mississippi, at the point where it flows from the lake, to be 172 feet wide, not having lost half the width it had at Sandy Lake, although in this distance it is diminished by the volume of its Leech Lake tributary, which the northwest agents informed Lieutenant Pike, in 1806, to be its largest tributary. I had reached it ten days earlier in the season than Governor Cass, having been exactly one day less in traversing the long line of intervening country from Sault de Ste. Marie. I proceeded directly to Grand Isle, the residence of a Chippewa band numbering 157 persons. This island was found to have a fertile soil, where they had always raised the zea maize. Its latitude is 47° 25′ 23″. Not only had I reached this point ten days earlier in the month than the expedition of 1820, but it was found that the state of the water on these summits was very favorable to their ascent. Ozawindib,‡ the Chippewa chief, said that his hunting-grounds embraced the source of the Mississippi, but that canoes of the size and burden which I had could not ascend higher than the *Pemidjegumaug*, or Queen Anne's Lake. I determined to encamp my extra men permanently on this island, with the heavy canoes,

---

* This is an anagram composed of the names of Schoolcraft, Cass, and Pike, the geographical discoverers, in reversed order, of the region.

† Beltrami.

‡ This name is derived from *osawau*, yellow; *winisis*, hair, and *kundiba*, bone of the forehead or head.

provisions, and baggage, leaving the camp in charge of Louis Default, a trusty man, of the *metif* class, well acquainted with the Indian language, who had been a guide in 1820, and to make explorations, in the lightest class of Indian canoes, provisioned for an *élite* movement. Lieutenant Allen also determined to encamp the United States soldiers of the party, leaving them under a sergeant. To give each gentleman of the party an opportunity of joining in this movement, it was necessary to procure five hunting canoes, which were of no greater capacity than to bear one *sitter** and two paddlers.

Ozawindib and his companions produced these canoes at an early hour on the following morning, and having, at my request, drawn a map of the route, embarked himself as the guide to the party. We left the island before it was yet daylight. The party now consisted of sixteen persons, including three Chippewas and eight *engagees*. The Mississippi enters this lake through a savanna, on its extreme western borders, after performing one of those evolutions through meadow lands so common to its lower latitudes; after reaching to within fifty yards of the lake, it winds about, through a natural meadow, for many miles before its debouchure. The chief, who was familiar with this feature, carried me to a fifty yards portage, by which we saved some miles of paddling. We reached the Mississippi at a place where it expands into an elongated lake, for which I heard no name, and which I called Lake Andrúsia.† After passing through this, the river appeared very much in size and volume as it had on the outlet below Cass Lake. It winds its way through the same species of natural meadows, during which there is but little current. On ascending this channel but a short distance, the river is found to display itself in a second lake—which the natives call Pamitascodiac‡—which, in general appearance and character, may be deemed the twin of Lake Andrúsia. On its upper margin, a tract of prairie land appears, of a sandy character, bearing

---

* The term "sitter," which is a northwest phrase in common use, is equivalent to the Canadian word *bourgoise*.

† From Andrew Jackson, at that time President of the United States.

‡ This word appears to be a derivation from *pemidji*, across, *muscoda*, a prairie, and *ackee*, land.

scattered pines. This appears to be the particular feature alluded to by the Indian name. About four miles above this lake, and say fifteen from Cass Lake, the rapids commence. It was eight o'clock A. M. when we reached this point, and we had then been four hours in our canoes from the Andrúsia portage. These rapids soon proved themselves to be formidable. Boulders of the geological drift period are frequently encountered in ascending them, and the river spreads itself over so considerable a surface that it became necessary for the bowsmen and steersmen to get out into the shallows and lead up the canoes. These canoes were but of two fathoms length, drew but a few inches water, and would not bear more than three persons. It was ten o'clock when we landed, on a dry opening on the right shore, to boil our kettle, and prepare breakfast. So dry, indeed, was the vegetation here, that the camp-fire spread in the grass and leaves, and it required some activity in the men to prevent its burning the baggage. There were ten of these rapids encountered before we reached the summit, or plateau, of Lake Pemidjegumaug, which is the *Lac Traverse* of the French. These were called the Metóswa rapids, from the Indian numeral for ten.

The term *Lac Traverse* has been repeated several times by the Canadian French, in our northwestern geography; being prominently known in the Upper Mississippi for a handsome sheet of water, connecting the St. Peter's, or Minnesota River, with Red River of Hudson's Bay; and as the Indian name, though very graphic, is not euphonious, I named it Queen Anne's Lake.* It is a clear and beautiful sheet of water, twelve miles in length, from east to west, and six or seven broad, with an open forest of hard wood. It is distant forty-five miles from Cass Lake, and lies at an elevation of fifty-four feet above that lake, and of 1,456 feet above the Gulf of Mexico. The latitude is 47° 28′ 46″. The peculiarity recognized by the Indian name of Pemidjegumaug, or Crosswater, is found to consist in the entrance of the Mississippi into its extreme south end, and its passage through or across part of it, at a short distance from the point of entrance. Another feature of its topography consists of its connection, by

---

* In allusion to an interesting period of British history, in its influences on America.

a lively channel of less than a mile's length, with another transverse lake of pure waters, to which I applied the name of Washington Irving. These features are shown by the subjoined
sketch.

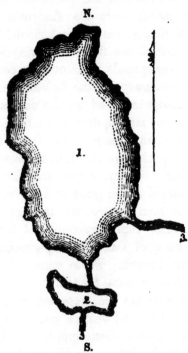

1. Queen Anne's Lake.   2. Washington Irving's Lake.   8. Mississippi River.

CHAPTER XXII.

Ascent of the Mississippi above Queen Anne's Lake—Reach the primary forks of the river—Ascend the left-hand, or minor branch—Lake Irving—Lake Marquette —Lake La Salle—Lake Plantagenet—Encamp at the Naiwa rapids at the base of the Height of Land, or Itasca Summit.

A SHORT halt was made on entering Queen Anne's Lake, to examine an object of Indian superstition on its east shore. This consisted of one of those water-worn boulders which assume the shape of a rude image, and to which the Chippewas apply the name *Shingabawassin*, or image-stone. Nothing artificial appeared about it, except a ring of paint, of some ochreous matter, around the fancied neck of the image.* We were an hour in crossing the lake southwardly from this point, which would give a mean rate of five miles. At the point of landing, stood a small, deserted, long building, which Ozawindib informed me had been used as a minor winter trading station. I observed on the beach at this spot some small species of unios, and, at higher points on the shore, helices. We here noticed the passenger pigeon. The forest exhibited the elm, soft maple, and white ash. Proceeding directly south from this spot a short distance, we entered the Mississippi, which was found to flow in with a broad channel and rapid current. This channel Lieutenant Allen estimated to be but one hundred yards long, at which distance we entered into a beautiful little lake of pellucid water and a picturesque margin, spreading transversely to our track, to which I gave the name of Irving. Ozawindib held his way directly south through this body of water, striking the river again on its opposite shore. We had proceeded but half a mile above

---

* An object of analogous kind was noticed, during the prior expedition of 1820, at an island in Thunder Bay of Lake Huron. *Vide* p. 55.

this lake, when it was announced that we had reached the primary forks of the Mississippi. We were now in latitude 47° 28′ 46″. Up to this point, the river had carried its characteristics in a remarkable manner. Of the two primary streams before us, the one flowing from the west, or the Itascan fork, contributes by far the largest volume of water, possessing the greatest velocity and breadth of current. The two streams enter each other at an acute angle, which varies but little from due south, as denoted in the diagram.

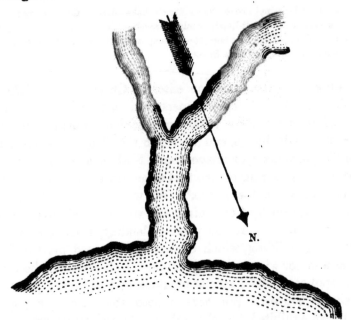

Primary forks of the Mississippi River, in lat. 47° 28′ 46″.

Ozawindib hesitated not a moment which branch to ascend, but shooting his canoe out of the stronger current of the Itascan fork, entered the other. His wisdom in this movement was soon apparent. He had not only entered the shallower and stiller branch, but one that led more directly to the base of the ultimate summit of Itasca. This stream soon narrowed to twenty feet. We could distinctly descry the moving sands at its bottom; but its diminished velocity was apparent from the intrusion of aquatic plants along its shores. It was manifest also from the forest vegetation, that we were advancing into regions of a more alpine flora. The branches of the larches, spruce, and gray pines, were clothed with lichens and floating moss to their very tops, denoting an atmosphere of more than the ordinary humidity. Clumps of gray willows skirted the margin of the stream.

It was found that the river had made its utmost northing in Queen Anne's Lake. From the exit from that point, the course was nearly due south, and from this moment to our arrival at the ultimate forks, which cannot exceed a mile and a half or two miles, it was evident why the actual source of this celebrated river had so long eluded scrutiny. We were ascending at every curve so far *south*, as to carry the observer out of every old line of travel or commerce in the fur trade (the sole interest here), and into a remote elevated region, which is never visited indeed, except by Indian hunters, and is never crossed, even by them, to visit the waters of the Red River—the region in immediate juxta-position north. This semi Alpine plateau, or height of land for which we were now pushing directly, is called in the parlance of the fur trade *Hauteurs de Terre*. It was evident that we were ascending to this continental plateau by steps, denoted by a series of rapids, presenting step by step, in regular succession, wide-spread areas of flat surface spotted with almost innumerable lakes, small and large, and rice-ponds and lagoons. Thus, after surmounting the step of the Packagama Falls, we enter on a wide and far stretching plateau which embraces the great area of Leech Lake, and its numerous lacustrine beds. This step or plateau may, in the descending order of the Mississippi, be called the fifth plateau, and is, by barometrical observation, 1,856 feet above the Gulf of Mexico. The next, or fourth step, is that of the plateau of Cass Lake, caused chiefly by the lively waters of the Leech Lake, the Upper Red Cedar, and the Winnepek outlets. The Cass Lake level extends west of this lake to the foot of the Metoswa rapids. This is forty-six feet above the Leech Lake level. The third plateau, on which the Mississippi spreads itself, is that of the Queen Anne summit, which is elevated by the Metoswa rapids sixty-four feet above the former. We had now entered on this third plateau, on which we found the river flowing with a just perceptible current, and frequently expanding itself in small lakes. On the first of these, after ascending the left hand, or minor fork, I bestowed the name of Marquette; and on

the second, that of La Salle. We proceeded beyond these to
a third lake of larger dimension, which the Chippewas call
Kubba-Kunna, or the Rest in the Path, being the site of crossing
of one of their noted land-trails; I named it Lake Plantagenet.
Lt. Allen deemed this lake ten miles long and five wide. At a
point a short distance above the head of this lake, we encamped
at a late hour. It was now seven o'clock P.M., and we had been
in our canoes sixteen hours, and travelled fifty-five miles. It
was not easy to find ground dry enough to encamp on, and while
we were searching for it, rain commenced. We had pushed through
the ample borders of the Scirpus lacustris and other aquatic plants,
to a point of willows, alders, and spruce and tamarack, with
pinus banksiana in the distance. The ground was low and wet,
the foot sinking into a carpet of green moss at every tread.
The lower branches of the trees were dry and dead, exhibiting
masses of flowing gray moss. Dampness, frigidity, and gloom
marked the dreary spot, and when a camp fire had been kindled
it threw its red glare around on strange masses of thickets and
darkness, which might have well employed the pencil of a Michael
Angelo. Tired and overwearied men are not, however, much
given to the poetic on these occasions, and they addressed themselves at once to the pacification of that uneasy organ, the
stomach. Travelling with men who strangely mix up two foreign
languages, one falls insensibly into the same jargon habits, of
which I convicted myself of a notable instance this evening. I
had on landing and pushing into the forest, laid a green morocco
portfolio on the branches of a little spruce, and could not find it.
Kewau bemuases, I said to one of the men, *en petite chose ver, mittig
onsing?* Have you not seen a small green roll in a sapling? not
recollecting that the middle clause of the sentence, though in
regimen with the Ojibwa, could have only been construed by one
familiar both with the Canadian French and the Algonquin.
Such, however, proved to be the case, and he soon handed me
the missing portfolio.

I observed, as the crews of the several canoes threw down their
day's game before the cook, there was a species of duck, the anas
canadensis, I think, which had a small unio attached to one of its
mandibles, having been engaged in opening the shell at the
moment it was shot. With every aid, however, from the tent and

the tea-kettle, and our cook's art in spitting ducks, the night here, in a gloomy and damp thicket, just elevated above the line of the river-flags, and quite in the range of the frogs and lizards, proved to be one of the most dreary and forlorn. It was felt that we were no longer on the open Mississippi, but were winding up a close and very serpentine tributary, nowhere over thirty feet wide, which unfolded itself in a savanna, or bog, bordered closely with lagoons and rice ponds. Indian sagacity, it was clear, had led Ozawindib up this tributary as the best, shortest, and easiest possible way of reaching to, and surmounting the Itasca plateau, but it required a perpetual use of hand, foot, paddle, and pole; nor was there a gleam of satisfaction to be found in anything but the most intense onward exertion. Besides, I had agreed to meet the Indians at the mouth of the Crow-Wing River on the 24th of July, and that engagement must be fulfilled.

At five o'clock the next morning (12th) we were on our feet, and resumed the ascent. The day was rainy and disagreeable. There was little strength of current, but quite a sufficient depth of water; the stream was excessively tortuous. Owing to the sudden bends, we often frightened up the same flocks of brant, ducks, and teals again and again, who did not appear to have been in times past much subjected to these intrusions. The flora of this valley appeared unfavorable. Dr. Houghton has reported a new species of malva and some five or six other species or varieties from the general region, but these have not, I think, been elaborately described. The localities of the known species of fauna might be marked by the occurrence, on this fork, of the cervus virginianus, which had not been seen after leaving the Sandy Lake summit till after getting above the primary forks, which flow from the south and west.

We toiled all day without intermission from daybreak till dark. The banks of the river are fringed with a species of coarse marsh-land grass. Clumps of willows fringe the stream. Rush and reed occupy spots favorable to their growth. The forest exhibits the larch, pine, and tamarack. Moss attaches itself to everything. Water-fowls seem alone to exult in their seclusion. After we had proceeded for an hour above Lake Plantagenet, an Indian in the advance canoe fired at and killed a deer. Although fairly shot, the animal ran several hundred yards. It then fell dead.

The man who had killed it brought the carcass to the banks of the river. The dexterity with which he skinned and cut it up, excited admiration. He gave the *moze*, which I understood to mean the hide and feet, to my guide, Ozawindib. Signs of this animal were frequent along the stream. But we were impelled forward by higher objects than hunting. It was, indeed, geographical and scientific facts that we were hunting for. To trace to its source an important river, and to fix the actual point of its origin, furnished the mental stimulus which led us to care but little where we slept or what we ate.

When the usual hour for breakfast arrived, the banks of the river proved too marshy to land, and we continued on till a quarter past twelve P. M., before a convenient landing could be made. After this recruit to stomach and spirits, the men again pushed forward, threading the stream as it wound about in a savanna, seldom halting more than a few minutes at a time. Frequently, a shot was fired at the numerous water-fowl, so abundant on these waters. Sometimes a small unio or anadonta was picked up from the shores; occasionally a plant pulled up, for the botanical press. Nowhere was the water found too shallow for our canoes, which were only embarrassed at some points by the density of vegetable tissue. Rain showers were encountered during the whole of the day, the equilibrium of the atmosphere being disturbed by rolling, cumulous clouds, which often poured down their contents with little warning, and without, indeed, driving us from our canoes. For, on these occasions, where a fixed point is to be made, and the showers are not anticipated to be long or heavy, it is better to travel in the rain and submit to the wetting, than to attempt landing. Neither can the meal of dinner be stopped for. At length, at half-past five o'clock in the evening, we came to the base of the highlands of the Itasca or Hauteurs de Terre summit. The flanks of this elevation revealed themselves in a high, naked precipice of the drift and boulder stratum, on the immediate margin of the stream which washed against it. Our pilot, Ozawindib, was at the moment in the rear; halting a few moments for him to come up, he said that we were within a few hundred yards of the Naiwa rapids, and that the portage around them commenced at this escarpment. We had seen no rock of any species, in place, thus far.

A general landing was immediately made at the foot of the hill, and as the five canoes came up the baggage was prepared in bundles and packages for being carried, the canoe-paddles and poles securely tied in bundles, and the canoes lifted from the water and dried in the sun to make the transportation of them as light as possible, and mended and pitched wherever they leaked. It was found that the whole baggage, canoes and all, could be arranged for eleven back-loads, this being the precise number of our carriers, white and red; and being ready, Ozawindib led the way, having a single canoe for his share, and he was soon followed by the whole line, each one of our sitters falling in this line, charged with the particular instrument of his observation, or record of it. The hill was steep, and the footing soft and yielding in the crumbling diluvion, and the scene, as the party struggled up the ascent, presented quite a study for the picturesque. Lieutenant Allen carried his canoe-compass, which I had had mounted by an artisan of Detroit; Dr. Houghton grasped his hortus siccus under his arms; Mr. Johnston, our interpreter, had his pipe and fowling-piece, and Mr. Boutwell had wellnigh lost his pocket-bible and notes, while staying himself against the treacherous influence of a steep sand cliff. While the party thus took their way over the hill to cross a peninsula of a mile or two, and strike the river above the junction of the Naiwa River, I went to observe the rapids. The river, at this point, is forced through a narrow gorge, where the water descends with loud murmuring over a series of rapids, which form a complete check to navigation. The portage is two miles. I judged the entire descent of the channel, from the beginning to the terminus of the portage, to be forty-eight feet. Boulders of the peculiar northern sienite, highly charged with hornblende, and of trap-rock, or greenstone, quartz, and sandstone, were scattered over this elevation, and mixed with the more finely comminuted portions of the same rocks, and of amygdaloids and schistose fragments. Among these, I observed some specimens of the zoned agate, which identifies the stratum with the extensive drift formation of the upper Mississippi. It would seem that extensive amygdaloidal strata formerly extended over these heights, which have been broken down by the fierce and general rush of the oceanic currents of the north, which once manifestly swept over these elevations.

Darkness fell as we reached an elevation overlooking the river above the Naiwa Rapids, and after some deliberation as to the spot where we should suffer less annoyance from mosquitos, I proceeded to the lower part of the valley near the river, and set up my tent there for the night. On questioning Ozawindib of the Naiwa River, he informed me that it was a stream of considerable size, and that it originated in a lake on a distant part of the plateau, which was infested with the copper-head snake; hence the name. Mr. Allen's estimate of this day's journey was fifty-two miles. We had reached the second, or Assawa plateau of the Mississippi, which is, barometrically, seventy-six feet above the Queen Anne summit, and now had but one more to surmount.

CHAPTER XXIII.

The Expedition having reached the source of the east fork in Assawa Lake, crosses the highlands of the Hauteurs de Terre to the source of the main or west fork in Itasca Lake.

THE next morning (18th) a dense fog prevailed. We had found the atmosphere warm, but charged with water and vapors, which frequently condensed into showers. The evenings and nights were, however, cool, at the precise time of the earth hiding the sun's disk. It was five o'clock before we could discern objects with sufficient distinctness to venture to embark. We found the channel of the river strikingly diminished on getting above the Naiwa. Its width is that of a mere brook, running in a valley half a mile wide. The water is still and pond-like, the margin being encroached on by aquatic plants. It presents some areas of the zizania palustris, and appeared to be the favorite resort for several species of duck, who were continually disturbed by our progress. After diligently ascending an hour and a half, or about eight miles, the stream almost imperceptibly began to open into a lake, which the Indians called Assawa, or Perch Lake. Its borders are fringed with the *monomin* of the Chippewas, or wild rice, and several of the liliaceous water plants. The water is transparent when dipped up and viewed by the light, but from the falling of leaves and other carbonaceous fibre to the bottom, it reflects a sombre hue. We were just twenty minutes in passing through it, denoting a length of perhaps two miles, and a width of half a mile. Our course through it was directly south. Ozawindib, who took the advance, entered an inlet, but had not ascended it far, when he rested on his paddles, and exclaimed *o-omah mekun-nah*, here is the path, or portage. We had, in fact, traced this branch of the river into its utmost sources. It was seven o'clock in the morning. We were surrounded by what the natives term *azhiskee*, or mire, broad-leaved plants extending over the surface of the water, in which I recognized a diminutive species of yellow pond-lily. There was no mode of reaching dry land but by stepping into this yielding azhiskee. The water was rather tepid. After wading about fifty yards the footing became more firm, and we soon began to ascend a slight elevation. Some traces of an Indian trail appeared here, which led to an opening in the thicket, where vestiges of the bones of birds, and old camp-poles, indicated the prior encampment of Indians.

I had now traced this branch of the Mississippi to its source, and was at the south base of the inter-continental highlands, which give origin to the longest and principal branch of the Mississippi. To reach its source it was necessary to ascend and cross these. Of their height, and the difficulty of their ascent, we knew nothing. This only was sure, from the representation of the natives, that it could be readily done, carrying the small bark canoes we had thus far employed. The chief said it was thirteen *opugidjiwenun*, or putting-down-places, which are otherwise called *onwaybees*, or rests. From the roughness of the path, not more than half a mile can be estimated to each *onwaybee*. Assawa Lake is shown, by barometric measurement, to be 1,532 feet above the Gulf. Having followed out this branch to its source, its very existence in our geography becomes a new fact.

While the baggage and canoes were being carried to the spot of our encampment, a camp-fire was kindled and the cook busied himself in preparing breakfast. The canoes were then carefully examined and repaired, and the baggage parted into loads, so as to permit the whole outfit and apparatus to be transported at one trip. These things having been arranged, and the breakfast dispatched, we set forward to mount the highlands. Ozawindib having thrown one of the canoes over his shoulders, led the way, complaisantly, being followed by the entire party.

The prevailing growth at this place is thick bramble, spruce, white cedar, and tamarak. The path plunges at once into a marshy and matted thicket, which it requires all one's strength to press through—then rises to a little elevation covered with white cedar, and again plunges into a morass strewed with fallen and decayed logs, covered with moss. From this the trail emerges on dry ground. Relieved from the entanglement about our feet, we soon found ourselves ascending an elevation of the drift stratum, consisting of oceanic sand, with boulders. On the side of this eminence we enjoyed our first *onwaybee*. The day had developed itself clear and warm, and glad indeed were we to find the chief had put down his canoe, and by the time we reached had lit his pipe. The second onwaybee brought us to the summit of this elevation; the third to the side of a ridge beyond it; the fourth to another summit; in fine, we found ourselves crossing a succession of ridges and depressions, which seemed to have owed their original outlines to the tumultuous waves of some mighty ocean, which had once had the mastery over the highlands. Trail there was often none. The day being clear, the chief, however, held his course truly, and when he was turned out of it by some defile, or thicket, or bog, he again found his line at the earliest possible point. In one of the depressions, we crossed a little lake in the canoes; in another, we followed the guide on foot, through and along the border of a shallow lake, to avoid the density of the thickets.

Ripe strawberries were brought to me at one of our onwaybees. I observed the diminutive rebus nutkanus on low grounds. The common falco was noticed, and the Indians remarked tracks of the deer, not, however, of very recent date. The forest growth is small, by far the most common species being the scrubby pinus banksianus, exhibiting its parasitic moss. The elevated parts of the route were sufficiently open, with often steep ascents. Over these sienite and granite, quartz and sandstone boulders were scattered. Every step we made in crossing these sandy and diluvial elevations, seemed to inspire renewed ardor in completing the traverse. The guide had called the distance, as we computed it, about six, or six and a half miles. We had been four hours upon it, now clambering up steeps, and now brushing through thickets, when he told us we were ascending the last elevation, and I kept close to his heels, soon outwent him on the trail, and got the first glimpse of the glittering nymph we had been pursuing. On reaching the summit this wish was gratified. At a depression of perhaps a hundred feet below, cradled among the hills, the lake spread out its elongated volume, presenting a scene of no common picturesqueness and rural

16

beauty. In a short time I stood on its border, the whole cortege of canoes and pedestrians following; and as each one came he deposited his burden on a little open plat, which constituted the terminus of the Indian trail. In a few moments a little fire threw up its blaze, and the pan of *pigieu*, or pine pitch, was heated to mend the seams of the bark canoes. When this was done, they were instantly put into the lake, with their appropriate baggage; and the little flotilla of five canoes was soon in motion, passing down one of the most tranquil and pure sheets of water of which it is possible to conceive. There was not a breath of wind. We often rested to behold the scene. It is not a lake overhung by rocks. Not a precipice is in sight, or a stone, save the pebbles and boulders of the drift era, which are scattered on the beach. The waterfowl, whom we disturbed in their seclusion, seemed rather loath to fly up. At one point we observed a deer, standing in the water, and stooping down, apparently to eat moss,

The diluvial hills inclosing the basin, at distances of one or two miles, are covered with pines. From these elevations the lands slope gently down to the water's edge, which is fringed with a mixed foliage of deciduous and evergreen species. After passing some few miles down its longest arm, we landed at an island, which appeared to be the only one in the lake. I immediately had my tent pitched, and while the cook exerted his skill to prepare a meal, scrutinized its shores for crustacea, while Dr. Houghton sought to identify its plants. While here, the latter recognized the mycrostylis ophioglossoides, physalis lanceolata, silene antirrhina, and viola pedata. We found the elm, lynn, soft maple, and wild cherry, mingled with the fir species.

An arm of the lake stretches immediately south from this island, which receives a small brook. Lieutenant Allen, who estimates the greatest length of the lake at seven miles, drew the following sketch of its configuration. (See p. 243.)

The latitude of this lake is 47° 13' 35".* The highest grounds passed over by us, in our transit from the Assowa Lake, lie at an elevation of 1,695 feet. The view given of the scene in the first

---

* By the report of Governor Stevens (June, 1854), the selected pass for the contemplated railroad through the St. Mary to the Columbia valley is in 47° 30', where there is but little snow at any time, and rich pasturage for cattle. The phenomena of the climates of our northern latitudes are but little understood.

volume of my *Ethnological Researches*, p. 146, is taken from a point north of the island, looking into the vista of the south arm of the

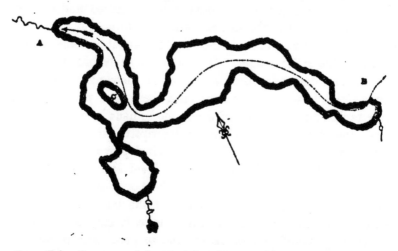

Itasca Lake, the source of the Mississippi River, 3,160 miles from the Balize.
A. Mississippi River.   B. Route of expedition to the Lake.   C. Schoolcraft's Island.

lake.   I inquired of Ozawindib the Indian name of this lake; he replied *Omushkös*, which is the Chippewa name of the Elk.[*] Having previously got an inkling of some of their mythological and necromantic notions of the origin and mutations of the country, which permitted the use of a female name for it, I denominated it ITASCA.[†]

[*] The Canadian French call this animal *la Biche*, from *Biche*, a hind.
[†] This myth is further alluded to, in the following stanzas from the *Literary World*, No. 387:—

STANZAS.

ON REACHING THE SOURCE OF THE MISSISSIPPI RIVER IN 1832.[‡]

L.

Ha! truant of western waters!   Thou who hast
So long concealed thy very sources—flitting shy,
Now here, now there—through spreading mazes vast
Thou art, at length, discovered to the eye
In crystal springs, that run, like silver thread,
From out their sandy heights, and glittering lie

[‡] Narrative of an Expedition to Itasca Lake. Harpers. 1834.   1 vol. 8vo. p. 307.

The line of discovery of the Mississippi, explored above Cass
Lake, taking the east fork from the primary junction, as shown
by Mr. Allen's topographical notes, is one hundred and twenty-
three miles.*   This is the shortest and most direct branch.   The
line by the Itascan or main branch of it is, probably, some twenty
or twenty-five miles longer.   It is evident, as before intimated,
that the river descends from its summit in plateaux.   From the
pseudo-alpine level of the parent lake, there is a principal and
minor rapids, for the former of which the Indians have the appropriate name of *Kakabikons*, which is a descriptive term for a cascade
over rocks or stones.   Then the river again deploys itself in a lake
and a series of minor lakes on the same level, and this process is
repeated, until it finally plunges over the horizontal rocks at St.
Anthony's Falls, and displays itself, for the last time, in Lake
Pepin.   Commencing with the latter lake, it may be observed
for the purposes of generalization, and to give definite notions
rather of its hydrography than geology, that there are nine plateaux, of which Governor Cass, in 1820, explored six.   The other
three, beginning at his terminal point, have now been indicated.
The heights of these are given, barometrically.   The distances
travelled are given from time.   The annexed diagram of these
plateaux, extending to the Pakagama summit, will impress these
deductions on the eye.

Within a beauteous basin, fair outspread
  Hesperian woodlands of the western sky,
As if, in Indian myths, a truth there could be read,
And these were tears, indeed, by fair Itasca shed.

II.

To bear the sword, on prancing steed arrayed;
  To lift the voice admiring Senates own;
To tune the lyre, enraptured muses played;
  Or pierce the starry heavens—the blue unknown—
These were the aims of many sons of fame,
  Who shook the world with glory's golden song.
I sought a moral meed of less acclaim,
  In treading lands remote, and mazes long;
And while around aerial voices ring,
  I quaff the limpid cup at Mississippi's spring.   H. R. S.

---

* Mr. Nicollet, who ascended the same fork in 1836, makes the distance twelve
miles more. *Vide* Ex., Doc. No. 237.

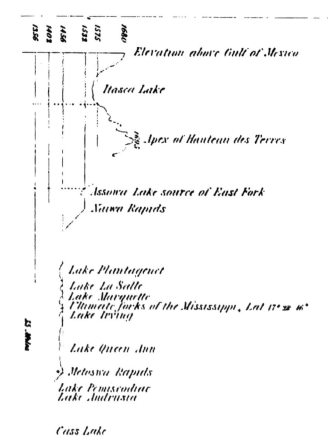

Elevation above Gulf of Mexico

Itasca Lake

Apex of Hauteau des Terres

Assowa Lake source of East Fork
Nuwa Rapids

Lake Plantagenet
Lake La Salle
Lake Marquette
Ultimate forks of the Mississippi, Lat 47° 28′ 16″
Lake Irving

Lake Queen Ann

Metoswa Rapids
Lake Pemisconduc
Lake Andrusia

Cass Lake

Lake Winnepek

Little Winnepek Lake

Leech Lake

LACUSTRINE PLATEAUX OF THE SOURCE OF THE MISSISSIPPI RIVER.

The length of the Mississippi, from the Gulf of Mexico, pursuing its involutions, may be stated to be three thousand miles. By estimates from the best sources made, respectively, during the expeditions of 1820 and 1832, it is shown to have a winding thread of three thousand one hundred and sixty miles. Taking the barometrical height of Itasca Lake at fifteen hundred and seventy-five feet, it has a mean descent of a fraction over six inches per mile. As one of the most striking epochs in American geography, we have known this river, computing from the era of Marquette's discovery to the present day (July 18, 1832), but one hundred and fifty-nine years—a short period, indeed! How rich a portion of the geology of the globe lies buried in the flora and fauna of the tertiary, the middle or secondary, and the palæozoic eras of its valley, we have hardly begun to inquire. It will, *doubtless*, and, so far as we know, *does*, contribute evidences to the antiquity and mutations of the earth's surface, conformably to the progress of discoveries in other parts of the globe. The immense basins of coal, found in the middle and lower parts of its valley, prove the same gigantic epoch of its flora which has been established for the coal measures of Europe,* and sweep to the winds the jejune theory that the continent arose from a chaotic state, at a period a whit less remote than the other quarters of the globe. While the large bones of its later eras, found imbedded in its unconsolidated strata, prove how large a portion of its fauna were involved in the gigantic and monster-period.

* Entire trees are often found imbedded in its rocks of the middle era, as is evidenced by an individual of the juglans nigra, of at least fifty feet long, in the River De Plaine, valley of the Illinois. *Vide* Appendix.

CHAPTER XXIV.

Descent of the west, or Itascan branch—Kakabikons Falls—Junction of the Chemaun, Paniddiwin, or De Soto, and Allenoga Rivers—Return to Cass Lake.

ITASCA LAKE lies in latitude twenty-five seconds only south of Leech Lake, and five minutes and eleven seconds west of the ultimate northerly point of the Mississippi, on the Queen Anne summit; it is a fraction over twelve minutes southwest of Cass Lake. The distance from the latter point, at which discovery rested in 1820, is, agreeably to the observations of Lieutenant Allen, one hundred and sixty-four miles.

On scrutinizing the shores of the island, on which I had encamped, innumerable helices, and other small univalves, were found; among these I observed a new species, which Mr. Cooper has described as planorbis companulatus.* There were bones of certain species of fish, as well as the bucklers of one or two kinds of tortoise, scattered around the sites of old Indian camp fires, denoting so many points of its natural history. Amidst the forest-trees before named, the betula papyracea and spruce were observed. Directing one of the latter to be cut down, and prepared as a flagstaff, I caused the United States flag to be hoisted on it. This symbol was left flying at our departure. Ozawindib, who at once comprehended the meaning of this ceremony, with his companions fired a salute as it reached its elevation.

Having made the necessary examinations, I directed my tent to be struck, and the canoes put into the water, and immediately embarked. The outlet lies north of the island. Before reaching it, we had lost sight of the flagstaff, owing to the curvature of the shore. Unexpectedly, the outlet proved quite a brisk brook, with a mean width of ten feet, and one foot in depth. The water is as

* Appendix.

clear as crystal, and we at once found ourselves gliding along, over a sandy and pebbly bottom, strewed with the scattered valves of shells, at a brisk rate. Its banks are overhung with limbs and foliage, which sometimes reach across. The bends are short, and have accumulations of flood-wood, so that, from both causes, the use of the axe is often necessary to clear a passage. There was also danger of running against boulders of black rock, lying in the margin, or piled up in the channel. As the rapid waters increased, we were hurled, as it were, along through the narrow passages, and should have descended at a prodigiously rapid rate, had it not been for these embarrassments to the navigation. Its course was northwest. After descending about ten miles, the river enters a narrow savanna, where the channel is wider and deeper, but equally circuitous. This reaches some seven or eight miles. It then breaks its way through a pine ridge, where the channel is again very much confined and rapid, the velocity of the stream threatening every moment to dash the canoe into a thousand pieces. The men were sometimes in the water, to guide the canoe, or stood ever ready, with poles, to fend off. After descending some twenty-five miles, we encamped on a high sandy bluff on the left hand.

The next morning (14th), we were again in our canoes before five o'clock. The severe rapids continued, and were rendered more dangerous by limbs of trees which stretched over the stream, threatening to sweep off everything that was movable. We had been one hour passing down a perfect defile of rapids, when we approached the Kakabikons Falls. *Kakábik,*[*] in the Chippewa, means a cascade, or shoot of water over rocks. *Ons* is merely the diminutive, to which all the nouns of this language are subject. How formidable this little cataract might be, we could not tell. It appeared to be a swift rush of water, bolting through a narrow gorge, without a perpendicular drop, and Ozawindib said it required a portage. Halting at its head, for Lieut. Allen to come up, his bowsman caught hold of my canoe, to check his velocity. It had that effect. But, being checked sud-

---

[*] Kakábik. *Abik* is a rock. The prefixed syllable, *Kak*, may be derived from *Kukidjewum*, a rapid stream. *Ka* is often a prefix of negation in compound words, which has the force of a derogative.

denly, the stern of his canoe swung across the stream, which per·
mitted the steersman to catch hold of a branch. Thus stretched
tensely across the rapid stream, in an instant the water swept over
its gunwale, and its contents were plunged into the swift current.
The water was about four feet deep. Allen and his men found
footing, with much ado, but his canoe—compass, apparatus, and
everything, was lost and swept over the falls. He grasped his
manuscript notes, and, by feeling with his feet, fetched up his
fowling-piece; the men clutched about, and managed to save the
canoe. Fortunately, I had a fine instrument to replace the lost
compass, though wanting the nautical rig of the other.

We made a short portage. Two of the canoes, with Indian
pilots, went down the rapids, but injured their canoes so much as
to cause a longer delay than if they had carried them by land.
Below this fall, the river receives a tributary on the right hand,
called the *Chemaun*, or Ocano. It contributes to double its
volume, very nearly, and hence its savanna borders are enlarged.
Conspicuous among the shrubbery on its shores are the wilding
rose and clumps of the salix. The channel winds through these
savanna borders capriciously. At a point where we landed for
breakfast, on an open pine bank on the left shore, we observed
several copious and clear springs pouring into the river. Indeed,
the extensive sand ranges which traverse the woodlands of the
Itasca plateau are perfectly charged with the moisture which is
condensed on these elevations, which flows in through a thousand
little rills. On these sandy heights the conifera predominate.

The physical character of the stream made this part of our
route a most rapid one. Willing or unwilling, we were hurried
on; but, indeed, we had every desire to hasten the descent. At
four o'clock P. M., we came to the junction of the Piniddiwin,* or
Carnage River, a considerable tributary on the left. On this
river, which originates in a lake, on the northeastern summit of
the Hauteur des Terres, I bestowed the name of De Soto. It has
also a lake, called Lac la Folle, at the point of its junction with
the Mississippi, whose borders are noted for the abundant and

---

* From the term *Iah-pininiddewin*, an emphatic expression for a place of carnage,
so called from a secret attack made at this place, in time past, by a party of Sioux,
who killed every member of a lodge of Chippewas, and then shockingly mangled
their bodies.

vigorous growth of wild rice, reeds, and rushes. It is called Monomina,* by the Chippewas. By this accession, the width and depth of the river are strikingly increased. The Indian reed first appears at this spot.

While passing through this part of the river, I observed a singular trait in the habits of the onzig duck, which, on being suddenly surprised by the traveller, affects for the moment to be disabled; flapping its wings on the water, as if it could not rise, in order to allow its brood, who are now (July) unfledged, to escape, when the mother instantly rises from the water, and wings her flight vigorously. We observed, sailing above the marshy areas of this fork, the falco furcatus, the feathers of which are much esteemed by the Indians, for this is considered a brave species, as its habit is to seize serpents by the neck, who twist themselves around its elongated body, while it flies off to some convenient perch to devour them. The deer is also noticed along the Itascan fork. Ozawindib landed a little below the junction of the Chemaun, to fire at one of them, which he discovered grazing at some distance; but, although he carefully landed and crept up crouchingly, he failed in his shot, either from the distance or some other cause. Immediately, he put a fresh charge of powder in his gun, and threw in a bullet, unwadded, and fired again before the animal had made many leaps, but it held its way.

We descended about eighteen miles below the Piniddiwin, and landed to encamp. The day's descent had been an arduous one. Lieut. Allen estimated it at seventy-five miles. We had now fairly followed the Mississippi out of what may be denoted its Alpine passes. All its dangerous rapids had been overcome. It was now a flowing stream of sixty feet wide. Immediately on landing, one of the Indians captured an animal of the saurian type, called *ocaut-e-kinabic*,† eight inches in length, striped blue, black, and white, with four legs of equal length. The colors were very vivid.

Having reached a part of the stream which could be safely navigated, I resolved to re-embark after supper, and continue the descent by night. We were now about fifteen miles above the

---

* From *Monominakauning*, place of wild rice.
† From *ocaut*, a leg, and *kinabic*, a snake.

primary forks. Lieut. Allen determined to remain till daylight, in order to trace the river down to the point at which it had been left in the ascent. Nothing of an untoward nature occurred. A river of some size enters, on the left hand, about six miles below the saurian encampment, which originates in a lake. This stream, for which I heard no name, I designated *Allenoga*, putting the Iroquois local terminal in *oga* to the name of the worthy officer who traced out the first true map of the actual sources of the Mississippi.* We passed the influx of the east fork, about half-past one A. M. on the 15th, traversed the Lake of Queen Anne, and descended the whole series of the Metoswa rapids, to Lake Andrusia, by the hour of daybreak, and reached the island of my primary encampment, in Cass Lake, at nine o'clock in the morning. We had been eleven hours and a half in our canoes, from the time of re-embarkation at the camp above Allenoga. Lieut. Allen did not rejoin us till six o'clock in the afternoon. He estimated the entire distance, *out* and *in*, at 290 miles, it being 125 miles to Itasca Lake, and, as before intimated, 165 miles from thence to Cass Lake. He estimates the length of the Mississippi, above the Falls of St. Anthony, at 1,029 miles. Taking the distance from the Gulf of Mexico to the Falls at 2,200 miles,† this would give to this stream a development of 8,229 miles, which exceeds my prior estimates more than fifty miles.

---

* Lieut.-Col. James Allen, U. S. A. This officer graduated at West Point in 1825. After passing through various grades, he was promoted to a captaincy of infantry in 1887. He was lieutenant-colonel and commandant of the battalion of Mormon volunteers in the Mexican war, which was raised by his exertions, and died at Fort Leavenworth, on the Missouri, on the 23d of August, 1846.

† Doc. No. 287.

CHAPTER XXV.

The expedition proceeds to strike the source of the great Crow-Wing River, by the Indian trail and line of interior portages, by way of Leech Lake, the seat of the warlike tribe of the Pillagers, or Mukundwa.

HAVING, while at Sandy Lake, summoned the Indians to meet me in council at the mouth of the *L'aile de Corbeau*, or Crow-Wing River, on the 20th of July, no time was to be lost in proceeding to that place. The 15th, being the Sabbath, was spent at the island, where the Rev. Mr. Boutwell addressed the Indians. The next day, I met the Cass Lake band in council, and, having finished that business, rewarded the Indians for their services and canoes on the trip to Itasca Lake, distributed the presents designed for them, replied to a message from Nezhopenais of Red Lake, and invested Ozawindib with the President's largest silver medal and a flag, and was ready by 10 o'clock A. M. to embark. Dr. Houghton employed the time to complete his vaccinations. I rewarded Mr. Default for taking charge of my camp during the journey to Itasca Lake. As well to shorten the line of travel as to visit an entirely unexplored section of the country, I resolved to pursue the Indian trail and line of interior portages from Cass to Leech Lake, and from the latter to the source of the great Crow-Wing fork.

Passing southwardly across the lake, between Red Cedar and Garden Islands, we have a prolonged bay running deep into the land, toward the south. This bay is in the direct line to Leech Lake; and as it had been crossed on the ice in January, 1806, by Lieutenant Pike, in his adventurous and meritorious journey of exploration, I called it Pike's Bay. It was twelve o'clock, meridian, when we debarked at its head. The portage commenced on the edge of an open pine forest, interspersed with scrub oak. The path is deeply worn, in the sand-plain, and looks as if it had been trod by the Indians for centuries. I observed, as we passed along, the alum root, hyacinthus, and sweet fern, with the ledum latifolium, vaccinium dumosum, and more common species of pine plains. The pinus resinosa assumes here a larger size, and the Indians pointed out to me markings and pictographs drawn with charcoal, and covered with the resin of the tree, which were made by the Indian tribe who preceded them in the occupancy of the sources of the Mississippi. This must have been, if I rightly apprehend their history, prior to A. D. 1600. That such markings should be preserved by the pitch, which sheds the rain, is, however, probable. They were of the totemic character, *i. e.* relating to the exploits or achievements of groups of families, in which the individual actor sinks his specific in the generic family or clan name. Antiquities of this character are certainly a new feature in Indian history. Letters have perfectly preserved the landing of Cartier at the mouth of the St. Lawrence in 1534. Pictography here records, that certain clans had killed bears and taken human scalps before that time. And the fact is deeply important in shedding light on Indian history and character; for the killing of deers and bears, and the taking of human scalps, is precisely what these tribes are doing at the present time. In the three hundred years' interval, they have made no mental progress. The Chippewa is just as fierce to-day, in hunting a Dacota scalp, as the Dacota is in hunting a Chippewa scalp. The conquering tribe has, however, pushed the Dacotas nearly one thousand miles down the Mississippi.

> " Talk of your Hannibals, Napoleons, and Alps,
> My glory," quoth the feathered hunter, " is in scalps."

After following the deeply indented path nine hundred and fifty yards, we reached a small lake which disclosed, as we passed it, patches of a dark, coarse, mossy-like substance at its bottom. On reaching down with their paddles, the men brought up a singular species of aquatic plant with coral-shaped branches. After crossing this lake, the pine plain resumed its former character. There was then a shallow bog of fifty or sixty yards. The rest of the path consists of an arid sand plain, which is sometimes brushy, but generally presents dry, easy travelling. We had walked four thousand one hundred yards, or about two and a half miles, when we reached an elongated body of clear living water, having its outflow into Leech Lake. Embarking on this, we crossed it, and entered a narrow stream, winding about in a shaking savanna, where it was found difficult to veer the large five-fathom canoes in which we now travelled. This tortuous stream was joined by a tributary from the right, and at no remote distance, entered an elongated duplicate body of water, named by the Indians *Kapuka Sagatawag*, or the Abrupt Discharges.[*] Below the junction of these lakes, which appear to be outbursts from the Hauteur de Terre range, the stream is a wide-flowing river. Its shores abound in sedge, reeds, and wild rice. The last glimpses of daylight left us as this broad river entered into Leech Lake. Moonlight still served us, as we began the traverse of this spreading sea, but it soon became overcast, and it was intensely dark before we reached the recurved point of land of the principal chief's village. It was now ten o'clock at night, and it was eleven before the military canoes, under Lieutenant Allen, came up. In the morning a salute was fired by the Indians, who welcomed us. Aishkebuggekozh,[†] or the Flat-mouth, the reigning chief, invited me to breakfast. As this chief exercises a kind of imperial sway over the adjacent country, it was important to respect him. Having sent a dish of hard bread before me, I took my interpreter and went to his residence. I found him living in a tenement built of logs, with two rooms, well floored and roofed, with two small glass windows. At one end of the breakfast-room were extended his flags, medals, and warlike paraphernalia. In the centre of the floor, a large mat of rushes, or Indian-woven *apukwa* was spread, and upon this the breakfast and breakfast things were arranged in an orderly manner. There were teacups, teaspoons, plates, knives and forks, all of plain English manufacture. A salt-cellar contained salt and pepper mixed in unequal proportions. There were just as many plates as expected guests. A large white fish, boiled, and cut up in good taste, occupied a dish in the centre. There was a dish of sugar made from the acer saccharinum. There were no stools, or chairs, but

* From the word *puka*, abrupt phenomenon, and the verb *saugi*, outflowing.

† From *Aishenagoss*, countenance, and *kozh*, a bill of a bird, or snout of an animal. The word is appropriately translated *guelle* by the Canadians.

small apukwa mats were spread for each guest. I observed the
dish of hard bread, which came opportunely, as there was no
other representative form of bread. The chief sat down at the head
of his breakfast, in the oriental fashion. Imitating his example,
I sat down with a degree of repose and nonchalance, as if this
had been the position I had practised from childhood. His empress—Equa,* sat on one side, near him, to pour out the tea, but
neither ate nor drank anything herself. Her position was also
that of the oriental custom for females; that is, both feet were
thrown to one side, and doubled beside her.† The chief helped
us to fish and to tea, taking the cups from his wife. He was dignified, grave, yet easy, and conversed freely, and the meal passed
off agreeably and without a pause, or the slightest embarrassment. This was, perhaps, owing in part to my having been acquainted with him before, he having visited me at my agency at
Sault Ste. Marie in 1828, and sat as a guest at my own table.
Nor, in a people so loath to give their confidence as the Indian, is
the fact undeserving of mention, of general affiliation to the tribe,
caused by my marriage with a granddaughter of the ruling chief
of Lake Superior, a lady of refinement and intelligence, who was
the child of a gentleman of Antrim, Ireland, where she was edu-
cated.

On rising to leave, I invited him to a council, at my tent,
which was ordered to assemble at the firing of the military. It
is not unimportant to observe, that, in preparing to set out on this
expedition into the Indian country, at a time when the Blackhawk had raised the standard of revolt on Rock River, and the
tribes of the Upper Mississippi were believed to be extensively
in his views, I had caused my canoe, after it had been finished in
most perfect style of art known to this kind of vessel, to be
painted with Chinese vermilion, from stem to stern. Ten years'
residence among the tribes, in an official capacity, had convinced
me that fear is the controlling principle of the Indian mind, and
that the persuasions to a life of peace, are most effectively made
under the symbols of war. To beg, to solicit, to creep and cringe

---

* *Equa*, a female; it is not, appropriately, the term of wife, for which the vocabulary has a peculiar term, but is generally employed in the sense of woman.

† I have observed this to be the universal custom among all the aboriginal females of America. They never part the feet.

to this race, whether in public or private, is a delusive, if not a fatal course; and though I was told by one or two of my neighbors that it was not well, on this occasion, to put my canoe in the symbolic garb of war, I did not think so. I carried, indeed, emphatically, messages of peace from the executive head of the Government, and had the means of insuring respect for these messages, by displaying the symbol of authority at the stern of each vessel, by an escort of soldiery, and by presents, and the services of a physician to arrest one of the most fatal of diseases which have ever afflicted the Indian race. But I carried them fearlessly and openly, with the avowed purpose of peace. The canoe, itself, was an emblem of this authority, and, like the *oriflamme* of the Mediæval Ages, cast an auspicious influence on my mission - over these bleak and wide summits, lakes, and forests, inhabited alone by fierce and predatory tribes, who acknowledged no power but force. Long before I had reached the sources of the Mississippi, St. Vrain, my fellow agent, had been most cruelly murdered at his agency, and General Scott, with the whole disposable army of the United States, had taken the field at Chicago.

Lieut. Allen paraded his men that morning with burnished arms. We could not, jointly, in an emergency, muster over forty men, of whom a part were not reliable in a mêlée, but arranged our camp in the best manner to produce effect. Effect, indeed, it required, when the hour of the council came. Not less than one thousand souls, men, women, and children, surrounded my tent, including a special deputation from the American borders of Rainy Lake. Of these, two hundred were active young warriors, who strode by with a bold and lofty air, and glistening eyes, often lifting the wings of my tent, to scan the preparations going forward. Aishkebuggekozh entered the council area, having in his train Majegabowi, the man who had led the revolt in the Red River settlement of Lord Selkirk, and who had tomahawked Gov. Semple, after he fell wounded from his horse. This association did not smack of peaceful designs. The chief, Aishkebuggekozh, himself, has the countenance of a very ogre. He is over six feet high, very brawny, and stout. That feature of his countenance from which he is named Flat-mouth, consisting of a broad expansion and protrusion of the front jaws, between the long incision of the mouth, reminds one much of a bull-dog's jaw. He held in his hand, suspended by ribbons, five silver medals, smeared with vermilion, to symbolize blood.

A person not familiar with Indian symbols, might deem such signs alarming. I knew him to be very fond of using these symbols, and, indeed, a man who never made a speech without them; and I had the fullest confidence that, while he aimed to produce the fullest effect upon his listening, but less shrewd tribe of folks, and upon all, indeed, he never dreamed of an act which should bring him into conflict with the United States. Like Blackhawk, who was now exciting and leading the tribes at lower points to war, he had, from his youth, been in the British interests. He displayed a British flag at his breakfast, and three of his medals were of British coinage, but he was a man of far more comprehensive mind and understanding than Blackhawk.

Having been, as a government agent, the medium of the agreement of the Chippewas and Sioux in fixing on a boundary line for their respective territories at the treaty of Prairie-du-Chien, in 1825, I made that agreement, on the present occasion, the basis of my remarks, for their preserving in good faith the stipulations of that treaty, and of renewing the principles of it in the points where they had since been broken and violated. I concluded by assuring them of the friendship of the United States, of which my visit to this remote region must be deemed proof, and of the sincerity with which I had communicated the words of the President. The presents were then delivered and distributed.

Aishkebuggekozh, or the Guelle Plat, replied, with much of the skill and force of Indian oratory. He began by calling the attention of the warriors to his words; he then turned to me, thanking me for the presents. He said that he had been present when Pike visited this lake in 1806. He pointed with his fingers across the lake, to the Ottertail Point, where the old trading-house of the British Northwest Company had stood. "You have come," he continued, "to remind us that the American flag is now flying over the country, and to offer us counsels of peace. I thank you. I have heard that voice before, but it was like a rushing wind. It was strong, but soon went. It did not remain long enough to choke up the path. At the treaty of Prairie-du-Chien, it had been promised that whoever crossed the lines, the long arms of the President should pull them back; but, that very year, the Sioux attacked us, and they have killed my people almost every year since. I was myself present when they fired on a peaceful delegation, and killed four Chippewas under the walls of Fort Snelling. My own son—my *only* son—has been killed. He was basely killed, without an opportunity to defend himself." A subordinate here handed him, at his request, a bundle of small sticks. "This," handing them to me, "is the number of Leech Lake Chippewas killed by the Sioux since the treaty of Prairie-du-Chien." There were forty-three sticks.

He then lifted up a string of silver medals, smeared with vermilion. "Take notice, they are bloody. I wish you to wipe the blood off. I cannot do it. I find myself in a war with this people, and I believe it has been intended by the Creator that we should be at war with them. My warriors are brave [looking significantly at them]; it is to them that I owe success. But I have looked for help where I did not find it."*

* It is hoped, hereafter, to give further sketches of this interview, and of this chief's life and character.

17

CHAPTER XXVI.

LEECH LAKE is a large, deep, and very irregularly-shaped body
of water. It cannot be less than twenty miles across its extreme
points. I requested the chief to draw its outlines, furnishing a
sheet of foolscap. He began by tracing a large ellipsis, and then
projecting large points and bays, inwardly and outwardly, with
seven or eight islands, and that peculiar feature, the Kapuka
Sagotawa, which I apprehend to originate in gigantic springs.
The following eccentric figure of the lake is the result.

This lake has been the seat of the Mukundwa, or Pillagers,
from early days. The date of their occupancy is unknown. The
French found them here early in the seventeenth century, when
they began to push the fur trade from Montreal. They were the
advance of the Algonquin group, who, when they had reached
the head of Lake Superior, proceeded still towards the west and
northwest. Two separate bodies assumed the advance in this
migratory movement, one of which went from the north shore,
at the old Grand Portage, north-northwest, by the way of the
Rainy Lakes, and the other went northwest from Fond du Lac.
The former soon earned for themselves the title of Killers, or
Kenistenos,* and speak a distinct dialect; the other, whose
language continued to be, with little variations, good Odjibwa,
acquired in a short time the name of Takers, or Mukundwa. The
Kenistenos advanced, through the Great Lake Winnepeck, and
up its inflowing waters, to the Portage du Trait, of the great
Churchill or Missi-nepi (much water) River, where they sent up

* Called by the French *Cress.*

a skinned frog, in derision of the feebler Athapasca race, whom they here encountered. *Mackenzie's Voyages*, p. lxxiii. *Hist. Fur*

Leech Lake.—*a*, Rush Bay; *b*, Leech Lake River; *c*, Three Points; *d*, Boy's River; *e*, Bear Island; *f*, Pelican Island; *g*, Two Points; *h*, Ottertail Point; *i*, Chippewa Village; *j*, Sugar Point; *k*, Carp River; *l*, Old N. W. House; *m*, Goose Island; *n*, Encampment, July 16; *o*, Trading House Am. P. Co.; *p*, Flatmouth's House; *q*, Chippewa Village; *r*, Encampment, July 17; *s, s*, Route to Crow-Wing River; *t*, Sandy Point; *u*, Big Point; *v*, Sandy Bay River.

Trade.   The Odjibwas were led from Chegoimegon, in Lake Superior, by two noted chiefs, called Nokay and Bainswah, under whom they drove the Sioux from the region of Sandy Lake and the source of the Mississippi. (*Ethnological Researches*, vol. ii. p. 135.)

Another party of this Algonquin force, which conquered the country lying round the sources of the Mississippi, proceeded through the Turtle River to Red Lake, and thence descended into the valley of the Red River of Hudson's Bay, where their descendants still reside. Large portions of these mingled with the Canadian stock, forming that remarkable people called Boisbrules. These advanced parties pressed into the buffalo plains, along the Rivers Assinabwoin and Saskatchawine, which is the ultimate western area of the spread of the Algonquin language. And to this migration the Blackfeet are believed to be indebted for the intermixture of this language which exists, and which Mr. Gallatin has erroneously supposed to arise from original elements, in the Blackfeet tongue.

This lake yields in abundance the corregonus albus, a fish which is unknown to the Mississippi, and which delights only, it appears, in very limpid and cold waters.

I found the population living at this lake to be eight hundred and thirty-two souls, under three chiefs, the Guelle Plat, Nesia, or the Elder Brother, and Chianoquet, or the Big Cloud, the latter of whom is exclusively a war chief. Having dined these chiefs at my tent, and finished my business, and the vaccinations and very numerous cases of odontalgia being got through with, I directed my canoes to be put in the water, with the view of going a few miles down the shore, in order to get a quiet night's encampment, and be ready for an early start on the morrow. It was near the hour of sunset before we could embark. Aiskebuggekozh came down to the boat to take leave of me. He was dressed, on this occasion (having been in Indian costume all the morning), in a blue military frock coat, with scarlet collar and cuffs, white underclothes, a ruffled shirt, shoes and stockings, and a citizen's hat. He was accompanied by Nesia and other followers, and it appeared to me if there ever was a person who had popular and undisputable claims to imperial sway, notwithstanding this poor taste in costume, it was he.

We went about five miles in the general direction towards

the source of the L'ail de Corbeau, and encamped. Dr. Houghton, who had been left behind with Lieut. Allen, to complete the vaccinations, rejoined me about seven o'clock. Guelle Plat had promised to send me guides, to cross the country to the Crow-Wing River, early the next morning (18th), but, as they did not arrive, I proceeded across the arm of the lake for the main shore without them. After reaching it, some time was spent in searching for the commencement of the portage path. It was found to lie across a dry pine plain. The Canadians, who are quick on finding the trail of a portage, wanted nothing more, but pushed on, canoes and baggage, without any further trouble about the Indian guides. A portage of 1,078 yards brought us to the banks of a small, clear, shallow lake, called Warpool, which had a very narrow, tortuous outlet, through which the men, with great difficulty, and by cutting away acute turns of the bank with their paddles, made way to force the canoes into Little Long Lake, which we were twenty-four minutes in crossing. The outlet from this lake expanded, at successive intervals, into three pond-like lakes, redolent with the nymphæ valerata; the series terminating in a fourth lake, lying at the foot of elevated lands, which was called the Lake of the Mountain. At the head of the latter, we debarked on a shaking bog. At this spot commences the portage *Plé*, which lies over a woodless and bleak hill. It is short and abrupt, and terminates on the banks of a deep bowl-shaped lake, where we took breakfast at twelve o'clock. We were now at the foot of elevated lands. Here began the mountain portage, so called. Its extent is, first, nine hundred and ten yards, terminating on the shores of a little lake, without outlet, called the Lake of the Isle. There is then a portage of 1,960 yards to another mountain lake, without outlet. We were now near the apex of the summit between Leech Lake and the source of De Corbeau. Another portage of one onwaybee or about a thousand yards, partly through a morass, carried us quite across this summit, and brought us out on elevated and highly beautiful grounds overlooking the Kaginogumaug, or Longwater Lake, which is the source of the Crow-Wing River. Here we encamped (18th).

There is no rock stratum seen in place, on the De Corbeau summit. Its surface is purely composed of geological drift and boulders. The journey had been a very hard and fatiguing one for the men, who were on the push and trot all day, embarking and debarking continually on lakes, or scrambling, with their burdens and canoes, over elevations or through morasses. It was particularly severe on the soldiers, who are ill-prepared for this kind of toil.

The chief Guelle Plat, with some companions of the Mukandwa band, had overtaken us, at the Lake of the Isle, and came and encamped beside us. I invited him to sup with us, and the evening was passed in conversing with him on various topics. I found him a man of understanding and comprehensive views, who was well acquainted with the history of his people. It was twelve o'clock before these conversations ended, when he got up to go to his camp fire. With him there sat Majegabowee,* a tall, gaunt, and savage-looking man of Red River, who scarcely uttered a word, but sat a silent listener to the superior powers of conversation and reflection of his chief. But I could not look at this person without a sense of horror, when I reflected that in him I beheld the murderer of Gov. Semple, of the Hudson's Bay Territory, a circumstance which I have previously adverted to, while at Leech Lake.†

Bidding adieu to the Leech Lake chief the next morning at sunrise (4 h. 45 m.), after giving him a lancet, with directions to vaccinate any of his people who had been overlooked, I embarked on the Kaginogamaug. This is a beautiful lake, with sylvan shores and crystal water, some four or five miles long. We were just forty minutes, with full paddles, in passing it. The outlet is narrow, and overhung with alders. The width is not over six feet, with good depth, but the turns are so sudden, and the stream so thickly overhung with foliage, that the use of the axe and the paddle as an excavator were often necessary. It then expands into a lake, called Little Vermilion, which is fringed with a growth of birch and aspen, with pines in the distance. Its outlet is fully doubled in width, and we had henceforth no more embarrassment in descending. This outlet is pursued about eight

---

* The Fore-standing man. From the verb *maja*, to go, *ninabow*, I stand, and *izzea*, a person or man.

† For an account of this transaction, *vide* Reports of the Disputes between the Earl of Selkirk and the Northwest Company, at the assizes held at York, Upper Canada, Oct. 1818. 1 vol. 8vo. pp. 664. Montreal, Casie & Mower, 1819.

miles. I noticed the tamarack on its banks, and the nymphæ odorata, scirpus lacustris, and Indian reed on the margin. It expands into Birch Lake, a clear sheet, about one mile long, with pebbly bottom, interspersed with boulders. A short outlet, in which we passed a broken fish-dam, connects it with Lac Plè. This lake is about three and a half miles long, exhibiting a portion of prairie on its shores, interspersed with small pines. From it, there is a portage to Ottertail Lake, the eastern source of Red River. This is the common war road of the Mukundwa against the Sioux.

On coming out of Lac Plè, freshwater shells began to show themselves, chiefly species of naiades, a feature in the natural history of this stream which is afterwards common; but I observed none of much size, and they are often greatly decorticated. Four or five miles lower, we entered Assowa Lake, and about a mile and a half further, Lac Vieux Desert, or Old Gardon Lake, so called from the remains of a trading station, where we halted for breakfast. On resuming the descent, just twenty minutes were required, with vigorous strokes of the paddle, to pass it. It has an outlet about two miles long, when the stream again expands into a lake of considerable size, which we called Summit Lake. Thus far, we had been passing on a geological plateau of the diluvial character, extending southwest. But from this point the course of the river veers, at first towards the east and northeast, and, after a wide circuit, to the southeast, and eventually again to the southwest. From this point, rapids begin to mark its channel. The river, consequently, assumes a velocity which, while it hurries the traveller on, increases his danger of running his frail bark against rocks or shoals. We had been driven down this accelerated channel two hours and fifteen minutes, when it expanded into a sheet called Long Rice Lake. This is some three miles in length, and, at a very short distance below it, the river again expands into a considerable lake, which, from the circumstance of Lieut. Allen having circumnavigated it, I called Allen's Lake. He found it the recipient of a small river from the north. It is, apparently, the largest of this series of river lakes below the Kaginogumaug. While crossing it, we experienced a very severe and sudden tempest of wind and rain, accompanied by most severe and appalling peals of thunder and vivid lightning. Broad ribbons of fire, in acute angles, appeared to rend the skies. Before the shore could be reached, the tempest had subsided, so sudden was its development. A short distance below this, the river makes its tenth evolution, in the shape of a lake, on which, as my Indian maps gave no name, I bestowed the name of *Illigan.**

* From *ininᵉⁱg*, men, and *sagiegan*, lake, signs of a war party having been discovered at this place. In this derivative, the usual transition of *n* to *l* of the old Algonquin is made.

CHAPTER XXVII.

Complete the exploration of the Crow-Wing River of Minnesota—Indian council—Reach St. Anthony's Falls—Council with the Sioux—Ascent and exploration of the River St. Croix and Misakoda, or Broulé, of Lake Superior—Return of the party to St. Mary's Falls, Michigan.

AT Illigan Lake, large oaks and elms appear in the forest; its banks are handsomely elevated, and the whole country puts on the appearance of being well adapted to cultivation. We landed to obtain a shot at some deer, which stood temptingly in sight, and were impressed with the sylvan aspect of the country. While in the act of passing out of the lake in our canoes, a small fire was observed on shore, with the usual signs of its having been abandoned in haste by Indians, who had been lying in ambush. Every appearance seemed to justify such a conclusion, and it was evident a party of Sioux had been concealed waiting the descent of Chippewas, but, on observing our flag, and the public character of the party, they hastily withdrew. Our men, knowing the perfidious and cruel character of this tribe, were evidently a good deal alarmed at these signs. We had been one hour in our canoes, descending the river with the double force of current and paddles, when the river was found again expanded, and for the eleventh and last time, in a lake, which the natives call *Kaitchebo Sagatowa*, meaning the lake through one end of which the river passes. As this is not a term, however graphic, which will pass into popular use, I named it Lake Douglas, in allusion to a former companion in explorations in the northwest.[*] Ten miles below this lake, the river receives its first considerable tributary in Shell River, the Aisisepi of the Chippewas, which flows in from the right, from the slope of the Hauteurs des Terres, near the Ottertail Lake. Below this tributary, the Crow-Wing is nearly doubled in width,

---

[*] Professor D. B. Douglas.

and there is no further fear of shallow water. We held on our way for a distance of fourteen miles below the point of junction, and encamped on the right hand bank at eight o'clock P.M. It had rained copiously during the afternoon, and everything in the shape of kindling stuff had become so completely saturated with moisture, that it was quite an enterprise in the men to light a camp-fire. Lieut. Allen did not reach our encampment this night, having been misled in Allen's Lake, and, being driven ashore by the tempest, he encamped in that quarter. Presuming him to be in advance, I had pushed on, to a late hour, and encamped under this impression.

The next morning (20th), we set off from our camp betimes, and, having now a full flowing river, made good speed. The river passes for a dozen or more miles through a willowy low tract, on issuing from which there begins a series of strong rapids. Twenty-four of these rapids were counted, which were called the Metunna Rapids. Lieut. Allen estimates that they occupy thirty miles of the channel of the river. Below these rapids, the river extends to a mean width of three hundred feet. At this locality we were overtaken by Mr. Allen, at about two o'clock in the afternoon, and were thus first apprised of the fact that he had been all the while in our rear instead of in front.

Twenty miles below the Metunna Rapids, Leaf River flows in from the right, by a mouth of forty yards wide. This stream originates in Leaf Lake, and is navigable sixty miles in the largest craft used by the traders.* The volume of the Crow-Wing River is constantly increased in width and velocity by these accessions, which enabled us fearlessly to make a large day's journey. We encamped together after sunset, on an elevated pine bank, having descended ninety miles.

The 21st, we were early in motion, the river presenting a broad rushing mass of waters, every way resembling the Mississippi itself. On reaching within twenty miles of its mouth, we passed, on the right bank, the mouth of the Long Prairie River,† a prime tributary flowing from the great Ottertail slope, which has

---

* The angle of country above Leaf River, on the Crow-Wing, has been proposed as a refuge for the Menomonee tribe, of Wisconsin, for whom temporary arrangements, at least, are now made, on the head of Fox River, of that State.

† This river has been assigned as the residence of the Winnebago Indians. It is the present seat of the United States agency, and of the farming and mechanical establishment for that tribe.

been, time out of mind, the war road between the Chippewas and Sioux; and between this point and the confluence coming in we passed, on the left bank, the confluence of the Kioshk, or Gull River, through which there is a communication, by a series of portages, with Leech Lake.*

From head to foot, we had now passed through the valley of the De Corbeau River, without finding in it the permanent location of a single Indian. We had not, in fact, seen even a temporary wigwam upon its banks. The whole river lies, in fact, on the war road between the two large rival tribes of the Chippewas and Sioux. It is entered by war parties from either side, decked out in war-paints and feathers, who descend either of its tributaries, the Leaf and Long Prairie Rivers. The Mukundwa descends the main channel from the Kaginogumaug Lake in canoes. On reaching the field of ambush, these canoes are abandoned, and the parties, after an encounter, haste home on foot.

From this deserted and uninhabited state of the valley we were the more surprised, as noon drew on, to descry an Indian canoe ascending the river. It proved to be spies on the look-out, from the body of Chippewas encamped at the mouth of the river, agreeably to my invitation at Sandy Lake. After mutual recognitions, and learning that we were near the mouth of the river, we resumed our descent with renewed spirit, and soon reached its outflow into the Mississippi, and crossed it to the point at which the Indians had established their camp. We were received with yells of welcome. It occupied an eminence on the east bank of the Mississippi, directly opposite to the mouth of the De Corbeau.† The site was marked by a flag hoisted on a tall

---

* Mr. J. J. Nicolet pursued this route in 1836, on his visit to the sources of the Mississippi. *Vide* Senate Doc. No. 237. Washington, D. C., 1843.

† CROW-WING RIVER.—This stream is the largest tributary of the Mississippi above the falls of St. Anthony. It enters the Mississippi in lat. 46° 15′ 50″, 180 miles above the latter, and 145 miles below Sandy Lake. Government first explored it, in 1832, from its source in Lake Kaginogumaug to its mouth, and an accurate map of its channel, and its eleven lakes, was made by Lieut. Allen, U. S. A., who accompanied the party as topographer. It is 210 miles in length, to its source in Long Lake. The island, in its mouth, is about three miles long, and covered with hard-wood timber. The whole region is noted for its pine timber; the lands lie in gentle ridges, with much open country; a large part of it is adapted to agriculture, and there is much hydraulic power. It is navigable at the lowest stages of water, about 80 miles, and by small boats to its very source.

staff. The Indians fired a salute as we landed, and pressed down to the shore, with their chiefs, to greet us. They informed me that by their count of sticks, of the time appointed by me at Sandy Lake, to meet them at this spot, would be out this day, and I had the satisfaction of being told, within a short time of my arrival, that the canoe, with goods and supplies, from Sandy Lake, was in sight. The Indians were found encamped a short distance above the entrance of the Nokasippi* River, which is in the line of communication with the Mille Lac and Rum River Indians. I found the latter, together with the whole Sandy Lake Band, encamped here, awaiting my arrival. They numbered 280 souls, of whom 60 were warriors.

A council was immediately summoned, to meet in front of my tent, at the appointed signal of the firing of the military; the business of my mission was at once explained, the presents distributed, and the vaccinations commenced. Replies were made at length, by the eldest chief, Gros Guelle, or Big Snout; by Soangekumig, or the Strong Echoing Ground; by Wabogeeg, or the White Fisher; and by Nitumegaubowee, or the First Standing Man. The business having been satisfactorily concluded, the vaccination finished, and having still a couple of hours of daylight, I embarked and went down the Mississippi some ten or fifteen miles, to a Mr. Baker's trading-house at Prairie Piercie.

At this place, I remained encamped, it being the Sabbath day, and rested on the 22d, which had a good effect on the whole party, engaged as it had been, night and day, in pushing its way to accomplish certain results, and it prepared them to spring to their paddles the more cheerfully on Monday morning. Indeed, it had been part of my plan of travel, from the outset, to give the men this rest and opportunity to recruit every seventh day, and I always found that they did more work in the long run, from it. I had also engaged them, originally, not to drink any ardent spirits, promising them, however, that their board and pot should be well supplied at all times. And, indeed, although I had frequently travelled with Canadian canoemen, I never knew a crew who worked so cheerfully, and travelled so far, per diem, on the mean of the week, as these six days' working canoemen.

* From *Noka*, a man's name, and *seebi*, a river.

At Mr. Baker's, 170 miles above St. Anthony's Falls, I found a stray number of a small newspaper, and first learned the state of the Sauc and Fox war. The chief, Blackhawk, had crossed the Mississippi, to enter the Rock River valley; had murdered Mr. St. Vrain, the United States agent, sustained a conflict with the Illinois militia, under Major Stillman, fled to Lake Gushkenong, on the head of Rock River, and drawn upon his movement the United States army, leaving, at last accounts, Generals Atkinson and Dodge in pursuit of him.

Having struck the Mississippi at the point where the prior narrative describes it (*vide* Chap XII.), it becomes unnecessary to give details of my descent to St. Anthony's Falls. Leaving Prairie Piercie on the 23d, two days were employed in the descent to Fort Snelling. I found Captain Wm. R. Jouett in command, who received me with courtesy and kindness, and offered every facility, in the absence of Mr. Talliaferro, the United States Indian Agent, for laying the object of my mission before the Sioux. He had received no very recent intelligence of the progress of the Sauc war, in addition to that which I had learned at the mouth of the De Corbeau; although he was in the habit of sending a mail boat or canoe twice a month to Prairie du Chien.*

On the 25th, being the day after my arrival, I met the assembled Sioux, in council, at the Agency House, the commanding officer being present, and having finished that business, and finding the Sioux wholly unconnected with, and disapproving the proceedings of Blackhawk and his adherents, I embarked early the next morning on my return to Lake Superior. I reached the mouth of the River St. Croix, at three o'clock P. M. on the 26th, and having entered the sylvan sheet of Lake St. Croix, ascended it to within a few miles of its head, and encamped. Lieut. Allen did not reach my camp, but halted for the night some seven or eight miles short.†
This lake is one of the most beautiful and picturesque sheets of

---

* It was not till some time after my return to St. Mary's that I learned of the overthrow of the chief and his army, and his being taken prisoner at the battle of the Badaxe, on the 14th of August, 1832.

† United States soldiers are not adapted to travelling in Indian canoes. Comparatively clumsy, formal, and used to the comforts of good quarters and shelter, they flinch under the activities and fatigue of forest life, and particularly of that kind of life and toil, which consists in the management of canoes, and the carrying

water in the West, being from two to three miles wide, and some four-and-twenty or thirty in length.* The next morning I reached the head of the lake after a couple of hours of travel, and, by a diligent and hard day's work, during which we passed between perpendicular walls of sonorous trap-rock, reached and encamped at the falls of St. Croix, at eight o'clock in the evening.† We were now about fifty miles from the line of the Mississippi River. For the last few miles, there had been either a very strong current or severe eddies of water, around angular masses of trap-rock; and we were encamped at the precise foot of the falls, where the river, narrowed to some fifty feet, breaks its way through trap-rock, falling some fifty feet in the course of six hundred yards. We had been carried, at a tangent, from the great Mississippi series of the silurian period, beginning at St. Anthony's Falls, to the vitric formations of trap and greenstone of the Lake Superior

forward canoes and baggage over bad portages, and conducting these frail vessels over dangerous rapids and around falls. No amount of energy is sufficient on the part of the officers to make them keep up, on these trips, with the gay, light, and athletic *voyageur*, who unites the activity and expertness of the Indian with the power of endurance of the white man. Lieut. Allen deserves great credit, as an army officer, for urging his men forward as well as he did on this arduous journey, for they were a perpetual cause of delay and anxiety to me and to him. They were relieved and aided by my men at every practicable point; but, having the responsibility of performing a definite duty, on a fixed sum of money, with many men to feed in the wilderness, it was imperative in me to push on with energy, day in and day out, and to set a manful example of diligence, at every point; and, instead of carping at my rapidity of movement, as he does in his official report of the ascent of the St. Croix, he having every supply within himself, and being, moreover, in a friendly tribe, where there was no danger from Indian hostilities, he should not have evinced a desire to control my encampments, but rather given his men to understand that he could not countenance their dilatoriness.

* It is, at this time, a part of the boundary between the State of Wisconsin and the Territory of Minnesota, and is the site of several flourishing towns and villages. On its western head is the town of Stillwater, the seat of justice for Washington County, Minnesota. This town has a population of 1,500 inhabitants, containing a court house, several churches, schools, printing offices, a public land office, and territorial penitentiary, with stores, mills, &c. Hudson is a town seated on its east bank, at Willow River, being the seat of justice for St. Croix County, Wisconsin. It contains a United States land-office, two churches, and 94 dwellings, besides stores and mills. Steamboats freely navigate its waters from the Mississippi.

† FALLS OF ST. CROIX.—A thriving post town is now seated on the Wisconsin side of these falls in Polk County, Wisconsin, which contains several mills, at which it is estimated, four millions of feet of pine lumber are sawed annually. It is at the head of steamboat navigation of St. Croix River.

system, and were now to ascend a valley, in which a heavy diluvial drift and boulder stratum rested on this broken and angular basis.* On reaching the summit of the St. Croix, there are found vast plateaux of sand, supporting pine forests; and on descending the Misakoda, or Brulé of Fond du Lac, the sandstone strata of that basin are again encountered. This ascent was rendered arduous, from the low state of the water. I reached Snake River on the 80th, had an interview with the Buffalo chief (Pezhikee) and his subordinates; finding the population 800, with thirty-eight half breeds. The men, while here, cut their feet, treading on the trap-rock debris, in the mouth of the river. The distance thence to Yellow River is about thirty-five miles, which we accomplished on the 31st, by eight o'clock in the morning, having found our greatest obstacle at the Kettle Rapids, which discloses sharp masses of the trap-rock. The river, in this distance, receives on its right, in the ascent, the Aisippi, or Shell River, which originates in a lake of that name, noted for its large unios and anadontas.

At Yellow River, I halted to confer with the Indians in front of a remarkable eminence called Pokunogun, or the Moose's Hip. This eminence is not, however, of artificial construction. This river, with its dependencies of Lac Vaseux, Rice Lake, and Yellow Lake, contains a Chippewa population of three hundred and eighty-two souls. We observed here the unio purpureus, which the Indians use for spoons, after rubbing off the alatæ and rounding the margin. We also examined the skin of the sciurus tredacem striatus of Mitchill.

We reached the forks of the St. Croix about two o'clock P. M. The distance from Yellow River is about thirteen miles; it required five and a half hours to accomplish this. The water was, indeed, so low, that the men had often to wade; and, on reaching this point, we were to lose half its volume, or more, for the Namakagun† fork, which enters here, carries in more than half the quantity of water.

---

* Vide Owen's Geological Report, for the first attempt to delineate the order of the various local and general formations. Philada., Lippincott & Co., 1852.

† From nama, a sturgeon, and kagun, a yoke or wier. I explored this stream in 1831, having reached it after ascending the Mauvais or Maskigo of Lake Superior. Vide Personal Memoirs: Lippincott, Grambo, & Co., 1851.

I found the chief Kabamappa and his followers encamped at the forks, awaiting my arrival, who received me with a salute. He disclaimed all connection with the movement of the Black-hawk. He stated facts, however, which showed him to be well acquainted with the means which that chief had used to bring the Indians into an extensive league against the United States. He readily assented to the measures proposed to the upper bands, for bringing the Sioux and Chippewas into more intimate and permanent relations of peace and friendship.

With respect to the ascent of the St. Croix, in the direction of the Brulé, his exclamation was *iskutta-iskutta*, meaning it is dried up, or there is no water. Dry the channel, indeed, looked, but by leading the canoes around the shoals, all the men walking in the water, and picking out channels, we advanced about seven miles before the time of encampment. The next morning (Aug. 1) a heavy fog detained us in our encampment, till five o'clock, when we recommenced the ascent of a similar series of embarrassments from very low water, rapid succeeding to rapid, till two o'clock P. M., when we reached the summit of a plateau, and found still water and comparatively good navigation. Five hours canoeing on this summit brought us to Kabamappa's village at the Namakowágon, or sturgeon's dam, where we encamped. The chief gave us his population at 88 souls, of whom 28 were men, including the minor chief, Mukudapenas,* and his men. We had now got above all the strong rapids, and proceeded from our encampment at four o'clock, A. M., on the 2d. The river receives two tributaries, from the right hand, on this summit, namely, the Buffalo and Clearwater, and, at the distance of about ten miles above the Namakowágon, is found to be expanded in a handsome lake of about six miles in extent, called Lake St. Croix. This is the source of the river. We were favored with a fair wind in passing over it, and having reached its head debarked on a marshy margin, and immediately commenced the portage to the Brulé, or Misakoda River.†

* From *mukuda*, black, and *penaica*, a bird, the name of the rail.

† From *misk*, red or colored, *mamecda*, a plain, and *auk*, a dead standing tree, as a tree burned by fire or lightning. From the French translation of the word, by the phrase *Brulé*; the Indian meaning is clearly shown to be burnt, scorched, or parched—a term which is applied to motifs of the mixed race.

I had now reached the summit between the St. Croix and Lake Superior. The elevation of this summit has not been scientifically determined; but from the great fall of the Brulé, cannot be less than 600 feet. The length of the Brulé is about 100 miles, in which there are 240 distinct rapids. Some of these are from eight to ten feet each. Four of them require portages, at which all the canoes are discharged. The river itself, on looking down it, appears to be a perfect torrent, foaming and roaring; and it could never be used by the traders at all, were it not that it had abundance of water, being the off-drain for an extensive plateau of lakes and springs. To give an adequate idea of this foaming torrent, it is necessary to conceive of a river flowing down a pair of stairs, a hundred miles long.

The portage from the St. Croix to it begins on marsh, ascending in a hundred yards or so, to an elevated sandy plain, which has been covered, at former times, with a heavy forest of the pinus resinosa; that having been consumed, there is left here and there a dry trunk, or *auk*, as the Indians call it. The length of the portage path is 3,850 yards, or about two miles. At this distance, we reach a small, sandy-bottomed brook, of four feet wide and a foot deep, of most clear crystalline cold water, winding its way, in a most serpentine manner, through a boggy tract, and overhung with dense alder bushes. It is a good place to slake one's thirst, but appears like anything else than a stream to embark on, with canoes and baggage. Nobody but an Indian would seem to have ever dreamed of it. Yet on this brook we embarked. It was now six o'clock in the evening. By going a distance below, and damming up the stream, a sufficient depth of water was got to float the canoes. The axe was used to cut away the alders. The men walked, guiding the canoes, and carrying some of the baggage. In this way we moved slowly, about one mile, when it became quite dark, and threatened rain. The voyageurs then searched about for a place on the bog dry enough to sleep on, and came, with joy, and told me that they had found a kind of bog, with bunches of grassy tufts, which are called by them *tete de femme*. The very poetry of the idea was something, and I was really happy, amid the intense gloom, to rest my head, for the night, on these fair tufts. The next morning we were astir as soon as there was light enough to direct our steps. After

18

a few miles of these intricacies, we found a brisk and full tributary, below which, the descent is at once free, and on crossing the first narrow geologic plateau, the rapids begin; the stream being constantly and often suddenly enlarged, by springs and tributaries from the right and left. To describe the descent of this stream, in detail, would require graphic powers to which I do not aspire, and time which I cannot command. We were two days and a part of a night in making the descent, with every appliance of voyageur craft. It was after darkness had cast her pall over us, on the evening of the 4th of August, before we reached still water. The river is then a deep and broad mass of water, into which coasting vessels from the Lake might enter. Some four miles from the foot of the last rapids, it enters the Fond du Lac of Lake Superior. Some time before reaching this point, we had been apprised of our contiguity to it, from hearing the monotonous thump of the Indian drum; and we were glad, on our arrival, to find the chief, Mongazid,* of Fond du Lac, with the military barge of Lieut. Allen, left at that place on our outward trip, which he had promised to bring down to this point.

Having thus accomplished the objects committed to my trust, and rejoined the track described in my prior narrative, I rested here on the next day (5th), being the Sabbath; and then proceeded through Lake Superior, to my starting-point at Sault de Ste Marie.†

* From *mong*, a loon, and *osid*, his foot. The name is in allusion to the track of the bird on the sand.

† On passing through Lake Superior, I learned from an Indian the first breaking out of Asiatic cholera in the country, in 1882, and the wide alarm it had produced.

APPENDIX.

No. 1.

THE EXPEDITION TO THE SOURCES OF THE MISSISSIPPI IN 1820.

I. OFFICIAL REPORTS OF THE EXPEDITION OF 1820.

1. DEPARTMENTAL REPORTS.

I. Announcement of the Return of the Expedition. By Hon. Lewis Cass.

II. General Report to the Department of War. By Hon. Lewis Cass.

III. Further Explorations of Western Geography recommended. By Hon. Lewis Cass.

IV. Personal Testimonial on the close of the Expedition. By Hon. Lewis Cass.

2. TOPOGRAPHY AND ASTRONOMY.

V. Results of Observations for Latitudes and Longitudes during the Expedition of 1820. By David B. Douglass, *Capt. Engineers, U. S. A.*

3. MINERALOGY AND GEOLOGY.

VI. Report on the Copper Mines of Lake Superior. By Henry R. Schoolcraft.

VII. Observations on the Mineralogy and Geology of the country embracing the sources of the Mississippi River and the Great Lake Basins. By Henry R. Schoolcraft.

VIII. Report in reply to a Resolution of the U. S. Senate on the Value and Extent of the Mineral Lands on Lake Superior. By Henry R. Schoolcraft.

IX. Rapid Glances at the Geology of Western New York, beyond the Rome summit, in 1820. By Henry R. Schoolcraft.

X. A Memoir on the Geological Position of a Fossil Tree in the secondary rocks of the Illinois. Albany: E. & E. Hosford, pp. 18, 1822. By Henry R. Schoolcraft.

4. BOTANY.

XI. List of Plants collected by Capt. D. B. Douglass at the sources of the Mississippi River. This paper has been published in the 4th vol. p. 56 of Silliman's Journal of Science. By Dr. John Torrey.

5. ZOOLOGY.

XII. A Letter embracing Notices of the Zoology of the Northwest, addressed to Dr. Mitchell on the return of the Expedition. By Henry R. Schoolcraft.

(1.) FRESH-WATER CONCHOLOGY.

XIII. Species of Bivalves collected by Mr. Schoolcraft and Capt. Douglass in the Northwest. Published in the 6th vol. Amer. Journ. of Science, pp. 120, 259. By D. H. Barnes.

XIV. Fresh-water Shells collected by Mr. Schoolcraft in the valleys of the Fox and Wisconsin Rivers. American Philosophical Transactions, vol. 5. By Mr. Isaac Lea.

(2.) FAUNA: ICHTHYOLOGY: REPTILIA.

XV. Summary Remarks respecting the Zoological Species noticed in the Expedition. By Dr. Samuel L. Mitchell.

XVI. Mus Busarius. Medical Repository, vol. 21, p. 248. By Dr. Samuel L. Mitchell.

XVII. Sciurus Tredecem Striatus. Med. Rep. vol. 21. By Dr. Samuel L. Mitchell.

XVIII. Proteus of the Lakes. Am. Journ. Science, vol. 4. By Dr. Samuel L. Mitchell.

6. METEOROLOGY.

XIX. Memoranda on Climatic Phenomena, and the distribution of Solar Heat, in 1820. By Henry R. Schoolcraft.

7. INDIAN LANGUAGES AND HISTORY.

XX. A Pictographic mode of communicating ideas by the Northwestern Indians. By Hon. Lewis Cass.

XXI. Inquiries respecting the History, &c. of the Indians of the United States. Detroit, 1822. By Hon. Lewis Cass.

XXII. A Letter on the Origin of the Indian Tribes of America, and the Principles of their Mode of uttering Ideas. By Dr. J. M'Donnell, Belfast, Ireland.

XXIII. Difficulties of studying the Indian Tongues of the United States. Schoolcraft's Travels in the Central Portions of the Mississippi Valley, p. 381. By Dr. Alexander Wolcott, Jr.

XXIV. Examinations of the Elementary Structure of the Odjibwa-Algonquin Language. First paper. By Henry R. Schoolcraft.

XXV. A Vocabulary of the Odjibwa-Algonquin. By Henry R. Schoolcraft.

APPENDIX.

DETROIT, September 14, 1820.

SIR: I am happy to be enabled to state to you that I reached this place four days since, with some of the gentlemen who accompanied me on my late tour, after a very fortunate journey of four thousand miles, and an accomplishment, without any adverse accident, of every object intrusted to me. The party divided at Green Bay, with a view to circumnavigate Lake Michigan; and I trust they may all arrive here in the course of a week.

As soon as possible, I shall transmit to you a detailed report upon the subject.

Since my arrival, I have learned that Mr. Ellicott, professor of mathematics, at the military academy, is dead. I cannot but hope that the office will not be filled until the return of Captain Douglass. I do not know whether such an appointment would suit him; but from my knowledge of his views, feelings, and pursuits, I presume it would. And an intimate acquaintance with him during my tour enables me to say that in every requisite qualification, as far as I can judge, I have never found a man who is his superior. His zeal, talents, and acquirements are of the first order, and I am much deceived if he do not soon take a distinguished rank among the most scientific men in our country. His situation as an assistant professor to Colonel Mansfield, and his connection with the family of Mr. Ellicott, furnish additional reasons why he should receive this appointment.

Very respectfully, sir,

I have the honor to be

Your obedient servant,

LEWIS CASS.

Hon. J. C. CALHOUN, *Secretary of War.*

II.

DETROIT, October 21, 1820.

SIR: I had the honor to inform you some time since that I had reached this place by land from Chicago, and that the residue of the party were daily expected. They arrived soon after, without accident, and this long and arduous journey has been accomplished without the occurrence of any unfavorable incident.

I shall submit to you, as soon as it can be prepared, a memoir respecting the Indians who occupy the country through which we passed; their numbers, disposition, wants, &c. It will be enough at present to say, that the whole frontier is in a state of profound peace, and that the remote Indians, more particularly, exhibit the most friendly feelings towards the United States. As we approach the points of contact between them and the British, the strength of this attachment evidently decreases, and about those points few traces of it remain. During our whole progress but two incidents occurred which evinced in the slightest degree, an unfriendly spirit. One of these was at St. Mary's, within forty-five miles of Drummond's Island, and the other within thirty miles of Malden. They passed off, however, without producing any serious result.

It is due to Colonel Leavenworth to say, that his measures upon the subject of the outrage committed by the Winnebago Indians, in the spring, were prompt, wise, and decisive. As you have long since learned, the murderers were soon surrendered; and so impressive has been the lesson upon the minds of the Indians, that the transaction has left us nothing to regret, but the untimely fall of the soldiers.

In my passage through the Winnebago country, I saw their principal chiefs, and stated to them the necessity of restraining their young men from the commission of acts similar in their character to those respecting which a report was made by Colonel Smith. I have reason to believe that similar complaints will not again be made, and I am certain that nothing but the intemperate passions of individuals will lead to the same conduct. Should it occur, the act will be disavowed by the chiefs, and the offenders surrendered with as much promptitude as the relapsed state of the government will permit.

The general route which we pursued was from this place to Michilimackinac by the southern shore of Lake Huron. From thence to Drummond's Island and by the River St. Mary's to the Sault. We there entered Lake Superior, coasted its southern shore to Point Kewena, ascended the small stream, which forms the water communication across the base of the point, and, after a portage of a mile and a half, struck the lake on the opposite side. Fifty miles from this place is the mouth of the Ontonagan, upon which have been found large specimens of copper.

We ascended that stream about thirty miles, to the great mass of that metal, whose existence has long been known. Common report has greatly magnified the quantity, although enough remains, even after a rigid examination, to render it a mineralogical curiosity. Instead of being a mass of pure copper, it is rather copper imbedded in a hard rock, and the weight does not probably exceed five tons, of which the rock is the much larger part. It was impossible to procure any specimens, for such was its hardness that our chisels broke like glass. I intend to send some Indians in the spring to procure the necessary specimens. As we understand the nature of the substance, we can now furnish them with such tools as will effect the object. I shall, on their return, send you such pieces as you may wish to retain for the Government, or to distribute as cabinet specimens to the various literary institutions of our country. Mr. Schoolcraft will make to you a detailed report upon this subject, in particular, and generally upon the various mineralogical and geological objects to which his inquiries were directed. Should he carry into effect the intention, which he now meditates, of publishing his journal of the tour, enriched with the history of the facts which have been collected, and with those scientific and practical reflections and observations, which few men are more competent to make, his work will rank among the most important accessions which have ever been made to our national literature.

From the Ontonagon we proceeded to the Fond du Lac, passing the mouths of the Montreal, Mauvais, and Brulé Rivers, and entered the mouth of the St. Louis, or Fond du Lac River, which forms the most considerable water communication between Lake Superior and the Mississippi.

The southern coast of the lake is sterile, cold, and unpromis-

ing. The timber is birch, pine, and trees of that description
which characterize the nature of the country. The first part of
the shore is moderately elevated, the next, hilly, and even moun-
ainous, and the last a low, flat, sandy beach. Two of the most
sublime natural objects in the United States, the Grand Sable and
the pictured rocks, are to be found upon this coast. The former
is an immense hill of sand, extending for some miles along the
lake, of great elevation and precipitous ascent. The latter is an
unbroken wall of rocks, rising perpendicularly from the lake to
the height of 300 feet, assuming every grotesque and fanciful
appearance, and presenting to the eye of the passenger a spectacle
as tremendous as the imagination can conceive, or as reason itself
can well sustain.

The emotions excited by these objects are fresh in the recollection of us all; and they will undoubtedly be described, so that
the public can appreciate their character and appearance. The
indications of copper upon the western part of the coast, are
numerous; and there is reason to suppose that silver, in small
quantities, has been found.

The communication by the Montreal with the Chippewa River,
and by the Mauvais and Brulé Rivers with the St. Croix, is
difficult and precarious. The routes are interrupted by long,
numerous, and tedious portages, across which the boats and all
their contents are transported by the men. It is doubtful whether their communication can ever be much used, except for the
purposes to which they are now applied. In the present state of
the Indian trade, human labor is nothing, because the number of
men employed in transporting the property is necessary to conduct the trade, after the different parties have reached their destination, and the intermediate labor does not affect the aggregate
amount of the expense. Under ordinary circumstances, and for
those purposes to which water communication is applied in the
common course of civilized trade, these routes would be abandoned. From the mouth of the Montreal River alone to its source,
there are not less than forty-five miles of portage.

The St. Louis River is a considerable stream, and for twenty-
five miles its navigation is uninterrupted. At this distance, near
an establishment of the Southwest Company, commences the
Grand Portage about six miles in length, across spurs of the

Porcupine ridge of mountains. One other portage, one of a mile and a half, and a continued succession of falls, called the Grand Rapids, extending nine miles, and certainly unsurmountable except by the skill and perseverance of the Canadian boatmen, conduct us to a comparatively tranquil part of the river. From here to the head of the Savannah River, a small branch of the St. Louis, the navigation is uninterrupted, and after a portage of four miles, the descent is easy into Lake au Sable, whose outlet is within two miles of the Mississippi.

This was until 1816 the principal establishment of the British Northwest Company upon these waters, and is now applied to the same purpose by the American Fur Company.

From Lac au Sable, we ascended the Mississippi to the Upper Red Cedar Lake, which may be considered as the head of the navigation of that river. The whole distance, 850 miles, is almost uninhabitable. The first part of the route the country is generally somewhat elevated and interspersed with pine woods. The latter part is level wet prairie.

The sources of this river flow from a region filled with lakes and swamps, whose geological character indicates a recent formation, and which, although the highest table-land of this part of the Continent, is yet a dead level, presenting to the eye a succession of dreary uninteresting objects. Interminable marshes, numerous ponds, and a few low, naked, sterile plains, with a small stream, not exceeding sixty feet in width, meandering in a very crooked channel through them, are all the objects which are found to reward the traveller for the privations and difficulties which he must encounter in his ascent to this forbidding region.

The view on all sides is dull and monotonous. Scarcely a living being animates the prospect, and every circumstance recalled forcibly to our recollection that we were far removed from civilized life.

From Lac au Sable to the mouth of the St. Peter's, the distance by computation is six hundred miles. The first two hundred present no obstacles to navigation. The land along the river is of a better quality than above; the bottoms are more numerous, and the timber indicates a stronger and more productive soil. But near this point commence the great rapids of the Mississippi, which extend more than two hundred miles. The river flows over a rocky bed, which forms a continuous succession of rapids, all of which are difficult and some dangerous. The country, too, begins here to open, and the immense plains in which the buffaloes range approach the river. These plains continue to the Falls of St. Anthony.

They are elevated fifty or sixty feet above the Mississippi, are destitute of timber, and present to the eye a flat, uniform surface, bounded at the distance of eight or ten miles by high ground. The title of this land is in dispute between the Chippewas and Sioux, and their long hostilities have prevented either party from destroying the game in a manner as improvident as is customary among the Indians. It is consequently more abundant than in any other region through which we travelled.

From the post, at the mouth of the St. Peter's, to Prairie du Chien, and from that place to Green Bay, the route is too well-known to render it necessary that I should trouble you with any observations respecting it.

The whole distance travelled by the party between the 24th of May and the 24th of September exceeded 4,200 miles, and the journey was performed without the occurrence of a single untoward accident sufficiently important to deserve recollection.

These notices are so short and imperfect that I am unwilling to obtrude them upon your patience. But the demands upon your attention are so imperious, that to swell them into a geographical memoir would require more time for their examination than any interest which I am capable of giving the subject would justify.

I propose hereafter to submit some other observations to you in a different shape.

Very respectfully, sir,

I have the honor to be

Your obedient servant,

LEWIS CASS.

Hon. J. C. CALHOUN, *Secretary of War.*

III.

Copy of a letter from Gov. Lewis Cass to Hon. John C. Calhoun, Secretary of War, dated

DETROIT, September 20, 1820.

SIR: In examining the state of our topographical knowledge,
respecting that portion of the Northwestern frontier over which
we have recently passed, it occurs to me that there are several
points which require further examination, and which might be
explored without any additional expense to the United States.

The general result of the observations made by Capt. Douglass,
will be submitted to you as soon as it can be prepared. And I
believe he will also complete a map of the extensive route we
have taken, and embracing the whole of the United States,
bounded by the Upper Lakes and by the waters of the Mississippi,
and extending as far south as Rock Island and the southern extremities of Lakes Michigan and Erie. The materials in his possession are sufficient for such an outline, and he is every way
competent to complete it. But there are several important
streams, respecting which it is desirable to procure more accurate
information than can be obtained from the vague and contradic-
tory relations of Indians and Indian traders. The progress of
our geographical knowledge has not kept pace with the extension
of our territory, nor with the enterprise of our traders. But I
trust the accurate observations of Captain Douglass will render a
resort to the old French maps for information respecting our own
country entirely unnecessary.

I beg leave to propose to you, whether it would not be proper
to direct exploring parties to proceed from several of our frontier
posts into the interior of the country, and to make such observations as might lead to a correct topographical delineation of it.
An intelligent officer, with eight or ten men, in a canoe, would
be adequate to this object. He would require nothing more than
a compass to ascertain his course, for it is not to be expected that
correct astronomical observations could be taken. In ascending
or descending streams, he should enter in a journal every course
which he pursues, and the length of time observed by a watch.
He should occasionally ascertain the velocity of his canoe, by
measuring a short distance upon the bank, and should also enter

in his journal his supposed rate of travelling. This, whenever it
is possible, should be checked by the distance as estimated by
traders and travellers. By a comparison of these data, and by a
little experience, he would soon be enabled to ascertain with
sufficient precision, the length of each course, and to furnish
materials for combination, which would eventually exhibit a perfect view of the country. I do not know any additional expense
which it would be necessary to encounter. An ordinary compass
is not worth taking into consideration. A necessary supply of
provisions, a small quantity of powder, lead, and tobacco, to present occasionally to the Indians, and a little medicine, are all the
articles which would require particular attention. Officers
employed upon such services should be directed to observe the
natural appearances of the country; its soil, timber, and productions; its general face and character; the height, direction, and
composition of its hills; the number, size, rapidity, &c., of its
streams; its geological structure and mineralogical products; and
any facts which may enable the public to appreciate its importance in the scale of territorial acquisitions, or which may serve
to enlarge the sphere of national science.

It is not to be expected that officers detached upon the duties
can enter into the detail of such subjects in a manner which their
importance would render desirable. But the most superficial
observer may add something to the general stock; and to point
their inquiries to specific objects, may be the means of eliciting
facts, which in other hands may lead to important results. The
most important tributary stream of the Upper Mississippi is the
Saint Peter's. The commanding officer at the mouth of that river
might be directed to form an expedition for exploring it.

It is the opinion of Captain Douglass, and it is strongly forti-
fied by my personal observation, and by the opinion of others,
that Lieut. Talcott, of the Engineers, now at the Council Bluffs,
would conduct a party upon this duty in a very satisfactory
manner. He might ascend the St. Peter's to its source, and from
thence cross over to the Red River, and descend the stream to
the 49th parallel of latitude, with directions to take the necessary
observations upon so important a point.* Thence up that branch
of the Red River, interlocking with the nearest water of the

* This is the origin of Major Long's second expedition.

Mississippi, and down this river to Leech Lake. From this lake, there is an easy communication to the River de Corbeau, which he could descend to the Mississippi, and thence to St. Peter's.*

The St. Croix and Chippewa Rivers, entering the Mississippi above and below the Falls of St. Anthony, might, in like manner, be explored by parties from the same post.* The former interlocks with the Mauvais and Brulé Rivers, but a descent into Lake Superior would not probably be considered expedient, so that the party would necessarily ascend and descend the same stream.*

The Chippewa interlocks with the Montreal and Wisconsin Rivers, and consequently the same party could ascend the former and descend the latter stream.

A party from Green Bay might explore Rocky River from its source to its mouth.

A correct examination of Green Bay and of the Menomonie River might be made from the same post.

The St. Joseph and Grand River, of this peninsula, could be examined by parties detached from Chicago.

It is desirable, also, to explore the Grand Traverse Bay, about sixty miles south of Michilimackinac, on the east coast of Lake Michigan.

These are all the points which require particular examination. Observations made in the manner I have suggested, and connected with those already taken by Captain Douglass, would furnish ample materials for a correct chart of the country.

It is with this view that it might be proper, should you approve the plan I have submitted to you, to direct, that the reports of the officers should be transmitted to Captain Douglass, by whom they will be incorporated with his own observations, and will appear in a form best calculated to promote the views which you entertain upon the important subject of the internal geography of our country.

IV.

DETROIT, October 8, 1820.

SIR: On the eve of separating from my associates in our late tour, I owe it to them and to myself, that I should state to you my opinion respecting Captain Douglass and Mr. Schoolcraft.

I have found them, upon every occasion, zealous in promoting

* Explored by the preceding narrative in 1831-1838.

the objects of the Expedition, indefatigable in their inquiries and observations, and never withholding their personal exertions. Ardent in their pursuit after knowledge, with great attainments in the departments of literature to which they have respectively devoted themselves, and with powers which will enable them to explore the whole field of science, I look forward with confidence to the day when they will assume distinguished stations among our scientific men, and powerfully aid in establishing the literary fame of their country.

Should any object of a similar character again require similar talents, I earnestly recommend their employment. Whoever has the pleasure of being associated with them, will find how easily profound acquirements may be united with that urbanity of manners, and those qualities of the heart, which attach to each other those who have participated in the fatigues of a long and interesting tour.

Very respectfully, sir,

I have the honor to be

Your obedient servant,

LEWIS CASS.

Hon. JOHN C. CALHOUN, *Secretary of War.*

---

2. TOPOGRAPHY AND ASTRONOMY.

Topographical materials were collected by Capt. Douglass, U. S. A., for a map of the northwestern portions of the United States, embracing the complete circumnavigation of the great lake basins, and accurate delineations of the sources of the Mississippi, as low down as the influx of the River Wisconsin. Being provided with instruments from the Military Academy of West Point, astronomical observations were made at every practical point over the vast panorama traversed by the Expedition. A line of some four thousand miles of previously unexplored country was visited; his notes and memoranda for a topographical memoir were full and exact; and they were left, I am informed, in a state of nearly perfect elaboration, accompanied by illustrations, and many drawings of scenery. Having written to his family recently, for the astronomical observations, they were transmitted by his son in a letter, of which the following is an extract:—

GENEVA, JUNE 28, 1854.

DEAR SIR: I inclose you herewith, on another page, the results of my father's observations of latitude and longitude, so far as I have been able to collect them. His calculations indicate great pains and labor to obtain accurate results. They are too voluminous to copy. I trust, however, that I have been as particular as was necessary in the inclosed memoranda. If anything else is wanting, I should like you to inform me.

I am, sir, with great respect,

Your obedient servant,

MALCOLM DOUGLASS.

V.

Results of Observations for Latitude and Longitude during the Expedition of 1820. By DAVID B. DOUGLASS, Capt. Engineers, U. S. A.

Mean latitude of Detroit
- By 3 sets of observations at Cunningham's Island, 1819, and reduced by exact measurement on the Boundary Bay
- By 1 set of observations at Gibraltar Island (Put-in Bay), taken, like the preceding, in 1819, and reduced as before
- By 1 set of observations taken on Sugar Island, and reduced as before
- By mean results of 2 sets of observations—May 17 and 21, 1820
- By mean observation, Sept. 29, 1820

$42° 19' 20''$

Mean longitude of Detroit, by 6 sets of observations, May 17 and 19, 1820 . . . . 82 39 00

Latitude of Presque Isle, Lake Huron, June 5, 1820 45 19 45

Latitude of Mackinaw, by 4 sets of observations, June 7 and 11, 1820, by meridian observations, Sept. 12, 1820 . . . . . . 45 50 54

Height of Fort Holmes. From the water to the brow of the hill near Robin-

19

son's Folly, nearly on a level with
Fort Mackinaw 115.8
Thence to the top of the block H of Fort
Holmes 260.9

Total height, 376.7 feet

Longitude of Mackinaw, by several sets of observations, Sept. 12, 1820 . . . . . 84° 28' 40"

Mean latitude of Sault de St. Marie, June 16, 1820 46 26 45

Latitude of Turtle Camp, on Lake Superior, June
22—primitive bluff (Granite Point.—S.) . . 46 41 15

Latitude of Keweena Camp 47 02 80

Mean latitude of Sandy River, July 4, 1820 . 46 55 24

Mean longitude (by 25 observations for degrees,
and 25 observations for time). In time, 6 h. 8 m.
48 sec. In degrees 90 57 00

Latitude of the gallais* on the Grand Portage of
St. Louis, July 6, 1820 . . . . 46 89 84

Latitude of camp at head of Grand Portage, July
8, 1820 . . . . 46 41 07

Latitude of camp at west end of Savanna Portage 46 51 47†

Mean latitude of Sandy Lake post, from observations, July 16 and 25 . . . . , 46 45 85

Mean longitude of Sandy Lake post, from 4 sets
of observations, July 15 and 16 . . . 93 21 80

Latitude of Wolverine Camp, July 23, 1 day from
Sandy Lake 47 4 15

Latitude of halting-place above forks of Leech
River on the Mississippi, July 20 . . . 47 24 00†

Latitude of camp at Lake Winnipec, July 20 . 47 80 56

Latitude of halting-place near first return camp,
July 21 . . . . 47 27 10

Latitude of return camp, near the above, same
day 47 26 40

Latitude of camp at Buffalo hunting-ground, above
Pe-can-de-quaw Lake, July 28 and 29 . . 46 00 00

Breadth of river at camp on the Buffalo Plain,
148 yards.

* *Galet*, in the Canadian patois, means a smooth, flat rock.—H. R. S.
† A little doubtful.

Latitude of halting-place between the Great Falls and St. Francis River	45°	25'	43"
Breadth of river at camp above Falls of St.'Anthony, 200 yards.			
Mean latitude of Fort St. Anthony, new site, July 31, by 5 sets of observations	44	53	20
Mean longitude of Fort St. Anthony, new site, July 31, by 3 sets of observations	92	55	45
Latitude of Fort Prairie du Chien, Aug. 6 and 7 .	43	03	19*
Latitude of Fox and Ouisconsin Portage, Aug. 14 and 15, 43° 42' 36"; say	43	42	00
Latitude of camp near mouth of River De Loup, Aug. 17	44	6	44
Latitude of Fort Howard, Green Bay, Aug. 21 .	44	31	38
Longitude of Fort Howard (some error), probably between 87° 45' 30" and	87	46	00
Latitude of camp at Sturgeon Portage, Lake Michigan, Aug. 23	44	47	43
Latitude of camp 3 miles north of the Manetowag, Aug. 24	44	13	47
Latitude of camp south of the Sheboyegan, Aug. 25	43	41	26
Latitude of camp at Milwaukie, Aug. 26 . .	43	01	35
Mean latitude of Fort Dearborn, Chicago, by 6 sets of equal altitudes, Aug. 31, and meridian altitude	41	54	06
Mean longitude of Fort Dearborn, 3 sets of observations. In time, 5 h. 50 m. 8 sec. In degrees	87	32	30
Longitude of Detroit, calculated from above .	82	54	53
Latitude of camp near head of Lake Michigan, Aug. 31 and Sept. 1	41	38	48
Mean latitude of the extreme south point of Lake Michigan, 4 sets of observations and meridian observation	41	37	28
Latitude of camp next north of the St. Joseph's, near Kekalamazo, Sept. 3	42	32	16
Latitude of camp at Maskegon River, Sept. 4 .	43	13	41

* Or 20".

Latitude of camp near Point aux Salles, Lake
   Michigan, Sept. 5   .   .   .   .   .   .   44° 5' 17"
Latitude of camp at Grand Traverse Bay, Lake
   Michigan, Sept. 7   .   .   .   .   .   .   45 34 24

---

8. MINERALOGY AND GEOLOGY.

VI.

Report on the Copper Mines of Lake Superior.   By HENRY R.
                    SCHOOLCRAFT.

To the Hon. JOHN C. CALHOUN, *Secretary of War.*

VERNON (Oneida County, N. Y.), November 6, 1820.

SIR: I have now the honor to submit such observations as
have occurred to me, during the recent expedition under Gov.
Cass, in relation to the copper mines on Lake Superior; reserving, as the subject of a future communication, the facts I have
collected on the mineralogy and geology of the country explored
generally.

The first striking change in the mineral aspect of the country
north of Lake Huron, is presented near the head of the Island of
St. Joseph, in the River St. Mary, where the calcareous strata of
secondary rocks are succeeded by a formation of red sandstone,
which extends northward to the head of that river at Point Iroquois, producing the falls called the *Sault de Ste. Marie*, fifteen
miles below; and thence stretching northwest, along the whole
southern shore of Lake Superior, with the interruptions noted, to
Fond du Lac.

This extensive stratum is perforated at various points by up-
heaved masses of sienitic granite and trap, which appear in elevated points on the margin of the lake at Dead River, Keweena
Point, Presque Isle, and the Chegoimagon Mountains. It is
overlaid, in other parts, by a stratum of gray or neutral-colored
sandstone, of uncommon thickness, which appears in various
promontories along the shore, and, at the distance of ninety miles
from Point Iroquois, constitutes a lofty perpendicular and caverned wall, upon the water's edge, called the Pictured Rocks.

So obvious a change in the geological character of the rock strata, in passing from Lake Huron to Lake Superior, prepares the observer to expect a corresponding one in the imbedded minerals and other natural features—an expectation which is realized during the first eighty leagues, in the discovery of various minerals. The first appearances of copper are seen at Keweena Point, two hundred and seventy miles beyond the Sault de Ste. Marie, where the debris and pebbles along the shore of the lake contain native copper disseminated in particles varying in size from a grain of sand to a mass of two pounds' weight. Many of the detached stones of this Point are also colored green by the carbonate of copper, and the rock strata exhibit traces of the same ore. These indications continue to the River Ontonagon, which has long been noted for the large masses of native copper found upon its banks, and about the contiguous country.

This river is one of the largest of thirty tributaries, mostly small, which flow into the lake between Point Iroquois and Fond du Lac. It originates in a district of mountainous country intermediate between the Mississippi River and lakes Huron and Superior. After running in a northern direction for about one hundred and twenty miles, it enters the latter at the computed distance of fifty miles west of the portage of Keweena, in north latitude 46° 52′ 2″, according to the observations of Capt. Douglass. It is connected, by portages, with the Monomonee River of Green Bay, and with the Chippewa River of the Mississippi. At its mouth there is a village of Chippewa Indians of sixteen families, who subsist chiefly on the fish taken in the river. Their location, independent of that circumstance, does not appear to unite the ordinary advantages of an Indian village of the region.

A strip of alluvial land of a sandy character extends from the lake up the river three or four leagues, where it is succeeded by hills of a broken, sterile aspect, covered, chiefly, with a growth of pine, hemlock, and spruce. Among these hills, which may be considered as lateral spurs of the Porcupine Mountains, the copper mines, so called, are situated, at the computed distance of thirty-two miles from the lake, and in the centre of a region characterized by its wild, rugged, and forbidding appearance. The large mass of native copper lies on the west bank of the river, at the water's edge, at the foot of an elevated bank, part of which appears to have slipped into the river, carrying with it the mass of copper, together with detached blocks of sienitic granite, trap-rock, and other species common to the soil at that place.

The copper, which is in a pure and malleable state, lies in connection with serpentine rock, one face of which it almost completely overlays. It is also disseminated in masses and grains throughout the substance of the rock. The surface of the metal, unlike most oxidable metals which have been long exposed to the atmosphere, presents a metallic brilliancy, which is probably attributable to the attrition of the semi-annual floods of the river.

The shape of the rock is very irregular; its greatest length is three feet eight inches; its greatest breadth, three feet four inches, with an average thickness of twelve inches. It may, altogether, contain eleven cubic feet.* It exceeds, in size, the great mass of native iron found some years ago on the banks of Red River, in Louisiana. I have computed the weight of metallic copper in the rock at twenty-two hundred pounds, which is about one-fifth of the lowest estimate made of it by former visitors. Henry, who visited it in 1766, estimated its weight at five tons. The quantity may, however, have been much diminished since its discovery, and the marks of chisels and axes upon it, with the discovery of broken tools, prove that portions have been cut off and carried away. Notwithstanding this reduction, it may still be considered one of the largest and most remarkable bodies of native copper on the globe, and is, so far as known, only exceeded in weight by a specimen found in a valley in Brazil, weighing twenty-six hundred and sixty-six Portuguese pounds. Viewed as a subject of scientific interest, it presents illustrative proofs of an important character. Its connection with a rock which is foreign to the immediate section of country where it lies,† indicates a removal from its original bed; while the intimate connection of the metal and matrix, and the complete envelopment of masses of the copper by the rock, point to a common and cotemporaneous origin, whether that be referable to volcanic agency or water. This conclusion admits of an obvious application to the beds of serpentine and other magnesian rock found in other parts of the lake.

* This copper rock now (1854) lies in the yard of the War Office at Washington.

† A locality of serpentine rock has since been discovered at Presque Isle, on Lake Superior.

Several other large masses of native copper have been found, either on this river or within the basin of the lake, at various periods since the country has been known, and taken into different parts of the United States and of Europe. A recent analysis of one of these specimens, at the University of Leyden, proves it to be native copper in a state of uncommon purity, and uncombined with any notable portion of either gold or silver.

A mass of copper, weighing twenty-eight pounds, was discovered on an island in Lake Superior, eighty miles west of the Ontonagon. It was taken to Michilimackinac and disposed of. The War Department was formerly supplied with a specimen from this mass, and the analysis above alluded to is also understood to have been made from a portion of it. A piece weighing twelve pounds was found at Winnebago Lake. Other discoveries of this metal have been made, within the region, at various times and places.

The existence of copper in the region of Lake Superior appears to have been known to the earliest travellers and voyagers.

As early as 1689, the Baron La Hontan, in concluding a description of Lake Superior, adds: "That, upon it, we also find copper mines, the metal of which is so fine and plentiful that there is not a seventh part lost from the ore."—*New Voyages to North America*, London, 1703.

In 1721, Charlevoix passed through the lakes on his way to the Gulf of Mexico, and did not allow the mineralogy of the country to escape him.

"Large pieces of copper are found in some places on its banks [Lake Superior], and around some of the islands, which are still the objects of a superstitious worship among the Indians. They look upon them with veneration, as if they were the presents of those gods who dwell under the waters. They collect their smallest fragments, which they carefully preserve, without, however, making any use of them. They say that formerly a huge rock of this metal was to be seen elevated a considerable height above the surface of the water, and, as it has now disappeared, they pretend that the gods have carried it elsewhere; but there is great reason to believe that, in process of time, the waves of the lake have covered it entirely with sand and slime. And it is certain that in several places pretty large quantities of this metal have been discovered without being obliged to dig very deep. During the course of my first voyage to this country, I was acquainted with one of our order (Jesuits) who had been formerly a goldsmith, and who, while he was at the mission of Sault de Ste. Marie used to search for this metal, and made candlesticks, crosses, and censers of it, for this copper is often to be met with almost entirely pure."—*Journal of a Voyage to North America.*

In 1766, Captain Carver procured several pieces of native copper on the shores of Lake Superior, or on the Chippewa and St. Croix Rivers, which are noticed in his travels, without much precision, however, as to locality, &c. He did not visit the southern shores of Lake Superior, east of the entrance of the Brulé, or Goddard's River, but states that virgin copper is found on the Ontonagon. Of the north and northeastern shores, he remarks: "That he observed that many of the small islands were covered with copper *ore*, which appeared like beds of copperas, of which many tons lay in a small space."—*Three Years' Travels, &c.*

In 1771 (four years before the breaking out of the American Revolution), a considerable body of native copper was dug out of the alluvial earth on the banks of the Ontonagon River by two adventurers, of the names of Henry and Bostwick, and, together with a lump of silver ore of eight pounds' weight, it was transported to Montreal, and from thence shipped to England, where the silver ore was deposited in the British Museum, after an analysis had been made of a portion of it, by which it was determined to contain 60 per cent. of silver.

These individuals were members of a company which had been formed in England for the purpose of working the copper mines of Lake Superior. The Duke of Gloucester, Sir William Johnson, and other gentlemen of rank were members of this company. They built a vessel at Point aux Pins, six miles above the Sault Ste. Marie, to facilitate their operations on the lake. A considerable sum of money was expended in explorations and digging. Isle Maripeau and the Ontonagon were the principal scenes of their search. They found silver, in a detached form, at Point Iroquois, fifteen miles above the present site of Fort Brady.

"Hence," observes Henry, "we coasted westward, but found nothing till we reached the Ontonagon, where, besides the detached masses of copper formerly mentioned, *we saw much of the same metal imbedded in stone.*

"Proposing to ourselves to make a trial on the hill, till we were better able to go to work upon the solid rock, we built a house, and sent to the Sault de Ste. Marie for provisions. At the spot pitched upon for the commencement of our operations, a green-colored water, which tinges iron of a copper color, issued from the hill, and this the miners called *a leader.* In digging, they found frequent masses of copper, some of which were of three pounds' weight. Having arranged everything for the accommodation of the miners during the winter, we returned to the Sault.

"Early in the spring of 1772, we sent a boat-load of provisions, but it came back on the 20th day of June, bringing with it, to our surprise, the whole establishment of miners. They reported that, in the course of the winter, they had penetrated forty feet into the face of the hill, but, on the arrival of the thaw, the clay, on which, on account of its stiffness, they had relied, and neglected to secure it by supporters, had fallen in. That, from the detached masses of metal which, to the last, had daily presented themselves, they supposed there might be ultimately reached a body of the same, but could form no conjecture of its distance, except that it was probably so far off as not to be pursued without sinking an air shaft. And, lastly, that the work would require the hands of more men than could be fed in the actual situation of the country.

"Here our operations, in this quarter, ended. The metal was probably within our reach, but, if we had found it, the expense of carrying it to Montreal must have exceeded its marketable value. It was never for the exportation of copper that our company was formed, but always with a view to the silver, which it was hoped the ores, whether of copper or lead, might in sufficient quantity contain."—*Travels and Adventures of Alexander Henry.*

[In the summer of 1832, being detained by head winds at the mouth of Miner's River, on Lake Superior, I observed the names of several persons engraved on the sand rock, but much obliterated by the water's dashing over the rock. Tradition represents that Henry's miners were detained there, and that they made explorations of the river, which is named from the circumstance.

The stream is a mere brook, coming over the shelving sand rock, which is a part of the precipitous range of the Pictured Rocks.]

Sir A. Mackenzie passed through Lake Superior, on his first voyage of discovery, in 1789. He remarks: "At the River Tennagon (Ontonagon) is found a quantity of virgin copper. The Americans, soon after they got possession of the country, sent an agent thither; and I should not be surprised to hear of their employing people to work the mine. Indeed, it might be well worthy the attention of the British subjects to work the mines on the north coast, though they are not supposed to be so rich as those on the south." — *Voyages from Montreal through the Continent of North America.*

It is difficult to conceive what, however, is apparent, from the references of Dr. Franklin to the subject, that the supposed mineral riches of Lake Superior had an important bearing on the discussions for settling the ultimate northern boundary of the United States. The British ambassadors had, it seems, from an old map which is before me, claimed a line through the Straits of Michilimackinac and the Illinois and Mississippi rivers, to the Gulf of Mexico.

The attention of the United States Government appears first to have been turned toward the subject during the administration of President John Adams, when the sudden augmentation of the navy rendered the employment of copper in the equipment of ships an object of moment. A mission was therefore authorized to proceed to Lake Superior, of the success of which, as it has not been communicated to the public, nothing can, with certainty, be stated; but from inquiries which have been made during the recent expedition, it is rendered probable that the actual state of our Indian relations, at the time, arrested the advance of the officer into the region where the most valuable beds of copper were supposed to exist, and that the specimens transmitted to Government were procured through the instrumentality of some friendly Indians, employed for the purpose.

Such are the lights which those who have preceded me in this inquiry have thrown upon the subject, all of which have operated in producing public belief in the existence of extensive copper mines on Lake Superior. Travellers have generally coincided that the southern shore of the lake is most metalliferous, and that the Ontonagon River may be considered as the seat of the principal mines. Mr. Gallatin, in his report on the state of American manufactures in 1810, countenances the prevalent opinion, while it has been reiterated in some of our literary journals, and in the numerous ephemeral publications of the times, until public expectation has been considerably raised in regard to them.

Under these circumstances, the recent expedition under Gov. Cass entered the mouth of the Ontonagon River on the 27th of June, having coasted along the southern shore of the lake from the head of the River St. Mary. We spent four days upon the banks of that stream, in the examination of its mineralogy, during which the principal part of our party was encamped at the mouth of the river. Gov. Cass, accompanied by such persons as were necessary in the exploration, proceeded, in two light canoes, to the large mass of copper which has already been described. We found the river broad, deep, and gentle for a distance, and serpentine in its course; then becoming narrower, with an increased velocity of current, and, before reaching the Copper Rock, full of rapids and difficult of ascent. We left our canoes at a point on the rapids, and proceeded on foot, across a rugged tract of country, around which the river formed an extensive semicircle. We came to the river again at the locality of copper. In the course of this curve the river is separated into two branches of nearly equal size. The copper lies on the right-hand fork, and it is subsequently ascertained that this branch is intercepted by three cataracts, at which the river descends over precipitous cliffs of sandstone. The aggregate fall of water at these cataracts has been estimated at seventy feet.

The channel of the river at the Copper Rock is rapid and shallow, and filled with detached masses of rock, which project above the water. The bed of the river is upon sandstone, similar to that under the Palisades on the Hudson. The waters are reddish, a color which they evidently owe to beds of ferruginous clay. The Copper Rock lies partly in the water. Other details in the geological structure and appearance of the country are interesting; but they do not appear to demand a more particular consideration in this report.

During our continuance upon this stream, we procured from an Indian a separate mass of copper weighing nearly nine pounds, which will be forwarded to the War Department. This specimen is partially enveloped with a crust of green carbonate of copper. Small fragments of quartz and sand adhere to the under side, upon which it would appear to have fallen in a liquid state. Several smaller pieces of this metal were procured during our excursion up the Ontonagon, or along the shores of the lake east of this stream.

It may be added that discoveries of masses of native copper, like those of gold and other metals, are generally considered indicative of the existence of mines in the neighborhood. The practical miner regards them as signs which point to larger bodies of the same metals, in the earth, and he is often determined by discoveries of this nature in the choice of the spot for commencing his labors. The predictions drawn from such evidence are more sanguine in proportion to the extent of the discovery. They are not, however, unerring indications, and appear liable to many exceptions. Metallic masses are sometimes found at great distances from their original repositories; and the latter, on the contrary, sometimes occur in the earth, or imbedded in rock strata, where there have been no great external discoveries.

From all the facts, which I have been able to collect on Lake Superior, and after a full deliberation upon them since my return, I have drawn the following conclusions:—

1. That the diluvial soil along the banks of the Ontonagon River, extending to its source, and embracing the contiguous region, which gives origin to the Monomonee River of Green Bay, and to the Wisconsin, Chippewa, and St. Croix Rivers of the Mississippi, contains very frequent, and several extraordinary masses of native, or metallic copper. But that no body of this metal, which is sufficiently extensive to become the object of profitable mining operations, has yet been found at any particular place. This conclusion is supported by the facts adduced, and, so far as theoretical aids can be relied upon, by an application of those facts to the theories of mining. A further extent of country might have been embraced, along the shores of Lake Superior, but the same remark appears applicable to it.

2. That a more intimate knowledge of the mineralogical resources of the country, may be expected to result in the discovery of valuable ores of copper, in the working of which occa-

sional masses and veins of the native metal, may materially enhance the advantages of mining. This inference is rendered probable by the actual state of discoveries, and by the geological character of the country.

These deductions embrace all I have to submit on the mineral geography of the country, so far as regards the copper mines. Other considerations arise from the facilities which the country may present for mining—its adaptation to the purposes of agriculture—the state and disposition of the Indian tribes, and other topics which a design to commence metallurgical operations would suggest. But I have not considered it incumbent upon me to enter into details upon these subjects. It may, in brief, be remarked that the remote situation of the country does not favor the pursuit of mining. It would require the employment of a military force to protect such operations. For, whatever may be their professions, the Indian tribes of the north possess strong natural jealousies, and in situations so remote, are only to be restrained from an indulgence in malignant passions, by the fear of military chastisement.

In looking upon the southern shore of Lake Superior, the period appears distant, when the advantages flowing from a military post upon that frontier, will be produced by the ordinary progress of our settlements—for it presents but few enticements for the agriculturalist. A considerable portion of the shore is rocky, and its alluvions are, in general, of too sandy and light a character for profitable husbandry. With an elevation of six hundred and forty-one feet above the Atlantic, and drawing its waters from territories situated north of the forty-sixth degree of north latitude, Lake Superior cannot be represented as enjoying a climate favorable to the productions of the vegetable kingdom. Its forest trees are chiefly those of the fir kind, mixed with varieties of the betula, lynn, oak, and maple. Meteorological observations indicate, however, a warm summer, the average observed heat of the month of June being 69. But the climate is subject to a long and severe winter, and to sudden transitions of the summer temperature. We saw no Indian corn among the natives.

A country lacking a fertile soil, may still become a rich mining country, like the county of Cornwall in England, the Hartz Mountains in Germany, and a portion of Missouri, in our own country.

But this deficiency must be compensated by the advantages of geographical position, a contiguous or redundant population, partial districts of good land, or a good market. To these, the mineral districts of Lake Superior can advance but a feeble claim, while it lies upwards of three hundred miles beyond the utmost point of our settlements, and in the occupation of savage tribes whose hostility has been so recently manifested.

Concerning the variety, importance, and extent of its latent mineral resources, I think little doubt can remain. Every fact which has been noticed tends to strengthen the belief that future observations will indicate extensive mines upon its shores, and render it an attractive field of mineralogical discovery. In the event of mining operations, the facilities of a ready transportation of the crude ores to the Sault de Ste. Marie, will point out that place as uniting, with a commanding geographical position, superior advantages for the reduction of the ores, and the general facilities of commerce. At this place, a fall of twenty-two feet, in the river, in the distance of half a mile, creates sufficient power to drive hydraulic works to any extent; while the surrounding country is such as to admit of an agricultural settlement.

I accompany this report with a geological sketch of a vertical section of the left bank of the Mississippi at St. Peter's, embracing a formation of native copper. This formation was first noticed by the officers of the garrison, who directed the quarrying of stone at this spot. The masses of copper found are small, none exceeding a pound in weight.

I have the honor to be, sir,

With great respect,

Your ob't servant,

HENRY R. SCHOOLCRAFT.

VII.

Observations on the Geology and Mineralogy of the Region embracing the Sources of the Mississippi River, and the Great Lake Basins, during the Expedition of 1820. Illustrated with Geological Profiles, and Numerous Diagrams and Views of Scenery. By HENRY R. SCHOOLCRAFT, U. S. Geol. and Minera. Exp.

To the Hon. JOHN C. CALHOUN, *Secretary of War,*

WASHINGTON, April 2, 1822.

SIR: I have the honor, herewith, to submit the general report of my observations on the geology and mineralogy of the region visited by the recent expedition to the sources of the Mississippi River. I transmitted to the Department on the 6th of November, 1820, a report on the existence of Copper Mines in the Basin of Lake Superior, together with specimens of the native metal, which were politely taken charge of at Albany by General Stephen Van Rensselaer, M. C. Will it be consistent with the views of the Department to print these reports?

I have the honor to be, sir,

Very respectfully,

Your obedient servant,

HENRY R. SCHOOLCRAFT.

REPLY.

WAR DEPARTMENT, April 6, 1822.

SIR: I have received your interesting report on the geology and mineralogy of that section of the western country embraced by the late expedition of Gov. Cass; and, although I have not had it in my power, as yet, to peruse it with attention, I will see you, at any time you please, on the subject of your letter respecting it.

I am, sir,

Respectfully,

Your obedient servant,

J. C. CALHOUN.

Mr. HENRY R. SCHOOLCRAFT.

SIR: Agreeably to your appointment as a member of the expedition to explore the sources of the Mississippi, by the way of the Lakes, I proceeded to join the party organized for that purpose at Detroit, by His Excellency Lewis Cass. Diurnal notes were kept of the changes in the geological features of the regions visited; of the mineralogy of the country; and of such facts as could be ascertained, with the means at command, to determine its general physical character and value.*

I have heretofore reported to you the facts and appearances which indicate the existence of the ores of copper, and of valuable deposits of copper in its native form, in the basin of Lake Superior—a point which constituted one of the primary objects to which my attention was called—and I now proceed to state such particulars in the topics confided to me as fell within my observation.

In generalizing the facts, it must be observed that the expedition had objects of a practical character relative to the number, disposition, and feelings to be learned respecting the Indian tribes; that the transit over large portions of the country was necessarily rapid; and that few opportunities of elaborate or long-continued observations occurred at any one point. The topography was committed to a gentleman who is every way qualified for that topic, who was well supplied with instruments, and who will do ample justice to that department. I make these remarks to prepare you for a class of observations which are necessarily technical, and quite imperfect, and to which it is felt that it will not be an easy task to impart a high degree of interest, whatever may have been the anticipations.

To prepare the mind to appreciate the account which I give of changes and developments in the physical structure of the country, it may be observed that the American continent has experienced some of the most striking mutations in its structure *at* and *north* of the great chain of lakes. That chain is itself rather the evidence of disruptions and upheavals of formations, which give its northern coasts, to some extent, the character of ancient—very

---

* The two geological profiles of the Mississippi Valley and the Lake Basins accompanying the original are here omitted; as, also, most of the illustrative views of scenery which accompanied the original.

ancient—volcanic areas of action. These lakes form—except Erie and Ontario—the general boundaries between the primitive and secondary strata. But, however striking this fact may, at particular localities, appear—such as at the Straits of St. Mary, of which the east and west shores are, geologically, of different construction—yet nothing in the grand phenomena of the whole region visited is so remarkable as the boulder stratum, which is spread, generally, from the north to the south. Some of the blocks of rock are enormous, and would seem to defy any known cause of removal from their parent beds; others are smaller, and have had their angles removed, and far the greater number of these transported boulders are quite smooth and rounded by the force of attrition. This drift stratum has been tossed and scattered from its northern latitudes over the surface of the limestones and sandstones of the south. It is mixed with the diluvial soils, in Michigan and elsewhere; but it is evident that, in its diffusion south, the heavier pieces have settled first, while comparatively minute boulders have been carried *over* or dropped in the plains and prairies of Ohio, Illinois, and more southerly regions. Nobody, with an eye to geology, can mistake the heavy boulder deposits which mark the southern shores of Huron, and become still more abundant on the St. Mary's, the shores of Lake Superior, and along the channels of the River St. Louis and the Upper Mississippi.

Lake Superior has been the central theatre of volcanic upheavals; but they must have operated at very remote periods, for there is not only no evidence of existing volcanic fires, but the heavy debris everywhere bespeaks long intervals of quietude, and slow elementary degradation. Some of the upheavals were made after the deposition of the sandstone rocks, which are, as at the foot of the Porcupine Mountains, raised up to stand nearly vertical; while other districts of the granitic rock, as at Granite Point, had been elevated before the deposition of the sandstone rock, which is accurately adjusted to its asperities, and remains quite horizontal.

The granitical series of strata, which is apparent in northern New York in the Kayaderasseras Mountains, and at the Thousand Islands of the St. Lawrence, reappear on the north shores of Huron and Superior, underlie the bed of the latter, and rise up

20

in the rough coast between the Chocolate River and Kewaiwenon, cross the Mississippi at the Petite Roche, above the Falls of St. Anthony, and put out spurs as low down as the source of the Fox, the St. Croix, and the head of the St. Peter's Rivers.

These glimpses of some of the leading points in the geological structure of the regions visited, will enable you to follow my details more understandingly. These details begin at Detroit. From this place the expedition passed, by water, along the southern shores of Lakes St. Clair, Huron, and Superior, to the Fond du Lac; thence, up the River St. Louis, to the Savanne summit. Thence we proceeded across the portage to Sandy Lake, which has an outlet into the Mississippi, and followed up the latter, through the lesser Lake Winnipek, to the entrance of the Turtle River, in Cass, or upper Red Cedar Lake, which is laid down by Pike in north latitude 47° 42' 40".* The state of the water was unfavorable to going higher.

From this point, which formed the terminus of the expedition, we descended the Mississippi, making portages around the Falls of Pekagama and St. Anthony, to Prairie du Chien. An excursion was made by me down the Mississippi to the mineral district of Dubuque. We ascended the Wisconsin, to the portage into the Fox River, and traced the latter down to its entrance into Green Bay. At this point, the expedition separated; a part proceeding north, through the bay, to Michilimackinac, and a part going south, along the west shores of Lake Michigan, to Chicago, the latitude of which is placed by Capt. Douglass in 41° 54' 06". At this place, a further division took place. Dr. Wolcott, having reached his station, remained. Governor Cass proceeded across the peninsula of Michigan to Detroit on horseback, leaving Capt. Douglass and myself to complete the survey of Lake Michigan. We rejoined the northern party detached at Green Bay, under Mr. Trowbridge and Mr. Doty, at Michilimackinac; and, after repassing the southern coast of Lakes Huron and St. Clair, reached Detroit.

Topographically, a very wide expanse of wilderness country had been seen. The entire length of route computed to have

* Pike's Expedition. This observation is corrected by Capt. Douglass to 47° 27' 10": the point of observation being, however, a few miles south.

been traversed, exceeds four thousand miles, in the course of which we had crossed nineteen portages, over which all the baggage and canoes were conveyed on the shoulders of men. We encountered actual resistance from the Indians at only one point.[*] I kept my journals continually before me, and had my pencil in hand every morning as soon as it was light enough to discern objects. I began my geological observations at Detroit.

This ancient city, founded by the French in 1701, stands upon an argillaceous stratum, which is divided, topographically, into an upper and lower bank. Wherever this clay has been examined by digging, it discloses pebbles of various species of rock, denoting it, as far as these extend, at least, to be a part of the great drift stratum.

In digging a well near the old Council House, in the northeast part of the city, the top soil appeared to be less than two feet. The workmen then passed through a stratum of blue clay, of eight or ten feet, when they struck a vein of coarse sand, six or eight inches in thickness, through which the water entered profusely. The digging was carried through another bed of blue clay, twenty or twenty-two feet in depth, when the men reached a stratum of fine yellow sand, into which they dug three feet and stopped, having found sufficient water. The whole depth of the well is thirty-three feet. The water is clear and rapid. No vegetable or other remains were found, and but few primitive pebbles.

In another well, situated near the centre of the town, the depth of which is twelve feet, the top soil was found to be two feet and a half; then a bed of gravel, seven feet; a vein of blue clay, eight inches, and the residue a whitish-blue clay, very compact and hard; a copious supply of water having been found. The water is, however, slightly colored, and is of a quality called hard.

In some places, this clay drift yields balls of iron pyrites, which renders the water unpalatable. At what depth the rock would be struck, if the excavation were continued, can only be conjectured. A well has been dug, a short distance below the city, upwards of sixty feet, chiefly through clay and gravel, without reaching the rock; but abraded fragments of granite and hornblende rocks were thrown from the greatest depths.

* *Vide* Narrative Journal.

308 APPENDIX.

The bed of the river opposite the city has been stated to consist of limestone rock, but without any proof or much probability. From the fact of its affording a good anchorage to vessels, I am inclined to think that it is wholly composed of clay and gravel.

DETROIT FLUVIATILE CLAY.—The argillaceous stratum of Detroit extends along both banks of the river to its head; passes around the shores of Lake St. Clair, and up the River St. Clair to Fort Gratiot—a distance of seventy miles. In this distance there are some moderate elevations and depressions in the surfaces of the soil, but no very striking changes in its general character and composition. The boulder stratum is prominent at Gros Point, at the foot of Lake St. Clair, where the shore exhibited some heavy blocks of granite, and other foreign rock.

ST. CLAIR FLATS OF PLASTIC CLAY.—At the mouth of the River St. Clair, the current is divided into several channels, and spread over a considerable tract of low ground, which is covered with grasses and aquatic plants. These channels have worn their way through beds of tough blue clay, called the flats, over which there is sometimes not over seven feet eight inches of water in the ship channel. They consequently form an impediment to commerce. The depth is, however, always increased in the spring season, when twelve inches more may be generally relied on. Frequently, during the droughts of summer, a change of wind, and its steady continuance for some time, will allow ships to pass without lighters. The permanent removal of this bar is, however, an object of national importance, which cannot but be felt, as the tonnage of the lakes increases.

ANCIENT DUNE; A BURIED FOREST.—The principal spot where the lands, in the immediate vicinity of the water, assume any considerable or abrupt elevation, is included between Black River of the St. Clair and Lake Huron. Here the outlet of the lake, which is rapid, washes the base of a ridge, or ancient dune, elevated fifty or sixty feet above the water. Fort Gratiot occupies the upper part of this elevation. The lower part consists of the blue clay stratum, corresponding in character with that found in the wells of Detroit. It is overlaid by a deposit of sand, forming two-thirds of the entire height. This elevation is crowned with a light forest of oak and other species. At the line of junction between the sand and clay, a number of trees are seen to be horizontally imbedded, projecting their roots and trunks in a striking manner above the water. These trees, on inspection, are merely preserved, not petrified. They appear to have been exposed to view, in modern times, by the wearing away of the bank. Certainly, none of the old travellers mention them.

The mode of this formation may be clearly seen. Winds, at some ancient period, have been the agent of blowing the sands, as they were washed up by the lake, and redepositing them on part of a prostrated forest, resting directly on the clay stratum. The trees, thus buried in dry sand, have been preserved. In process of time, the river encroached upon these antique beds, exposing them to view. There are also antique fresh-water shells found in similar positions near this spot. No rock is, thus far, found *in sitû* in ascending the lakes. The old surface of the country is wholly of diluvial formation, except where it shows lake action.

HURON COAST FROM FORT GRATIOT TO MICHILIMACKINAC.— About two hundred and thirty miles lie stretched out between these two points. Lake Huron charms the eye, with the view of its freshness and oceanic expanse. But the entrance is without rock scenery, and the student of its geology must be a patient gleaner along its shores. Long coasts of sand and gravel extend before the eye, and they are surmounted, at a moderate elevation, with a dense foliage, which limits the view of its structure to a narrow line. Portions of this coast are heavily loaded with the primitive debris* from the North. These are found, in some places, in heavy masses, but all are more or less abraded, showing that they have been transported from their original beds. In one of these, I observed crystals of staurotide.

The first section of this coast reaches from Fort Gratiot to Point aux Barques, a distance of about seventy-five miles. Nearly midway lies the White Rock, a very large boulder of whitish-gray semi-crystalline limestone, lying off the shore about half a mile, in water of about one and a half fathom's depth. It is the effect of gulls lighting upon this rock, and not the intensity of the color of the stone, that has originated the name—

---

* In 1824, an Indian brought me a specimen of native silver found on this part of the coast. It was imbedded in a boulder of mixed granite and steatite.

which is a translation of the *Róche Blanche* of the older *voyageurs*. The Detroit clay-formation still characterizes the coast.

FIRST EMERGENCE OF ROCK, IN PLACE, ABOVE THE SURFACE.— We are passing, in this section, along and near to the outcrop of the secondary strata of the peninsula, but these strata are covered with a heavy deposit of diluvial clays, sands, and pebble drift. The first emergence of fixed rocks, above the line of the drift, occurs after passing Elm Creek in the advance to Ship Point (*Pointe àux Barques*). It is a species of coarse gray, loosely compacted sandstone, in horizontal layers. This rock continues to characterize the coast to and around the Ship Point promontory into Saganaw Bay. It possesses a few fossil remains of corallines; but the rock is not of sufficient compactness and durability for architectural purposes. It is conjectured to be one of the outlying series of the coal measures, of which this coast exhibits, further on, other evidences.

SAGANAW BAY.—The phenomena of this large body of water, which is some sixty miles long, appear to indicate an original rent in the stratification, having its centre of action very deep. If the peninsula of Michigan be likened to a huge fish's head, this bay may be considered as its open mouth. We crossed the inner bay from Point aux Chenes, where it is estimated to be twenty miles across.* The traverse is broken by an island, to which the Indians, with us, applied the name of Sha-wan-gunk.† It is composed of a dark-colored limestone, of dull and earthy fracture and compact structure. It presents broken and denuded edges at the water level. I observed in it nodular masses of chalcedony and calc. spar. The margin of the island bears fragments of the boulder stratum.

HIGHLANDS OF SAUBLE.—On crossing the bay, these highlands present themselves to view in the distance. They are the northeastern verge of the most elevated central strata of the peninsula. Their structure can only be inferred from the formations along

---

* Ships make the traverse where it is sixty miles wide.

† The reason of this name I did not learn. It is apparently the same name as that bestowed on a mountain range in Orange and Ulster Counties, New York, lying south of the Catskills, where it is sometimes called, for short, Shongum. The meaning is, evidently, something like South-land-place. The local *unk* may be translated hill, island, continent, &c. &c.

the margin of the lake, extending by Thunder Bay and Presque Isle, and the Isles of Bois Blanc and Round Island to Michili-mackinac. At Thunder Bay, the compact limestone of the Saga-naw Islands reappears, and is constantly in sight from this point to Presque Isle. It exists in connection with bituminous shale, at an island in Thunder Bay. It is of a dark carbonaceous cha-racter on the main opposite Middle Island, at a point which is called by the Indians *Sho-sho-ná-bi-kó-king*, or Place of the Smooth Rock. I noticed at this point the cyathophyllum helianthoides in abundance, and easily detached them from the rock. The more compact portions of this formation in the approach to Presque Isle, disclosed the ammonite, two species of the gorgonia, and the fragment of a species of chambered shell, whose character is inde-terminate.

Much of the coast was footed, as the winds were adverse, and its debris thus subjected to a careful scrutiny. Wherever the limestone was broken up or receded from the water, long lines of yellow beach-sand and lake-gravel, including members of the erratic block stratum, intervened. In some localities, local beds of iron sand occur.

MICHILIMACKINAC.*—The approach to this island was screened from our view by the woody shores and forests of Bois Blanc, an island of some twelve miles in length lying off the main land; and the view of it first burst upon us in the narrow channel be-tween it and Round Island. It is a striking geological monument of mutations. Here the calcareous rock, which had before exhi-bited itself in low ledges along the shore is piled up in masses, which reach an extreme altitude of three hundred and twelve feet. About two hundred feet of this elevation is precipitous on its south, east, and west edge. A hundred feet or more is piled up on its centre, part rock and part soil, in a crowning shape. The highest part of this apex, which is surmounted by the ruins · of Fort Holmes, consists of the drift stratum,.among which are boulders of sienite, and other foreign rocks. A locality of these abraded boulder-rocks, near the Dousman farm, is worthy of a visit from all who take an interest in the phenomena of boulders dis-

* The name, as pronounced by the Indians, is Mich-en-i-mack-in-ong, meaning Place of Turtle Spirits, a notion of their mythology. It was anciently deemed a sacred spot, or one where Monetoes revealed themselves.

persed over the continent. The fishermen represent the water around this island to be eighty fathoms in depth. Yet, across these waters, to the utmost altitude of the island, these blocks of foreign rock have been transported. No force capable of effecting this is now known. And the argument of their having been transported on cakes of ice, in the nascent periods of the globe, is rendered stronger by these appearances than any geological proofs which I have yet seen.

DISTINCTIVE CHARACTER OF THE MACKINAC LIMESTONE.—Nothing appears so completely to puzzle the observer as the first glance at this rock. It is different in appearance from the calcareous rocks, to which my attention has heretofore been called in Western New York, and in Missouri and Illinois. The difficulty is to find a point of comparison. I walked entirely around the island, partly in water, the northern shores being comparatively low. There appeared to be three layers. The first, which rises up from the depths of the lake, scarcely, if at all, reaches the water level. Upon this is superimposed a vesicular rock, of which the vesicles are filled with carbonate of lime in the state of agaric mineral. By exposure to the air, this substance readily decomposes, and assumes an almost limey whiteness, and sometimes a complete pulverulent state. The reticular, or vesicular lines, by which the mass is held together, are thus weakened, and large masses of the craggy parts fall, and assume the condition of debris at the water's edge. Some conditions of the reticulated filaments are covered with minute crystals of cal. spar; others of minutely crystallized quartz. There appear, at other localities, in low positions, layers of quartz in the condition of a coarse bluish, flinty, striped agate. The entire stratum appears to be a reproduced mass, which is plainly denoted, if I mistake not, by some imbedded masses of an elder lime-rock. The whole stratum is too shelly and fissured to be of value for economical purposes. It yields neither quicklime nor building stone.

Fort Mackinac is erected on the summit of this stratum. The two objects of curiosity, called the Arched Rock, and the point called Robinson's Folly, are evidences of this tendency of the cliffs to disintegration. The superior stratum which constitutes the nucleus of the Fort Holmes' summit, contains more silex, diffused throughout its structure. It is, however, of a loose, though hard and shelly character; and has, in the geological mutations of the island been chiefly demolished and washed away. The monumental mass of this period of demolition, called the Sugar Loaf, is a proof that it contained, either by its shape, or otherwise, a superior power of resisting these means of ancient prostration. Striking as it now appears, this is the simple story which it tells. Its apex is probably level, or nearly so, with the Fort Holmes's summit. Over the whole island, after these demolitions, the drift stratum was deposited.

The German geognosts apply the term *mushelkalk*, to this species of calcareous rock. It is, apparently, the magnesian limestone of English writers.

ANCIENT WATER LINES.—Such marks appear on the most compact parts of the cliffs, denoting the water to have stood, during the ancient boundaries of the lake, at higher levels.

LAKE ACTION.—It is known that strong currents set into the Straits of Michilimackinac, and out of it, from Lake Michigan, at this point. The fishermen, who set their nets at 'four hundred feet in the waters, often bring up, entangled in their nets, large compact masses of limestone, which have been fretted into a kind of lacework, by the rotatory motion of little pebbles and grains of sand, kept in perpetual motion by the water at the bottom of the lake.

ORGANIC IMPRESSIONS.—There are cast up among the lake debris of this island, casts of some species of orthocaratites, ammonites, and madrepores, which appear to be derived from the calcareous rocks in place in the basin of Lake Huron. But the rock strata of the island itself appear to be singularly destitute of these remains. The only species which I have noticed, is one that was thrown up from a well attempted to be dug, on the apex of Fort Holmes, by the British troops, while they held possession of the island in 1813, 1814, and 1815. But this is uniformly fragmentary. It has the precise appearance of the head of a trilobite, but never reveals the whole of the lateral lobes, nor any of the essential connecting parts. It is silicious.

GYPSEUS FORMATION.—Evidences of the extension of this formation to this vicinity were brought to my notice; in consequence of which I visited the St. Martin's Islands, which belong to the Mackinac group. Masses of gypsum were found imbedded in the soil, both of the fibrous and compact variety. These islands are low diluvial formations. Similar masses are found on Goose Island; and the mineral has been found at 'Point St. Ignace on the main land.

Taken in connection with the discovery of this mineral, at a subsequent part of the journey on Grand River, the indications of the series of the saline group of rocks, so prevalent in the Mississippi Valley, are quite clear up to this extreme point, which is, however, very near the northern verge of this group.

HONEYCOMBED ROCKS.—As evidences of existing lake action, it has already been mentioned that the fishermen bring up, from great depths in the straits, pieces of compact limestone, completely fretted and excavated by small pebbles, which are kept in motion by the strong currents which prevail at profound depths. The process of their formation by these currents is such, as in some instances to give the appearance of cellepores,, and analogous forms of organic life. I have seen nothing in these carious forms which does not reveal the mechanical action of these waters.

PSEUDOMORPHIC FORMS.—Amongst the limestone debris, of recent date, found on these shores, are pieces of rock which have an appearance as if they had been punctured with a lancet, or blade of a penknife. These incisions are numerous, and from their regularity, appear to have been moulded on some crystals which have subsequently decayed. Yet, there are difficulties in supposing such to have been the origin of these small angular orifices.

Whenever these masses are examined by obtaining a fresh fracture, they are found to consist of the compact gray and semigranular rock of the inferior Mackinac group, but in no instance of the vesicular or silicious varieties. These blocks appear to be identical in character with the White Rock, before noticed.

NORTH SHORE OF LAKE HURON.—The next portion of the country examined was that of the north shores of the lake, extending from Michilimackinac to Point Detour, the west Cape of the Straits of St. Mary's, a distance computed to be forty miles. The calcareous rock, such as it appears in the inferior stratum of Mackinac, extends along this coast. The first three leagues of it, consist of an open traverse across an arm of the lake. Goose

Island offers a shelter to the voyager, which is generally embraced. It consists of an accumulation of pebbles and boulders on a reef, with a light soil, resting on the lower limestone. It does not, perhaps, at any point, rise to an elevation of more than eight or ten feet above the water. Outard Point, a short league, or rather three miles further, exhibits the same underlying formation of rock, which is found wherever solid points put out into the lake, during the entire distance. The chain of islands called Chenos, extends about twenty miles, and affords shelter during storms to boatmen and canoemen, who are compelled to pass this coast. Large masses of the rock, with its angles quite entire, lie along parts of the shore, and appear to have been but recently detached. The intervals between these blocks and points of coast, are formed of the loose sand and pebbles of the lake, which are more or less affected by every tempest. The only organic remains and impressions are drift-specimens, which have been driven about by the waves, and are abraded. Broken valves of the anadonta, occasionally found in similar positions, denote that this species exists in the region, but that the outer localities of the coast are entirely unfavorable to their growth.

DRUMMOND ISLAND.—This island, now in the possession of British troops, who removed from Michilimackinac in 1816, is the western terminus of the Manatouline chain. We did not visit it, but learn from authentic sources, that it is a continuation of the nether Mackinac limestone—and that the locality abounds in loose petrifactions, which appear to have belonged to an upper stratum of the rock, now disrupted.*

STRAITS OF ST. MARY'S.—These straits, and the river which falls into their head, connect Lakes Huron and Superior. They appear to occupy the ancient line of junction between the great calcareous and granitic series of rocks on the continent. The limestone, which has been noticed along the north shore of the

---

* Dr. John Bigsby, in a memoir read before the London Geological Society, has described and figured several of these. In a memoir by Charles Stokes, Esq., of London, read before this Society in June, 1837, some of its most striking fossils are figured and described, with references to the prior discoveries of Dr. Bigsby, Captain Bayfield, and Dr. Richardson. Six new species of the Arctinoceras, and five of the Huronia, Ormoceras, and Orthoceras, are figured and described in the most splendid manner. This memoir is essential to all who would understand its fossil history, and that of the North generally.

Huron from Michilimackinac, and which continues, with interruptions of water only, from Detour to Drummond Island, and
the Manatoulines, is to be noticed up the straits as high as Isle a
la Crosse, where the last locality of a pure carbonate of lime
appears to occur. The island of St. Joseph is chiefly primitive
rock, and its south end is heavily loaded with granitic, porphy-
ritic, and quartz boulders. The north shores of the river, opposite and above this island, are entirely of the granitic series,
which continues to Gros Cape of Lake Superior. On reaching
the *Nebeesh,*[*] or Sailor's Encampment Island, sandstone rocks of
a red color present themselves, and are found also on the American side of the river, and continue to characterize it to the Falls,
or Sault de Ste. Marie,[†] and to Point Iroquois and Isle Parisien
in Lake Superior.

The Sault of St. Mary's is *upon* and *over* this red sandstone.
The river makes several successive leaps, of a few feet at a time,
in its central channel, falling, altogether, about twenty-two feet in
half a mile. This gives it a foaming appearance, and the volume
pours a heavy murmur on the ear.[‡] It is, of course, a complete
interruption to the navigation of vessels, which can, however,
come to anchor near its foot, while barges may be pushed up,
empty, on the American shore. The water-power created by
such a change of level, is such as must commend the spot, at a
future period, to manufacturers, lumbermen, and miners. The
foot of these falls is heavily incumbered, both with masses of the
disrupted sand-rock§ and granitic and conglomerate boulders.

RED SANDSTONE OF LAKE SUPERIOR.—That this is the old red
sandstone, may be inferred simply from the fact that, although

* Strong water.
† Reached somewhere about 1641, by the French missionaries.
‡ In 1825, Lieutenant Charles F. Morton, U. S. A., sent to my office a mass of
this red sand rock, of about twelve inches diameter, perfectly round and ball-shaped,
which he had directed one of the soldiers to pick up, in an excursion among the
islands of the lower St. Mary's. This ball was a monument of that physical throe
which had originally carried this river through the sandstone pass of St. Mary's,
having been manifestly rounded in what geologists have called "a pocket hole" in
the rock at the falls, and afterwards carried away, with the disrupted rocks, down
the valley.

§ The Indians call it *Pauwateeg* (water leaping on the rocks), when speaking of
the phenomenon, and *Pawating*, when referring to the place of it.

deposited originally in horizontal beds, its position has been disturbed in many localities.

PLASTIC CLAY STRATUM OF THE LAKES.—The northern extremity of Muddy Lake—a sheet of water some twenty miles in length—is the head of the straits, and the beginning of the River St. Mary's. This sheet of water has the property of being rendered slightly whitish, or turbid, by continuous winds. Its bottom appears to be formed of the same plastic blue clay which obstructs the passage of vessels of large draft on the St. Clair flats, and forms an impediment of a similar kind in this river in Lake George. This stratum seems to be the result of causes not now in operation. If dredged through, or excavated, there is no reason to suppose it would again accumulate; for the waters of the lake are clear and pure, and carry down no deposit of the kind. These clay deposits remain to attest physical changes which are past. They denote the demolition of formations of slate in the upper regions, which have been broken down and washed away when the dominion of the waters was far more potential than they now are.

This formation is favorable to the growth of some species of fresh-water shells. I observed several species of the anadonta and the plenorbis, and think, from the broken valves, that research would develop others.

PORPHYRY AND CONGLOMERATE BOULDERS.—A formation of red jasper, in common white quartz, exists, in the bed of intersection, on the southeastern foot of Sugar Island. The fragments of jasper are of a bright vermil red, quite opaque, and have preserved their angles. I had observed fragments of the formation along the shores of the lower part of the straits, and even picked up some specimens, entirely abraded, however, on the south shores of the Huron, between the White Rock and Michilimackinac—a proof of the course of the drift.

The granitic conglomerates appear quite conclusive, one would think, of the results of fusion. The attraction of aggregation would seem inadequate to hold together such diverse masses. In these curious and striking masses we see the red feldspathic granite, black and shining hornblende rock, white fatty quartz, and striped jasper, held together as firmly, and polished by attrition as completely, as if they were—what they are not—the results of crystallization in this aggregate form.

ERRATIC BLOCK GROUP.—Wherever, in fact, the geologist sets his foot, on the shores of the upper lakes, he finds himself on the great drift stratum, and cannot but revert to that era when waters, on a grander scale, swept over these plains, and the lakes played rampantly over wider areas.*

BASIN OF LAKE SUPERIOR.—We entered this island sea as if by a kind of geological gate, in which the sandstone cliffs of Point Iroquois, on the one hand, stand opposite to the granitical hills of Gross Cape on the other.

In order to conceive of its geology, it may subserve the purposes of description to compare it to a vast basonic crater. The rim of this crater has been estimated, by Sir Alexander Mackenzie, at fifteen hundred miles. The primitive formations of Labrador and Hudson's Bay coasts come up, so as to form the eastern and northern sides of the rim, around which they stand in cliffs of sienitic greenstone and hornblendic rocks, in some places a thousand feet high. On its south and southwest shores, this formation of the elder class of rocks forms also a considerable portion of the coast; as in the rough tract of Granite Point, the Porcupine and Iron River Mountains, and the primitive tract west of Chegoimegon, or Lapointe. It will serve to denote the broken character of this rim, if we state that the entire plain of the lake, running against and fitting to this rim, was originally filled up with the red, gray, and mottled sandstone, which gave way and fell in at localities west of the great Keweena Peninsula, converting its bottom into an anteclinal axis.

Volcanic action, to which this disturbance in its westerly

---

* During a subsequent residence of eleven years at this point, the excavations made on both sides of the river, in digging wells, canals made by the military, &c., fully demonstrated the truth of this general observation. In these positions, it was evident that some greatly superior force of watery removal, such as does not now exist, had heaped together particles of similar matters, according to laws which govern moving, compacted masses of water, leaving clay to settle according to the laws of diffused clay, sand of sand, and pebbles and boulders of pebbles and boulders. In their change and redeposit, gravity has evidently been the primary cause, modified by compressed currents, attraction, and probably those secret and still undeveloped magnetic and electric influences which exist in connection with astronomical phenomena. That the earth's surface, "standing out of the water and in the water," has been disrupted and preyed upon by oceanic power, no one, at this day of geological illumination, will deny.

bearings may be attributed, appears to have thrown up the trap-rocks of the Pic, of the Porcupine chain, of the Isle Royal group, and other trap islands, and the long peninsula of Keweena. This system of forces appears to have spent itself from the northeast to the southwest. The shocks brought with them the elements of the copper and other metallic bodies which characterize the trap-rock. They exhausted their power, on the American side, west of the granitic tract of Chocolate and Dead Rivers, and the Totosh and Cradle-Top Mountains. The most violent disturbance took place at the west of the Keweena Peninsula, and thence it was propagated in the direction of the higher Ontonagon, the Iron, and the Montreal rivers.

This disturbance of the level of the sandstone produced undulations, which are observable on the St. Mary's, where the variation from a level is not more than eight or ten degrees. They left portions of it—as between Isle au Train and the Firesteel River—undisturbed; and they threw other portions of it—as between Iron and Montreal rivers—almost completely on their edges.

The entire north shore from Gargontwa to the old Grand Portage, inclusive of the Michepicotin and Pic regions, cannot be particularly alluded to, as that part of the coast was not visited; but the accounts of observers represent it as consisting of trap-rocks. Without the application of such forces, it appears impossible to understand the geology of this lake, or to account for the sectional and disturbed formations.

The lake itself, whose depth is great, and which has an extreme length of about 500 miles, by an extreme width of some 180, is endowed with powerful means of existing elemental action. This consists almost entirely of the force of its winds and long, sweeping waves. Its bottom may, in this light, be looked upon as an immense mortar or triturating apparatus, in which its sandstones, trap-boulders, and pebbles are driven about and comminuted. This power has greatly changed its configuration, and the process of these mutations is daily going on.

It is only by such a power of geological action that we can account for the powerful demolitions and inroads which it has made upon some parts of its southern borders. The coasts of the Pictured Rocks, which have a prominent development of about 12 to 15 miles, consist in horizontal strata of coarse gray sandstone, of little cohering power. The effect of waves beating upon rocks is to communicate a curved line. This has operated to excavate numerous and extensive caves into the coast. These, after reaching hundreds of feet, have in some cases united. The effect is to isolate portions of the coast, and to leave it in fearful pinnacles, having many of the architectural characters of Gothic or Doric ruins.

The portion of coast immediately west of Grand Marrais is scarcely less unique. It denotes the effect of the prostrating power of the lake in another way. The sandstone of parts of the coast, ground down into yellow sand by this vast machinery, is lifted up by the winds as soon as it reaches the point of dryness, and heaped up into vast dunes. Standing trees are buried in these tempests of sand, and its effect is, for about nine miles along the coast, to present, at an elevation of several hundred feet, a scene of arid desolation, which can only be equalled by the Arabic deserts.

A dyke of trap seems once to have extended from the north shore to Point Keweena; but, if so, it has been prostrated, and its contents—veins and deposits, silicious and metallic—scattered profusely around the shores of the lakes. A cause less general is hardly sufficient to account for the wide distribution of fragments of the copper veins and vein-stones which have so long been noticed as characters of this lake. The basal remains of this antique dyke form the peninsula of Keweena. The tempests beating against this barrier from the northwest, have ripped up terrific areas from the solid rock, and left its covering, amygdaloid and rubblestones, in fantastic patches upon the more solid parts, or constituting islands in front of them.

STRUCTURE OF ITS SOUTHERN COAST.—The estimated distance from Sault Ste. Marie to Fond du Lac is a fraction over 500 miles. The sandstone, as it appears in the Falls of the St. Mary's, does not appear to be entirely level. It exhibits an undulation of about 8° or 10°, dipping to west-northwest. Two instances of this waved stratification of the Lake Superior sandstone deserve notice. The first terminates at the intersection of red sand rock at la Point des Grande Sables with the beginning of the horizontal strata of the Pictured Rocks. We again observe an inclination of the strata of a few degrees at Grand Island, which is more ingfish River, and appears to dip at Isle aux Trains, about twenty miles northeast. The scenery is peculiarly soft and pleasing in passing the Huron Islands, a granitic group, and directing the view, as in the sketch, to the coast and the rough granitical hills rising behind Huron Bay. The strata are level, as shown above, around the Bay of Presque Isle and Granite Point, and continue so, resting on the roots of the granitical tract of the *Tötosh*, or Schoolcraft, and Cradletop Mountains, and at Point aux Beignes, and Keweena Bay. This level position of the rock is preserved to the south cape of the shallow bay of the Bete Gre, on the north, at which the trap-dykes of the peninsula first begin; and so continues after passing that rugged coast of the vitreous series of that remarkable point, to and beyond Eagle River and Sandy Bay, in the approach to the portage of the Keweena.

The same horizontality is observed on the headland west of it, and upon all the points and headlands to Misery and Firesteel Rivers and the mouth of the Ontonagon. The trap-dyke of Keweena crosses this river about ten miles, in a direct line, inland.

At Iron River, we observe a stratum of compact gray grauwacke, over the hackly bed of which that river forces its way during the spring months, and stands in tanks and pools during the summer. On reaching the foot of the Porcupine Mountains, the sandstone, which is here of a dark chocolate color, with quartz pebbles of the bigness of a pigeon's egg, and organic remains of palæozoic type, is found to be tilted up into nearly a vertical position, as shown in the sketch. The grauwacke reappears, in a most striking manner, at the Falls of Presque Isle River, where the whole mass of water precipitated from the highlands drops into a vast pot-hole, a hundred feet wide and perhaps twice that depth. The whole upper series of rocks, from the Porcupine Cliffs west to the Montreal River, is a conglomerate. At the Falls of the Montreal, the river drops over the vertical edges of the red sandstone. Beyond the Bay of St. Chares, at Lapointe Chegoimigon, masses of sienitic mountains arise, which have their apex near La Riviere de Fromboise.

The Islands of the Twelve Apostles, or Federation Group, appear to be all based on the sienitic or trap, with overlying red sandstone; which latter again reappears on the point of the entrance into Fond du Lac Bay, and marks its southern coast, till

21

near the entrance of the Brulé, or Misakoda River, as seen in the illustration beneath. Shores of sand then intercept its view to the entrance of the River St. Louis, and up its channel to its first rapids, about eighteen miles, where the red sandstone again appears, as the first series of the Cabotian Mountains.

SERPENTINE ROCK.—At the nearest point north of Rivier du Mort is a headland of this rock, jutting out from the granitical formation. Lapping against it, at the mouth of the river, is a curious formation of magnesian breccia. The serpentine rock appears, in nearly every locality examined, to be highly charged with particles of chromate of iron. It may be expected to yield the usual magnesian minerals.* Its position is between the Carp River and Granite Point, in the Bay of Presque Isle, or rather Chocolate River, for that river pours into this bay by far the largest quantity of water.†

ANCIENT DRIFT-STRATUM.—In the intervals between the points and headlands, where the rock formation is exposed by streams or gorges, the drift, or erratic boulder stratum, is found. Such is its position beneath the sand-dunes of the Grandes Sables, and in the elder plains and uplands, stretching with interruptions on the coast from the head of the Mary's valley to that of the St. Louis. The edge of this formation is composed of the sand and loose pebbles and boulders of the lake. Mighty as are the existing causes of action of the lake in beating down and disrupting strata of every kind, and in reproducing alluvial lands and dunes, they are weak and local when compared to the causes which have spread these ponderous boulders, and drift masses over latitudes and longitudes which appear to be limited only by the leading elevations of the continent. That oceanic torrents of water, suddenly heaped on the land, and wedged into compactness and power now unknown to it, is after all, the most plausible theory of the dispersion of this formation, and this theory avoids the necessary local one of the glacial dispersion which presupposes a very low temperature over the whole surface of the globe.

* In 1831, in making some explorations of this rock with gunpowder, I found the serpentine in a crystalline state, of a beautiful deep-green color, but appearing as if the crystallization was pseudomorphous.

† The extensive iron mines of Marquette County, Upper Michigan, are now worked in this vicinity.

KAUGWUDJU.*—This imposing mass of the trap-rocks is the highest on the southern shores of Lake Superior. The following outlines of it are taken from a point on the approach to the Ontonagon River, about forty miles distant.

They rise to their apex about thirty miles west of that stream, in north lat. 46° 52' 2", as observed by Captain Douglass. They are distant three hundred and fifty miles from St. Mary's. In a serene day they present a lofty outline, and were seen by us from the east, at the distance of about eighty miles. The Indians represent them to have a deep tarn, with very imposing perpendicular walls, at one of the highest points. If Lake Superior be estimated at six hundred and forty feet above the Atlantic, as my notes indicate, its peaks are higher than any estimates we have of the source of the Mississippi, and are, at least, the highest elevations on this part of the continent. The granitical tract of the St. Francis, 'Missouri,† and of the quartz high lands of Wachita, Arkansas, the only two known primitive elevations between the Rocky and Alleghany chains, are far less elevated.

I have now taken a rapid glance at the formations along the southern shore of the lake between St. Mary's and Fond du Lac; but have passed by some features which may be thought to merit attention.

EXISTING LAKE DRIFT.—The gleaner among the rock debris of this lake has a field of labor which is not dissimilar to that of the fossilist. If he has not, so to say, to put joint to joint, to establish his conclusions, he has a mineralogical adjustment to make every way as obscure. A boulder of sienite, or a mass of sandstone, or grauwacke, may be easily referred to a contiguous rock. But when the observer meets with species which are apparently foreign to the region, he is placed in a dilemma between the toil of an impossible scrutiny and the danger of an unlicensed conjecture.

Among the more common masses which may be assigned a locality within the compass of the lake, are granites, sienites, hornblendes, greenstones, schists, traps, grauwackes, sandstones, porphyries, quartz rocks, serpentines, breccias, amygdaloids, am-

* Porcupine Mountains. From *kaug*, a porcupine, and *wudju*, mountain.
† *Vide* my view of the lead mines, in the Appendix to " Scenes and Adventures in the Ozark Mountains."

phiboles, and a variety of masses in which epidote and horn-blende are essential constituents. With these, the coast mineralogist must associate, in place or out of place, agates, chalcedonies, carnelians, zeolite, prehnite, calcareous spar, crystalline quartz, amethystine quartz, coarse jaspers, noble serpentine, iron-sand, iron-glance, sulphate of lead, chromate of iron, native copper, carbonate of copper, and various species of pyrites. These were, at least, my principal rewards for about eighteen days' labor, in scrutinizing, at every possible point, its lengthened and varied coasts.

CUPREOUS FORMATION.—The whole region, above Grand Island at least, appears to have been the theatre of trap-dykes, and an extensive action from beneath, which brought to the surface the elements of the formation of copper veins. These have not been much explored; but, so far as observation goes, there are evidences which cannot be resisted, that the region contains this metal in various shapes and great abundance. I refer to my report of the 6th of November, 1820, for evidences of a valuable deposit of this metal in the valley of the Ontonagon River, and at other points. I found the metal in its native state at various other localities, and always under physical evidences which denoted its existence, in the geological column of the lake, in quantity. These indications were confined almost exclusively to the area intervening between the peninsula of Keweena, and La Pointe Chegoimegon, a distance of about one hundred and fifty miles. Of this district, the two extremities would make the Ontonagon Valley about the centre.* A profile of one of the detached pieces, found in the Ontonagon Valley, and forwarded to you by Mr. Van Rensselaer, is herewith given.

VITRIC BOULDERS.—Among the debris of Lake Superior are masses of trachyte, and also small pieces of the sienitic series, in which the red feldspar has a calcined appearance, the quartz being, at the same time, converted into a perfectly vitreous texture. Similar productions, but not of the same exact character, exist on the sandy summits of the Grande Sable. These

---

* I would also refer, for subsequent information, to my report of the 1st of October, 1822, made in compliance to a resolution of the Senate, and printed in the Executive Documents of that year, No. 365, 17th Congress, 2d session.

exhibit an exterior of glistening cells or orifices: it may be possible that they have been produced by fusion; but I think not. The smooth cells appear like grains of sand hurled by the winds over these bleak dunes. I have brought from that locality a single specimen of pitchstone, perfectly resinous, bleak and shining.

LA POINTE CHEGOIMEGON.—A sketch of these islands, as given in the Narrative, denotes that their number is greatly underrated, and will serve to show the configuration of a very marked part of the Superior coast. It must, hereafter, become one of the principal harbors and anchoring-ground for vessels of the lake.

VALLEY OF THE ST. LOUIS RIVER.—The St. Louis River takes its rise on the southern side of the Hauteur des Terres, being the same formation of the drift and erratic block stratum which gives origin, at a more westerly point, to the Mississippi. Its tributaries lie northwest of the Rainy Lakes. Vermilion Lake, a well-known point of Indian trade, is a tributary to its volume, which is large, and its outlet rushes with a great impetus to the lake. At what height its sources lie above Lake Superior, we can only conjecture. It was estimated to have a fall of two hundred and nine feet to the head of the Portage aux Coteaux, and may have a similar rise above.

By far its most distinguishing feature is its passage at the Grand Portage through the Cabotian Mountains. We entered it at Fond du Lac and pursued up its channel through alluvial grounds, in which it winds with a deep channel about nineteen or twenty miles to the foot of its first rapids. This point was found one mile above the station of the American Fur Company's trading-house. Here we encountered the first rock stratum, in the shape of our old geological acquaintance, the old red sandstone of Lake Superior. It was succeeded in the first sixteen miles, in the course of which the river is estimated to fall two hundred feet—most of it in the first twenty-nine miles—by trap, argillite, and grauwacke. Through these barriers the water forces its way, producing a series of rapids and falls which the observer often beholds with amazement. The river is continually in a foam for nine miles, and the wonder is that such a furious and heavy volume of water should not have prostrated everything before it. The sandstone, grauwacke, and the argillite, the latter

of which stands on its edges, have opposed but a feeble barrier;
but the trap species, resisting with the firmness, as it has the color
of cast-iron, stand in masses which threaten the life and safety of
everything which may be hurled against them. I found a loose
specimen of sulphuret of lead and some common quartz in place
in the slate rock, a vein of clorite slate, and a locality of coarse
graphite, to reward my search.

The Portage aux Coteaux, which is over the basetting edges of
the argillite, will give a lively idea of the effects of this rock
upon the feet of the loaded voyageurs.

The sandstone is last seen near the Galley on the Nine Mile
Portage. Above the Knife Portage, some eight miles higher,
vast black boulders of hornblendic and basaltic blocks, are more
frequent; and these masses are observed to be more angular in
their shapes than the boulders and blocks of kindred character
encountered on the shores of Lakes Superior and Huron. There
is a vast sphagnous formation, which spreads westwardly from
the head of the Coteau Portage, and gives rise to the remote tributaries of Milles lac and Rum River. Much of this consists of
what the Indians term *muskeeg*, or elastic bog. Hurricanes and
tempests have made fearful inroads upon areas of its timber, and
it is seldom crossed, even by the Indians. This tract lies east of
the summit of sandhills and drift, which environ Sandy Lake, the
Komtaguma of the Chippewas. The portage of the Savanna
River, a tributary of the St. Louis, is the route pursued by persons with canoes; there is no other species of water craft adapted to this navigation. But wherever crossed, this swamp-land
tract imposes labor and toil which are of no ordinary cast. It is
the equivalent of the argillite which has been broken down and
disintegrated, forming beds of clay soil which are impervious to
the water, and we may regard this ancient slate formation of
the true source of the St. Lawrence tributaries, as the remote
origin of those extensive beds of an argillaceous kind, which
exist at many places in the lower lakes and plains.

Immediately west of the Savanna Portage, the Komtagama
summit is reached. This summit consists wholly of arid pebble
and boulder drift of the elder period. It exhibits evidences of
broken-down amygdaloids, which not only furnish a part of its
pebbles, but also of the contents of this stratum, in numerous

agates and other subspecies of the quartz family which are found scattered over the surface. This is, in fact, the origin of that extensive diffusion of these species, which is found in the valley of the Upper Mississippi, as at Lake Pepin, &c., and which has even been traced, in small pieces, as low as St. Louis and Herculaneum in Missouri.* We may conclude that the ancient sandstones, slates, and rubblestone, and amygdaloids, of which traces still remain, were swept from the summit of the Mississippi by those ancient floods which appear to have diffused the boulder drift from the North.

SANDY LAKE.—The first view of this body of water was obtained from one of those eminences situated at the influx of the west Savanna River.

This lake is bounded, on its western borders, by the delta of the Mississippi; its outlet is about two miles in length. We here first beheld the object of our search. The soil on its banks is of the richest alluvial character. From this point, dense forests and a moderately elevated soil, varying from three or four to fifteen feet, confined the view, on either side, during more than two days' march. On the third day after leaving Sandy Lake, at an early hour, we reached the Falls of Pakágama. Here the rock strata show themselves for the first time on the Mississippi, in a prominent ledge of quartz rock of a gray color. Through this formation the Mississippi, here narrowed to less than half its width, forces a passage. The fall of its level in about fifty rods may be sixteen or eighteen feet. There is no cascade or leap, properly so called, but a foaming channel of extraordinary velocity, which it is alike impossible to ascend or descend with any species of water craft. It lies in the shape of an elbow. We made the portage on the north side.

PAKÁGAMA SUMMIT.—The observer, when he has surmounted the summit, immediately enters on a theatre of savannas, level to the eye, and elevated but little above the water. Vistas of grass, reeds, and aquatic plants spread in every direction. On these grassy plains the river winds about, doubling and redoubling on itself, and increasing its cord of distance in a ratio which, by the most moderate computation, would seem extravagant. On those

---

* *Vide* View of the lead mines.

plateaux, and the small rivers and lakes connected with them, the wild rice reaches the highest state of perfection.

Our men toiled with their paddles till the third day, through this unparalleled maze of water and plants, when we reached the summit of the Upper Red Cedar or Cass Lake, where we encamped. In this distance no rock strata appeared, nor any formation other than a jutting ridge of sand, or an alluvial plain. Plateau on plateau had, indeed, carried us from one level or basin to another, like a pair of steps, till we had reached our extreme height.

CASS LAKE BASIN.—From estimates made, this lake is shown to lie at thirteen hundred and thirty feet above the Atlantic.* This is a small elevation, when we consider it as lying on the southern flank of the transverse formation which forms the connecting link with the Rocky Mountains. A rise or a subsidence of this part of the continent to this amount, would throw the Hudson's Bay and Arctic waters down the Mississippi valley. The scenery of its coasts is in part arenaceous plains, and in part arable land, yielding corn to the Indians.

SOURCES OF THE MISSISSIPPI.—In order to understand the geology of this region, it is necessary to premise, that the St. Lawrence, the Hudson's Bay, and the Mexican Gulf waters are separated by a ridge or watershed of diluvial hills, called the Hauteur des Terres, which begins immediately west of the basin of the Rainy Lakes and Rainy Lake River. This high ground subtends the utmost sources of the Mississippi, and reaches to the summit of Ottertail Lake, where it divides the tributaries of the Red River of Lake Winnepec from those of the Des Corbeau, or Great Crow-Wing River.

Within this basin, which circumscribes a sweep of several hundred miles, there appears to have been deposited, upon the trap and primary rocks which form its nucleus, a sedimentary argillaceous deposit, capable of containing water. Upon this, the sand and pebble drift reposes in strata of unequal thickness, and the sand is often developed in ridges and plains, bearing species of the pine. The effect has been, that the immense amount of

---

* Agreeable to barometric observations made in 1836, by Mr. Nicollet, its true altitude is found to be 1,402 feet above the Gulf of Mexico. Its latitude, by the same authority, is 47° 25′ 28″.

vapor condensed upon these summits, and falling in dews, rains, and snows, being arrested by the impervious subsoil of clay, has concentrated itself in innumerable lakes, of all imaginable forms, from half a mile to thirty miles long. These are connected by a network of rivers, which pour their redundancy into the Mississippi, and keep up a circulation over the whole vast area. The sand plains often resting around the shores of these lakes create the impression of bodies of water resting on sand, which is a fallacy. Some of these bodies of water are choked up, or not well drained, and overflow their borders, forming sphagnous tracts. Hence the frequent succession of arid sand plains, impassable muskeegs, and arable areas on the same plateaux. Every system of the latter, of the same altitude, constitutes a plateau. The highest of these is the absolute source of the Mississippi waters. The next descending series forms another plateau, and so on, till the river finally plunges over St. Anthony's Falls.

In this descending series of plateaux, the Cass, Leech Lake, and Little Lake Winnipec form the third and fourth levels.

In descending the Mississippi below the Pakágama, the first stratum of rock, which rises through the delta of the river, occurs between the mouth of the Nokasippi and Elm Rivers, below the influx of the Great De Corbeau. This rock, which is greenstone trap, rises conspicuously in the bed of the stream, in a rocky isle seated in the rapid called—I know not with what propriety—the BIG FALLS, or *Grande Chute*. The precipitous and angular falls of this striking object decide that the bed of the stream is at this point on the igneous granitical and greenstone series. This formation is seen at a few points above the water, until we pass some bold and striking eminences of shining and highly crystalline hornblendic sienite, which rises in the elevation called by us Peace Rock, on the left bank, near the Osaukis Rapids. This rock lies directly opposite to the principal encampment on the 27th of July, which was on an elevated prairie on the west bank. To this point a delegation of Sioux had ascended on an embassy of peace from Fort Snelling to the Chippewas, having affixed on a pole what the exploring party called a bark letter, the ideas being represented symbolically by a species of picture writing, or hieroglyphics. In allusion to this embassy, this locality was called the

Peace Rock. This rock is sienite. It is highly crystalline, and extends several miles. Its position must be, from the best accounts, in north latitude about 44° 30'. From this point to Rum River, a distance of seventy miles, no other point of the intrusion of this formation above the prairie soil was observed.

INTRODUCTION OF THE PALÆONTOLOGICAL ROCKS.—After passing some fifty miles below this locality there are evidences that the river, in its progress south, has now reached the vicinity of the great carboniferous and metalliferous formations, which, for so great a length, and in so striking a manner, characterize both banks of the Mississippi below St. Anthony's Falls. About nine or ten miles before reaching these Falls, this change of geological character is developed; and on reaching the Falls the river is found to be precipitated, at one leap, over strata of white sandstone, overlaid by the metalliferous limestone. The channel is divided by an island, and drops in single sheets, about sixteen to eighteen feet, exclusive of the swift water above the brink, or of the rapids for several hundred yards below. This sandstone is composed of grains of pure and nearly limpid quartz, held together by the cohesion of aggregation. If my observations were well taken it embraces, sparingly, orbicular masses of hornblende. It is horizontal, and constitutes, in some places, walls of stratification, which are remarkable for their whiteness and purity. This sandstone is overlaid by the cliff limestone, the same in character, which assumes at some points a silicious, and at others, a magnesian character. It is manifestly the same great metalliferous rock which accompanies the lead ore of Missouri and mines of Peosta or Dubuque. There rests upon it the elder drift stratum of boulders, pebble, and loam, which marks the entire valley. This latter embraces boulders of quartz and hornblende rock, along with limestones and sandstones. It is overlaid by about eighteen inches of black alluvial carbonaceous mould.

From St. Anthony's Falls the river is perpetually walled on either side with those high and picturesque cliffs which give it so imposing and varied an appearance, and its current flows on with a majesty which seems to the imagination to make it rejoice in its might, confident of a power which will enable it to reach and carry its name to the ocean in its unchanged integrity.

ST. PETER'S RIVER AND VALLEY.—The importance, fertility,

and value of this tributary have particularly impressed every member of the party. Its position as the central point of the Sioux power, and its border position to the Chippewas, the representative tribe of the great Algonquin family, render it now a place of note, which fully justifies the policy of the department in establishing a military post at the confluence of the river; and the importance cannot soon pass away, in the progress of the settlement of the Mississippi Valley.* It is the great route of communication with the valley of the Red River of the North, and the agricultural and trading settlements of Lord Selkirk in that fertile valley, and its complete exploration by a public officer is desirable, if not demanded.†

Of its geological character but little is known, and that connects it with both the great formations which have been noticed as succeeding each other at the great Peace Rock. That the granitical formation reaches it at a high point is probable, from the large reported boulders. The Indians bring from the blue earth fork of it, one of their most esteemed green and blue argillaceous pigments, of which the coloring matter appears to be carbonate of copper. They also bring from the Coteau des Prairie, probably Carver's "shining mountains," specimens of that fine and beautiful red pipe stone, which has so long been known to be used by them for that purpose. This mineral is fissile, and moderately hard, which renders it fit for their peculiar ripe sculptures. I found small masses of native copper in the drift stratum at the mouth of this stream, on the top of the cliffs on the Mississippi, opposite the mouth of the St. Peter's.

CRYSTALLINE SAND ROCK.—This stratum reveals the same crystalline structure which is so remarkable in the sandstone caves, near the Potosi road, in the county of St. Genevieve, Missouri; and the sand obtained from it, like that mineral, would probably fuse, with alkali, in a moderate heat, and constitute an excellent material for the manufacture of glass. It is also, like the Missouri sandstone, cavernous. In both situations, these caves appear to be due to water escaping through fissures of the rock, where its cohesion is feeble, carrying it away grain by grain.

* Thirty years has made it the centre of the new territory of Minnesota, which has now entered on the career of nations.

† This object was accomplished by an expedition by Major L. Long, in 1828.

In stopping at one of these caves, about twelve miles below St. Peter's, we found this cause of structure verified by a lively spring and pond of limpid water flowing out of it.

VALLEY OF THE ST. CROIX.—This river originates in an elevated range of the elder sand and pebble drift, which lies on the summit between the Mississippi system of formations, and the Lake Superior basin. It communicates with the Brulé, which is "Goddard's River" of Carver, and with the Mauvaise or Bad River of that basin. Specimens of native copper have been found on Snake River, one of its tributaries.*

GEOLOGICAL MONUMENTS.—In descending the river for the distance of about one hundred miles below St. Anthony's Falls, my attention was arrested, on visiting the high grounds, by a species of natural monuments, which appear as if made by human hands seen at a distance, but appear to be the results of the degradation and wasting away, on the Huttonian theory, of all but these, probably harder, portions of the strata.

LAKE PEPIN.—This sheet commends itself to notice by its extent and picturesque features. It is an expansion of the river, about twenty-four miles long, and two or three wide. Both its borders and bed reveal the drift stratum, and the observer recognizes here, boulders of the peculiar stratification which has, in ancient periods, characterized the high plateaux about the sources of the river. Such are its hornblendic, sienite, quartz, trap, and amygdaloid pebbles, and that variety of the quartz family which assumes the form of the agate and other kindred species. Moved as these materials are annually, lower and lower, by the impetus of the stream, other supplies, it may be inferred, are still furnished by the shifting sand and gravel bars from above. The mass must submit to considerable abrasion by this change, and the diminished size of the drifted masses become a sort of measure of the distance at which they are found from their parent beds.

CHIPPEWA RIVER.—This stream is the first to bring in a vast mass of moving sand. Its volume of water is large, which it gathers from the high diluvial plains that spread southwest of the Porcupine Mountains, and about the sources of the Wis-

---

* This river was explored by me in 1831 and 1832, in two separate expeditions in the public service, accounts of which have been published in 1831 and 1832, of which abstracts are given in the preceding pages.

consin, the Montreal, and the St. Croix Rivers, with which it originates.

TROMPELDO (*Le Montaine des Tromps d'Euux*).—This island mountain stands as if to dispute the passage of the Mississippi, whose channel it divides into two portions. Distinct from its height, which appears to correspond with the contiguous cliffs, and in the large amount of fresh debris at its base, it presents nothing peculiar in its geology.

PAINTED ROCK.—This vicinity is chiefly noted for its large and fine specimens of fresh-water shells.

WISCONSIN.—Like the Chippewa, this stream brings down in its floods, vast quantities of loose sand, which tend to the formation of bars and temporary islands. It originates in the same elevated plains, and bespeaks a considerable area at its sources, which must be arid. It is a region, however, in which lakes and rice lands abound, and it may, in this respect, be geologically of the same formation as the higher plateaux of the Mississippi, above the Sandy Lake summit. Its sides produce many species to enrich our fresh-water Conchology.

LEAD MINES OF PEOSTA AND DUBUQUE.—In my researches into the mineral geography of Missouri, in 1818 and 1819, I had explored a district of country between the rivers Merrimak and St. Francis, and on the Ozarks, which revealed many traits which it has in common with the Upper Mississippi. There, as here, the mineral deposits appear to be, in many cases, in a red marly clay, whether the clay is overlaid by the calcareous rock or not. There, as here, also, the limestone and sandstone strata are perfectly horizontal. The leads of ore appear, in this section, to be followed with more certainty, agreeable to the points of the compass; but this may happen, to some extent, because the practice of mining on individual account, with windlass and buckets, in the Missouri district, has led common observers to be more indifferent to exact scientific methods. To say that the digging, at these mines, is equally, or more productive, is perhaps just. Capital and labor have been rewarded in both sections of the country, in proportion as they have been perseveringly and judiciously expended.

I found much of the ore, which is a sulphuret, at Dubuque's Mines, lying in east and west leads. These leads were generally

pursued in caves, or, more properly, fissures in the rock.   In one
of the excavations which I visited, the digging was continued
horizontally under the first stratum of rock, after an excavation
had been made perpendicularly, through the top soil and calcareous rock, perhaps thirty feet.   The ore is a broad-grained cubical galena, easily reduced, and bids fair very greatly to enhance
the value and resources of this section of the West.

Similar mines exist at Mississinawa, and the River Au Fevé,*
both on the eastern or left bank of the Mississippi. And a system of leasing or management, such as I have suggested for the
Missouri mines, appears equally desirable.

QUARTZ GEODES.—The amount of silex in the cliff limestone is
such, in some conditions of it, as to justify the term silico-calcareous.   This condition of the rock at the passage of the Mississippi through the Rock River and Des Moines Rapids, is such as
to produce a very striking locality of highly crystalline quartz
geodes, which accumulates in the bed of the stream.   Many of
these geodes are from a foot to twenty-two inches in diameter,
and on breaking them they exhibit resplendent crystals of limpid
quartz.   Sometimes these are amethystine; in other cases they
present surfaces of chalcedony or cacholong.   The latter minerals,
if obtained from the rock, and before unduly hardening by exposure, would probably furnish a suitable basis for lapidaries.

INTERMEDIATE COUNTRY IN THE DIRECTION TO GREEN BAY.—
There is a line which separates, on the north, the granitical and
trap region from the metal-bearing limestone, and its supporting
sandstone.   This formation of the elder series of rocks, having
been traced to the south shore of Lake Superior, and having
been seen to constitute the supporting bed of the alluviums and
diluviums of the Upper Mississippi, above the Peace Rock, it
may subserve the purpose of inquiry to trace this line of junction
by its probable and observed boundaries.

The line may be commenced where it crosses the Mississippi,
at the Peace Rock, and extended to the St. Croix, the falls of
which are on the trap-rock, to the sources of the Chippewa at
Lac du Flambeau, and the Wisconsin near Plover Portage.   The
source of Fox River runs amid uprising masses of sienite, and
this formation appears to pass thence northeasterly, across the

* GALENA has subsequently been made the capital of these mines.

Upper Menominee, to the district of the Totosh and Cradle-Top Mountains, west of Chocolate River, on the shores of Lake Superior.

I observed the crystalline sandstone and its overlying cliff limestone, along the valley of the Wisconsin, where ancient excavations for lead ore have been made. There is an entire preservation of its characters, and no reason occurs why its mineralogical contents should not prove, in some positions, as valuable as they have been found in Missouri, or in the Dubuque district west of the Mississippi.

On reaching the Wisconsin Portage, the limestone is found to have been swept by diluvial action, from its supporting sand rock. Such is its position not far north of the highest of the four lakes, and again at Lake Puckway, in descending the Fox River; consequently, there are no lead discoveries in this region. On coming to the calcareous rock, which is developed along the channel of the river, below Winnebago Lake, it appears rather to belong to the lake system of deposits. Its superior stratum lies in patches, or limited districts, which appear to have been left by drift action. Petrefactions are found in these districts, and the character of the rock is dark, compact, or shelly. The lower series of deposits, such as they appear at the Kakala Rapids, at Washington Harbor, in the entrance to Green Bay, and in the cliffs north of Sturgeon Bay and Portage, are manifestly of the same age and general character as the inferior stratum of Michilimackinac and the Manatouline chain.

BASIN OF LAKE MICHIGAN.—This basin, stretching from the north to the south nearly four hundred miles, lies deeply in the series of formation of limestones, sandstone, and schists, to which we apply the term of the Michilimackinac system. Its north and west shores are skirted from Green Bay to a point north of the Sheboygan, with the calcareous stratum. At this point, the ancient drift, the lacustrine clay of Milwaukie and the prairie diluvium of Chicago, constitute a succession, of which the surface is a slightly waving line of the most fertile soils.

Among the pebbles cast ashore at the southern head of this lake I observed slaty coal. It seems, indeed, the only one of the lakes which reaches south into the coal basin of Illinois. If the level at which coal is found on the Illinois were followed through, it

would issue in the basin of the lake below low-water mark.
Digging for this mineral on the Chicago summit, promises indeed
not to be unsupported by sound hypothesis.

After passing Chicago, of which a sketch is added, the sands
which begin to accumulate at the Konamik, the River du Chemin,
and the St. Joseph's River,* appear in still more prominent
ridges, skirting the eastern coasts to and beyond Grand River.
These sands, which are the accumulations of winds, are cast on
the arable land, much in the manner that has been noticed at the
Grand Sable on Lake Superior, and reach the character of striking dunes at the coast denominated the Sleeping Bear. The
winds which periodically set from the western shore, produce
continual abrasions of its softer materials, and are the sole cause
of these intrusive sand-hills. Pent up behind them, the water is
a cause of malaria to local districts of country, and many of the
small rivers upon this side are periodically choked with sand. The
sketch transmitted of this bleak dune-coast (omitted here), as it is
seen at the mouth of Maskigon Lake, will convey a false idea of
the value of this coast, even half a mile from the spot where the
surf beats. It is designed to show the air of aridity which the
mere coast line presents. The stratification regains its ordinary
level and appearance before reaching the Plate or Omicomico
River, and the peninsula of the Grand Traverse Bay, and the
settlements of the Ottawa Indians on Little Traverse Bay, afford
tracts of fertile lands. Point Wagonshonce consists of a stratum
of limestone of little elevation, which constitutes the southeast
cape of the strait. Here a lighthouse is needed to direct the
mariner.

LAKE HURON.—Notices of this sheet of water have been given
in our outward voyage. It appears rather as the junction of
separate lakes which have had their basins fretted into one another,
than as one original lake. Michigan is connected with it through
the Straits of Michilimackinac. The Georgian Bay, north of the
Manatouline chain, seems quite distinct. The Saganaw Bay is an
element of another kind. The Manitouline chain separates the
calcareous and granitic region, and its numerous trap and basaltic
islands towards the north shore, of which there are many thou-

---

* The subjoined petrifaction of a leaf, apparently a species of betula, was obtained on this river. See *ante*, p. 206.

sands, denote that it has been the scene of geological disturbance of an extraordinary kind.

ULTERIOR CONCLUSIONS.—In taking these several views of the geological structure of the Northwest—of the Lake Superior basin, and of the valleys of the St. Louis River—the region about the Upper Mississippi, its striking change at the Falls of St. Anthony—and the valleys of the Wisconsin and Fox Rivers, and the basins of Lakes Michigan and Huron, I am aware of the temerity of my task. Allowance must, however, be made for the rapidity of my transit over regions where the question was often the safety and personal subsistence of the party. A very large and diversified area was passed over in a short time. At no place was it possible to make elaborate observations. A thousand inconveniences were felt, but they were felt as the pressure of so many small causes impeding the execution of a great enterprise. A sketch has been made, which, it is hoped, will reveal something of the physical history and lineaments of the country. These glimpses at wild scenes, heretofore hid from the curious eye of man, have been made, at all points, with the utmost avidity. I have courted every opportunity to accumulate facts, and I owe much to the distinguished civilian who has led the party so successfully through scenes of toil and danger, not always unexpected, but always met in a calm, bold, and proper spirit, which has served to inspire confidence in all; to him, and to each one of my associates, I owe much on the score of comity and personal amenity and forbearance; and I have been made to feel, in the remotest solitudes, how easy it is to execute a duty when all conspire to facilitate it.

The views herein expressed are generalized in two geological maps (hereto prefixed), which, it is believed, will help to fix the facts in the mind. They exhibit the facts noticed, in connection with the theory established by them, and by all my observations, of the construction of this part of the continent.

The mineralogy of the regions visited is condensed in the following summary, drawn from my notes, which, it is believed, constitutes an appropriate conclusion to this report.

With the exception of one species, namely, the ores of copper, the region has not proved as attractive in this department as I

22

found the metalliferous surface of Missouri. There are but few traces of mining, and those of an exceedingly ancient character, in the copper region of Lake Superior. The excavations in search of lead ore on the Upper Mississippi do not date back many years, but the indications are such as to show that few countries, even Missouri, exceed them in promises of mineral wealth.

I have employed the lapse of time between the termination of the exploration and the present moment, to extend my mineralogical observations to some parts of the Mississippi Valley which were not included in the line of the expedition, but which were visited in the following year, in the service of the Government, namely, the Miami of the Lakes, and Wabash Valleys, the Cave in Rock Region in Lower Illinois, and the Valley of the River Illinois. The whole is concentrated in the following notices :—

Tabular View of Minerals observed in the Northwest.

I. ORES.

Genera.	Species.	Subspecies.	Varieties.
Copper	Native copper.		
	Green carbonate of copper		Fibrous. / Compact.
Lead	Sulphuret of lead		Common.
Zinc	Sulphuret of zinc		Blende.
Iron	Sulphuret of iron		Common. / Radiated. / Spheroidal. / Cellular. / Hepatic.
	Magnetic oxide of iron		Iron sand.
	Specular oxide of iron.	Micaceous.	
	Red oxide of iron		Ochrey. / Scaly. / Compact.
	Brown oxide of iron		Ochrey.
Silver.			

(Genus bracket: METALLIC MINERALS)

II. EARTHS AND STONES.

Genus.	Species.	Varieties.

SILICIOUS MINERALS

Quartz
- Common quartz
  - Milky.
  - Radiated.
  - Tabular.
  - Greasy.
  - Granular.
  - Arenaceous.
  - Pseudomorphous.
  - Amethystine.
- Amethyst.
- Ferruginous quartz
  - Yellow.
  - Red.
- Prase.
- Chalcedony
  - Common.
  - Cacholong.
  - Carnelian.
  - Sardonyx.
  - Agate.
- Hornstone.
- Jasper
  - Common.
  - Striped.
  - Red.
- Heliotrope.
- Opal . . . . . Common.

Silicious slate
- Common.
- Basanite.

Petrosilex.

Mica
- Common.
- Gold yellow.

Schorl
- Common.
- Indicolite.

Feldspar . . . . . Common.

Prehnite . . . . . Radiated.

Hornblende
- Common.
- Actynolite.

Woodstone
- Mineralized wood.
- Agatized wood.

CALCAREOUS MINERALS

Carbonate of lime
- Calcareous spar
  - Crystallized.
  - Lamellar.
- Granular limestone
- Compact limestone
  - Common.
  - Earthy.
- Agaric mineral
  - Common.
  - Fossil farina.
- Concreted carbonate of lime
  - Oolite.
  - Calcareous sinter
    - Stalactite.
    - Stalagmite.
  - Calcareous tufa.
  - Pseudomorphous carbonate of lime.
- Marl . . . . . Ludus helmontii.

Sulphate of lime . Gypsum
- Fibrous.
- Granular.
- Granularly foliated.
- Earthy.

Fluate of lime . . . . Fluorspar.

	Genus.	Varieties.
	Argillaceous slate	{ Argillite. / Bituminous shale.
	Chlorite	Chlorite slate.
	Staurotide.	
ALUMINOUS MINERALS	Clay	{ Potters' clay. / Pipe clay. / Variegated clay. / Blue sulphated clay. / Green sulphated clay. / Opwagunite.
MAGNESIAN MINERALS	Serpentine	Common serpentine.
	Steatite	Steatite.
	Asbestus	Com. asbestus.
BARYTIC MINERALS	Sulphate of barytes	Lamellar.
STRONTIAN	Sulphate of strontian	Foliated.

III. COMBUSTIBLES.

BITUMINOUS MINERALS	Bitumen	{ Petroleum. / Maltha. / Asphaltum.
	Graphite	Granular graphite.
	Coal	Slate coal.

IV. SALTS.

SODA	Muriate of soda	{ Native salt. / Salt springs.
	Alkaline sulphate of alumina .	Alum.

a. *Metallic Minerals.*

1. COPPER.

This metal is frequently found, in detached masses, in the diluvial soil along the southern shore of Lake Superior, and in the high and barren tract included between Lakes Huron, Michigan, and Superior, and the Mississippi River, as general boundaries. Thus, it has been found upon the sources of the Menomonie, Wisconsin, Chippewa, St. Croix, and Ontonagon Rivers, but most constantly, and in the greatest quantity, upon the latter. There are many localities known only to the aborigines, who appear to set some value upon it, and have been in the habit of employing the most malleable pieces in several ways from the earliest times. It occurs mostly in detached masses, resting upon, or imbedded in, diluvial soil. These masses, which vary in size, are sometimes connected with isolated fragments of rock. Such is the geognostic position of the great mass of native copper upon the banks of the Ontonagon, which has been variously estimated to weigh from two to five tons. This extraordinary mass is situated at the base of a diluvial precipice composed of reddish loam and mixed boulders and pebbles of granite, greenstone, quartz, and sandstone and diallage rocks. The nearest strata, in situ, are red sandstone, grauwacke, and greenstone trap. A company of miners was formerly employed in searching for copper mines upon the banks of this river. They dug down about forty feet into the diluvial soil, at a spot where a green-colored water issued from the hill. In sinking this pit, several masses of native copper were found, and they discovered, as their report indicates, the same metal "imbedded in stone." But the enterprise was abandoned, in consequence of the falling in of the pit.

At Keweena Point, on Lake Superior, I found native copper along the shore of the lake, constituting small masses in pebbles, and, in one instance, in a mass of several pounds' weight, which was found in the Ontonagon Valley. I also observed the green carbonate of copper, in several places, in the detritus. The strata of this point appear to be charged with this mineral, particularly in its native forms. Hardly a mass of the loose rock is without some trace of the metal, or its oxides or salts. It would be difficult, on any known principles, to resist the testimony which is offered, by every observer, to favor the idea that extensive and very valuable mines exist. The whole lake shore, from this peninsula to the Montreal River, is replete with these evidences.

There are indications that this mineral pervades the rocks and soils, in a radius of one hundred and fifty miles or more, south and west of this central point. It has been discovered at the sources of the Menominee, Chippewa, Montreal, and St. Croix, and even at more distant points.

At St. Peter's, in digging down for the purpose of quarrying the rock, about eighteen inches depth of dark alluvium was passed; then a deposit of diluvial soil, with large fragments of limestone, greenstone, quartz rock, &c., about six feet; and, lastly, one foot of small pebbles, &c., constituting the copper diluvium. No large mass was found; nor any veins in the rock.

2. LEAD.

The only ore of lead known to exist within the limits to which these remarks are confined, is the sulphuret. In the year 1780, Peosta, a woman of the Misquakee, or Fox tribe of Indians, discovered a lead mine upon the west banks of the Mississippi, at the computed distance of twenty-five leagues below Prairie du Chien, which the Indians, in 1788, gave Julian Dubuque a right to work. This permission was partially confirmed by the Baron de Carondelet, Governor of Louisiana, in 1796. No patent was, however, issued; but Dubuque continued to prosecute the mining business to the period of his death, which happened in 1810, when the mines were again claimed by the original proprietors.

The ore is the common sulphuret of lead, or galena, which Dubuque stated to have yielded him seventy-five per cent. in smelting in the large way. He usually made from 20,000 to 40,000 pounds per annum.

I made a cursory visit to these mines, and found them worked by the Fox Indians, but in a very imperfect manner. They cover a considerable area, commencing at the mouth of the Makokketa River, sixty miles below Prairie du Chien. Traces of the ore are found, also, on the east bank of the Mississippi at several points. It occurs disseminated in a reddish loam, resting upon limestone rock, and is sometimes seen in small veins pervading the rock; but it has been chiefly explored in diluvial soil. It generally occurs in beds having little width, and runs in a direct course towards the cardinal points. They are sometimes traced into a crevice of the rock. At this stage of the pursuit, most of the diggings have been abandoned. Little spar or crystalline matrix is found in connection with the ore. It is generally enveloped by a reddish, compact earth, or marly clay. Occasionally, masses of calcareous spar occur; less frequently, sulphate of barytes, green iron earth, and ochrey brown oxide of iron. I did not observe any masses of radiated quartz, which form so conspicuous a trait in the surface of the metalliferous diluvion of the mining district of Missouri.

Sufficient attention does not appear to have been bestowed, by mineralogists, upon the metalliferous soil of the Mississippi Valley. It is certainly very remarkable that such vast deposits of lead ore, accompanied by veins of sulphate of barytes, calc spar, and other crystallized bodies, should be found in alluvial beds; and it would be very interesting to ascertain whether any analogous formations exist in Europe, or in any other part of the earth's surface. It is one of the most striking features of this deposit, that the ore, spars, &c., do not appear as the debris of older formations, and have no marks of having been worn or abraded, like those extraneous masses of rock which are very common in the alluvial soil of our continent. The lead ore and accompanying minerals appear to have been crystallized in the situations where they are now found. We should, perhaps, except from this remark the species of lead called *gravel ore* by the miners, which is in rounded lumps, and is never accompanied by spars.

Sulphuret of lead is also found near the spot where the small River Sissinaway enters the Mississippi, and two leagues south of it, upon the banks of the River Aux Fevre, at both of which places considerable quantities have been raised, and continue to be raised, for the purposes of smelting, by the Fox and Sac tribes of Indians. At these places, it is most frequently connected with a gangue of heavy spar and calcareous spar, with pyrites of iron. I procured from a trader, at Dubuque, several masses of galena crystallized in cubes and octahedrons.

In descending the Upper Mississippi, a specimen of galena was exhibited to me, by a Sioux Indian, at the village of the Red Wing, six miles above Lake Pepin, said to have been procured in that vicinity. Galena is also reported to have been discovered in several places on the south side of the Wisconsin River, and these localities may be entitled to future notice, as furnishing important hints.

8. Zinc.

The sulphuret of zinc (black blende) is found disseminated in limestone rock along the banks of Fox River, between the post of Green Bay and Winnebago Lake. Although frequently seen in small masses, no body of it is known to exist. I also found blende, in small, orbicular masses of calcareous marl, along the east shore of Lake Michigan, between the Rivers St. Joseph and Kikalemazo.

4. Iron.

This mineral is distributed, in several of its forms, throughout the region visited, although but little attention has yet been directed to its exploration. In the basin of Lake Superior it exists, in valuable masses, in the form of a magnetic oxide, on the coasts of the lake between Gitchi Sebing (Great River), called by the French Chocolate River, and Granite Point. Specimens from Dead River (Riviere du Morts) and Carp River, the Namabin of the Indians, in this district, denote the latter to be the chief locality. It is the iron glance, and occurs in mountain masses.

Sulphuret of Iron.—This variety is found, in limited quantities, in a state of crystallization, in clay beds, on the west shore of Lake Michigan, between Milwaukie and Chicago. It is frequently in the form of a cube or an octahedron. Some of the crystals are in lumps of several pounds' weight, with a metallic lustre. Often the masses, on being broken, are found radiated, sometimes cellular, and occasionally irised.

Iron Sand.—The breaking-up and prostration of the sandstone and other sedimentary formations, along the shores of lakes Michigan, Huron, and Superior, liberates this ore in considerable quantities. It arranges itself, on the principle of its specific gravities, in separate strata along the sandy shores, where it invariably occupies the lowest position at and below the water's edge. The shores of Fond du Lac, on Lake Superior, may be particularly mentioned as an abundant locality.

Micaceous Oxide of Iron.—In detached mass, among the debris of the River St. Louis and of Fond du Lac. It exists in veins in the clay slate which characterizes the banks of this river.

Ochrey Red Oxide of Iron (Red ochre)—Is produced near a spot called the Big Stone, on the head of the River St. Peter's. It is said to occur in a loose form, in a stratum of several inches thick, lying below the soil of a level dry prairie or plain. The Sioux Indians, who employ it as a paint, make this statement. The color of a portion given to me by them is of a bright red; and a considerable proportion of the mass is in a state of minute division. Particles of quartz are occasionally mixed with it. This ore of iron is also represented to be found in the prairies north of

Gros Point, along the west shore of Lake Michigan, between Mil-waukie and Chicago.

Ochrey red oxide of iron occurs on the shores of Big Stone Lake, at the source of the St. Peter's River. A large spring rises from a level, dry plain, a few feet beyond which the mineral occurs. The Indians, who employ it as a pigment, take it up with their knives. The stratum is about eight inches thick, but just below the surface it is mixed with common earth. The spring of water is pure and unadulterated.

5. SILVER.

The belief in the existence of silver ore in the region of the lakes, and particularly on Lake Superior, seems to have early prevailed. So much confidence was placed in the reports of its existence, that Henry tells when a company was formed in Eng-land for exploring the copper mines of Lake Superior (A. D. 1771), they were impelled to the search more from an expectation of the silver, which it was hoped would be found in connection with it, than from the copper.*

b. *Silicious Minerals.*

1. QUARTZ.

This interesting species being distributed in its numerous va-rieties throughout the region visited, I shall confine my notices to a few localities.

Subs. 1.—*Common Quartz.*

Occurs in the form of large water-worn masses along the shores of Lakes Huron, Michigan, and Superior. Also, in veins in the granite of Lake Superior, and in the argillite of St. Louis River. These localities all consist of the opaque varieties, with a slight

---

* This metal has subsequently (namely, in 1844) been found to constitute a per-centage in the native copper of the Eagle River mines of Lake Superior. Traces of it were found in a mass of native copper found on the shores of Keweena Lake, by Mr. Moliday, in 1826. A mass of pure silver was discovered in a boulder in the drift of Lake Huron, west of White Rock, in 1824. These discoveries induce the belief that this element will be found to be extensively present in the eventual metallurgic operations of the Lake Superior basin.

degree of translucence in some places. It exists in mass at Huron Bay, Lake Superior, and in fragments of red jasper on Sugar Island, St. Mary's River.

1. *Radiated Quartz.*—In detached masses on the Grange, and also at the rapids of the River Desmoines, on the Upper Mississippi. At the Grange, the crystals, which are usually minute, sometimes possess a cinnamon color, or pass into a variety of crystallized ferruginous quartz.

2. *Tabular Quartz.*—In small, flattened masses along the shores of Lake Pepin. These masses are transparent, or only translucent. Their color is generally white, but sometimes yellow. They appear to be closely allied to chalcedony.

3. *Greasy Quartz.*—In detached masses along the shores of Lake Superior.

4. *Granular Quartz.*—At the Falls of Puckaiguma, on the Upper Mississippi, in large, compact beds rising through the soil. Also, in some conditions of the cliffs commencing at the Falls of St. Anthony, Carrer's Cave, &c.

5. *Arenaceous Quartz.*—This is sometimes the condition of fine, even-grained, translucent sand rock of the preceding localities. Valuable as an ingredient of glass.

6. *Pseudomorphous Quartz.*—On the shores of Lake Pepin, occasionally. These masses appear to have taken their crystalline *impress* from rhomboidal crystals of carbonate of lime.

7. *Amethystine Quartz.*—In the trap-rock of Lake Superior.

Subs. 2.—*Amethyst.*

This mineral occurs most frequently in the condition of amethystine quartz, in hexahedral prisms, lining the interior of geodes, in the bed of the River Desmoines, and on the Rock Rapids, in the channel of the Mississippi. The crystals which I have examined are generally limpid, with a high lustre, and of a pale violet color. Sometimes the tinge of color approaches to a full red, or is only apparent in the summit of the crystal. These geodes are sometimes eight or ten inches in diameter, with a rough and dark-colored exterior, often so nearly spherical as to resemble cannon *balls.* Some of the finest specimens I have observed from this locality are preserved in the museum of Gov. Clarke, at St. Louis, Missouri.

In amorphous masses, of a deep-red, brown, or yellowish-red color, along the southern shore of Lake Superior. Likewise, crystallized, in very minute hexagonal prisms, terminated by six-sided pyramids, of a reddish color, on the summit and declivities of the Grange de Terre.

Subs. 4.—*Prase.*

In the drift of Lake Superior. Its color is a light green and not fully translucent. It possesses a hardness and a lustre intermediate between waxy and resinous.

Subs. 5.—*Chalcedony.*

1. *Common Chalcedony.*—In globular or reniform masses imbedded in trap-rock, on the Peninsula of Keweena, Lake Superior. It is found sometimes in association with other quartz minerals. Its color is white or gray, sometimes veined or spotted with red. Also, constituting the interior lining of geodes at the rapids of Rock Island and the River Desmoines. These geodes, on breaking, often present a mammillary surface. In the form of translucent fragments, with a highly conchoidal fracture, among the debris of the shores of Lake Pepin. These fragments possess an extremely delicate texture, color, and lustre.

2. *Cacholong.*—Some loose fragments of this mineral exist along the west shore of Lake Michigan, between Green Bay and Chicago. These fragments possess small cavities studded over with very minute and perfect crystals of quartz.

3. *Carnelian.*—This mineral occurs in fragments in the debris of Lake Superior; also, in the amygdaloid; also, around the shores of the Upper Mississippi. Its color is various shades of red, or yellowish red, sometimes spotted or clouded, fully translucent, and occasionally presenting a considerable richness and beauty. Most commonly, the fragments are too small to be applied to the purposes of jewelry. Sometimes it is seen in very regular spheroidal masses, which contain a nucleus of radiated quartz. Some of the specimens would be considered as sardonyx.

4. *Agate.*—Is found with the preceding. It is more frequently found in larger masses, in the rock, which are sometimes spher-.

oidal, reniform, or globular. These agates are chiefly arranged in concentric layers, which are white, red, yellow, &c., according to the colors of the different varieties of chalcedonies, carnelians, &c., of which they are composed. A close inspection would also separate them into several varieties—as onyx, agate, dotted agate, &c.

Subs. 6.—*Hornstone.*

In nodular or angular masses, imbedded in the secondary limestone of the west shores of Green Bay; and in the beds of argillaceous white clay strata of Cape Girardeau, of Missouri. Also, on the hills of White River, Arkansas.

Subs. 7.—*Jasper.*

1. *Common Jasper.*—In detached fragments, yellow, in the drift of Lake Superior.

2. *Striped Jasper.*—With the preceding. Most commonly, these specimens consist of alternate bands of red and black, or brown.

3. *Red Jasper.*—In quartz rock, Sugar Island, River St. Mary's, Michigan. Masses of this mineral have been met in situ.

Subs. 8.—*Heliotrope.*

A fine specimen of this mineral, now before me, was procured at the mouth of the Columbia River, Oregon. It is in the form of an Indian dart. Its color is a deep uniform green, variegated with small spots of red; those parts which are green being fully translucent, the others less so, or nearly opaque. This beautiful mineral is represented to have been in common use by the Indian tribes of the Northwest Coast, for pointing their arrows, previous to the introduction of iron among them. It differs chiefly from the dotted jaspers of Lake Michigan, in its translucence and green color.

Subs. 9.—*Opal.*

Common opal occurs as a constituent of agate, along with chalcedony rarely, in the drift on the south shore of Lake Superior.

2. SILICIOUS SLATE.

1. *Common.*—In subordinate beds, in the argillite of the River St. Louis, northwest of Lake Superior.

2. *Basanite (Touchstone).*—In detached fragments in the drift on Lake Superior, and along the banks of the Upper Mississippi generally.

8. PETROSILEX.

In large isolated masses in the bed of the Illinois River, on the shallow rapids between the junction of the Fox and Vermilion Rivers. It is mostly arranged in stripes or circles of white, gray, yellow, &c., resembling certain jaspers, or approaching sometimes to hornstone. The bed of the Illinois River, at this place, is a species of gray sandstone. Also, in detached fragments, on the south shore of Lake Superior, intimately mixed with prehnite. In regard to the latter, Professor Dewey, of Williamstown College, writes me: "I have received from Dr. Torrey, a curious mixture of petrosilex and prehnite, in imperfect radiating crystals, which was sent him by you and collected at the West. He did not tell me the name, but examination showed what it was. The association is singularly curious." The locality of this mineral is Keweena Point, Lake Superior.

4. MICA.

Occurs rarely in the granite of Lake Superior. It is found in place on the Huron Islands. Also, in minute folia, in the alluvial soil of the Upper Mississippi. A beautiful aggregate, consisting of plates of gold-yellow mica, connected with very black and shining crystals of schorl, has been dug up from the alluvial soil of the Island of Michilimackinac.

5. SCHORL.

1. *Common Schorl.*—In crystals, in boulders of granite, at Green Bay.
2. *Tourmaline.*—With the preceding.

6. FELDSPAR.

As an ingredient in the granite of Huron Islands, Lake Superior. Also, in detached masses of granite along the west shores of Lake Michigan. Also, in the form of prismatic crystals of a light-green color, in the rolled masses of hornblende, porphyry, greenstone, and epidotic boulders of Lakes Huron, Michigan, and Superior.

7. Prehnite.

This mineral occurs at Keweena Point, on Lake Superior. It is found in connection with isolated blocks of amygdaloid, of primitive greenstone, and of petrosilex. Sometimes native copper, and carbonate of copper, are also present in the same specimen. In some instances, a partial decomposition has taken place, converting its green color into greenish-white, or perfect white, and rendering it so soft as to be cut with a knife. Sometimes the grains or masses of native copper are interspersed among the prehnite, and slender threads of this metal occasionally pass through the aggregated mass of greenstone, prehnite, &c., so that, on breaking it, the fragments are still held together by these metallic fibres.

8. Hornblende.

1. *Common Hornblende.*—Occurs as a constituent of the hornblende rocks near Point Chegoimegon, Lake Superior. Also, at the Peace Rock, on the Upper Mississippi, and in certain granite aggregates, and rolled masses of porphyries, &c., around the shores of Lakes Huron, Michigan, and Superior.

2. *Actynolite.*—In slender, translucent, greenish crystals, pervading rolled masses of serpentine, on the west shores of Lake Michigan.

9. Woodstone.

1. *Mineralized Wood.*—In bed of the River Des Plaines, Illinois.

2. *Agatized Wood.*—This variety of fossil wood is found along the alluvial shores of the Mississippi and of the Missouri.

c. *Calcareous Minerals.*

1. Carbonate of Lime.

Of a substance so universally distributed throughout the western country, it will not be necessary to give many localities, and these will be principally confined to its crystalline forms.

Subs. 1.—*Calcareous Spar.*

Crystallized Calcareous Spar.—This mineral occurs, in minute rhomboidal crystals, in the calcareous rock of the Island of

Michilimackinac. Sometimes these crystals fill cavities or seams of the rock, or are studded over the angular surfaces of masses of vesicular limestone of that island. I also found this mineral at Dubuque's mines, and in small crystals in the metalliferous limestone bordering the Fox River, between the post of Green Bay and Winnebago Lake, where it is associated with iron pyrites and blende.

Subs. 2.—*Compact Limestone.*

In proceeding northwest of Detroit, this mineral is first observed, in situ, on an island in Lake Huron. It is afterwards found to be the prevailing rock along the south and southwest shores of Lake Huron. In many places, it incloses fossil remains. Sometimes it is *earthy,* as at Bay De Noquet, a part of Green Bay, on Lake Michigan, where it contains very perfect remains of the terrebratula. (Parkinson.) In other places, no remains whatever are visible, and the structure is firm and compact; or even passes, by a further graduation, into transition-granular, of which, it is believed, the west shores of Lake Michigan afford an instance. It is most commonly based upon sandstone, which also contains, in many places, the fossil organized remains of various species of crustaceous animals, and of vegetables, sometimes, coal, &c.

Subs. 3.—*Agaric Mineral.*

This mineral substance occurs in crevices and cavities in the calcareous rock of the Island of Michilimackinac, Michigan.

Subs. 4.—*Concrete Carbonate of Lime.*

1. *Calcareous Sinter.*—In the form of *stalactites* and *stalagmites,* in a cave situated near Prairie du Chien, on the Upper Mississippi.

2. *Calcareous Tufa.*—A remarkable formation of tufa is seen on the east banks of the Wabash River, near Wynemac's Village, about ten miles above the junction of the Tippecanoe. It extends for several miles, and is deposited to the thickness of thirty or forty feet above the water, forming cliffs which are covered with alluvial soil and sustain a growth of forest trees. The precise points of its commencement and disappearance were not observed. The structure is cellular or vesicular, and resembles, in some places, a coarse dried mortar. It is very light, and possesses a

white color in inferior situations, but the surface is somewhat colored by fallen leaves and other decaying vegetation. It imbeds fluvatile shells and some vegetable remains, the species of which have not been ascertained. The opposite, or west side of the river consists of a kind of puddingstone, or caschalo, made up of pebbles of quartz, &c., cemented by carbonate of lime, of a yellow color and translucent. This beautiful aggregate is overlayed by a stratum, of fifteen or twenty feet in thickness, of diluvial soil. These localities fall within the limits of the State of Indiana; but on territories still occupied, if not owned, by the aborigines.

3. *Pseudomorphous Carbonate of Lime.*—This form of carbonate of lime occurs in Pope County, Illinois, a district celebrated for its fluorspar, lead, crystallized quartz, &c., and bearing the unequivocal marks of a secondary formation. Scattered in large masses over the soil, we observe compact limestone, with very perfect cubical, octahedral, or other regular cavities, which have manifestly originated from crystals of fluorspar. The most common *impress* of this kind appears to have resulted from two cubes variously joined—a form of appearance very common to the Illinois fluates. Some of these cubical cavities exceed three inches square; but in no case is any remaining portion of the spar in these cavities, or anywise connected with the fragments of limestone thus impressed, although, at the same time, the spar is very abundant in the alluvial soil where these curious limestones are found.

2. SULPHATE OF LIME.

Subs. *Gypsum.*

1. *Fibrous Gypsum.*—In the alluvial soil of the St. Martin's Islands, Lake Huron. The fibres are sometimes five or six inches in length, of a white color and delicate crystalline lustre. Sometimes these fibrous masses are partially colored yellow or brown, apparently from the clay, or mixed alluvion, in which they are imbedded.

2. *Granular Gypsum.*
3. *Granularly-Foliated Gypsum.* } With the preceding.
4. *Earthy Gypsum.*

Fluor-Spar.—On the United States Mineral Reserve, Pope County, Illinois. This locality is abundant, and the mineral readily and constantly to be obtained. I first obtained specimens in June, 1818, and afterwards visited it in July, 1821. It is disseminated in loose masses throughout the soil, and in veins in the calcareous rocks. The spot most noted and resorted *to*, and where the original discovery was made, is four miles west of Barker's Ferry, at Cave-in-Rock, on the banks of the Ohio, and about twenty-six miles, by the course of the river, below Shawneetown. It is situated in the midst of a hilly, broken region, called *the Knobs*, a tract of highlands intervening between the banks of the Ohio and the Saline. The distance of this range from north to south, or parallel with the course of the Ohio, cannot be stated. It probably extends from near the banks of the Wabash River to the Little Chain of Rocks. Its breadth—from Barker's Ferry, west, to Ensminger's, at the Saline, is about twenty miles. It thus separates, by a rocky border, the prairies of the Illinois from the current of the Ohio River. These knobs, wherever observed, bear the indubitable marks of secondary formation, and may be stated to consist, essentially, of compact limestone resting on sandstone. The sandstone is sometimes so much colored by iron, and by globular or irregular masses of iron stone, as to give that rock a very singular aspect. This may be particularly instanced in the mural front of the Battery rocks on the banks of the Ohio. Every part of this formation has more or less the appearance of a mineral country; and it is already known as the locality of ores of lead, iron, and zinc, of crystallized quartz, of opal, heavy spar, crystallized pyrites, and of very perfect fossil madrepores. In one place (near the head of Hurricane Island) this spar forms a very large and compact vein, dipping under the bed of the Ohio. Where the rock has been explored, it is found in connection with sulphuret of lead, but it has been mostly procured, because most easy of access, in the alluvial soil. I went out about half a mile west of the Ohio, where a new locality has been opened, and, in removing about five or six solid feet of earth, procured as many specimens as filled a box of fourteen inches square. None of these were more than two feet below the surface. One of these

23

specimens is an irregular octahedral crystal, eight inches in diameter. The color of these masses is various shades of blue, violet, or red, sometimes perfectly white or yellow; and the form most commonly assumed is a cube, sometimes truncated at two or more angles, or variously clustered. The external lustre of the crystals, raised from alluvial soil, is feeble, but quite brilliant when taken from veins and cavities in the rock. These spars from the alluvion do not appear to exist as rock debris, or fragments worn off from other formations, but as original deposits. There are no marks of attrition. They appear as much in place as the limestone rocks below. It should also be recollected that this mineral tract is terminated by one of the greatest and most valuable salt formations in the western country—that of the Illinois Saline.

Septaria: Ludus Helmontii.—This variety of calcareous marl is found, in orbicular or flattened masses, along the eastern shores of Lake Michigan, between the rivers St. Joseph's and Kalemazo. Its original situation appears to be the beds of marly clay which form the banks of Lake Michigan at these places, from which these masses have been disengaged by the waves, and left promiscuously among the washed and eroded debris of the shore. These masses are penetrated by numerous seams and lines of calcareous spar, sometimes radiating star-like, or intersecting each other irregularly. Occasionally, these seams are filled with sulphuret of zinc, and in these cases the spar, if any be present, is rose-colored.

d. Aluminous Minerals.

1. ARGILLACEOUS SLATE.

1. *Argillite*, or *Common Argillaceous Slate.*—Along the banks of the River St. Louis, at the Grand Portage, &c. It occurs in a vertical position, embracing veins, or subordinate beds, of grauwakke, milky quartz, chlorite slate, and silicious slate, &c. It is bounded on one side by red sandstone, and on the other by an extensive tract of diluvial soil.

2. *Bituminous Shale.*—In detached masses, along the shores of Lake Huron, between Fort Gratiot and Thunder Bay. It contains amorphous masses of iron pyrites, of a yellow color and metallic brilliancy, which soon tarnishes on exposure to the air.

2. Chlorite.

Chlorite Slate.—In subordinate strata in the argillite of the River St. Louis.

3. Staurotide.

In garnet-colored crystals, in detached blocks of mica-slate, in the drift of Lake Huron. These crystals consist of two intersecting six-sided prisms, truncated at both ends, forming the cross. They are nearly opaque, or feebly translucent on the fractured edge.

4. Clay.

1. *Plastic Clay.*—Very extensive beds of this clay are seen along the west shore of Lake Michigan, between Sturgeon Bay Portage and Chicago. Its color is generally a light blue, verging sometimes into deep blue or grayish-white. It is plastic in water, adheres strongly to the tongue, takes a polish from the nail, and emits an argillaceous odor when moistened or breathed upon. These beds of clay frequently contain iron pyrites, both in the crystallized and amorphous state.

2. *Pipe Clay.*—In the flats of the St. Clair and Lake George, Michigan. A bed of clay, apparently answering to this description, exists at White River, Lake Michigan. Its color is a grayish-white, verging to blue. It is very unctuous and adhesive when first raised, but acquires more or less of a meagre feel as it parts with its moisture, drying in firm and compact masses.

3. *Variegated Clay.*—On the banks of the River St. Peter's, Upper Mississippi. Neither the quantity in which it exists, nor the precise locality is known. Its color is white, variegated with stripes, spots, or clouds of red or yellow.

4. *Azure Blue Clay of St. Peter's.*—The locality of this substance, as communicated by the Indians, is the declivity of a hill, in the rear of the village of Sessitongs, one mile above the confluence of the Terre Blue River with the St. Peter's. It is found near the foot of this hill, between two layers of sandstone rock, in a vein about fifteen inches in thickness. This vein is elevated about twenty feet above the waters of the Terre Bleu, and does not extend far in the direction of the river. Having been resorted to by the Sioux Indians a long time, a considerable excavation has been made, but the supply is constant. The color of this mineral

substance (its distinguishing character) is an azure copper blue of more or less intensity. It is ductile and moderately adhesive, when first taken up, or when moistened with water, but acquires an almost stony solidity on drying. It is considerably adulterated with sand or particles of quartz. It parts with its moisture rapidly on exposure to the atmosphere, and dries without much apparent diminution of volume.

5. *Green Clay of St. Peter's.*—This differs little from the preceding, except in its color, which is a deep or verdigris green, admitting some diversity of shades. Its composition appears to be, essentially, alumina, silica, carbonate of copper, water, and iron.

6. *Opwagunite; Calamet Stone; Pipe Stone.*—The last of these terms is a translation of the first, which is Algonquin. Under these names, a peculiar kind of stone, which is much employed by the Indians for pipes, has been alluded to by travellers and geographers from the earliest times. It appears to be a variety of argillaceous wackke. Its color is most commonly a uniform dull red, resembling that of red chalk. Sometimes it is spotted with brown or yellow, but these spots are very minute, and the colors usually faint. It is perfectly opaque, very compact in its structure, and possessing that degree of hardness which admits its being cut or scraped with a knife, or sawed without injury to a common hand-saw, when first raised from the quarry; but it acquires hardness by exposure, and even takes a polish. But it is not capable of receiving a polish by the usual process of rubbing with grit-stone and pumice, these substances being too harsh for it. The Indian process is to scrape or file it smooth, and give it a polish by rubbing with the scouring rush. Its powder is a light red, and emits an argillaceous odor when wetted. This substance is procured at the Coteau des Prairie, intermediate between the sources of the St. Peter's and the Great Sioux Rivers. Some other places have been mentioned as affording this mineral, particularly a locality on the waters of Chippewa River; but the mineral procured here is chocolate-colored.

e. *Magnesian Minerals.*

1. SERPENTINE.

At Presque Isle Point, Lake Superior, common and precious, in isolated masses; also, in connection with, and imbedding native copper, along the southern shore of Lake Superior, at Ontonagon River, &c.

2. STEATITE.

At Presque Isle, near River au Mort, Lake Superior, in connection with the serpentine formation. Also, at the Lake of the Woods, of a black or very dark color, where it is employed by the Indians in carving pipes.

8. ASBESTOS.

Common Asbestos.—In serpentine and steatite, at Presque Isle Point, Lake Superior. Also, in minute veins, in detached masses of diallage and serpentine rocks, on the west shore of Lake Michigan. These veins are no more than a fourth of an inch in width, and the fibres of asbestos occur transversely. They are very flexible, and easily reducible into a flocculent mass.

f. *Barytic Minerals.*

SULPHATE OF BARYTES.

Lamellar Sulphate of Barytes.—In detached masses, imbedded in diluvial soil, at the mines of Peosta, or Dubuque, on the Upper Mississippi, where it is accompanied by sulphuret of lead, calcareous spar, &c. Also, at the Mine au Fevre (now Galena), and at the mouth of the Sissinaway River, on the east banks of the Mississippi, between Prairie du Chien and Fort Armstrong. Its colors are white or yellow, and it is frequently incrusted with a thin coat of yellow oxide of iron. It is most commonly opaque. The only translucent specimen seen was procured at Dubuque's mines.

g. *Strontian Minerals.*

SULPHATE OF STRONTIAN.

Foliated Sulphate of Strontian.—At Presque Isle (Wayne's Battle Ground), on the Maumee River, Wood County, Ohio. It occurs in veins and cavities, in compact limestone, most commonly in the form of flattened prisms. Its color is blue, frequently a very light or sky-blue, and the crystals are fully translucent, or even

transparent. In some instances, they appear to have suffered a
partial decomposition, and fall into fragments in the act of raising,
or are covered with a white powdery crust, frequently visible
only on the summits or terminating points of the prisms. The
same limestone yields crystallised calcareous spar. Both these
substances are abundant in the rocky banks and in the bed of
the Maumee. Also, on Grosse Isle, Detroit River, Michigan.

h. *Bituminous Minerals.*

1. Bitumen.

Petroleum.—Occurs in cavities, in loose fragments of limestone
rock, along the west shore of Lake Michigan, between Milwaukie
and Chicago. These masses of rock lie promiscuously among
fragments of quartz, granite, sandstone, fossil madrepores, &c.,
along the alluvial shore of the lake, and appear to have been
washed up from its bed. The petroleum is in a free and liquid
state; but, where it has suffered an exposure to the atmosphere,
it has acquired a stiff and tar-like consistence passing into *maltha.*
Not unfrequently, fragments of mineral coal are also found scattered along these shores, and there is reason to conclude that a
bituminous formation exists in the contiguous inferior strata
forming the basin of the lake.

2. Graphite.

Granular Graphite.—In a small vein, in the clay-slate of the
River St. Louis, at the head of the nine-mile portage. It is
coarse-grained and *gritty.*

3. Coal.

Slaty Coal.—The only spot where this mineral has been observed, in situ, is at La Charbonniére, on the west banks of the
Illinois River, at the computed distance of one hundred and
twenty miles south of the post of Chicago. It is here seen in
horizontal strata, not exceeding two or three inches in thickness,
interposed between layers of sandstone and shale. Breaking out
on the declivity of the bank of the river, where the overlaying
strata are constantly crumbling down, and thus obscuring the
seams, no very satisfactory examination could be made in a hasty

visit; but the nature and position of the rock strata and soils, and the general aspect of the country, do not justify the conclusion that the bed is of much thickness or extent. Valuable beds may be discovered, however, by exploring this formation. This coal has a shining black color, a slaty structure, inflames readily, burning with a bright flame. It is very fragile where exposed to the weather, falling into fine fragments. Hence, a very black color has been communicated to the contiguous and overlaying soil, which is manifestly more or less the result of disintegrated coal.

Detached fragments of coal, corresponding in mineral characters with the above, are occasionally found around the southern shores of Lake Michigan. The inference, as to the existence of coal around the shores of this lake, is obvious. And we are led to inquire: Does the La Charbonniére formation of coal exist in the sandstone and limestone strata forming the table-land between the Illinois River and Lake Michigan, and reappearing around the basin of the latter, but at such a depression below its surface as to elude observation? And, if so, does not this coal formation extend quite across the southern portion of the peninsula of Michigan? The secondary character of the region alluded to, so far as observed, the horizontal and relative position of the strata, and the general uniformity which is generally observed in the species and order of the coal measures, favor this suggestion.

i. Soda.

1. MURIATE OF SODA.

No traces of salt are known to have been discovered in those parts of the territory of the United States situated north of latitude 46° 31' (which is that of the Sault Ste. Marie) and *east* of the Mississippi River. The great secondary formations which pervade the western country cease south of this general limit, and with them terminate the salt springs, the gypsum beds, the coal measures, and other connected minerals which are generally found in association. It is one of the most important facts which the science of geology has contributed to the stock of useful information, that, in the natural order of the rocky and earthy deposits, muriate of soda always occupies a position contiguous to that of gypsum. This intimate connection between the sulphate of lime and the muriate of soda, enables us, by the discovery of the one, to predict, with considerable but not unerring certainty, the presence of the other. It adds weight to an observation first made among the salt formations of Europe, to find its general correctness corroborated by the relative position of these substances in the United States. These remarks will apply particularly to the salt formations of New York, and to some portions of the muriatiferous region of Virginia and the Arkansas.

There appears to be a salt formation extending from the northwest angle of the Ohio through Michigan, for a distance of two hundred to three hundred miles. It commences in the Seweekly country, passing around the Sandusky River of Lake Erie, where an extensive bed of granular gypsum has recently been discovered, and continues, probably, northwest, so as to embrace the Saganaw basin, and reach quite to the end of the peninsula, and embracing, perhaps, the Gypsum Islands of Lake Huron, ten miles northeast of Michilimackinac. All the brine springs and gypsum beds noticed in the region are situated in the line of this formation.

During the fall of 1821, a number of gentlemen at the Island of Michilimackinac united in the expenses of a tour for exploring the Skeboigon River, a stream which originates in the peninsula of Michigan, and flows into Lake Huron opposite the Island of Bois Blanc. The particular object of this party was to ascertain the precise locality of certain salt springs reported to exist upon that stream. They proceeded to the places indicated, and examined several springs more or less impregnated with salt, but reported that, owing to the jealousy and hostility of those bands of Indians who were found upon that stream, they were not enabled fully to accomplish the object in view.

There are several salt springs reported to exist near the Indian village of Wendagon, on the Sciawassa River, and others on the Titabawassa River, the principal tributaries of the Sagana. Little is, however, known respecting these springs, but the water is represented to be so strongly impregnated, that the Indians manufacture from it all the salt necessary for their villages.

Grand River Valley has also been mentioned among the localities of salt water and gypsum rocks.

Hints may thus be derived of value to the future commerce of the country. Scarcely any of the new states are without indications of the existence of salt. Every day is adding to the number of localities.

In the region *west* of the Mississippi, I was informed that salt occurs, in the crystallized form, in the territories of the Yanktons, who inhabit the flat country at the sources of the River St. Peter's. In certain parts of these plains, the salt exists on the surface. It is mixed with earth, in specimens brought to me, but crystallized in cubes, very imperfect, of a gray or grayish-white color. The Indians scrape it up from certain parts of the prairies or plains, where the salt water is prevented from draining off.

2. ALKALINE SULPHATE OF ALUMINA.

This salt exists, in the form of efflorescences, in the cavities and fissures of rocks along the southeast parts of the shores of Sagana Bay, Lake Huron, and in the argillaceous formations at Erie, on Lake Erie, Pennsylvania.

These positions embrace the principal localities of minerals noticed. In travelling rapidly through a remote wilderness, there was but little opportunity to explore off the track; and the whole observation was confined to the mere surface of the country, which is much obscured by diluvial and alluvial formations.

It will be seen that the region of Lake Superior has been a fruitful field for mineralogical inquiry, and it is one which invites further exploration. Its mineralogy affords a variety of interesting substances which are objects of scientific research, and it may be anticipated to be the future theatre of extensive mining operations. The country northwest of Lake Superior, and the Upper Mississippi north of the Falls of St. Anthony—consisting mostly of upheaved primitive rocks and the pebble-drift, or diluvial, formations—has furnished but few subjects of mineralogical remark.

The district of country between the Falls of St. Anthony and Prairie du Chien, in common with the more southern portions of the Mississippi Valley, partakes of all the interest which the mineral kingdom presents in a calcareous and metalliferous country of secondary formation. It has added considerably to my collection. It is probable the Rivers St. Peter's, St. Croix, and Chippeway would well reward exploration; but the mines of Dubuque partioularly invite a mineralogical survey. Their future importance cannot fail to be duly appreciated.

If the country has put on an aspect unfavorable to mineralogy, its geological features have been observed to sustain its interest.

Much of the interest growing out of the examination, for the first time, of the mineralogy and natural history of the country, is such as to commend itself, in an especial manner, to the consideration of men of science, and of associations devoted to scientific details, rather than the department of a government. To these former, nature is a storehouse of facts, and a perpetual anxiety is felt by this class of observers to know the range, not only of our rock formations, but of our plants, shells, fossils, and other classes of objects in our physical geography. Such desires I have endeavored, as far as my means permitted, to gratify. The fresh-water conchology of the lakes and rivers visited was often attractive, when other objects excited little interest. The species collected in this department have been referred to the New York Lyceum of Natural History.

With these remarks, the result of an arduous and interesting journey through a part of the continent hitherto unexplored, I have the honor to conclude my report, and to terminate the trust confided to me.

I am, sir, with respect,
Your obedient servant,
HENRY R. SCHOOLCRAFT,
Geologist, &c. of the Ex. Exp.

VIII.

A Report to the Senate of the United States, in Answer to a Resolution passed by this Body, respecting the Value and Extent of the Mineral Lands on Lake Superior.* By HENRY R. SCHOOLCRAFT.

SAULT STE. MARIE, October 1, 1822.

SIR: In reply to the inquiries, contained in a resolution of the Senate of the United States, respecting the existence of copper

* *To the Senate of the United States :—*

In compliance with a resolution of the Senate of the 8th May last, requesting "information relative to the copper mines on the southern shore of Lake Superior, mines in the region of Lake Superior, inclosed to me in a note from the War Department, dated 8th May, 1822, I have the honor to submit to you the following facts and remarks:—

1. In relation to "*the number, value, and position of the copper mines on the south shore of Lake Superior.*" The remote position of the country alluded to, the infrequency of communication, and the little reliance to be placed on information derived through the medium of the aborigines or of traders, who are wholly engrossed with other objects, presents an embarrassment at the threshold of this inquiry, which must be felt by every person who turns his attention to the subject. The information sought for demands a minute acquaintance with the natural features and mineral structure of the country, which can only be acquired by personal examination; and it is a species of research requiring more leisure, better opportunities, and a freer participation in personal fatigue, than usually falls to the share of tourists and travellers. Not only are those difficulties to be encountered which are inseparable from the collection of isolated facts in a new and unsettled country; but those, also, which are peculiar to the subject, connected as it is, at every stage of the inquiry, with the

their number, value, and position, the names of the Indian tribes who claim them, the practicability of extinguishing their titles, and the probable advantage which may result to the Republic from the acquisition and working these mines," I herewith transmit a report from the Secretary of War, which comprises the information desired in the resolution referred to.

JAMES MONROE.

WASHINGTON, 7th December, 1822.

DEPARTMENT OF WAR, 3d December, 1822.

The Secretary of War, to whom was referred the resolution of the Senate of the 8th May last, requesting the President of the United States "to communicate to the Senate, at the commencement of the next session of Congress, any information which may be in the possession of the Government, derived from special agents or otherwise, shewing the number, value, and position of the copper mines on the south shore of Lake Superior, the names of the Indian tribes who claim them, the practicability of extinguishing their title, and the probable advantage which may result to the Republic from the acquisition and working these mines," has the honor to transmit a report of Henry R. Schoolcraft, Indian agent at the Sault of Ste. Marie, on the copper mines in the region of Lake Superior, which contains all the information in relation to the subject in this department.

All which is respectfully submitted.

J. C. CALHOUN.

To the PRESIDENT OF THE UNITED STATES.

prejudices and superstitions of the Indian tribes. [B.] It can, therefore, excite little surprise that, after having been the theme of speculation for more than a century, and obtained the notice of several works of merit in Europe,* both the position and value of these mineral beds have continued to the present times to be but partially known. To ascertain more clearly their value and importance to the Republic were objects more particularly confided to me as a member of the expedition sent by the Indian Department, in the year 1820, to traverse and explore those regions. My report of the 6th of November of that year—a copy of which, marked A, is herewith transmitted—gives the result of that inquiry. After a lapse of two years, little can be added. Reflection and subsequent inquiry convince me that the facts advanced in that report will be corroborated by future observation. No circumstance has transpired which is calculated to prove that my suggestions with regard to the fertility and future importance of those mines are fallacious; on the contrary, all information tends to strengthen and confirm those suggestions. Specimens of pure and malleable copper continue to be brought in to me by the aborigines from that region, but it is not deemed necessary to particularize in this place the additional localities. It will be sufficient to observe, that the number of these new discoveries justifies the expectations that have been created respecting the metalliferous character of the region of the Ontonagon, and the south shore of Lake Superior. [C.]

I shall here add the result of an accurate analysis made upon a specimen of this copper at the mint of Utrecht, in the Netherlands, at the request of Mr. Eustis, minister plenipotentiary from the United States, who carried samples of the American copper to that country. The report of the inspector of the mint, which communicates the result of this analysis, has the following remarks upon the natural properties of this species of copper, and the mode of its production: "From every appearance, the piece of copper seems to have been taken from a mass that has undergone fusion. The melting was, however, not an operation of art, but a natural effect caused by a volcanic eruption. The stream of

---

* *Vide* Jameson's Mineralogy, Parkes's Chemical Catechism, Phillips's Elementary Introduction to Mineralogy.

lava probably carried along in its course the aforesaid body of copper, that had formed into one collection, as fast as it was heated enough to run, from all parts of the mine. The united mass was probably borne in this manner to the place where it now rests in the soil. The crystallized form, observable everywhere on the original surface of the metal that has been left untouched or undisturbed, leads me to presume that the fusion it has sustained was by a process of nature; since this crystallized surface can only be supposed to have been produced by a slow and gradual cooling, whereby the copper assumed regular figures as its heat passed into other substances, and the metal itself lay exposed to the air.

"As to the properties of the copper itself, it may be observed that its color is a clear red; that it is peculiarly qualified for rolling and forging; and that its excellence is indicated by its resemblance to the copper usually employed by the English for plating. The dealers in copper call this sort *Peruvian copper* to distinguish it from that of *Sweden*, which is much less malleable. The specimen under consideration is incomparably better than Swedish copper, as well on account of its brilliant color as for the fineness of its pores and its extreme ductility. Notwithstanding, before it is used in manufactures, or for the coining of money, it ought to be melted anew, for the purpose of purifying it from such earthy particles as it may contain. The examination of the North American copper, in the sample received from his excellency the minister, by the operation of the cupel and test by fire, has proved that it does not contain the smallest particle of silver, gold, or any other metal." It is a coincidence worthy of remark, that the suggestions offered by the assayer respecting the volcanic origin of these masses of copper, are justified by the leading features of the Porcupine Mountains, and by the melted granites found upon the heights called Grande Sables and Ishpotonga.

2. The second and third inquiries of the resolution relate to "*the names of the Indian tribes who claim the mines, and the practicability of extinguishing their title.*" By the treaty concluded at this post on the 16th of June, 1820, the Ojibwai* Indians cede to the United States four miles square of territory, bounded by

---

* For the different names applied to this tribe of Indians, see Appendix H.

the River St. Mary's, and including the portage around the falls.*
This is the most northerly point to which the Indian title has
been extinguished in the United States. The different bands of
Ojibwais possess all the country northwest of this post, extending through Lake Superior to the sources of the Mississippi,
where they are bounded by the Assennaboins, the Crees, and the
Chippewyans of the Hudson Bay colony. Their lands extend
down the Mississippi to the Sioux boundary, an unsettled line
between the junction of the River De Corbeau and the Falls of St.
Anthony. South of Lake Superior, they claim to the possessions
of the Winnebagoes, on the Onisconsin and Fox Rivers, and to
those of the Pottawatamies and Ottoways, on Lake Michigan.
The Wild Rice, or Monomonee Indians, are an integral part of
the Ojibwai nation, deriving their name from the great reliance
they place on the zizania aquatica as an article of food. They
live in small, dispersed bands between the Ojibwais of the lake,
and the Winnebagoes of Fox River. Those residing among the
Ojibwais speak the same language, but with many peculiarities
and corruptions on the waters of Green Bay. They claim the
respective tracts upon which they are located. These are, principally, the valleys of the Fox and Monomonee Rivers, and the
rice lands contiguous to the Fol. Avoine, Clam Lake, and Lac de
Flambeau, which lie on the table-lands between Lake Superior
and the Mississippi.

The right of soil to all that part of the Peninsula of Michigan
not purchased by the United States is divided between the Ojibwais and the Ottoways. The former claim all the shores and
islands of Lake Huron situated north of the Saganaw purchase,
except those in the vicinity of Michilimackinac and the St. Martin,
or Gypsum Islands, which were ceded by treaty on the 6th of
July, 1820.† Their territories continue north, through the River
St. Mary's, embracing the country on both banks, and the islands
in the river, saving Drummond's Island, which is garrisoned by
the British, and the Four Mile concession at the Sault or Falls,
now occupied by a detachment of the United States' army. It is
not deemed necessary to point out the limits of their territories

* Vide acts passed at the second session of the 16th Congress of the United States,
page 88.
† Vide acts passed at the second session of the 16th Congress, p. 91.

with more precision, or to pursue them into the Canadas, where they are also very extensive. It will sufficiently appear, from this outline, that the discoveries of copper on the south shore of Lake Superior are upon their lands. That some of these discoveries have been made upon, or will be traced to, the possessions of the North Monomonees, is also probable.

With respect to the practicability of extinguishing the Indian title, no difficulty is to be apprehended. Living in small villages, or tribes of the same mark, scattered over an immense territory, and often reduced to great poverty by the failure of game and fish, it is presumed there would be a disposition among their chiefs and head men to dispose of portions of it. Those districts which most abound in minerals, presenting a rough and rocky surface, are the least valuable to them as hunting-grounds; and the goods and annuities which they would receive in exchange must be vastly more important to them than any game which these mineral lands now afford.

3. *"The probable advantage which may result to the Republic from the acquisition and working of these mines."* How far metallic mines, situated upon the public domain, may be considered as a source of national wealth, and what system of management is best calculated to produce the greatest advantages to the public revenue, are inquiries which are not conceived to be presented for consideration in this place; nor should I presume to offer any speculations upon topics which have been so often discussed, and so fully settled. In applying axioms, however, to a species of productive industry, the results of which are so very various under various situations, great caution is undoubtedly necessary; and it must appear manifest, on the slightest reflection, how much the comparative value of metallic mines, equally fertile and productive, ever depends upon situation and local advantages. Dismissing, therefore, all questions of abstract policy, I shall here adduce a few facts in relation to the fertility of these mineral beds, and their position with respect to a market—points upon which their value to the nation must ultimately turn.

That copper is abundantly found on the south shore of Lake Superior has been shown. It is unnecessary here to add to, or repeat the instances of its occurrence, or to urge, from an inspection of the surface, the fertility of subterranean beds. All the facts which I possess in relation to this subject are before you, and you will assign to them such importance as they merit. It is a subject upon which I have bestowed some reflection and much inquiry, superadded to limited opportunities of personal observation, and the result has led me to form a favorable estimate of their value and importance. It is not only certain that a prodigious number of masses of metallic copper are found along the borders of the lake, but every appearance authorizes a conclusion that they are only the indications of near and continuous veins. Some of these masses are of unexampled size, and all present metallic copper in a state of great purity and fineness. Of its ductile and excellent qualities for the purposes of coinage and sheathing, the analysis of Utrecht leaves no doubt. It is true that a mistaken idea has prevailed among travellers and geographers respecting the weight of the great mass of copper on the Ontonagon River; but it is, nevertheless, of extraordinary dimensions, and I have endeavored to show, from their works, how these errors have originated, and that the metal is disseminated throughout a much greater extent of country, and in masses of every possible form and size. Until my facts and data can, therefore, be proved to be fallacious, I must be permitted to consider these mines not only fertile in native copper and its congenerous species, but unparalleled in extent, and to recommend them as such to the notice of the Government.

But, whatever degree of incertitude may exist respecting the riches of these mines, their situation with respect to a market can admit of no dispute. As little can there be concerning the advantages which this situation presents for the purposes of mining and commerce. Let us compare it with that of other mines, and appeal to acknowledged facts for the decision. The value of a coal mine, a stone quarry, or a gypsum bed, often arises as much from its situation as its fertility. But the proposition may be reversed with respect to a metallic mine, the value of which to the proprietor arises more from its fertility and less from its situation. Gold, silver, copper, tin, lead, &c., when separated from the matrix of the mine, are so valuable that they can bear to be transported a long journey over land, and the most distant voyage by water. Their worth in coined money, produce, or manufactures, is not fixed in the particular circles of country where they are dug up, but depends upon the seaboard market, and embraces all countries. The silver of Mexico and Peru circulates throughout Europe, and is carried to China. It is no objection to those mines that they are situated in the Cordilleras, or upon the high table-lands of the American continent, and must be carried a thousand miles upon the backs of mules to the seaside. The very discovery of those mines has rendered many poor silver mines of Europe of no value, although possibly situated in the environs of the best silver markets in the world. It is the fertility, and not the situation of such mines, that constitutes their chief value; and it is so with many of the coarser metals.

The tin of the Island of Banka, and the Peninsula of Siam in Asia, and the copper of Japan, find their way to Europe, and are articles of commerce in the United States. The cobalt of Saxony is sent to Pekin, and the platina of Choco, to all parts of the world. In all these instances, the fertility of the mines compensates for every disadvantage of situation. But this principle is not alone confined to mines of tin, copper, &c.; it even holds true of the heavy and bulky articles of iron, lead, and salt. The lead of Missouri finds a market at New York, Philadelphia, and Boston, and will be carried to Europe. It is no objection that it must be conveyed in wagons forty miles from the interior, and sent a voyage of 3,000 miles in steamboats and merchant ships. The great fertility of the mines counterbalances the disadvantages of its remote position from the market, and it is the price of the metal in the market which always regulates its price at the mines. The malleable iron of Sweden is consumed on the summits of the Alleghany, although its strata are replete with iron ore, which is worked at numerous forges along the rivers which proceed from each side of it. It is believed that the salt springs of Onondaga, from their copiousness alone, would supply a vast portion of the interior and seaboard of the United States with salt, even if the facilities of water carriage had not been presented by the Erie Canal. The value of such mines and minerals ever depends as much upon the abundance as upon the favorable position of them. It is far otherwise with quarries of stone, gypsum, marl, fossil coal, &c., whose contiguity to a good market establishes their value. No abundance of these articles would justify a land carriage of one hundred miles. They constitute a species of mining,

24

the profits and value of which increases in the ratio of the surrounding population, and as the country advances in improvements. But this advantage is far less sensibly felt, and cannot be considered essential to the successful working of mines of silver, copper, &c. Neither the remote position, therefore, of the Lake Superior copper mines, nor the want of a surrounding population, present objections of that force which would at first seem to exist; and it is confidently believed that, if their fertility is such as facts indicate, they may be opened and wrought with eminent advantage to the Republic. But let us examine their situation with respect to a market, and compare it with that of other mines of the same metal, and of some of the coarser metals, which bear a considerable land, and the most distant water carriage. To favor the inquiry, let it be granted for the moment that proximity of situation to a market, or free water carriage, are indispensable to the success and value of the most fertile mines.

Assuming the confluence of the Ontonagon River with Lake Superior (which is apparently the centre of the mine district) as the place where the metal is first to be embarked for market, it must be carried down the lake 300 miles to the Sault or rapids of St. Mary's. Here, if it is in barges, it may descend the rapids in perfect safety, as is the invariable practice of the traders on arriving with their annual returns of furs and skins from the north. If in vessels, it must be transferred either into boats or carts, and carried half a mile to the foot of the rapids, where it will again be embarked in vessels, and transported through the Lakes Huron, St. Clair, and Erie, and their connecting straits, to Buffalo, a distance of 650 miles. The progress made in the construction of the great canal which is to connect the lakes and Atlantic, is such as to leave no doubt upon any reasonable mind of the full completion of that work with the close of the year 1824. Through this channel, the transportation is to be continued in boats or barges, by a voyage of 353 miles, to the Hudson at Albany; thence a sloop navigation of 144 miles, which, for speed and freedom from risk, is perhaps unequalled in all America, takes it into the harbor of New York, making the entire distance, from the mouth of the Ontonagon, 1,447 miles. From New York it is distributed to our naval depots, and to the markets of Europe.

It is exchanged for the lead of Missouri, the iron of Sweden, or the silver of Mexico; and the same ready communication transports the return cargo to Buffalo, from whence the commerce is extended, by means of the lakes, throughout western New York, Pennsylvania, Ohio, Indiana, Illinois, Michigan, and the interminable regions of the north. Thus it is seen that, when the Erie Canal is completed, a free and direct water communication, from the mines to one of the best markets in America, will exist, in which the rapids of St. Mary's are the only interruption, and this is only an interruption to large vessels. Not only so, but the Ontonagon River may be ascended many miles with vessels of light burden, and thus the copper of Lake Superior, wafted from the heart of the interior, and from the base of the Porcupine Mountains, into the harbors of New York, Philadelphia, &c. Of this whole distance, 1,047 miles are now navigated by the largest class of river craft and lake schooners; the balance of the distance is the length of the Erie Canal. (See Note D.)

Let it be recollected that there are no mines of copper situated upon the margin of the sea, and that every quintal of sheet copper, bolts, nails, &c., which we receive from Great Britain, Russia, Sweden, or Japan, is transported a greater or less distance on turnpikes or canals, before it reaches the place of shipment. The richest copper mines of the Russian empire are seated on the summits of the Uralian Mountains; those of Fahlun, in Sweden, and Cornwall, in England, are scarcely more favored as to position; and, owing to a want of coal, all the ores raised at the latter are transported into Wales to be smelted.* But we need not resort to Europe for instances. All the lead raised at the fertile mines in Missouri is transported an average distance of forty miles in carts and wagons before it reaches the banks of the Mississippi. Steamboats take it to New Orleans, a distance, by the shortest computation, of 1,000 miles. But it must still pass through the Gulf of Mexico, and encounter the perils of the Capes of Florida, and a voyage of 2,000 miles along the coast of the United States, before it reaches its principal marts. The average cost of transporting a hundredweight of lead from Mine au Breton and Potosi to the banks of the Mississippi, during the year 1818,

* Silliman.

was seventy-five cents. The distance is thirty-six miles. The price of conveying the same quantity from the storehouses at Herculaneum and St. Genevieve to New Orleans, by steamboats, was seventy cents. The distance exceeds 1,000 miles. Hence, it costs more to transport a given quantity thirty-six miles by land than to convey it 1,000 by water. These rates have probably varied since, but the proportionate expense of land carriage, compared to that of water, will remain the same. A quintal of copper may, therefore, be transported from the mines of Superior to Buffalo or Lockport, in New York, for the same sum required to convey an equal quantity of lead from Potosi to St. Genevieve. If we consider the city of New York as the market of both, no hesitancy or doubt can be experienced as to the decided and palpable advantages possessed by the northern mines. It is only necessary to adduce these facts; the conclusions are inevitable. In every point of view, the distance of these mines from the market presents no solid objection to their being explored with profit to the nation.

Pig copper, which is the least valuable form in which this metal is carried to market, is now quoted in the Atlantic cities at 19 cents per pound; sheathing, at 27; brazier's, at 32. I have no data at hand to show the amount of these articles consumed in the United States, and for which we are annually transmitting immense sums to enrich foreign States. But those who best appreciate the advantages of commerce will readily supply the estimate. It would be an interesting inquiry to ascertain how much of the sums yearly paid for sheathing copper, bolts, nails, engravers' plates, &c., is contributed to the wealth of the respective foreign States who possess mines of this metal. We can look back to a period in the history of Great Britain, when that power did not contribute one pound of copper to the commerce of Europe. During a period of nine years, closing with the memorable year (in American history) of 1775, the produce of the copper mines of Cornwall was 2,650 tons of fine copper. (See Note E.) Since that time, the yearly returns of those mines exhibit a constant increase; and the copper mines of Great Britain are now the most valuable in the world. The amount produced by the mines of Cornwall and Devon, after deducting the charges of smelting, for the single year of 1810, was 969,376 pounds sterling.

(See Note F.)   The clear profits of the Dolgoath mine, one of the richest in Cornwall, for a period of five months, during the year 1805, was £18,000, which is at the rate of £43,200, or $192,000, per annum.   Next to Great Britain, the most considerable mines of Europe are those of Russia, Austria, Sweden, and Westphalia, as it was in 1808.   Of less importance are those of Denmark, France, Saxony, Prussia, and Spain.   The proportion in which the British mines exceed those of the most favored European nation is as 200,000 × 67,000. (See Note G.)

There is another consideration connected with this subject which is worthy of remark.   Should it be inquired what would be the effects of the purchase of these mines upon the condition of the Indian tribes, the reply is obvious.   It would have the most beneficial tendency.   They would not only profit by an exchange of their waste lands for goods, implements of husbandry, the stipulated services of blacksmiths, teachers, &c., but the intercourse would have a happy tendency to allay those bitter feelings which, through the instigation of the British authorities in the Canadas, they have manifested, and still continue to feel, in degree, towards the United States.   The measures which the President has recently directed to be pursued to assuage these feelings of hostility, and to induce them to cherish proper sentiments of friendship and respect, are already in a train of execution that bids fair for success.   Continued exertions, and the necessary and proper means, are all that seem necessary to confirm and complete the effect; and whatever measures have a tendency to increase the intercourse of American citizens with these "remote tribes," and to give them a true conception of the power and justice, and the pacific and benevolent policy of our Government, must favor and hasten such a result.

I have the honor to be, sir,
With the highest respect,
Your most obedient servant,
HENRY R. SCHOOLCRAFT,
U. S. Indian Agent at the Sault Ste. Marie.
Hon JOHN C. CALHOUN,
Secretary of War, Washington.

Notes.

(B.)

Among the numerous superstitions which the Indian tribes entertain, that respecting mines is not the least remarkable.   They are firmly impressed with a belief that any information communicated to the whites, disclosing the position of mines or metallic treasures situated upon their grounds, is displeasing to their manitos, and even to the Great Spirit himself, from whom they profess to derive every good and valuable gift; and that this offence never fails to be visited upon them in the loss of property, in the want of success in their customary pursuits or pastimes, in untimely death, or some other singular disaster or untoward event.   This opinion, although certainly not a strange one to be cherished by a barbarous people, is, nevertheless, believed to have had its origin in the transactions of an era which is not only very well defined, but must ever remain conspicuous in the history of the discovery and settlement of America.   It is very well known that the precious metals were the principal objects which led the Spanish invaders to penetrate into the interior of Mexico and Peru, and ultimately to devastate and conquer the country, to plunder and destroy its temples, and to tax and enslave its ill-fated inhabitants.   It is equally certain that, to escape these scenes of cruelty and oppression, many tribes and fragments of tribes, when further resistance became hopeless, fled towards the north, preferring the enjoyment of liberty and tranquillity upon the chilly borders of the northern lakes, to the pains of servitude in the mild and delightful valleys of Mexico, and the golden plains of the Incas.·   In this way, many tribes who originally migrated from the north, along the Pacific Ocean, to the Gulf of California, and thence over all New Spain, were returned towards the north over the plains of Texas and the valley of the Mississippi; those tribes nearest the scenes of the greatest atrocities always pressing upon the remoter and less civilized, who, in turn, pressed upon the nations less enlightened than themselves, and finally drove them into the unfrequented forests of the north.   Among these terrified tribes, the traditions of the Ojibwais affirm that their ancestors came, and that they originally dwelt in a country destitute of snows. Many tribes who now speak idioms of their language were left upon the way, and have since taken distinctive names. Among these, are the Pottawatamies, the Ottoways, &c. The latter formerly were, as they still remain, the agriculturists. The Miamis and Shawnees, whose languages bear some affinity, preceded them in their flight. The Winnebagoes, speaking a separate and original tongue, came later, and preserve more distinct traditions of their migration. All these tribes carried with them the strong prejudices and fixed hatred excited by the cruelty, rapacity, and cupidity of their European conquerors; and, above all, of that insatiable thirst for gold and silver which led the Spaniards to sack their towns, burn their temples, and torture their people. Cruelty and injustice of so glaring a character must have made upon their minds too deep an impression ever to be forgotten, or completely erased from their traditions. To that memorable epoch we must, therefore, look for the origin of that cautious and distrustful disposition which these tribes have since manifested with regard to the mines and minerals situated upon their lands; and the circumstance seems to offer an abundant excuse, if not a justification, for those prevarications and evasions which present a continual series of embarrassment to every person who seeks through their aid to develop the mineral resources, or describe the natural productions, of their territories. Hence, too, the cause why they are prone to imagine that all mineral or metallic substances obtained or sought upon their lands, are susceptible of being converted or *transmuted* into the precious metals.

(C.)

The following *additional* localities of native copper, derived from sources entitled to respect, and accompanied, in some instances, by specimens of the metal, may here be given:—

1. Grand Menou, or Isle Royal, Lake Superior. Captain ——, of the schooner ——, in the employ of the Hudson's Bay Company, on Lake Superior, describes this island as affording frequent masses of copper. While becalmed off its shores in the spring of 1822, and, afterwards, in coasting along the island for a distance of one hundred miles, his men frequently went ashore, and never failed to bring back with them lumps of metallic copper, which they found promiscuously scattered among the fragments of rock.

These were more abundant in approaching its southwestern extremity, where they unite in representing it to exist in a solid vein. Specimens of limpid quartz, chalcedony, and striped agate, were also brought to me from this island. [J. S. J. J.]

2. On the extremity of the great peninsula, called by the natives Meenaiewong, or Keweena Point, which forms so prominent a feature in the physiognomy of Lake Superior. It occurs in the detached form. [J. H. J. J.]

3. At Point aux Beignes, which is the east cape of the entrance into L'Ance Quewiwenon. A mass from this place was raised from the sandstone rock, which predominates there. [J. Y. B.]

4. At Caug Wudjieu, or the Porcupine Mountains, Lake Superior; in masses, enveloped with a green crust, along the banks of the Carp, or Neemaibee River, which originates in these mountains. [W. M. G. Y. J. J.]

5. On the banks of Lac Courterroile. This lake lies near the source of the River Broule, or Cawesacotai, which enters Lake Superior near La Pointe. It occurs in the alluvial soil, which is a kind of loamy earth, with pebbles intermixed, but of a rich quality, and timbered with beech and maple. It is found mostly in small, flat masses, more or less oxidated. [B. G. J. G. Y.]

6. In a vein on the shore of Lake Superior, between La Riviere de Mort and St. John's, a little to the west of Presque Isle. [J. J.]

7. On the northeast branch of the Ontonagon River. [J. H.]

8. In the precipitous bluffs called Le Portail, and the Pictured Rocks. A green matter oozes from the seams in these rocks, and forms a kind of stalactites, which is apparently a carbonate of copper. [G. Y.]

These localities embrace a range of more than two hundred miles along the south shore of Lake Superior, which proves how intimately this metal and its ores are identified with the rocks and the soil of that region.

(D.)

In all our calculations respecting the position and advantages of these mines, too much stress cannot be laid upon the facilities of the lake navigation. It is believed that a ton of merchandise, or a barrel bulk, can be transported through the lakes at the same rates that are paid in the coasting trade of the United States.

Nor is the risk greater. The best data which I can command, induce me to conclude that a quintal of copper can be conveyed from the place of shipment on Lake Superior, to the city of New York, for *one dollar*. The present price of transportation, for a barrel bulk, from Buffalo to Mackina, may be stated, on the average of freights, at 8*s.*, New York. The mean weight of a barrel bulk, taking flour as the standard, may be safely put down at 200 lbs. gross, being 50 cents per cwt. But it must be recollected that there is no return freight; and, consequently, that this sum covers the expenses not only of the outward and return voyage, but still leaves a profit to the owner. Messrs. Gray and Griswold, sutlers of the 2d regiment, paid 9*s.* 6*d.*, New York, per barrel bulk, from Buffalo to the Sault. This gives a result of 59 cents per cwt. But, if a return cargo could be obtained, one-half of this sum would afford an equal profit on the voyage; and it is believed that the article of bar copper could at all times be conveyed from the Sault to Buffalo for 20 cents per cwt. Being a very convenient species of ballast, it would oftentimes be taken in lieu of stone, and, consequently, cost no greater sum than the price of carrying it on board. But the facilities and cheapness of the lake navigation cannot, perhaps, be better illustrated than by stating the price of provisions at the post of St. Mary's, every article of which is carried from 800 to 700 miles through the lakes. The following statement of the assistant commissary has been politely furnished at my request:—

SAULT STE. MARIE, October, 1822.

DEAR SIR: Agreeably to your request, I send you a statement of the actual cost of subsistence stores furnished at this post for the use of troops at present making the military establishment, ordered by the Government to this place.

The prices of the several articles below enumerated are at a small advance on the stores of the settlers outside of the cantonment.

The expenses of subsisting, or rather of maintaining, a garrison at this place will be as small, if not less, per annum, than at any other frontier post in our country. The provisions for the soldier cost as little, I believe, as at any other post, and next year we shall be able to raise all the forage for the use of our beef cattle, and the horses and oxen of the quartermaster's department.

I am, dear sir, yours, &c.,

W. BICKER,

A. C. S. U. S. A.

Statement of the Cost of United States Subsistence Stores at the Sault de Ste. Marie, 1822.

	Cents.
Pork, per pound	$4\frac{1}{4}$
Flour, per pound	$1\frac{9}{10}$
Whiskey, per gallon	29
Fresh beef, per pound	$6\frac{1}{4}$
Vinegar, per gallon	22
Salt, per bushel	90
Soap, per pound	10
Candles, per pound	$20\frac{1}{2}$
Beans, per quart	$4\frac{7}{10}$

The total cost of a soldier's ration is 9 cents and 1 mill per diem.

WALTER BICKER,

A. C. S. U. S. A.

H. R. Schoolcraft, Esq., *U. S. I. Agent.*

(E.)

Statement of the Returns of Copper Ores Smelted at the Mines of Cornwall (Eng.) from 1726 to 1775.—[Rees's Cyclopedia.]

Periods.	Tons of ore.	Average price per ton.	Amount.	Annual quantity of fine copper.
1726 to 1735	64,800	£7 15 10	£473,500	700 tons
1736 to 1745	75,520	7 8 6	560,106	830 "
1746 to 1755	96,790	7 8 0	731,457	1,080 "
1756 to 1765	169,699	7 6 6	1,243,045	1,800 "
1766 to 1775	264,273	6 14 6	1,778,837	2,650 "

(F.)

Statement of the Produce of the Mines of Cornwall and Devon (Eng.) for a period of four years, ending with 1811.

		Tons of ore.			Fine copper.				Average standard per ton.	Annual amount after deducting charges of smelting.	
		Tons.	cwt	qrs.	Tons.	cwt.	qrs.	lbs.	£	£	s.
1808 {	Cornwall	78,484	2	1	7,118	5	1	17	} 107	781,348	16
	Devon	8,725	0	0	869	10	0	0			
1809 {	Cornwall	72,088	12	2	6,972	17	0	17	} 122	875,784	2
	Devon	8,210	0	0	865	1	0	8			
1810 {	Cornwall and Devon	} 80,238	14	8	7,006	18	2	5	141	969,876	19
1811 {	Cornwall and Devon	} 78,579	0	1	6,272	0	2	2	125	769,879	4

(G.)

Table of the Annual Quantity of Copper raised from the Earth in Different Countries, in Quintals—the Quintal valued at 100 lbs.

1. England 200,000
2. Russia 67,000
3. Austria, including Bohemia, Gallicia, Hungary, Transylvania, Styria, Carinthia, Carniola, Salzburg, and Moravia 60,000
4. Sweden 22,000
5. Westphalia, in 1808 . . . 17,229
6. States of Denmark 8,500
7. Bavaria, including the Tyrol . . 3,000
8. France 2,500
9. Saxony, in 1808 1,320
10. Prussia, as left by the treaty of Tilsit 337
11. Spanish European mines . . 309

Total, 382,186

(H.)

I shall here give the synonoma for this tribe of Indians, which appears to have been first recognized by the United States as an independent tribe by Wayne's treaty of 1795,* under the

---

* This fact is not stated in full confidence. I cannot refer to any authorities to prove that they were formally recognized by the United States before this very

name of Chipewa. This name has been retained in all subsequent treaties with them, not, however, without some discrepance
in the orthography. These variations are chiefly marked by the
introduction of the letter p at the beginning of the second syllable, or the vowel y annexed to the third; producing Chip-pe-wa,
Chip-pe-way, and Chip-e-way. The French missionaries and
traders, whose policy it was to discard the names of the aboriginal
tribes from their conversations, bestowed upon this tribe, at a
very early period, the *nom de guerre* of *Saulteurs*, or *Sauteurs*, from
the Sault or Falls of St. Mary's, which was the ancient seat of
this tribe—a name which is still retained by the Canadians, and
by many of the American traders. Among the early French
writers, they were also sometimes denominated *Outchipouas*.
There is as little uniformity among travellers and geographers.
Pinkerton, Darby, Morse, Carver, Mackenzie, and Herriot, either
employ the word according to the orthography of Wayne's treaty,
or with the modifications above noticed. The name of Chippewyans, employed by Mackenzie, relates to a tribe residing north
and west of the sources of the Mississippi, who speak a language
having no affinity, and are a distinct people. Henry, who was
well versed in the Chippewa language, also conforms to the
popular usage, but observes that the true name, as pronounced
by themselves, is Ojibwa.

Having taken pains to ascertain and fix the pronunciation of
this word, I have not hesitated to introduce it into my correspondence and official accounts; but I am aware of my great
temerity in so doing. Popular prejudices, and several of the
authorities above cited, stand opposed to the proposed innovation.
The continued use of the word "Chippewa" is also sanctioned by
a name entitled to conclusive respect. "I write the word in this
way," observes the Executive of Michigan, "because I am apprehensive the orthography is inveterately fixed, and not because I
suppose it is correct." Still, there are reasons for changing it.
Justice to this unfortunate race requires it. Since the popular

recent period. By the French and British governments they were known soon after
the first settlements at Quebec and Albany (A. D. 1608, 1614), and subsequently
treated with. A band of warriors from Chegoimegon, on Lake Superior, under the
command of Waub Ojeag, or the White Fisher, was present at the taking of Fort
Niagara by Sir W. Johnston in 1759.

apathy to their condition is such that every remembrance of their actual customs, manners, and traditions will probably perish with them, and their *name*, ere long, be all that is left, it is at least incumbent upon us to transmit *that* to posterity in its true sound— as the fathers and sachems pronounced it. If, then, there is an acknowledged error in this respect, shall we hesitate to correct it?

IX.

Rapid Glances at the Geology of Western New York, west of the Rome Summit, in 1820.*

ROCK FORMATIONS.—1. Assuming the area of the most eastwardly head of the Onondaga Valley, the Wood Creek, and the Rome Summit, and the valley of the Niagara, with an indefinite extent laterally, to form the limits of this inquiry; it is in coincidence with all known facts to say that it is a secondary region, consisting of the sedimentary and semi-crystalline strata, the lines of which are perfectly horizontal. Colored sandstone, generally red, forms the lowest observed stratum.

Wherever streams have worn deep channels, they either disclose this rock or its adjuncts, the grits, or silicious sinter. It is apparent in the chasm at Niagara Falls, about half a mile below

---

* At the time these sketches were written, no geological observations had been made on this field, which has, at subsequent periods, been so elaborately described; nor had the topic itself attracted much attention. I landed at New York, in the ship Arethusa, from New Orleans, in the summer of 1819, and published, in that city, in the fall of that year, an account of the lead-bearing rocks of Missouri, and their supporting white sandstones, which rest, in horizontal deposits, on the primitive formation of the St. Francis; bringing, at the same time, a rich collection of the mineralogy of that region, which soon became known in private cabinets. This became the cause of my employment, by the United States Government, to visit the alleged copper mines on Lake Superior, as a member of the expedition to the sources of the Mississippi. I left Oneida County, in the district remarked on, on the 10th of April of that year, and reached the banks of the Niagara River on the 29th of that month. On returning from the sources of the Mississippi, I entered the same region on the 17th of October, and reached Oneida on the 21st of the same month. Prior to my visit to the Great West, I had dwelt some three years— namely, 1809, '10, '11, '12—in Oneida and Ontario counties. These were the opportunities enjoyed, up to the period, for acquiring a knowledge of the geography and geology of the country. Mr. A. Eaton's *Index to Geology*, published early in 1820, embraces nothing extending to western New York.

the cataract. It is often seen on the surface of the country, or buried slightly beneath the soil. In color, hardness, and other characters, there is a manifest variety. But, considered as a "formation," no doubt can exist of its unity. Its thickness can only be conjectured, as no labor has, so far as we know, penetrated through it.

Judging from observations made in Cattaraugus County, in 1818, the coal measures have been completely swept from this area.

2. Next in point of altitude, is the series of dark, carbonaceous, shelly slate rock. The thickness of this formation, as indicated at Niagara, cannot be less than ninety feet. It is also often a surface-rock in the district, forming portions of the banks of lakes, streams, &c. It is characterized by organic remains of nascent species. Portions of it also disclose rounded masses of pre-existing rocks.

3. Last in the order of superposition, is the secondary limestone formation. It is, most commonly, of a dark, sedimentary aspect. It is not invariably so, but portions of it have a shining, semi-crystalline fracture. Shades of color also vary considerably, but it never, in the scale of colors, exceeds a whitish-gray. Viewed at different localities, the mass is either compact, fetid, shelly, or silicious. Much of it produces good quicklime. It is often rendered "bastard," as the phrase is, by argillaceous and earthy impurities. Organic impressions, and remains of sea shells and coarse corals are frequent. Encrinites give some portions of it the appearance of eyed or dotted secondary marble. The occurrence of a hard variety of hornstone, which is not flint, is apparently confined to the compact, fetid variety. This formation, like the two preceding, may be found to consist of separate strata. Localities, joinings, overlayings, substrata, mineral contents, organic species, &c., require observation. The following notices are added.

. GEOLOGICAL CHANGES.—The evidences which are furnished of ancient submersion, which has "changed and overturned" vast portions of the solid land, are neither few nor equivocal. They are seen as well in the rock strata as the alluvial soils. The most elevated hills and the lowest valleys are equally productive of the evidences of extensive changes. The whole aspect of the country seems to attest to the ancient dominion of water. But the most striking proof of its agency is, perhaps, found in the sea-shells, polypi, and crustacea, which are preserved, in their outlines, in solid strata. Some of these are most vivid in their shapes and ray-like markings, particularly the univalve shells.

A subsequent change, in the surface of the country, is indicated by the marks of attrition and watery action upon the faces of these rocks, in situations greatly elevated above the present water-levels. This action must, consequently, be referred to a period when extensive submersions, in the nature of lakes or semi-seas, existed; for there is no power in present lakes and streams, however swelled and reinforced by rains or melting snows, to reach even a moiety of the elevation of these ancient water-marks. It is to the era of these last submersions that we are encouraged, by evidences, to look, as the disturbing cause which has buried trees, leaves, and bones in alluvial soils.

ACTION OF WATER.—In examining some portions of the flat lands of Ontario County, such as the township of Phelps, there are strata of a fine sedimentary soil, such as might be expected to result from the settlings of water not greatly agitated. The bottoms of mill-ponds afford an analogous species of soil. In these level districts, there are also not unfrequently observed fields of bare flat rock, of the limestone species, which is checkered in its surface, conveying the idea of their having formed a flooring to some former lake. An appearance of this kind may be seen a few hundred yards from the meeting-house in Phelps. The rock, in this instance, is a carbonate of lime, and affords organic remains.

The Oak Openings, in Erie County, are a kind of natural meadows or prairies. Many suppose them to have been ancient clearings; but of this the Indians have no tradition, and the evidences of such a settlement are by no means satisfactory. In many places, on these extensive openings, there are naked and barren layers of calcareous rock, whose surface exhibits appearances analogous to those in Ontario. The limestone is, however, of a darker color, and contains numerous imbedded nodules of hornstone, and it emits a fetid odor on breaking.

In crossing the elevated calcareous highlands, between Danville and Arkport, in Steuben County, we perceive in the bluff rocks which bound the valley of the Conestoga River, at an elevation of perhaps two hundred feet above its bed, horizontal water-marks, deeply impressed upon the face of the rocks, as if the waters had formerly stood at that level; and it is impossible to resist the conviction, in travelling over this rugged district of country, that it has not been totally submerged by waters, which have been suddenly drawn off, but by gradual or periodical exhaustions, standing for many ages at different levels.

SLATE ROCKS.—These were, not inaptly, denominated "brittle slate," by Dr. Mitchell, in 1809. Brittleness is their pervading character; and it is owing to this quality, in a formation of great thickness, that the action of the water at Niagara Falls is of so very striking a character. There is no portion of the Niagara slate solid enough to be used for building stone. It is uniformly shelly, and exhibits, even in hand specimens, its reproduced character.* Those portions of the general formation which are solid constitute silicious slate. A locality of this variety may be seen at the Halfway House, eight miles east of Canandaigua.

SENECA LAKE.—This clear and picturesque lake has its bed in the secondary formations, and may be referred to as exhibiting localities of them. Its upper parts afford the compact limestone in quadrangular blocks. Large portions of its margin consist of the brittle carbonaceous slate. The shores, from the vicinity of Rose's Farm to Appletown, are little else but a continuous bank of the slate. On the opposite coast, it is also visible at various localities below the Crooked Lake inlet. Cashong Creek may be particularly referred to. A short ascent of its valley brings the spectator into a scene where the walled masses of slaty rock assume a character of grandeur. Among the recent portions which have been thrown into the valley, may be seen masses having large species of the stem-like organic remains, which indicate its newness as a formation. Here are also disclosed orbicular masses, and pebbles of other rocks, imbedded in the slate. These prove it to be—what its texture would, in other places, indicate—a secondary slate.

The order of position on the banks of this lake is the same as at Niagara; but the sandstone is not apparent above the water line. Its existence, in the bed of the lake, may be satisfactorily

---

* Appropriately pronounced a "secondary graywacke slate," by Mr. Eaton.

inferred, from the masses of yellow coarse sand which are driven up at the foot of the lake, and particularly around its outlet. When the winds prevail, the water is driven violently against this part of the shore. As it is an alluvial flat, they soon surmount the stated margin, and produce a partial inundation. On their recession, wreathes of sand remain.

DILUVIAL ELEVATIONS.—Bounding the alluvial plain of the Seneca outlet westward, there is a series of remarkable wave-like ridges, whose direction is parallel to that of the lake. On the declivity-stop of the first of these ridges, stands the village of Geneva, the buildings of which are thus displayed in an amphitheatric manner above the clear expanse of the lake. The substratum of these ridges is an argillaceous, compact soil of the eldest formation. Some parts of it are a stiff clay, and yield septaria; but there is no considerable portion of it, which has been examined, wholly destitute of primitive boulders and pebbles. Little doubt can remain but that it is the result of the broken-down slaty rock mixed with the extraneous and far-fetched primitive masses. They are conclusive of its diluvial character. I have attentively examined this formation, in the section of it exposed on the shores of the lake between the village of Geneva and Two-mile Point. All its solid, stony contents are piled along the margin of the lake, the soil being completely washed away. Granite, quartz, and trap pebble-stones and boulders, are here promiscuously strewn with recent debris. Over the argillaceous deposit is spread a mantle of newer soil, of unequal depth and character, which forms, exclusively, the theatre of farming and horticultural labors.

WHITE SPRINGS.—On the declivity of one of these parallel ridges, at the distance of two miles from the lake, is found an extensive bed of white marl. This deposit, which is on the estate of the late Judge Nicholas, covers many acres, and yields so copious a spring of pure water that it is sufficient, at the distance of about three hundred yards from its issue, to turn a gristmill. There are to be found in this bed of marl several species of helix and voluta. The marl is generally covered with an alluvial deposit of two feet in depth. The depth of the marl itself is unexplored. Is not this marl the result of decomposed sea shells?

25

BEDS OF QUARTZOSE SAND.—In certain parts of the Seneca Valley are found limited deposits of a white quartzose sand, in a state of comparative purity. This substance is capable of being readily vitrified by the addition of alkaline fluxes, and is thus converted into glass. Its existence, as a local deposit, beneath separate strata of alluvial soil, supporting a growth of trees and shrubs, is such as to render it probable that the present stream, in its exhausted state, could have had no agency in producing these deposits. If we are compelled to look to a former condition of the waters passing off through this valley, as affording the requisite power of deposit, we are then carried back to an era in the geology of the country which we must refer to, to account for by far the greater number of changes in all its recent soils. Indeed, wherever we examine these soils, out of the range comprehended between high and low-water mark, on any existing lake or stream, there will be found occasion to resort to the agency of more general and anterior submersions. A few localities may be appealed to.

FOSSIL WOOD.—In digging a well in the Genesee Valley, one mile east of the river (at Hosmer's), part of the trunk of a tree, of mature growth, was found at the depth of forty-one feet below the surface. The soil was a loose sand mixed with gravel. The position is more elevated than the flats, so called.

ANTLERS.—A large pair of elk's horns were discovered in an excavation made for the foundation of a mill at Clyde, in Seneca County. They were imbedded in alluvial soil, ten feet below the surface. This surface had been cleared of elm and other forest trees of mature growth. Near the same place, logs of wood were found at the depth of fourteen feet. These discoveries were made in the valley of Clyde River, which is formed by the junction of the Canandaigua Outlet with Mud Creek.

FROGS INCLOSED IN THE GEOLOGICAL COLUMN.—At Carthage, on the Genesee, twelve or fifteen frogs were found in excavating a layer of compact clay marl, about nine feet below the surface. The position is several hundred feet above the bed of the Genesee River, to which elevation no one, after viewing the spot, will deem it probable its waters could have reached, this side of the diluvian era.

A frog was dug out of the solid rock, at Lockport, Niagara

County, by the workmen engaged in excavating the canal. It was enveloped by the limestone which abounds in cavities filled with crystals of strontian and dog-tooth spar. It came to life for a few moments, and then expired. There was no aperture by which it could possibly communicate with the atmospheric air. The cavity was only large enough to retain it, without allowing room for motion.

The inclosure of animals of the inferior classes in the sedimentary strata, and even in the most solid substance of rock, is a fact which has been frequently noticed, without, however, any very satisfactory theory having been given of the process, at least to common apprehension. *Vide* Addenda, for some further notices of this kind.

FOSSIL VEGETATION.—A well was dug in the lower part of the village of Geneva, in 1820, which disclosed, at the depth of thirteen feet, the branches and buds of a cedar-tree. They were found lying across the excavation, and in the sides of it; and were in excellent preservation. No one could conjecture in what age they had been buried. But this discovery would seem to establish the position that the catastrophe occurred *in the spring.*

MADREPORES.—A madrepore, measuring eight inches in diameter, was found in the upland soil of Caledonia, Genesee County. Smaller specimens of the same species occur in that township. Madrepores of a large size have also been found imbedded in the soil, or lying on the surface, in various places in Cattaraugus and Alleghany counties. They are locally denominated petrified wasps' nests. The lands containing these loose fossil remains are contiguous to, or based on, secondary rocks at considerable elevations.

BOULDERS AND PRIMITIVE GRAVEL.—But the most abundant evidences of diluvial action are furnished by the masses of foreign crystalline rocks which are scattered, in blocks of various sizes, on the surface of the soil, or imbedded at all depths within it. Primitive rocks are foreign to the district, and these masses could not, therefore, have resulted from local disintegration. They must have been transported from a distance. They required not only an adequate cause for their removal, but one commensurate with the effects. Such a cause Cuvier supposes, in discussing the general question, may have existed in eruptions, or in the action of oceanic masses of water, operating at an ancient period.

The latter opinion appears to be generally adopted. Dr. Mitchell, in reference to northwestern boulders, attributes their distribution over secondary regions to the draining of interior seas or lakes. Mr. Hayden, in his *Geological Essays*, refers them to the action of oceanic currents setting "from north and east to south and west."

SUBORDINATE AND EQUIVALENT STRATA.—These constitute the most intricate subjects of reference. They are either adjuncts or residuary deposits of leading formations. But their order, as accompanying series, must sometimes be sought for by a previous determination of the formations themselves. Could we certainly know, for instance, that the sandstone of Western New York is or is not the true coal-sandstone, or the limestone is or is not the carboniferous limestone, it would at once direct to positive eras, and serve to impart confidence in the prediction of unknown deposits of an important character. But, in order to fix the formations, it is often the safest mode of procedure to employ the subordinate and local deposits as evidences of the character of the formations embracing them.

GYPSUM.—A stratum of gypsum of the plaster of Paris kind—that is, consisting of an admixture of the carbonate with the sulphate of lime—occurs on the banks of the Canandaigua outlet. It has been chiefly explored in the township of Phelps, Ontario. In visiting the principal bed (1820), I found the following order of deposits composing the banks of the outlet:—

1. Alluvial soil of a dark, arenaceous, and mellow character, having small stones of the primitive kind sparingly interspersed, two and a half to three feet. Cultivated in improved farms.

2. Shelly limestone, of an earthy, dull-gray color and loose texture, in layers, three feet.

3. Limestone of a more firm character, but still shelly, or rather slaty, fissile, and easily quarried, six feet. This stratum contains iron pyrites in a decomposed state. Also, nodular or kidney-shaped masses of what the quarrymen call *plaster-eggs*—apparently snowy gypsum.

4. Plaster of Paris, ten feet. This stratum yields granular, earthy, fibrous, and foliated gypsum. It is the first two varieties

which are quarried. In some places, the mass is firm enough to admit of blasting. In others, it is loose and veiny, and is readily broken up with iron bars and sledges. Portions of it appear to consist of a shelly limestone identical with No. 2. They are rejected in quarrying.

5. Limestone similar to No. 3, four feet.

At this depth it is covered by the waters of the outlet. How deep it extends is uncertain. The rapids at the village of Vienna are caused by shelving strata of this limestone.

There is a suite character in these strata which appears to constitute them a single deposit. The plaster-bed at Canasaraga exists in a ledge more elevated in reference to the local stream, and presents a broader section of the limestone. The shades of difference which are observable in its color and texture, do not appear to indicate a difference of geological era. Nor do appearances denote, for the calcareous formation which embraces these beds, much antiquity in the scale of secondary rocks.

SALIFEROUS RED CLAY-MARL.—Examinations, at various points, render it a probable supposition that the red clay-marl of western New York is the equivalent for the new red sandstone, in positions where the latter is—as it often is—wanting. It is extensively deposited in the upland soils, in the range of the salt rock and gypsum counties, from the summit grounds of Oneida County west. It may be seen in various stages of the decomposition. I have more attentively examined it on the upper parts of the Scanado* and Oneida creeks. Large areas of it exist in Westmoreland, Verona, and Vernon townships, and bordering the valley grounds of the Oneida reservation, and the northerly portions of Sullivan County. The existence of salt water might, apparently, be searched for with as much probability of success, in the district thus indicated, as at more westerly points.

COAL-FORMATION.—With a strong predisposition to regard our leading sandstone and limestone surface-formations as members of the "independent" or true coal-formation, inquiry has led me to relinquish the impression that they will, to any great degree, be found to yield this mineral. If the sandstone is—as facts indicate it to be—the new red or saliferous sandstone, it may be ex-

* Usually written Skenanodoah, but pronounced as above.

pected to yield thin seams of coal, in distant places, but no deposit
of this mineral which will reward exploration in this or its superincumbent series of rocks, the slates, limestones, &c.   It will result, that the coal-measures, properly so denominated, are a prior
deposit in the order of series; and, should they hereafter be found,
such a discovery must take place above the range of the sandstone, which is the basis rock at Niagara and Genesee Falls.

Having premised the character of the sandstone, all the series
occupying a position above it must derive their character, as
secondary deposits, from this.   The limestone cannot, therefore,
be a part of the carboniferous or "medial."   The slates, as shown
at Cashong, are fragmentary, and rather nearer slaty grau-
waks.   The arenaceous and calcareous upper deposits assume
nearly the position of the oolitic series, and, in fact, ought, in
some localities, to be regarded as equivalents.

WESTERN COAL-MINES.—Much of the data employed in these
inquiries is the result of previous examinations of the great coal
deposits in the Ohio Valley, and other parts of the western country.
Here we have the coal-sandstone and the slate clay, with slate,
&c., alternating with the coal-measures.   Such is the order of
deposits at the junction of the Alleghany and Monongahela, where
the formation is well developed, and where there exists, too, in
the elevated valley hills, several repetitions of the series.   The
zechstone, or compact limestone, which is a pervading rock in
the Mississippi Valley, occupies a position next above the great
Mississippi sandstone.*   It may always be distinguished from the
shelly, entrochal limestone of the Genesee,† by the absence of
gypsum and of the fetid odor emitted on fracture.

ALLEGHANY VALLEY.—A question of interest, in connection
with the extent of the Ohio Valley coal-formation, arises from
the attempt to fix the point to which this formation ascends the
Alleghany Valley—being the direct avenue into Western New
York.   I have examined this valley in its entire length between
Pittsburg and Olean, in Cattaraugus County, and have not been
able to observe that there are any evidences of its termination

---

* This formation cannot be called "red sandstone," from its being generally
white or gray, but appears to occupy the position of the "horizontal red sandstone"
among European rocks.

† The cornutiferous limerock of Mr. Eaton.

below the latter point. The general order and parallelism of strata remain the same. The coal stratum is apparently present. The qualities of the coal at Armstrong, and at various points below French Creek—the first primary fork of the river—are not distinguishable from the products of the Pittsburg galleries. Less search has been made above that point, but wherever the hills have been penetrated, they have—as at Brokenstraw—produced the bituminous coal. Above the Conawango Valley, which brings in the redundant waters of Chatauque Lake, the Alleghany discloses frequent rapids. The effect of parallelism upon the strata is to sink the coal-measures deeper as they ascend the Alleghany; and this cause may, in connection with the unexplored character of the country, be referred to in accounting for the absence of coal along this part of the line. The reappearance of traces of this mineral at Potato Creek, forty miles above Olean, is a proof, however, that the coal-formation extends to that point. This locality is a few miles within the limits of Pennsylvania. It occurs in a valley.

COAL IN WESTERN NEW YORK.—The coal-bed above Olean is south of the summit of the Genesee, and not remote from its primary source. The expectation may be indulged that the western coal-formation embraces portions of Cattaraugus and Alleghany or Steuben counties. The noted spring of naphtha, called Seneca Oil, is on Oil Creek in this county. As this substance, in the class of bitumens, is nearly allied to the coal series, it may be deemed favorable to the existence of the formation in the substrata.[*] Fragments of carbonized wood are frequently found in the large tracts of marine sand,[†] as well as in some of the mixed alluvions of these counties; and it needs but an examination, as cursory as it has fallen to my lot to make, of this portion of the country, to render it one of high geological interest, and to denote that the coal-measures probably extend into some portions of Western New York.[‡]

---

[*] These tracts bear a valuable growth of pines, which constitute the source of a profitable lumber trade with the Ohio Valley.

[†] This mineral oil also occurs in several of the lower tributaries of the Alleghany River, within the coal district.

[‡] A discovery of coal has been announced in Alleghany County, New York, as these sheets are going through the press, more than thirty years after these lines were penned.

ADDENDA.

Animals inclosed in Rock, &c.

TOADS.—In 1770, a toad was brought to Mr. Grignon inclosed in two hollow shells of stone; but, on examining it nicely, Mr. G. discovered that the cavity bore the impression of a shell-fish, and, of consequence, he concluded it to be apocryphal.

In 1771, another instance occurred, and was the subject of a curious memoir read by Mr. Guettard to the Royal Academy of Sciences at Paris. It was thus related by that famous naturalist:—

In pulling down a wall, which was known to have existed upwards of a hundred years, a toad was found without the smallest aperture being discoverable by which it could have entered. Upon inspecting the animal, it was apparent that it had been dead but a very little time; and in this state it was presented to the Academy, which induced Mr. Guettard to make repeated inquiries into the subject, the particulars of which will be read with pleasure in the excellent memoir we have just cited.

WORMS.—Two living worms were found, in Spain, in the middle of a block of marble which a sculptor was carving into a lion, of the natural color, for the royal family. These worms occupied two small cavities to which there was no inlet that could possibly admit the air. They subsisted, probably, on the substance of the marble, as they were the same color. This fact is verified by Captain Ulloa, a famous Spaniard, who accompanied the French academicians in their voyage to Peru to ascertain the figure of the earth. He asserts that he saw these two worms.

ADDER.—We read in the *Affiches de Provence*, 17 June, 1772, that an adder was found alive in the centre of a block of marble thirty feet in diameter. It was folded nine times round, in a spiral line. It was incapable of supporting the air, and died a few minutes after. Upon examining the stone, not the smallest trace was to be found by which it could have glided in or received air.

CRAWFISH.—Misson, in his *Travels through Italy*, mentions a crawfish that was found alive in the middle of a marble in the environs of Tivoli.

FROGS.—M. Peyssonel, king's physician at Guadaloupe, having

ordered a pit to be dug in the back part of his house, live frogs were found by the workmen in beds of petrifaction. M. P., suspecting some deceit, descended into the pit, dug the bed of the rock and petrifactions, and drew out himself green frogs, which were alive, and perfectly similar to what we see every day.

We are informed by the *European Magazine*, February 21, 1771, that M. Herissan inclosed three live toads in so many cases of plaster, and shut them up in a deal box, which he also covered with thick plaster. On the 6th of April, 1774, having taken away the plaster, he opened the box, and found the cases whole and two of the toads alive. The one that died was larger than the others, and had been more compressed in its case. A careful examination of this experiment convinced those who had witnessed it, that the animals were so inclosed that they could have no possible communication with the external air, and that they must have existed during this lapse of time without the smallest nourishment.

The Academy prevailed upon M. Herissan to repeat the experiment. He inclosed again the two surviving toads, and placed the box in the hands of the Secretary, that the Society might open it whenever they should think proper. But this celebrated naturalist was too strongly interested in the subject to rest satisfied with a single experiment; he made, therefore, the two following:—

1. He placed, 15 April, 1771, two live toads in a basin of plaster, which he covered with a glass case that he might observe them frequently. On the 9th of the following month, he presented the apparatus to the Academy. One of the toads was still living; the other had died the preceding night.

2. The same day, April 15, he inclosed another toad in a glass bottle, which he buried in sand, that it might have no communication with the external air. This animal, which he presented to the Academy at the same time, was perfectly well, and even croaked whenever the bottle was shook in which he was confined. It is to be lamented that the death of M. Herissan put a stop to these experiments.

We beg leave to observe upon this subject, that the power which these animals appear to possess of supporting abstinence for so long a time, may depend upon a very slow digestion, and, perhaps, from the singular nourishment which they derive from themselves. M. Grignon observes that this animal sheds its skin several times in the course of a year, and that it always swallows it. He has known, he says, a large toad shed its skin six times in one winter. In short, those which, from the facts we have related, may be supposed to have existed many centuries without nourishment, have been in a total inaction, in a suspension of life, or a temperature that has admitted of no dissolution; so that it was not necessary to repair any loss, the humidity of the surrounding matter preserving that of the animal, who wanted only the component parts not to be dried up, to preserve it from destruction.

The results of modern chemistry and philosophy have proved the number of elementary substances to be far greater than was admitted in the preceding century. And this discovery is progressive, and will probably go on a long time; after which, it is not improbable a new race of chemical and philosophical observers will spring up, who will be able to decompose many substances we now consider elementary, and thus again reduce the number of elements of which all external matter is composed. It would not be wonderful if posterity should reduce the number of elements even as low as the ancients had them. Such a result would throw new light on the mysterious and intricate connection which seems to exist between animal, vegetable, and mineral matter. We should then, perhaps, have less cause to wonder that toads, &c., are capable of supporting life in stone, that birds should exist in solid blocks of wood, &c.

But toads are not the only animals which are capable of living for a considerable length of time without nourishment and communication with the external air. The instances of the oysters and dactyles, mentioned at the beginning of this article, may be advanced as a proof of it. But there are other examples.—*European Magazine*, March, 1791.

A beetle, of the species called capricorn, was found in a piece of wood in the hold of a ship at Plymouth. The wood had no external mark of any aperture.—*European Magazine*.

A bug eat itself out of a cherry table at Williamstown, Mass. See an account of this phenomenon, by Professor Dewey, in the *Lit. and Philos. Repertory*.

These phenomena remind us of others of a similar nature and equally certain.

In a trunk of an elm, about the size of a man's body, three or four feet above the root, and precisely in the centre, was found, in 1719, a live toad, of a moderate size, thin, and which occupied but a very small space. As soon as the wood was cut, it came out and slipped away very alertly. No tree could be more sound. No place could be discovered through which it was possible for the animal to have penetrated, which led the recorder of the fact to suppose that the spawn from which it originated must, from some unaccountable accident, have been in the tree from the very moment of its first vegetation. The toad had lived in the tree without air, and, what is still more surprising, had subsisted on the substance of the wood, and had grown in proportion as the tree had grown. This fact was attested by M. Hebert, Ancient Professor of Philosophy at Caen.

In 1731, M. Leigne wrote to the Academy of Sciences at Paris an account of a phenomenon exactly similar to the preceding one, except that the tree was larger, and was an oak instead of an elm, which makes the instance the more surprising. From the size of the oak, M. Leigne judged that the toad must have existed in it without air or any external nourishment, for the space of eighty or a hundred years.

We shall cite a third instance, related in a letter the 5th Feb. 1780, written from the neighborhood of Saint Mexent, of which the following is a copy.

"A few days ago, I ordered an oak tree of a tolerable size to be cut down, and converted into a beam that was wanting for a building I was then constructing. Having separated the head from the trunk, three men were employed in squaring it to the proper size. About four inches were to be cut away on each side. I was present during the transaction. Conceive what was their astonishment when I saw them throw aside their tools, start back from the tree, and fix their eyes on the same point with a kind of amazement and terror. I instantly approached, and looked at that part of the tree which had fixed their attention. My surprise equalled theirs, on seeing a toad, about the size of a large pullet's egg, incrusted, in a manner, in the tree, at the distance of four inches from the diameter and fifteen from the root.

It was cut and mangled by the axe, but still moved. I drew it with difficulty from its abode, or rather prison, which it filled so completely that it seemed to have been compressed. I placed it on the grass; it appeared old, thin, languishing, decrepit. We afterwards examined the tree with the nicest care, to discover how it had glided in; but the tree was perfectly whole and sound."
—*European Magazine.*

BAT.—A woodman engaged in splitting timber for rail-posts in the woods close by the lake in Haming (a seat of Mr. Pringle's in Selkirkshire), lately discovered, in the centre of a large wild cherry tree, a living bat, of a bright scarlet color, which he foolishly suffered to escape, from fear, being fully persuaded it was (with the characteristic superstition of the inhabitants of that part of the country) a "being not of this world." The tree presented a small cavity in the centre, where the bat was inclosed, but is perfectly sound and solid on each side.—*N. Y. Lit. Journ. and Belles-Lettres Repository*, taken from the *London Semi-Monthly Magazine.*

SKULL IN WOOD.—A tenant of the Rev. J. Cattle, of Warwick, lately presented to him a part of the solid butt of an oak tree, containing within it the skull of some animal (unknown). It was in the part of the tree nine feet above the ground, and was perfectly inclosed in solid timber.—*N. Y. Lit. Journ. and Belles-Lettres Repository*, from *European Magazine.*

X.

A Memoir on the Geological Position of a Fossil-Tree in the Series of the Secondary Rocks of the Illinois.

The spirit of inquiry which has been excited in this country in regard to objects of natural history, while it has enlarged the boundaries of our knowledge of existing species, has directed some of its more valuable researches to those organized forms which have perished and become embalmed in the shape of petrifactions, in the body of solid rocks. A petrified tree of this kind has recently been discovered in the secondary* rocks

* This term is superseded, in geological discussions of the present day, by the term *silurian*, which embraces all strata of the era between the *palæozoic* and *tertiary* formations.

at the source of the Illinois River. Having recently visited this evidence of former changes in the flora of the West, I embrace the occasion, while my recollections are fresh, to give an account of it.

The tract of country separating the southern shores of Lake Michigan from the Illinois River, is a plat of table-land composed of compact limestone, based on floetz or horizontal sandstone. This formation embraces the contiguous parts of Illinois, and spreads through Indiana, Ohio, and the Peninsula of Michigan. It is overspread with a deposit of the drift era, covered with a stratum of alluvial soil, presenting a pleasing surface of prairies, forests, and streams. These features may be considered as peculiarly characteristic of the junction of the Rivers Kankakee and Des Plaines, which constitute the Illinois River. This junction is effected about forty miles south of Chicago.

The fossil in question occurs about forty rods above the junction of the Kankakee. The sandstone embracing it is deposited in perfectly horizontal layers, of a gray color and close grain. It lies in the bed of the Des Plaines. The action of this stream has laid bare the trunk of the tree to the extent of fifty-one feet six inches. The part at the point where it is overlaid in the western bank is two feet six inches in diameter. Its mineralization is complete. The trunk is simple, straight, scabrous, without branches, and has the usual taper observed in the living specimen. It lies nearly at right angles to the course of the river, pointing towards the southeast, and extends about half the width of the stream. Notwithstanding the continual abrasion to which it is exposed by the volume of passing water, it has suffered little apparent diminution, and is still firmly imbedded in the rock, with the exception of two or three places where portions of it have been disengaged and carried away; but no portion of what remains is elevated more than a few inches above the surface of the rock. It is owing, however, to those partial disturbances that we are enabled to perceive the columnar form of the trunk, its cortical layers, the bark by which it is enveloped, and the peculiar cross fracture, which unite to render the evidence of its ligneous origin so striking and complete. From these characters and appearances, little doubt can remain that it is referable to the species juglans nigra, a tree very common to the forest of the

Illinois, as well as to most other parts of the immense region drained by the waters of the Mississippi. The woody structure is most obvious in the outer rind of the trunk, extending to a depth of two or three inches, and these appearances become less evident as we approximate the heart. Indeed, the traces of organic structure in its interior, particularly when viewed in the hand specimen, are almost totally obliterated and exchanged, the vegetable matter being replaced by a mixed substance, analogous, in its external character, to some of the silicated and impure calcareous carbonates of the region. Like those carbonates, it is of a brownish-gray color and compact texture, effervesces slightly in the nitric and muriatic acids, yields a white streak under the knife, and presents solitary points, or facets, of crystals resembling calc spar. All parts of the tree are penetrated by pyrites of iron of a brass yellow color, disseminated through the most solid and stony parts of the interior, filling interstices in the outer rind, or investing its capillary pores. There are also the appearances of rents or seams between the fibres of the wood, caused by its own shrinkage, which are now filled with a carbonate of lime, of a white color and crystallized.

From an effect analogous to carbonization, the exterior rind and bark of the tree have acquired a blackish-hue, while the inclosing rock is of a light-gray color, characters which are calculated to arrest attention.

There is reason to conclude that the subject under consideration is the joint result, partly of the infiltration of mineral matter into its pores and crevices, prior to inclosure in the rock, and partly to the chemical action educed by the great catastrophe by which it was translated from its parent forest, and suddenly enveloped in a bed of solidifying sand.

At the time of my visit (August 13, 1821), the depth of water upon the floetz rocks forming the bed of the River Des Plaines, would vary from one to two feet; but it was at a season when these higher tributaries, and the Illinois itself, are generally at their lowest stage. Like most of the confluent rivers of the Mississippi and their tributaries, the Des Plaines is subject to great fluctuations, and during its periodical floods may be estimated to carry a depth of eight or ten feet of water to the junction of the Kankakee. At those periods, the water is also rendered turbid by the quantity of alluvial matter it carries down, and a search for this organic fossil must prove unsuccessful. But during the prevalence of the summer droughts, in an atmosphere of little humidity, when the waters are drained to the lowest point of depression, and acquire the greatest degree of transparency, it forms a very conspicuous trait in the geology of the stream, and no person, seeking the spot, can fail to be directed to it.

The sand-rock containing this petrifaction is found in a horizontal position, differing only with respect to hardness and color. The remains of fossil organized bodies in this stratum are not abundant, or have not been successfully sought. It is probable that future observations will prove that its organic conservata are chiefly referable to the vegetable kingdom. It is certain, that this inference is justified by the facts which are before me, and particularly by the characteristic appearances of the strata in the bed of the River Des Plaines, where the imbedded walnut is the representative of the ancient flora. At a short distance above, where the bed of the Des Plaines approaches nearer the summit level, limestone ensues, and continues from that point northward to the shores of Lake Michigan. In the vicinity of Chicago, where this limestone is quarried for economical purposes, it is characterized by the fossil remains of molluscous species.

Lake Erie lies at an elevation of five hundred and sixty-five feet above the Atlantic.*

There exists a water communication between the head of Lake Michigan, at Chicago, and the River Des Plaines, during the periodical rises of the latter, but its summer level is about seven feet lower, at the termination of the Chicago portage, than the surface of the lake. From this point to its junction with the Kankakee, a computed distance of fifty miles, the bed of the Des Plaines may be considered as having a mean southern depression of ten inches per mile, so that the floetz rocks at its mouth, lying on a level of forty-eight feet eight inches below the surface of Lake Michigan, have an altitude which cannot vary far from five hundred and fifty feet above the Atlantic. There are no mountains for a vast distance either east or west of this stream. It is a

* Public Documents relating to the New York Canals, with an Introduction, &c., by Colonel Haines.

country of plains, in which are occasionally to be seen alluvial hills of moderate elevation; but the most striking inequalities of surface proceed from the streams which have worn their deep-seated channels through it; and an oceanic overflow capable of covering the country, and producing these strata by deposition, would also submerge all the immense tracts of secondary and alluvial country between the Alleghany and the Rocky Mountains, converting into an arm of the sea the great valley of the Mississippi, from the Gulf of Mexico north to the Canadian Lakes. We find in the alluvial soil along the Illinois and Des Plaines blocks of granite, hornblende, and gneiss, of the drift stratum, exhibiting the same appearances of attrition, and of having been transported from their parent beds, which characterize the secondary table-lands along the margin of the great American lakes, the prairies of Illinois, and the western parts of New York.

There is nothing, perhaps, in the progress of modern science, which has tended to facilitate geological research so much as the study and investigation of fossil organic remains. They teach, with unerring lights, how extensively the ancient flora and fauna of this continent have been prostrated, leaving their exact impressions, in all their minuteness, in the newly-formed stratifications. That these impressions, fresh and vivid as we find them, should mark the eras of depositions and crystallization of rocks from the suspension of their elements in water, is the observation of Werner, and it is to him we owe the elements of the Neptunian hypothesis. His general recognition of the epochs of the primitive, transition, and secondary rocks, appears too probable not to commend itself to adoption with regard to all strata which can be conceived to be the products of watery menstrua.

But it remained for Werner, who was the first to perceive an order in strata, also to point out the important application of fossil organic bodies in elucidating their eras, and the natural order of their superposition.

To adopt the words of Dr. Thomas Cooper:—

"There appears to be a series of strata, or, as Werner calls them, formations, that may be considered as surrounding the nucleus of the earth. The first formed, or lowest series, always preserve the same situation to each other, except where occasional eruptions, or circumstances not of a general nature, make a variety in their situations. These strata are not only the deepest, but they are also the highest that are observable in the crust of the earth; forming the tops of the highest mountains. They are characterized by an appearance of crystallization, and by containing no remains of organic matter, animal or vegetable. The strata or formations that in general constitute this first, deepest, highest, and crystallized series, are granite, gneiss, mica-slate, clay-slate, primitive greenstone, granular limestone, serpentine, porphyry, and sienite. These formations are so generally found, and in the same situations as incumbent upon or subtending each other relatively, that they may be considered as universal. Their crystallized appearance shows that their particles have either been dissolved or very finely suspended in water, so that the attraction of crystallization has been free to operate; that this water has been deep, so that the lowermost parts of it have not been much agitated during the crystallization, which would otherwise have been more confused than it is; and, indeed, the oldest formations are the best crystallized. A part of the water covering the nucleus must have been taken up, as water of crystallization, in the primitive formations. When these were deposited, there were no vegetables formed; of course, no animals; nay, even the sea was unpeopled, for there is no trace of any organic remains in these strata. Even the belemnites, the asteriæ, the echini, the entrochi, the most simple forms of oceanic animal life, do not occur until the transition strata appear. Hence the propriety of denominating these formations *primitive*.

"By processes of nature, besides the consumption of water by the new crystallized masses, to us unknown, the waters appear to have diminished. The highest parts of the primitive formations became the shores to the water superincumbent on their bases and middle regions; the simplest forms of oceanic animals came into existence; the mosses and lichens of high latitude would generally occupy the surface of the primitive strata, gradually decomposed by the alternate action of air and water after many ages. During this period, while the strata were in a state of *transition* from the chaotic to the habitable state, other deposits would gradually be made from the waters, now decreased in quantity, and take their place below the summits of the primitive range. Those summits being exposed to the action of the atmo-

26

sphere, of rains, of frost probably, and to the action also of the waters with their contents still incumbent on the earliest strata, would furnish masses and particles washed away, which would mingle with the deposits of the transition series. This series, therefore, will exhibit appearances of mechanical and chemical intermixture of earths and stones, such as are found in the silicious porphyries, the graywackes, the silicious and argillaceous hornblende rocks, the elder red sandstone, &c. During the period when these transition formations were deposited, there would be no land animals, for there would be no vegetables for them to feed upon. There would be no vegetables unless some few lichens, mosses, or ericas, that would find foothold upon the slight decomposition that, after the lapse of some ages, would take place on the surface of the primitive rocks. The sea only would be peopled, and that but sparingly; for, in that mass of muddy water, none but the lowest and most inferior grades of animal life, and such as do not inhabit deep water, could exist. Hence, we find the transition formations contain in their substances some belemnites, asteriæ, entrochi, echini, &c., but no organized vegetable substance except, very rarely, in the latest rocks of this series, and no remains whatever of terrestrial animals. Indeed, in the high latitudes of the outgoings or summits of the primitive strata, very few vegetables, even at the present day, can live. No vegetation fit for animal life could take place until the transition, and most of the next series of *secondary* or *floetz* formations had subsided. These would occupy lower and lower situations, till a rich soil, from every kind of intermixture of earth mechanically deposited, would afford a proper temperature of region, and an easily decomposed soil, wherein vegetables could grow.

"Next to the transition series, come the *secondary*, or, as the German mineralogists call them, the *floetz* rocks; so called, because they appear to be more floated or horizontal, though I confess the appellation does not appear to me peculiarly appropriate. These strata consist principally of sandstone, limestone—sometimes fetid from bituminous impregnations, sometimes shelly—secondary greenstone, graphite, coal, gypsum, rock salt. I have observed that the Alpine heights of the primitive mountains could at no time furnish much food. The same remark, but in a less degree, will apply to the transition range; the low and kindly climates occupied by the secondary series. The soft and decomposable nature of these depositions would furnish the true theatre of vegetable life, and, until these regions were filled with vegetables, the race of animals could not have been produced; for on what could they subsist? Graminivorous animals, therefore, must have succeeded the various forms of vegetable existence; and carnivorous, the graminivorous. The vegetable matter imbedded in the substance of the secondary strata will consist of the remains of vegetables that grow in the transition strata; and the animal remains will consist chiefly of such animals as were produced in the early stages of animal existence, particularly the smaller aquatic animals; and, of these, chiefly shell-fish, as shells are not so soon decomposed as mere animal substance."

It is to the latter class of depositions—to the secondary series—that we must refer the sandstone of the River Des Plaines, in which we find a walnut, of mature growth, enveloped by, and imbedded in the rock, in the most complete state of mineralization; and, since all geological writers who subscribe to the Neptunian theory are constrained to employ the agency of oceanic depositions of different eras, in explaining the structure of the earth's surface, it is one of the most obvious and important conclusions, to be drawn from the fact that such submersions and depositions of rock matter have taken place subsequent to the existence of forests of mature growth, and that the rock strata and beds composing the exterior of the earth are the result of different geological epochs, and of successive subsidences of chaotic matter—positions which have been so severely attacked and so often denied, particularly by the disciples of the Huttonian school, that it is not without a feeling of lively interest, I communicate a discovery which appears so conclusive on the subject.

Considerations arising from the frontier position of the country, and the infrequency of the communication, have also induced me to draw from incidental sources, a corroboration of the facts advanced.

In a letter to Governor Cass, of Michigan, dated September 17, 1821, I made the following observations on the subject under review:—

"I consider the petrified tree discovered during our recent journey up the Illinois, so extraordinary an object in the natural history of the country, and calculated to lead to conclusions so important to the science of geology, that I am anxious to avail myself of your concurrent testimony as to the fact of the existence of the tree in a mineralized state, and the natural appearances of the spot where it lies imbedded. I feel the more solicitude on this subject, as I am aware that any description of this phenomenon which I may be induced to communicate to the public, will be received with a degree of caution and scrutiny which it is the province of the naturalist to exercise whenever any discovery is announced affecting the existing theories of the natural sciences, or tending to increase the volume of facts upon which their advancement and perfection depend. I am aware, also, that whatever degree of caution and vigilance it may be proper to exercise to prevent errors from mingling with the sound doctrines of the physical and other sciences, still more care and circumspection is requisite in examining facts which affect the progress of geology."

I quote an extract from Governor Cass's reply on the subject:—

"The appearance of the wood and bark indicates that it was a black walnut, the juglans nigra of our forests. We computed its original diameter, at the place where it is concealed in the earth, to have been three feet, and at the other end eighteen inches. The texture of the wood, and the bark and knots, are nearly as distinct as in the living subject, and the process of decay had not commenced previous to the commencement of this wonderful conversion. Every part of the mass which we could examine is solid stone, and readily yields fire by the collision with steel.

"When we visited the spot, the water of the river was at the lowest stage; but there was no part of the tree within some inches of the surface. The rocky bed of the stream was formed round and upon it. We raised from it pieces of the rock, which were evidently *in situ,* and which had been formed upon the tree posterior to the period of its deposit in its present situation. This rock is a species of sandstone, whose characteristic features must be well known to you.

"There are no mineralized substances of vegetable origin in the vicinity of this specimen, nor are there any appearances which indicate that its present condition has been caused by any peculiar property in the waters of the Des Plaines."

ADDENDA.

The publication of the foregoing memoir led to several letters being addressed to the author on topics connected with it. Some of these were from gentlemen eminent in science or politics, whose opinions are entitled to the highest respect. Extracts are given from such only as introduce new data, either of fact or opinion.

GEOLOGICAL THEORIES.—Professor Dewey, of Williams College, observes: "A friend has just lent me your 'Memoir on a Fossil-Tree.' Though the account is very interesting, I do not perceive its exact bearing on the Neptunian and Plutonian hypotheses. The fault is doubtless in me, and you will excuse my remarks and set me right. I had supposed the Huttonians and Wernerians did not dispute about the manner in which the *secondary* rocks were formed. Macculloch, and others before him, led me into this opinion, though it may be erroneous. But Bakewell, who is referred to as authority in *Rees's Cyclopœdia*, says, p. 131: 'Geologists are agreed that secondary rocks have been formed by the agency of water.' If this be so, they would agree generally with the account of Dr. Cooper respecting the formation of petrifactions, and especially those of vegetables, and the fossil-tree would be treated of in a similar manner by both."

Hutton's original hypothesis, and not the modifications of it introduced by the Neptu-Vulcanists, were adverted to in reply. Subsequently, Professor Dewey writes:—

"I was greatly obliged by your letter in various respects, and I write you now to make my acknowledgments for it, as well as to maintain the correctness of your notions on the Huttonian hypothesis. As you had seen a Scotch mineralogist directly from the mint of Playfair, I had every reason to suppose you had received correct views of Playfair's notions on the subject. I have been led, therefore, to examine the matter, and, as I may have set you on the search, I wish to prevent your continuing it on my account, or from what I wrote.

"Playfair's Illustrations I have never seen. Occasional extracts, or allusions to its points, have fallen in my way. But I have before me a very full abstract of Hutton's paper on the subject, from the *Transactions of the Royal Society of Edinburgh*. It is from the very paper in which he announces his hypothesis. In that paper he mentions that the consolidation of all the hard crust of the globe has been effected by *heat* and *fusion*, extending it to secondary as well as primitive rocks, and mentioning particularly Spanish marble, shell limestone, oolite, and chalk.

"This operation of heat, he says, is exemplified by *chalk, which is to be found in all gradations, from marble to loose chalk.* This is his precise notion, but not his words. I had once looked at this paper before, and thought much of this theory; but this thought had been obliterated from my mind by thoughts advanced by others, as I thought in consistency with the sentence I quoted from Bakewell. At least, one objection to Hutton's views would be removed by modifying his theory in the manner it seems to be by Bakewell. Though Hutton does not think this to be necessary; for he appears to feel no difficulty in accounting for petrifactions of wood on his hypothesis, for he mentions that *we have many proofs of the penetration of flinty matter, in a state of fusion, in other bodies, such as insulated pieces of flint in chalk or sand, and fossil wood penetrated with silicious matter.*

"Still, the grand reasons of Hutton for employing heat as the agent of consolidation are opposed to the above modification of his theory. These reasons, as you know, are the insolubility of most mineral substances in water, and the disappearance of the water from the cavities of minerals which have been consolidated. The first is, indeed, the great one for Hutton; for the crystallization of salts in water, and the existence of liquids, in some cases, in the cavities of the most solid minerals, show well enough that the water might or might not disappear, as the circumstances were different.

"If the Huttonians maintain, as he did, the formation of petrifactions by heat, which consistency requires, I concede, indeed, to you that that fossil-tree stands as a grand monument of some different process; and yet, we can hardly suppose that they do not see great difficulty in the common notion on the subject. The rapidity with which the petrifactions must have taken place—a point well illustrated in Hayden's *Geological Essays*—seems to require some new notions on the subject. What these may be, I cannot tell; but I believe that neither of these two hypotheses will be adopted exclusively, half a century hence, on this point, or on geology generally. I think, with you, that our countrymen need illumination on the subject of Hutton's hypothesis, and I wish some one would attempt it."

TRAP-ROCKS OF EUROPE AND AMERICA.—"I suspect the greenstone of our country, when examined as it ought to be, will be found, in its geological relations, much to resemble the basalt of Europe; and that the same difficulties will attend it, on Werner's hypothesis, as now attend the basalt. Indeed, I know not how we can account for what Bakewell and Macculloch state on this hypothesis."

SANDSTONE OF VIRGINIA.—"I have seen a piece of a petrified tree, about eight inches through, found in the sandstone of Virginia, but could get none of it. The petrifaction was far finer than the stone in which it lay, and was, like it, silex."

SANDSTONE OF OHIO.—C. Atwater, Esq., in a letter to the author, observes:—

"I can assure you that the finding of whole trees in sandstone is nothing strange in this State. Some of these trees are imbedded in sandstone one hundred feet below the surface. Zanesville and Gallipolis are the best spots to find these fossils.

"There is no part of the tree but what I have in my cabinet, not excepting their leaves, fruit, and even fungi attached to them."

MOSAICAL HISTORY OF THE CREATION.—B. Irvine, Esq., in adverting to remarks on the Illinois fossil, observes:—

"They may yet awaken some ideas in the minds of the people on the wonders of physics—and I had almost said, the *slow miracles of creation;* for, if ever there was a time when matter existed not, it is pretty evident that *millions of years*, instead of six days, were necessary to establish order in chaos, let Cuvier, &c. temporize as they may. However, it is the humble allotment of the herd to believe or stare; it is the glory of intelligent men to inquire and admire."

The doctrine of materialism, adverted to by Mr. Irvine, it is the province of divines to controvert. One remark may be predicted on the biblical era of the six days. It is now believed to be generally conceded by eminent geologists and ecclesiastics, that the term "day," employed by the translators of the English version of the Scriptures, is used in Gen. ch. i. in a sense synonymous with "era" or "time," as it is emphatically used in Gen.

ch. ii. ver. 4. For an able exposition of the present views on this subject, see the *American Journal of Science*, vol. xxv. No. 1.

---

4. BOTANY.

XI.

A descriptive list of the plants collected on the expedition, drawn up by Dr. John Torrey, has been published in the fourth volume of the *American Journal of Science*. References to this standard work may be conveniently made by botanists.

---

5. ZOOLOGY.

No professed zoologist was attached to the expedition, the topic being left to such casual attention as members of it might find it convenient to bestow. Of the fauna of the region, it was not believed that there were any of the prominent species which were improperly classed in the *Systema Naturæ* of Linnæus. It was doubtless desirable to know something more particularly of the character and habitat of the American species of the reindeer (*C. sylvestris*) and hyena, or glutton. Perhaps something new was to be gleaned respecting the extent of the genera arctomys and sciurus, among the smaller quadrupeds, and in the departments of birds and reptilia. The mode of travel gave but little opportunity of meeting the larger species in their native haunts, but it afforded opportunities of examining the skins of the quadrupeds at the several trading stations, and of listening to the narrations of persons who had engaged in their capture.

In effect, the crustacea of the streams furnished the most constant and affluent subject for enlarging the boundaries of species and varieties. The collections in this department were referred to members of the Lyceum of Natural History at New York, and of the Academy of Natural Sciences at Philadelphia. The results of their examinations have been published in two of the principal scientific journals of the country. It had been originally proposed to republish these papers in this Appendix, together with that on the botanical collections, and some other topics; but the long time that has elapsed, renders it, on second thought, inexpedient. Distinct references to the several papers are given.

XII.

A Letter embracing Notices of the Zoology of the Northwest.
By HENRY R. SCHOOLCRAFT.

VERNON, N. Y., October 27, 1820.

DEAR SIR: I reached this place, on my return from the sources of the Mississippi River, on the 21st instant, having left the canal at Oneida Creek at four o'clock in the morning, whence I footed it three miles through the forest, by a very muddy road, to the ancient location of Oneida Castle, while my baggage was carried by a man on horseback.

The plan of the expedition embraced the circumnavigation of the coasts of Lakes Huron, Michigan, and Superior. From the head of the latter, we ascended the rapid River of St. Louis to a summit which descends west to the Upper Mississippi, the waters of which we entered about five hundred miles above the Falls of St. Anthony, and some three hundred miles above the ulterior point reached with boats by Lieutenant Pike in December, 1805.

From this point we ascended the Mississippi, by its involutions, to its upper falls at Pakagama, where it dashes over a rock formation. A vast plateau of grass and aquatic plants succeeds, through which it winds as in a labyrinth. On this plateau we encountered and passed across the southern Lake Winnipek. Beyond this, the stream appears to be but little diminished, unless it be in its depth. It is eventually traced to a very large lake called Upper Lac Ceder Rouge, but to which we applied the name of Cass Lake. This is the apparent navigable source of the river, and was our terminal point. It lies in latitude 47° 25′ 23″.

The whole of this summit of the continent is a vast formation of drift and boulders, deposited in steps. In descending it, we found the river crossed by the primitive rocks in latitude about 46°, and it enters the great limestone formation by the cataract of

St. Anthony's Falls, in latitude 44° 58' 40". We descended the river below this point, by its windings among high and picturesque cliffs, to the influx of the Wisconsin, estimated to be three hundred miles. Thence we came through the Wisconsin and Fox valleys to Green Bay, on an arm of Lake Michigan, and, having circumnavigated the latter, returned through Lakes Huron and St. Clair to Detroit. The line of travel is about four thousand two hundred miles. Such a country—for its scenery, its magnificence, and resources, and the strong influence it is destined ultimately to have on the commerce, civilization, and progress of the country—the sun does not shine on! Its topography, latitudes and longitudes, heights and distances, have been accurately obtained by Captain Douglass, of West Point, who will prepare an elaborate map and description of the country.

Personally, I have not been idle. If I have sat sometimes, in mute wonder, gazing on such scenes as the Pictured Rocks of Lake Superior, or the sylvan beauty and mixed abruptness of the Falls of St. Anthony, it has been but the idleness of admiration. I have kept my note-book, my sketch-book, and my pencil in my hands, early and late; nor have once, during the whole journey, transferred myself, at an early hour, from the camp-fire or pallet to the canoe, merely to recompose myself again to sleep. If the mineralogy or geology of the country often presented little to note, the scenery, or the atmosphere, or that lone human boulder, the American Indian, did. The evidences of the existence of copper in the basin of Lake Superior are ample. There is every indication of its abundance that the geologist could wish. Nature here has operated on a grand scale. By means of volcanic fires, she has infused into the trap-rocks veins of melted metal, which not inaptly represent the arteries of the human system; for wherever the broken-down shores of this lake are examined, they disclose, not the sulphurets and carbonates of this ore, but fragments and lumps of virgin veins. These, the winds and waves have scattered far and wide.

But what, you will ask, can be reported of its quadrupeds, birds, reptilia, and general zoology? Have you measured the height and length of the mastodon—"the great bull"—who the Indians told Mr. Jefferson resisted the thunderbolts, and leaped over the great lakes?* Truly, I beg you to spare me on this head. You are aware that we had no professed zoologist.

I herewith inclose you a list of such animals as came particularly under our notice. Imperfect as it is, it will give you the general facts. The dried and stuffed skins of such species as were deemed to be undescribed, or were otherwise worthy attention, will be transmitted for description. Among these is a species of squirrel, of peculiar character, from the vicinity of St. Peter's, together with a species of mus, a burrowing animal, which is very destructive to vegetation. This appears to be the hamster of Georgia. Of the larger class of quadrupeds, we met, in the forest traversed, the black bear, deer, elk, and buffalo. The latter we encountered in large numbers, about one hundred and fifty miles above the Falls of St. Anthony, about latitude 45°, on the east bank of the river. We landed for the chase, and had a full opportunity of observing its size, color, gait, and general appearance.

Great interest was imparted to portions of the tour by the ornithology of the country, and it only required the interest and skill in this line of a Wilson or an Audubon, to have not only identified, but also added to the list of species.†

The geological character of the country has been found highly interesting. The primitive rocks rise up in high orbicular groups on the banks of Lake Superior. The interstices between groups are filled up with coarse red, gray, or mottled sandstone, which lies, generally, in a horizontal position, but is sometimes waved or raised up vertically. Volcanic fires have played an important part here. I have been impressed with the fact that the granitical series are generally deficient in mica, its place being supplied by hornblende. Indeed, the rock is more truly sienite, very little true granite being found, and, in these cases, it is in the form of veins or beds in the sienite.

There have also been great volcanic fires and upliftings under the sources of the Mississippi. Greenstone and trap are piled up

---

* Notes on Virginia.

† The only addition to ornithology which it fell to my lot to make, was in the grosbeck family, and this occurred after I came to return to St. Mary's. Mr. Wm. Cooper has called the new species fringilia vespertina, from the supposition that it sings during the evening. The Chippewas call this species pauahkundamo, from its thick and penetrating bill.

in huge boulders. The most elevated rock, in place, on the sources of the Mississippi, is found to be quartzite. This is at the Falls of Pakagama. In coming down the Mississippi, soon after passing the latitude of 46°, the river is found to have its bed on greenstones and sienites, till reaching near to the Falls of St. Anthony, where the great western horizontal limestone series begins. To facilitate the study of the latter, opportunities were sought of detecting its imbedded forms of organic life, but their infrequency, and the rapid mode of our journeying, was averse to much success in this line without the boundaries of the great lake basins.

In the department of mineralogy, I have not as brilliant a collection as I brought from Potosi in 1819—but, nevertheless, one of value—the country explored being a wilderness, and very little labor having been applied in excavations. Among the objects secured, I have fine specimens of the various forms of native copper and its ores, together with crystallized sulphurets of lead, zinc, and iron; native muriate of soda, graphite, sulphate of lime, and strontian, and the attractive forms which the species of the quartz family assume, in the shore debris of the lakes, under the names of agate, carnelian, &c. The whole will be prepared and elaborately reported to the Department.

I found the freshwater shells of this region to be a very attractive theme of observation in places

> "Where the tiger steals along,
> And the dread Indian chants his dismal song;"

where, indeed, there was scarcely anything else to attract attention; and I have collected a body of bivalves, which will be forwarded to our mutual friend, Dr. Mitchell, for description. Indeed, the present communication is designed, after you have perused it, to pass under his eye. No one in our scientific ranks is more alive to the progress of discovery in all its physical branches. Governor Clinton, in one of his casual letters, has very happily denominated him the Delphic oracle, for all who have a question to ask come to him, and his scientific memory and research, in books, old and new, are such, that it must be a hard question indeed which he cannot solve.

Next to him, as an expounder of knowledge, you, my dear sir, as the representative of the *corps editorial*, take your place. For, if it is the writer of books who truly increases information, every decade's experience more and more convinces me that it is the editor of a diurnal journal who diffuses it, by his brief critical notices, or by giving a favorable or unfavorable impetus to public opinion.

I am expected, I find, to publish my private narrative of the expedition, to serve at least—if I may say so—as a stay to popular expectation, until the more matured results can be duly elaborated. I am taking breath here, among my friends, for a few days, and shall be greatly governed by your judgment in the matter, after my arrival at Albany.

I am, sir,

With sincere respect,

Your obedient servant,

HENRY R. SCHOOLCRAFT.

To NATHANIEL H. CARTER, Esq., Albany.

List of Quadrupeds, Birds, &c. observed.

The identification of species in this list, by giving the Indian name, is herein fixed.

ENGLISH NAME.	INDIAN (ALGONQUIN) NAME.	SCIENTIFIC NAME.
Buffalo,	Pe-zhík-i,*	Bos Americanus. *Gm.*
Elk,	Mush-kos,	Cervus Canadensis. *L.*
Deer (common),	Wa-wash-ká-shi,	Cervus Virginianus. *Gm.*
Moose,	Môz,	Cervus alces. *L.*
Black Bear,	Muk-wah,	Ursus Americanus. *Gm.*
Wolf (gray),	My-een-gan,	Canis vulpes. *L.*
Wolverine,	Gwin-gwe-au-ga,†	Ursus luscus. *L.*
Fox (red),	Waú-goosh,	Canis vulpes. *L.*
Badger,	Ak-kuk-o-jeesh,	Meles labradoria. *C.*
Fox (black),	Muk-wau-goosh,	Canis argenteus. *C.*
Muskrat,	Wau-zhusk,	Fiber vulgaris. *C.*
Martin,	Wau-be-sha-si,	Mustela mortes. *L. & B.*
Fisher,	O-jeeg,	Mustela Pennanti. *C.* Am. ed., app. v.
Beaver,	Am-ik,	Castor fiber. *B.*
Otter,	Ne-gik,	Lutra vulgaris. *L.*
Porcupine,	Kaug,	Hystrix cristata. *C.*

* This animal was found grazing the prairies on the east bank of the Mississippi, about latitude 45° 30′.

† Means under-ground drummer.

              APPENDIX.

ENGLISH NAME.	INDIAN (ALGONQUIN) NAME.	SCIENTIFIC NAME.
Raccoon,	Ais-e-bun (from *ais*, a shell, and *bun*, past tense),	Procyon lotor. *C.*
Hare,	Wau-bose,	Lepus Americanus. *Gm.*
Polecat,	She-kaug,	Mephites putorius. *Cu.*
Squirrel (red),	Ad-je-dah-mo,	Sciurus vulgaris. *C.*
Squirrel (ground or striped),	Ah-gwing-woos,	Sciurus striatus. *C.*
Squirrel (an apparently new species).		
Pouched Rat or Hamster,	No-naw-pau-je-ne-ka-si,	Mus busarius. *Shaw.*
Weasel,	Shin-gwoos,	Mustela vulgaris. *L.*
Mink,	Shong-waish-ke,	Mustela lutreola. *C.*
Jerboa, called the Jumping Mouse,*		Dipus. *C.*
Eagle (bald),	Mik-a-si,†	F. lucocephulus. *L.*
Fork-tailed Hawk,	Ca-niew,	F. furcatus. *L.*
Chicken Hawk,	Cha-mees,	F. communis. *C.*
Pigeon Hawk,	Pe-pe-ge-wa-zains,	F. columbarius. *Wilson.*
Raven,	Kaw-gaw-ge,	Corvus corax. *L.*
Crow,	On-daig,	C. corone. *L.*
Magpie,	Wau-bish-kau-gau-gi (White Raven),‡	C. pica. *L.*
Cormorant,	Kau-kau-ge-aheeb (Raven-duck),	P. carbe. *Brin.*
Pelican,	Shay-ta,	P. onocrotalus. *Illig.*
Goose,	Wa-wa,	An. anser. *L.*
Brant,	Ne-kuh,	An. bernicla. *Wilson.*
Duck (d. and m.),	Shee-sheeb (a generic term),	Anas.
Duck (saw-bill),	On-zig,	A. tadorna. *C.*
Duck (Red-head or Fall),	Misquon-dib,	A. rufus. *Gm.*
Duck (alewives),	Ah-ah-wa.	
Swan,	Wau-bis-si,	A. cygnus. *C.*
Heron,	Moosh-kow-e-si,	Ardea. *C.*
Plover,	Tchwi-tchwish-ke-wa,	Charadriûs. *C.*
Turkey,	Mis-is-sa,	Meleagris. *C.*
Blackbird,	Os-sig-in-ok,	The red-winged species.
Rail,	Muk-ud-a-pe-nais,	
Jay (blue),	Dain-da-si,§	Garrulus. *C.*
Whippoorwill,	Paish-kwa,	Caprimulgas. *L.*
Robin,	O-pee-chi.	T. migratorius. *L.*

* Found at Lapointe, Lake Superior.

† This is a generic term for the eagle family. It is believed the kanieu, or black eagle, is regarded by them as the head of the family. The feathers of the falco furcatus are highly valued by warriors.

‡ The meaning is white raven.

§ The term is from *dain-da*, a bullfrog.

ENGLISH NAME.	INDIAN (ALGONQUIN) NAME.	SCIENTIFIC NAME.
Kingfisher,	Me-je-ge-gwun-a,	Alcedo. *C.*
Pigeon,	O-mee-mi,	Columba emigratoria.
Partridge,	Pe-na,*	Tetrao. *C.*
Crane,	Ad-je-jawk,	Crane family.
Gull,	Ky-aushk,	Gull family.
Woodpecker,	Ma-ma,	Picus. *C.*
Snipe,	Pah-dus-kau-unzh-i,	Scolipax. *C.*
Owl,	Ko-ko-ko-o,†	} Generic terms for the
Loon,	Mong,	} species.
Mocking-bird (seen as far north as Michilimacinac),		T. polyglotis. *Wilson.*
Sturgeon,	Na-ma,	Acipenser. *L.*
Sturgeon (paddle-nose),	Ab-we-on-na-ma,	Acipenser spatularia. *C.*
Whitefish,	Ad-ik-um-aig‡ (means deer of the water).	
Salmon trout,	Na-ma-gwoos,	} Salmo. *L.*
Trout (speckled),	Na-zhe-ma-gwoos,	}
Carp,	Nam-a-bin,	Denotes the red fin.
Catfish,	Miz-zi,	Silurus. *C.*
Bass,	O-gau.	The striped species.
Tulibee,	O-dŏn-a-bee (wet-mouth).	
Eel,	Pe-miz-zi (a specific term).	A specific term.
Snake,	Ke-ná-bik (a generic),	} Ophidia. *C.*
Snake,	A species supposed peculiar,	}
Turtle (lake),	Mik-e-nok,	} Chelonia. *C.*
Turtle (small land),	Mis-qua-dais,	}

PHILOLOGICAL NOTE.—Three of these fifty-seven terms of Indian nomenclature are monosyllables, and twenty-four dissyllables. The latter are compounds, as in *muk-wah* (black animal), and *wau-bose* (white little animal); and it is inferable that all the names over a single syllable are compounds. Thus, aisebun (raccoon), is from *ais*, a shell, and the term past tense of verbs in *bun*.

XIII.

Species of Bivalves collected in the Northwest, by Mr. Schoolcraft and Captain Douglass, on the Expedition to the Sources of the Mississippi, in 1820. By D. H. BARNES.

This paper, by which a new impulse was given to the study of our freshwater conchology, and many species were added to the list of discoveries, was published in two papers, to be found in

* This is the prairie grouse of the West.
† The name is generic for the owl family.
‡ This term arises from *adik*, a reindeer, and *gumaig*, waters.

the pages of *Silliman's American Journal of Science*, vol. vi. pp. 120, 259.

XIV.

Freshwater Shells collected in the Valleys of the Fox and Wisconsin, in 1820, by Mr. Schoolcraft. By ISAAC LEA, Member American Philosophical Society.

A description of these shells, in which several new species are established, was published by the ingenious conchologist, Mr. I. Lea, of Philadelphia, in the *Transactions of the American Philosophical Society*, vol. v. p. 37, Plate III., &c.

XV.

Summary Remarks respecting the Zoology of the Northwest noticed by the Expedition to the Sources of the Mississippi in 1820. By Dr. SAMUEL L. MITCHELL.

The squirrel [from the vicinity of the Falls of St. Anthony], is a species not heretofore described, and has been named *sciurus tredecem striatus*, or the federation squirrel. (A.)

The pouched rat, or *mus busarius*, has been seen but once in Europe. This was a specimen sent to the British Museum from Canada, and described by Dr. Shaw. But its existence is rather questioned by Chev. Cuvier. Both animals have been described, and the descriptions published in the 21st vol. of the *Medical Repository*, of New York, pp. 248, 249. The specimens [from the West] are both preserved in my museum. Drawings have been executed by the distinguished artist Milbert, and forwarded by him, at my request, to the administrators of thé King's Museum, at Paris, of which he is a corresponding member. My descriptions accompany them. The animals are retained as too valuable to be sent out of the country. [B.]

The paddle-fish is the *spatularia* of Shaw, and *polydon* of Lacepede. It lives in the Mississippi only, and the skeleton, though incomplete, is better than any other person here possesses. It is carefully preserved in my collection.

The serpent is a species of the ophalian genus anguis, the oveto of the French, and the blind worm of the English. The loss of the tail of this fragile creature renders an opinion a little dubious;

but it is supposed to be *opthiosaureus* of Dandrige, corresponding to the *anguis ventralis* of Linnæus, figured by Catesby.

The shells afford a rich amount of an undescribed species. The whole of the univalves and bivalves received from Messrs. Schoolcraft and Douglass have been assembled and examined, with all I possessed before, and with Mr. Stacy Collins's molluscas brought from the Ohio. Mr. Barnes is charged with describing and delineating all the species not contained in Mr. Say's *Memoir of the Productions of the Land and Fresh Waters of North America.* The finished work will be laid before the Lyceum, and finally be printed in Mr. Silliman's *New Haven Journal.* The species by which geology will be enriched will amount, probably, to nine or ten. (C.) We shall endeavor to be just to our friends and benefactors.   S. L. MITCHELL.

For Gov. CASS.

Notes.

(A.)

An animal similar, in some respects, has been subsequently found on the Straits of St. Mary's, Michigan, a specimen of the dried skin of which I presented to the National Institute at Washington; but, from the absence of the head bones and teeth, it is not easy to determine whether it is a sciurus, or arctomys.

(B.)

The duplicature of the cheeks of this animal having been extended *outwardly* in drying the skin, was left in its rigid state, giving it an unnatural appearance, which doubtless led to the incredulity of Cuvier when he saw the figure and description of Dr. Shaw. Dr. Mitchell was led to a similar error of opinion, at first, as to the natural position of these bags; but afterwards, when the matter was explained to him, corrected this mistaken notion.

(C.)

By reference to the descriptions of Mr. Barnes and Mr. Lea, recited above, the number will be seen to have exceeded this estimate.

XVI.

Mus Busarius. Vide *Medical Repository,* vol. xxi. p. 248.

27

XVII.

Sciurus Tredecem Striatus. *Medical Repository*, vol. **xxi.**

XVIII.

Proteus. *American Journal of Science*, vol. iv.

---

6. METEOROLOGY.

XIX.

Memoranda of Climatic Phenomena and the Distribution of Solar Heat in 1820. By HENRY R. SCHOOLCRAFT.

The influence of solar heat on the quantity of water which is discharged from the great table lands which give origin to the sources of the Mississippi was such, during the summer months of 1820, that, on reaching those altitudes in latitude but a few minutes north of 47°, on the 21st of July, it was found impracticable to proceed higher in tracing out its sources. Attention had been directed to the phenomena of temperatures, clouds, evaporations, and solar influences, from the opening of the year, but they were not prosecuted with all the advantages essential to generalization. Still, some of the details noticed merit attention as meteorological memoranda which may be interesting in future researches of this kind, and it is with no higher view that these selections are made.

Observations made at Geneva, N. Y.

1820.	7 A. M.	1 P. M.	7 P. M.	REMARKS.
April 20 . . .	64°	78°	60°	Clear.
" 21 . . .	62	74	61	Clear.
" 22 . . .	65	78	66	Clear.
" 23 . . .	60	69	59	Clear.
" 24 . . .	59	70	61	Clear.
" 25 . . .	54	64	55	Clear.
" 26 . . .	55	67	54	Cloudy, with rain.
" 27 . . .	50	60	51	Rainy.
" 28 . . .	64	...	...	Clear.

Observations made at Buffalo, N. Y.

1820.				8 A. M.	2 P. M.	REMARKS.
April 30	.	.	.	43°	60°	Clear.
May 1	.	.	.	49	64	Clear.
" 2	.	.	.	45	68	Clear.
" 3	.	.	.	44	65	Clear.
" 4	.	.	.	46	79	Cloudy.
" 5	.	.	.	40	68	Cloudy, with rain.
" 6	.	.	.	44	...	Cloudy.

These places are but ninety miles apart, yet such is the influence of the lake winds on the temperature of the latter position, that it denotes an atmospheric depression of temperature of 5°. At the same time, the range between the maximum and minimum was exactly the same.

Observations made at Detroit.

1820.	8 A. M.	12 M.	6 P. M.	REMARKS.	WIND.
May 15,	50°	61°	51°	Fair.	N. E.
" 16,	49	62	50	Fair.	N. E.
" 17,	50	64	51	Fair.	N. E.
" 18,	52	64	60	Fair.	N. E.
" 19,	60	68	60	Fair.	N. E.
" 20,	64	68	63	Fair.	N. E.
" 21,	67	82	66	Fair.	S. W.
" 22,	64	88	82	Fair.	S. W.
" 23,	72	84	76	Cloudy, some rain.	W. N. W.
" 24,	58	64	...	Cloudy.	N. W.

The average temperature of this place for May is denoted to be some five or six degrees higher while the wind remained at N. E., but on its changing to S. W. (on the 21st), the temperature ran up four degrees at once. As soon as it changed to N. W. (on the 24th), the thermometer fell from its range on the 21st fourteen degrees.

The uncommon beauty and serenity of the Michigan autumns, and the mildness of its winters, have often been the subject of remark. By a diary of the weather kept by a gentleman in Detroit, in the summer and fall of 1816, from the 24th of July to the 22d of October, making eighty-nine days, it appears that

57 were fair,

12 cloudy, and

20 showery and rainy.

By a diary kept at the garrison of Detroit (Fort Shelby), agreeable to orders from the War Department, from the 15th of Nov. 1818, to the 28th of Feb. 1819, making 105 days,

40 of them are marked "clear,"
40 "cloudy,"
13 "clear and cloudy," and
12 "cloudy, with rain or snow."

By Fahrenheit's thermometer, kept at the same place, and under the same direction, it appears that the medium temperature of the atmosphere was agreeable to the following statement:—

	7 A. M.	2 P. M.	9 P. M.	Average.	Lowest deg.	Highest deg.
Nov. 13 to 30,	41°	47°	41°	43°	31°	58°
December,	22	29	25	25	2	50
January,	30	31	30	30	10	58
February,	29	39	31	33	8	58

Prevailing winds, S. W. and N. W.

Observations on Lake and River St. Clair, Michigan.

1820.	6 A. M.	8 A. M.	12 M.	2 P. M.	6 P. M.	8 P. M.	REMARKS.
May 24,	...	...	...	...	...	51°	
" 25,	47°	56°	56°	...	46°	...	Clear. Wind N. W.
" 26,	...	52	53	56°	45	...	Clear. Wind N. W.
" 27,	...	54	55	...	...	44	Clear. Wind N. W.

Temperature of the Water of Lake and River St. Clair.

May 25,	at 6 A. M., 49°	at 12 M., 54°	
" 26,	at 8 A. M., 55	at 2 P. M., 55	
" 27,	at 8 A. M., 54	at 12 M., 55	at 8 P.M., 50°

Observations on Lake Huron.

	5 A. M.	6 A. M.	8 A. M.	9 A. M.	11 A. M.	12 M.	1 P. M.	2 P. M.	3 P. M.	5 P. M.	6 P. M.	7 P. M.	8 P. M.	Average temp.	REMARKS.
May 28	..	..	54°	..	..	53°	..	..	..	..	..	..	44°	51°	Clear. Wind N. W.
" 29	..	..	44	..	..	..	..	70°	..	..	53°	..	..	55	Clear in the morning; in the afternoon high wind from N. W. with thunder and lightning.
" 30	..	..	..	46	..	53	..	..	..	..	48°	..	..	49	Clear. Wind high; N. W.
" 31	..	..	..	54°	..	..	55°	..	54°	..	..	48	..	53	
June 1	46°	..	..	57	61°	..	..	..	..	..	54	..	..	54	
" 2	..	..	..	..	..	55	..	..	..	..	50	..	..	52½	
" 3	..	50°	..	..	..	..	..	61	..	..	47	..	..	52½	
" 4	..	52	..	51	..	..	..	..	49°	..	45	..	..	49	Cloudy, with rain. Wind strong; N. W.
" 5	..	48	..	..	..	57	..	..	..	..	44	..	..	49½	Flying clouds. Wind strong; N. W.
" 6	..	..	49	..	..	57	..	..	..	..	46	..	..	50½	Clear. Wind strong; N. W.
														51 5-10	Average temperature of the air.

Water at Lake Huron.

				Average.
May 28,	at 5 A. M., 55°	at 12 A. M., 58°	at 7 P. M., 56°	56°
" 29,	at 7 A. M., 54	at 12 A. M., 60	at 7 P. M., 68	59
June 1,	at 5 A. M., 42	at 11 A. M., 52	at 7 P. M., 44	40
" 3,	at 6 A. M., 46	at 2 P. M., 56	at 8 P. M., 46	47
" 6,	at 8 A. M., 50	at 12 A. M., 52	at 6 P. M., 49	50½

Observations at Michilimackinac and on the Straits of St. Mary's.

1820.	6 A. M.	8 A. M.	9 A. M.	1 P. M.	3 P. M.	7 P. M.	9 P. M.	Average.	WEATHER.	WIND.	
June 7	...	...	59°	61°	...	...	59°	59½°	Clear.	W. N. W.	
" 8	...	...	59	...	64°	...	59	60	Clear.	W. N. W.	
" 9	...	...	58	...	...	58°	...	52½	Cloudy with rain.	W.	
" 10	...	55°	...	...	60	...	54	56	Cloudy with rain.	W.	
" 11	...	52	...	...	54	...	51	52	Clear.	S. E.	
" 12	...	54	...	55	...	...	52	53	Clear.	S. E.	
" 13	53°	...	...	63	...	...	58	58	Fair.	S. W.	
" 14	55	...	...	73	...	...	57	61	Cloudy.	S. W.	
" 15	...	66	...	...	68	62	...	65	Clear.	S. W.	} St. Mary's
" 16	...	52	70	82	...	66	...	69	Clear.	S. W.	
" 17	...	58	...	...	82	...	78	74	Clear.	S. W.	
" 18	56	...	...	76	...	...	68	66	Cloudy; rain.	N. W.	

The chief conclusion to be drawn, is the extreme fluctuations of winds and temperatures, in these exposed positions on the open lakes.

Observations on Lake Superior.

1820.	4 A. M.	5 A. M.	6 A. M.	7 A. M.	8 A. M.	9 A. M.	10 A. M.	11 A. M.	12 A. M.	1 P. M.	2 P. M.	3 P. M.	4 P. M.	5 P. M.	6 P. M.	7 P. M.	8 P. M.	9 P. M.	10 P. M.	Average temp.	REMARKS.
June 19	..	..	64	..	..	..	..	..	..	78	..	..	..	..	72	..	..	..	..	70½	Stormy and rain. Wind N. W.
" 20	..	72	..	..	..	..	..	..	75	..	..	..	..	68	71	..	..	..	..	71½	Stormy and rain. Wind N. W. Hurricane at night.
" 21	..	..	..	..	..	65	..	..	..	..	..	..	..	70	..	..	50	..	..	62	Calm.
" 22	..	..	55	..	..	..	..	..	..	63	..	..	..	49	..	..	..	..	..	55¼	Clear. Wind light from N. W.
" 23	..	65	..	..	..	..	68	..	..	..	..	..	..	..	..	70	..	..	67¼	Clear. Wind S. E.
" 24	..	..	..	58	..	..	74	..	..	..	..	60	63	..	..	..	..	..	63	Clear. High wind, N. W.
" 25	..	..	..	60	..	62	..	76	..	..	..	..	53	..	..	62½	Clear. Wind N. W.
" 26	..	..	..	..	69	..	..	83	..	..	..	..	68	..	..	73	Rainy. Wind W. N. W.
" 27	..	..	..	68	..	..	71	..	..	..	69	..	..	..	60	Clear. Wind E. N. E. (Fair!)
" 28	..	..	..	74	..	..	91	..	..	..	74	..	..	..	79½	Sky clear. Wind N. W.
" 29	..	..	..	79	..	..	94	..	..	..	86	..	..	88	Clear. Wind N. W.
" 30	..	..	..	76	..	84	..	..	..	..	60	..	73	Clear. Wind N. W.
July 1	54	..	..	61	..	..	75	..	80	68	..	..	67½	Misty. Wind light at N. N. W.
" 2	70	..	..	75	..	..	76	..	65	65	..	70	Clear. Wind W. S. W.
" 3	..	70	..	..	..	..	66	..	52	61	..	65	Cloudy, mist, and rain. Wind S. S. W.
" 4	..	..	57	..	61	..	..	..	..	..	..	..	Wind S. S. W.

Temperature of Lake Superior.

						Lake. average.
June 20,	at 6 P. M., 55°					55°
" 21,	at 10 A. M., 60	at 6 P. M., 56°	at 9 P. M., 56°			57
" 22,	at 6 A. M., 56	at 3 P. M., 54				55
" 23,	at 5 A. M., 52	at 12 A. M., 56	at 10 P. M., 64			57
" 24,	at 6 P. M., 54	at 7 P. M., 51				53
" 25,	at 7 A. M., 67	at 11 A. M., 66	at 9 P. M., 68			60
" 26,	at 9 A. M., 56	at 8 P. M., 57				56
" 27,	at 8 A. M., 57	at 6 P. M., 62				60
" 28,	at 8 A. M., Superior 62°	at 6 P. M., Lake 72 }				67
	Ontonagon 54	River 71 }				
" 29,	at 8 A. M., Lake 64					61
	River 68	at 1 P. M., River 76	at 7 P. M., 75°			
" 30,	at 8 P. M., River 74					
July 1,	at 8 A. M., 61	at 2 P. M., 65	at 6 P. M., 66			64
" 2,	at 4 A. M., 63	at 11 A. M., 64	at 2 P. M., 68	at 9 P. M., 62		64
" 3,	at 6 A. M., 62	at 3 P. M., 60	at 9 P. M., 58			60
" 4,	at 7 A. M., 58					

It will be observed that the fluctuations of temperature noticed at lower points on the lake chain, about the latitude of Michilimackinac, have also characterized the entire length of Lake Superior. The atmosphere observed at three separate times, during twenty-four days, by Fahrenheit's thermometer, during the months of June and July, has varied from an average temperature of 62° to 88°, agreeable to masses of clouds interposed to the rays of the sun, and to shifting currents of wind, which have often suddenly intervened. Its waters, spreading for a length of five hundred miles from E. to W., observed during the same time by as many immersions of the instrument, has not varied more than two degrees below or above the average temperature of 55° in mere surface observations.

Observations on the Sources of the Mississippi River.

	5 A. M.	7 A. M.	8 A. M.	12 M.	2 P. M.	8 P. M.	9 P. M.	REMARKS.
July 17	...	...	...	76°	80°	79°	78°	Morning rainy, then fair.
" 18	...	...	51°	64	66	58	50	Fair.
" 19	...	...	46	68	70	55	...	Night rainy, morning cloudy, then fair.
" 20	...	...	60	80	84	75	...	
" 21	...	...	68	86	88	85	74	
" 22	...	...	78	88	90	77	...	Cloudy, some thunder.
" 28	...	...	70	82	88	78	...	Night and morning rain, afternoon thunder.
" 24	...	...	74	87	80	78	...	Fair.
" 25	...	...	...	...	85	74	...	Fair.
" 26	61°	...	...	...	81	61	...	Morning fair, evening cloudy and rain, clear.
" 27	62	...	...	...	80	75	...	Morning fair, evening fair.
" 28	62	...	...	...	76	61	...	Morning fair, rain in afternoon.
" 29	50	...	...	...	74	52	...	Clear.
" 30	...	60°	...	...	76	...	63	Wind N. W., weather clear.
" 31	...	65	...	...	81	...	69	Wind W., weather clear.
Aug. 1	...	67	...	...	88	70	...	Fair.
" 2	...	72	...	...	*	...	...	Fair.

Observations at St. Peter's (now Minnesota).

1820.	7 A. M.	2 P. M.	9 A. M.	WINDS.	WEATHER.
July 15,	61°	79°	64°	S.	Clear; fair.
" 16,	62	82	76	S.	Clear; rain towards morning.
" 17,	70	88	61	W.	Cloudy; rain, thunder and lightning.
" 18,	58	78	56	E.	Clear.
" 19,	59	80	64	S.	Cloudy; rain P. M.
" 20,	68	80	65	S.	Clear.
" 21,	69	84	72	S.	Clear.
" 22,	75	88	72	W.	Clear; cloudy P. M., rain, thunder and lightning during the night.
" 28,	78	86	70	W.	Clear, cloudy; rain and fair weather alternately.
" 24,	70	89	72	W.	Clear; calms.
" 25,	70	80	66	W.	Clear; high winds at night.
" 26,	68	82	64	W.	Clear; calm.
" 27,	72	78	62	W.	Clear.
" 28,	67	75	58	S. E.	Clear; fresh winds.
" 29,	60	71	54	N. E.	Clear.
" 30,	60	76	68	N. W.	Clear.
" 81,	65	81	69	W.	Clear.

* Broke instrument.

Meteorological Journal kept at Chicago by Dr. A. Wolcott.

1820.	Daylight.	9 A. M.	2 P. M.	9 P. M.	WIND.	WEATHER.
Jan. 1,	4°	11°	10°	0°	W. N. W.	Cloudy; light snow; first ice in the river, 14 inches thick; none in the lake.
" 2,	10	14	25	12	W. N. W.	Clear.
" 8,	4	9	18	14	W. S. W.	Clear.
" 4,	9	14	19	9	W.	Clear.
" 5,	9	5	4	10	W. N. W.	Clear.
" 6,	11	4	15	28	S. S. W.	Clear.
" 7,	86	86	89	86	S. W.	Cloudy.
" 8,	82	82	84	88	N. N. E.	Cloudy.
" 9,	82	88	86	84	N. E.	Cloudy.
" 10,	82	81	81	25	N. E.	Snow-storm.
" 11,	14	14	16	2	N.	Clear.
" 12,	17	15	2	12	S. S. W.	Clear.
" 18,	20	24	25	12	W. S. W.	Clear.
" 14,	14	15	15	15	N.	Snow-squalls.
" 15,	12	14	15	10	N. N. W.	Clear; lake covered with moving ice, as far as the eye can see.
" 16,	20	20	21	21	E. N. E.	Snow-storm.
" 17,	14	14	25	10	W. N. W.	Clear.
" 18,	14	18	15	6	W.	Cloudy.
" 19,	10	0	10	2	W. N. W.	Clear.
" 20,	6	12	25	18	W.	Clear.
" 21,	20	22	26	28	E. N. E.	Snow-storm.
" 22,	7	11	12	5	N. W.	Clear.
" 28,	20	4	0	8	W.	Clear.
" 24,	2	6	18	16	W.	Clear.
" 25,	4	8	9	7	W.	Clear.
" 26,	16	19	26	28	E. S. E.	Snow-storm.
" 27,	18	21	25	8	S. W.	Cloudy.
" 28,	8	1	11	10	W. N. W.	Clear.
" 29,	12	20	81	18	W.	Cloudy; ice 18 inches on river.
" 80,	6	6	4	5	W.	Clear.
" 81,	6	5	8	17	W. N. W.	Clear; snow 22 inches deep.
Feb. 1,	12	0	14	16	S. E.	Cloudy.
" 2,	22	25	29	20	E. N. E.	Snow-storm; ice 18¾ inches on river.
" 8,	10	7	9	7	W.	Clear.
" 4,	0	5	25	24	E. S. E.	Clear.
" 5,	80	86	40	40	S. W.	Clear.
" 6,	11	12	82	24	S.	Clear.
" 7,	28	88	42	80	W. S. W.	Clear.
" 8,	80	84	40	82	E.	Cloudy and mist; snow during the night fell six inches.
" 9,	80	84	84	81	E.	Clear.

1820.	Daylight.	9 A.M.	2 P.M.	9 P.M.	WIND.	WEATHER.
Feb. 10,	31	32	39	32	E.	Cloudy.
" 11,	28	32	38	34	S.	Clear.
" 12,	32	39	34	20	N. E.	Cloudy.
" 18,	12	22	39	32	W. S. W.	Clear.
" 14,	34	39	37	36	E.	Cloudy; some rain with thunder.
" 15,	36	38	39	36	E.	Cloudy; some rain with thunder.
" 16,	38	42	47	33	S. W.	Clear.
" 17,	27	27	28	22	W.	Light clouds.
" 18,	10	22	28	30	E.	Cloudy.
" 19,	32	36	46	24	W.	Clear.
" 20,	15	22	24	16	W.	Clear.
" 21,	8	20	37	38	S. W.	Clear.
" 22,	34	40	45	32	W.	Clear.
" 23,	28	37	46	36	S. W.	Cloudy; rain and hail with thunder.
" 24,	30	38	40	39	E.	Clear.
" 25,	44	50	59	54	S. W.	Clear.
" 26,	50	49	38	36	S. W.	Cloudy; tempest of wind with flurries of rain and hail.
" 27,	30	31	34	28	W. N. W.	Clear.
" 28,	20	28	30	39	S. E.	Clear.
" 29,	28	36	50	37	S. W.	Clear.
Mar. 1,	32	35	36	18	N. N. W.	Clear.
" 2,	8	15	25	20	N. N. W.	Clear.
" 8,	26	30	36	22	W. N. W.	Cloudy.
" 4,	19	28	42	36	S. W.	Clear.
" 5,	30	32	36	28	N. E.	Cloudy.
" 6,	13	19	25	14	N. N. W.	Clear.
" 7,	16	17	24	18	E. N. E.	Cloudy; light snow.
" 8,	17	24	23	21	N. E.	Cloudy.
" 9,	22	24	26	28	N. N. E.	Cloudy.
" 10,	24	26	31	24	N. N. E.	Cloudy.
" 11,	22	24	29	31	E. N. E.	Cloudy.
" 12,	28	32	38	32	E. S. E.	Cloudy; light snow.
" 18,	32	37	39	34	E. N. E.	Cloudy.
" 14,	32	36	36	33	E. N. E.	Cloudy; light snow.
" 15,	26	32	...	...		

Agreeable to a register kept at Council Bluffs during the month of January, 1820, the highest and lowest temperature at that place were, respectively, 36° and 22°, the month giving a mean of 17.89. Compared with the observed temperature, for the same month, at the following positions in the United States, both east and west of the Alleghanies, the Missouri Valley reveals the fact of its being adapted to the purposes of a profitable agriculture.*

	Mean temperature of the month.	Highest.	Lowest.
Council Bluffs	17.89°	86°	22°
Wooster	16.69	86	zero
Zanesville	25.84	42	zero
Marietta	28.42	45	zero
Chillicothe	82.48	48	10
Cincinnati	28.76	46	11
Jeffersonville	28.05	50	6
Shawneetown	82.91	52	8
Huntsville	86.48	62	12
Tuscaloosa	46.68	74	17
Cahaba	65.87	73	54
Ouachita	84.16	68	10
New Orleans	52.16	78	25
Portsmouth, N. H. . . .	19.81	40	4†
Washington City . . .	29.19	45	4

Council Bluffs, lat. 41° 45′, long. 19° 50′ W. of the capitol.
New Orleans, " 29 57 " 12 53 W. "
Portsmouth, " 48 05 " 6 10 E. "
Difference of lat. 18° 48′.　　　Difference of long. 26°.

Nor does it appear that the same quantity of snow falls in the Missouri Valley which is common east of the Alleghany Mountains. At the Council Bluffs, on the last of January, snow was but twelve inches deep; at the same period, it was three feet or more throughout the Eastern States.

A snow-storm fell over the middle and eastern latitudes of the United States, for the first time, during the autumn of the year (1820), in the first half of November. As a precursor to this, slight drifts and gusts of snow had showed themselves at Albany on the 25th, 26th, and 28th of October.‡

* In Europe, the mean annual temperature necessary for the production of certain plants is—

For the sugar-cane	67°
" coffee	64
" orange	68
" olive	54
" vine (vitis vinifera) . . .	51

† Below zero.
‡ Meteorological journal kept at the Albany Academy for October, 1820.

"MONTREAL, CANADA, October 28, 1820.—On Wednesday last, we had the first fall of snow this season. It commenced in the forenoon, and continued slightly during the remainder of the day. Although expected to disappear, the frosts in the nights have been pretty severe, and a considerable quantity still remains (Saturday) at the moment we are writing."

"SALEM, N. Y. October 31.—On Saturday last (27th), we had our first snow for the season. It fell during most of the fore-noon, and for an hour or two the atmosphere was quite filled with it. Some cool and shaded spots still remain whitened, though yesterday was one of our pleasant autumnal days, with a mild west wind."

Early Sleighing.—The *Burlington* (Vt.) *Sentinel* of the 27th ult. says: "On Tuesday night and Wednesday, the snow fell in this place about eight inches deep on the level. It is said to be twelve inches deep in some of the adjoining towns."—*October*, 1820.

At Philadelphia, it began on Saturday, 11th (morning), snow-storm from the east, and continued all day. At night a hurri-cane, accompanied by torrents of rain and snow, which did not subside until the 12th in the morning. Weather unsettled on the 13th.

At Worcester, a severe snow-storm, from northeast, on the 11th and 12th. On the 13th, snow was ten inches deep, the weather cold, and sleighing good.

Snow in Poughkeepsie fell twelve inches deep, and produced excellent sleighing.

At New Haven (Conn.), it began with snow, hail, and rain, on Saturday evening, 11th. The day before was wintery cold. The storm continued, without intermission, till Monday, 13th.

At Boston, it also began on Saturday, 11th, from the northeast, and fell six inches. On Sunday, rain and snow. Monday cold, and indifferent sleighing in the *streets.*—*Boston paper*, Nov. 14th.

In Vernon, Oneida County, it began on the 11th, in the even-ing, and continued, in all, till Monday, 13th, giving us snow, rain, hail, and wind, alternately. On the 15th, the snow, which lay six inches deep, began to thaw, and this was the beginning of our Indian summer.

The Buffalo papers, of November 14th, say that several vessels were lost in the gale and snow-storm, or driven ashore. The storm closed up on the 13th, at New York City; the wind at northwest, and very cold. The rain, snow, and hail which had fallen gave good sleighing a part of that day. These notices cover an area of about five hundred miles square, proving the universality of our autumnal phenomena.

Indian Summer.

This season appears to be produced by the settling of a thin azure vapor. It is supposed to arise from the partial decomposition of the foliage of the forest after the autumnal rains are past. "What is called the Indian summer," says an observer at Albany, "usually gives us fifteen or twenty days of uncommonly pleasant fall weather, commencing in the early part of October. The present season it set in as usual, and we had a week or ten days of very fine weather, when a northeast storm commenced, and continued for part of two days; within which time more rain is supposed to have fallen than during the whole of the preceding summer and fall. Most of the streams and springs were filled, and the Hudson River, in many places, overflowed its banks. It however again cleared off pleasant, and remained so till Tuesday evening, when another storm of rain commenced, which continued the whole night. In the morning, there was some fall of hail accompanying the rain, and about 8 o'clock a slight flurry of snow, and another on Thursday evening; since which the weather has set in cold, and has the appearance of the closing in of fall or the setting in of winter. We however expect to put off winter and cold weather for some time yet, and anticipate many pleasant days in November."

Indian summer, in Oneida, commenced on the 15th November. The weather had previously been cold, with snow and rain and a murky atmosphere.

Wednesday, Nov. 15. The snow, which lay six inches deep, began to thaw, and the sky was clear and sunny.

Thursday, " 16. Was a clear and pleasant day throughout; snow continued to melt.

Friday, " 17. The same, and smoky; warm sunshine; not a cloud to be seen; snow melts.

Saturday, " 18. The same.

Sunday,   Nov. 19. The same; full moon; cloudy, with wind in the evening; snow gone.

Monday,   " 20. The same; sky clear and warm.

Tuesday,   " 21. Weather cloudy; wind S. E.; prepares for a change; a little snow during the previous night, but melts from the roofs this morning; no sun appears.

Wednesday,   " 22. Cloudy, dull morning; rain afternoon; sun appeared a few moments about 4 P. M.

Thursday,   " 23. Cloudy, with alternate sunshine and rain.

Friday,   " 24. Clear and pleasant.

Saturday,   " 25. Clear and pleasant.

Dr. Freeman, of Boston, in one of his occasional sermons, employs the following poetic language in relation to this American phenomenon:—

"The southwest is the pleasantest wind which blows in New England. In the month of October, in particular, after the frosts which commonly take place at the end of September, it frequently produces two or three weeks of fair weather, in which the air is perfectly transparent, and clouds, which float in a sky of the purest azure, are adorned with brilliant colors. If at this season a man of an affectionate heart and ardent imagination should visit the tombs of his friends, the southwestern breezes, as they breathe through the glowing trees, would seem to him almost articulate. Though he might not be so wrapped in enthusiasm as to fancy that the spirits of his ancestors were whispering in his ear, yet he would at least imagine that he heard 'the still small voice' of God. This charming season is called the Indian Summer, a name which is derived from the natives, who believe that it is caused by a wind which comes immediately from the court of their great and benevolent God Cantantowan, or the Southwestern God; the God who is superior to all other beings, who sends them every blessing which they enjoy, and to whom the souls of their fathers go after their decease."

7. INDIAN HIEROGLYPHICS, OR PICTURE WRITING, LANGUAGES, AND HISTORY.

XX.

Pictographic Mode of Communicating Ideas among the Northwestern Indians, observed during the Expedition to the Sources of the Mississippi in 1820, in a Letter to the Secretary of War. By Hon. LEWIS CASS.

DETROIT, February 2, 1821.

SIR: An incident occurred upon my recent tour to the North-west, so rare in itself, and which so clearly shows the facility with which communications may be opened between savage nations, without the intervention of letters, that I have thought it not improper to communicate it to you.

The Chippewas and Sioux are hereditary enemies, and Charle-voix says they were at war when the French first reached the Mississippi. I endeavored, when among them, to learn the cause which first excited them to war, and the time when it commenced. But they can give no rational account. An intelligent Chippewa chief informed me that the disputed boundary between them was a subject of little importance, and that the question respecting it could be easily adjusted. He appeared to think that they fought because their fathers fought before them. This war has been waged with various success, and, in its prosecution, instances of courage and self-devotion have occurred, within a few years, which would not have disgraced the pages of Grecian or of Roman history. Some years since, mutually weary of hostilities, the chiefs of both nations met and agreed upon a truce. But the Sioux, disregarding the solemn compact which they had formed, and actuated by some sudden impulse, attacked the Chippewas, and murdered a number of them. The old Chippewa chief who descended the Mississippi with us was present upon this occasion, and his life was saved by the intrepidity and generous self-devotion of a Sioux chief. This man entreated, remonstrated, and threatened. He urged his countrymen, by every motive, to abstain from any violation of their faith, and, when he found his remonstrances useless, he attached himself to this Chippewa chief, and avowed his determination of saving or perishing with him. Awed by his intrepidity, the Sioux finally agreed that he should ransom the Chippewa, and he accordingly applied to this object all the property he owned. He then accompanied the Chippewa on his journey until he considered him safe from any parties of the Sioux who might be disposed to follow him.

I subjoin an extract from the journal of Mr. Doty, an intelligent young gentleman who was with the expedition. This extract has already been published, but it may have escaped your observation, and the incident which it describes is so heroic in itself, and so illustrative of the Indian character, that I cannot resist the temptation of transmitting it to you.

EXTRACT FROM MR. DOTY'S JOURNAL.—"The Indians of the upper country consider those of the Fond du Lac as very stupid and dull, being but little given to war. They count the Sioux their enemies, but have heretofore made few war excursions.

"Having been frequently reprimanded by some of the more vigilant Indians of the north, and charged with cowardice, and an utter disregard for the event of the war, thirteen men of this tribe, last season, determined to retrieve the character of their nation by making an excursion against the Sioux. Accordingly, without consulting the other Indians, they secretly departed, and penetrated far into the Sioux country. Unexpectedly, at night, they came upon a party of the Sioux, amounting to near one hundred men, and immediately began to prepare for battle. They encamped a short distance from the Sioux, and, during the night, dug holes in the ground into which they might retreat and fight to the last extremity. They appointed one of their number (the youngest) to take a station at a distance and witness the struggle, and instructed him, when they were all slain, to make his escape to their own land, and state the circumstances under which they had fallen.

"Early in the morning, they attacked the Sioux in their camp, who, immediately sallying out upon them, forced them back to the last place of retreat they had resolved upon. They fought desperately. More than twice their own number were killed before they lost their lives. Eight of them were tomahawked in the holes to which they had retreated; the other four fell on the field! The THIRTEENTH returned home, according to the directions he had received, and related the foregoing circumstances to his tribe. They mourned their death; but, delighted with the bravery of their friends, unexampled in modern times, they were happy in their grief.

"This account I received of the very Indian who was of the party and had escaped."

The Sioux are much more numerous than the Chippewas, and would have overpowered them long since had the operations of the former been consentaneous. But they are divided into so many different bands, and are scattered over such an extensive country, that their efforts have no regular combination.

Believing it equally consistent with humanity and sound policy that these border contests should not be suffered to continue; satisfied that you would approve of any plan of pacification which might be adopted, and feeling that the Indians have a full portion of moral and physical evils, without adding to them the calamities of a war which had no definite object, and no probable termination; on our arrival at Sandy Lake, I proposed to the Chippewa chiefs that a deputation should accompany us to the mouth of the St. Peter's, with a view to establish a permanent peace between them and the Sioux. The Chippewas readily acceded to this proposition, and ten of their principal men descended the Mississippi with us.

The computed distance from Sandy Lake to the St. Peter's is six hundred miles, and, as I have already had the honor to inform you, a considerable proportion of the country has been the theatre of hostile enterprises. The Mississippi here traverses the immense plains which extend to the Missouri, and which present to the eye a spectacle at once interesting and fatiguing. Scarcely the slightest variation in the surface occurs, and they are entirely destitute of timber. In this debatable land, the game is very abundant; buffaloes, elks, and deer range unharmed, and unconscious of harm. The mutual hostilities of the Chippewas and Sioux render it dangerous, for either, unless in strong parties, to visit this portion of the country. The consequence has been a great increase of all the animals whose flesh is used for food, or whose fur is valuable for market. We found herds of buffaloes quietly feeding upon the plains. There is little difficulty in approaching sufficiently near to kill them. With an eagerness which is natural to all hunters, and with an improvidence which always attends these excursions, the animal is frequently killed without any necessity, and no other part of them is preserved but the tongue.

There is something extremely novel and interesting in this pursuit. The immense plains, extending as far as the eye can reach, are spotted here and there with droves of buffaloes. The distance and the absence of known objects render it difficult to estimate the size or the number of these animals. The hunters approach cautiously, keeping to the leeward, lest the buffaloes, whose scent is very acute, should observe them. The moment a gun is fired, the buffaloes scatter and scour the field in every direction. Unwieldy as they appear, they move with considerable celerity. It is difficult to divert them from their course, and the attempt is always hazardous. One of our party barely escaped with his life from this act of temerity. The hunters, who are stationed upon different parts of the plain, fire as the animals pass them. The repeated discharge of guns in every direction, the shouts of those who are engaged in the pursuit, and the sight of the buffaloes at full speed on every side, give an animation to the scene which is rarely equalled.

The droves which we saw were comparatively small. Some of the party whom we found at St. Peter's, and who arrived at that place by land from the Council Bluffs, estimated one of the droves which they saw to contain two thousand buffaloes.

As we approached this part of the country, our Chippewa friends became cautious and observing. The flag of the United States was flying upon all our canoes, and, thanks to the character which our country acquired by the events of the last war, I found in our progress through the whole Indian country, after we had once left the great line of communication, that this flag was a passport which rendered our journey safe. We consequently felt assured that no wandering party of the Sioux would attack even their enemies, while under our protection. But the Chippewas could not appreciate the influence which the American flag would have upon other nations, nor is it probable that they estimated with much accuracy the motives which induced us to assume the character of an umpire.

The Chippewas landed occasionally to examine whether any
28

of the Sioux had recently visited that quarter. In one of these excursions, a Chippewa found in a conspicuous place, a piece of birch bark, made flat by being fastened between two sticks at each end, and about eighteen inches long by fifteen broad. This bark contained the answer of the Sioux nation to the proposition which had been made by the Chippewas for the termination of hostilities. So sanguinary has been the contest between these tribes, that no personal communication could take place. Neither the sanctity of the office, nor the importance of the message, could protect the ambassadors of either party from the vengeance of each other. Some time preceding, the Chippewas, anxious for the restoration of peace, had sent a number of their young men into these plains with a similar piece of bark, upon which they had represented their desire. The bark had been left hanging to a tree in an exposed situation, and had been found and taken away by a party of the Sioux.

The propositions had been examined and discussed in the Sioux villages, and the bark which we found contained their answer. The Chippewa who had prepared the bark for his tribe was with us, and on our arrival at St. Peter's, finding it was lost, I requested him to make another. He did so, and produced what I have no doubt was a perfect *fac-simile*. We brought with us both of these *projets*, and they are now in the hands of Capt. Douglass. He will be able to give a more intelligible description of them than I can from recollection, and they could not be in the possession of one more competent to the task.

The Chippewas explained to us with great facility the intention of the Sioux, and apparently with as much readiness as if some common character had been established between them.

The junction of the St. Peter's with the Mississippi, where a principal part of the Sioux reside, was represented, and also the American fort, with a sentinel on duty, and the flag flying. The principal Sioux chief is named the Six, alluding, I believe, to the bands or villages under his influence. To show that he was not present at the deliberations upon the subject of peace, he was represented upon a smaller piece of bark, which was attached to the other. To identify him, he was drawn with six heads and a large medal. Another Sioux chief stood in the foreground, holding the pipe of peace in his right hand, and his weapons in his left. Even we could not misunderstand that. Like our own eagle with the olive-branch and arrows, he was desirous of peace, but prepared for war. ·

The Sioux party contained fifty-nine warriors, and this number was indicated by fifty-nine guns, which were drawn upon one corner of the bark. The only subject which occasioned any difficulty in the interpretation of the Chippewas, was owing to an incident, of which they were ignorant. The encampment of our troops had been removed from the low grounds upon the St. Peter's, to a high hill upon the Mississippi; two forts were therefore drawn upon the bark, and the solution of this enigma could not be discovered till our arrival at St. Peter's.

The effect of the discovery of this bark upon the minds of the Chippewas was visible and immediate. Their doubts and apprehensions appeared to be removed, and during the residue of the journey, their conduct and feelings were completely changed.

The Chippewa bark was drawn in the same general manner, and Sandy Lake, the principal place of their residence, was represented with much accuracy. To remove any doubt respecting it, a view was given of the old northwest establishment, situated upon its shore, and now in the possession of the American Fur Company. No proportion was preserved in their attempt at delineation. One mile of the Mississippi, including the mouth of the St. Peter's, occupied as much space as the whole distance to Sandy Lake; nor was there anything to show that one part was nearer to the spectator than another; yet the object of each party was completely obtained. Speaking languages radically different from each, for the Sioux constitute one of three grand divisions into which the early French writers have arranged the aborigines of our country, while the Chippewas are a branch of what they call Algonquins, and without any conventional character established between them, these tribes thus opened a communication upon the most important subject which could occupy their attention. Propositions leading to a peace were made and accepted, and the simplicity of the mode could only be equalled by the distinctness of the representations, and by the ease with which they were understood.

An incident like this, of rare occurrence at this day, and throwing some light upon the mode of communication before the in-

vention of letters, I thought it not improper to communicate to
you. It is only necessary to add, that on our arrival at St. Peter's,
we found Col. Leavenworth had been as attentive and indefatigable
upon this subject, as upon every other which fell within the sphere
of his command.

During the preceding winter, he had visited a tribe of the
Chippewas upon this pacific mission, and had, with the aid of
the agent, Mr. Talliafero, prepared the minds of both tribes for a
permanent peace. The Sioux and Chippewas met in council, at
which we all attended, and smoked the pipe of peace together.
They then, as they say in their figurative language, buried the
tomahawk so deep that it could never be dug up again, and our
Chippeway friends departed well satisfied with the result of their
mission.

I trust that Mr. Bolvin, the agent at Prairie du Chien, has been
able before this to communicate to you a successful account of the
negotiation which I instructed him to open between the Sacs and
Foxes, forming one party, and the Sioux. Hostilities were carried on between these tribes, which, I presume, he has been able
to terminate.

We discovered a remarkable coincidence, as well in the sound
as in the application, between a word in the Sioux language and
one in our own. The circumstance is so singular that I deem it
worthy of notice. The Sioux call the Falls of St. Anthony HA HA,
and the pronunciation is in every respect similar to the same words
in the English language. I could not learn that this word was
used for any other purpose, and I believe it is confined in its
application to that place alone.* The traveller in ascending the
Mississippi turns a projecting point, and these falls suddenly appear before him at a short distance. Every man, savage or civilized, must be struck with the magnificent spectacle which opens
to his view. There is an assemblage of objects which, added to
the solitary grandeur of the scene, to the height of the cataract,

* Iha ha [iha-ikiha] are words given as equivalent to laugh, v. in Riggs's Dictionary of the Dakota language, published by the Smithsonian Institution in 1852.
Ihapi, n., is laughter. The letter h, with a dot, represents a strong guttural, resembling the Arabic Kha. Iha, by the same authority, is the lips or cover to anything; it is also an adverb of doubt. The vowel i has the sound of i in marine, or
e in me.

and to the eternal roar of its waters, inspire the spectator with awe and admiration.

In his *Anecdotes of Painting*, it is stated by Horace Walpole, that "on the invention of fosses for boundaries, the common people called them Ha Ha's! to express their surprise on finding a sudden and unperceived check to their walk." I believe the word is yet used in this manner in England. It is certainly not a little remarkable that the same word should be thus applied by one of the most civilized and by one of the most barbarous people, to objects which, although not the same, were yet calculated to excite the admiration of the observer.

Nothing can show more clearly how fallacious are those deductions of comparative etymology, which are founded upon a few words carefully gleaned here and there from languages having no common origin, and which are used by people who have neither connection nor intercourse. The common descent of two nations can never be traced by the accidental consonance of a few syllables or words, and the attempt must lead us into the regions of fancy.

The Sioux language is probably one of the most barren which is spoken by any of our aboriginal tribes. Colonel Leavenworth, who made considerable proficiency in it, calculated, I believe, that the number of words did not exceed one thousand. They use more gestures in their conversation than any Indians I have seen, and this is a necessary result of the poverty of their language.

I am well aware, that the subject of this letter is not within the ordinary sphere of official communications. But I rely for your indulgence upon the interest which you have shown to procure and disseminate a full knowledge of every subject connected with the internal condition of our country.

I am preparing a memoir upon the present state of the Indians, agreeably to the intimation in my letter of September last. I shall finish and transmit it to you as soon as my other duties will permit.

Very respectfully, sir,

I have, &c.,

LEWIS CASS.

Hon. JOHN C. CALHOUN,
Secretary of War.

XXI.

Inquiries respecting the History of the Indians of the United States.
By LEWIS CASS.

These queries were published at Detroit in separate pamphlets, about the era of 1822, and communicated to persons in the Indian country supposed to be capable of furnishing the desired information. The results became the topic of several critical disquisitions, which appeared in the pages of the *North American Review* in 1825 and 1826; disquisitions the spirit and tone of which created, as the reader who is posted up on the topic will remember, a sensation among philological and philosophical readers.

Whether we are most to admire the bold tone of inquiry assumed by Gen. Cass, the acumen displayed in the discussions, the eloquence of the language, or the general soundness of the positions taken, is the only question left for decision. Certainly, nobody can arise from the perusal of these papers without becoming wiser or better informed on the subjects discussed. The mere luxury of high-toned and eloquent language is a gratification to the inquirer. But he cannot close these investigations into a subject of deep historical and philological interest without feeling established in the principles of historic truth, or warmed in his literary ardor.

Prominent among the topics of the initial discussion, was the work of John Dunn Hunter, a singular adventurer in the Indian country, or, perhaps, an early captive, who, after wandering to the Atlantic cities, where his harmless inefficiency of character gained no favorable attention, found his way to London, where the booksellers concocted a book of travels from him, in which the United States is unscrupulously traduced for its treatment of the Indians. The scathing which this person and his book received arises from its having fallen in the way of the business journeys of the critic to visit some of the principal scenes referred to; and among others, the residence of John Dunn, of Missouri, after whom he professed to be named, who utterly denied all knowledge of the man or of his purported adventures.

The question of the authenticity of the Indian traditions of Mr. Heckewelder, derived from a single tribe, and that tribe telling

stories to salve up its own disastrous history, and the mere literary
capacities of the man to put his materials in order, is propounded
and examined in connection with the contemporary traditions and
languages of other tribes.   These traditions had been communicated to the Pennsylvania Historical Society, in 1816, and were
published under the special auspices of Mr. Duponceau, in 1819.
From the internal evidence of the letters themselves, the critic pro-
nounces them to be reproductions of Mr. Duponceau himself; and
it is an evidence of the aptness of this deduction to be told that
Mr. Gallatin admitted (*vide* my *Personal Memoirs*, p. 623), that the
letters of Mr. Heckewelder had all been rewritten previous to
publication.   It could no longer be a subject of admiration to
philologists, that from such imperfect sources of information, that
distinguished scholar should have pronounced the opinion that
the Delaware language rather exceeds than falls short of the Greek
and Latin in the affluence of syntactical forms and capacities of
expression. *Trans. Hist. and Lit. Com., Am. Philo. Soc.*, vol. i. p. 415.

XXII.

*A Letter on the Origin of the Indian Race of America, and the Prin-
ciples of their Mode of uttering Ideas; addressed to John Johnston,
Esq., late of St. Mary's Falls, Michigan.* By Dr. J. McDonnell,
of Belfast, Ireland.

BELFAST, April 16, 1817.

My Dear J.: I feel always as if I am guilty of some great crime,
in not writing to you.

An account came to Sir Joseph Banks, of very curious rocks,
with odd stripes and colors, having been seen, this last war, by
sailors on the lakes, I think on Lake Superior.*   Pray keep up
your thoughts to the geography of rocks.   I got some lately from
Bombay, exactly ditto with our Causeway.†

I shall ever regret the not having seen your daughter.   I think
it likely that mingling the European blood and character with
the Indian might bring out some superior traits of character.
Lest my letter should altogether fail of presenting any useful
point, I must put some questions to you that would be worth
something if answered.

* Most probably this idea arose from the very marked precipices of the coast
denominated Pictured Rocks.   H. R. S.
 † The Giant's Causeway, on the Coast of Antrim.

∟A man has published, in 1816, an octavo volume in Trenton (United States), the author's name Boudinot, to explain some things about the Indian nations, and, among other things, he fancies some resemblance between their languages and Hebrew. Baron Von Humboldt, a Prussian, was in Spanish America lately, and he found the natives had Hebrew opinions and usages, evidently things borrowed from Jewish doctrines. I don't want you to inquire much about their being of this extraction, but observe, for me, whether their languages have no pronouns, as one author, Colden, stated fifty years ago; and whether they are defective in the prepositions, as this Boudinot states; and whether those near you have any words, idioms, or traditions that are expressive of their early origin, or their connection with European nations.

In fact, I think you are better circumstanced, in most respects, than any other man that I ever heard of, to do something worth notice in that way; for, although you have not books, nor knowledge of many tongues, yet you could collect lists of great and radical words, expressed with proper letters, so that others could compare those words with Asiatic, and African, and European tongues, so as to enable mankind to judge of similitudes or dissimilitudes.

The words most apt to pervade different nations, and to pass from one people to another, are articles, pronouns, auxiliary verbs, prepositions; next to these, numerals; next to these, whatever terms are expressive of striking, useful, hurtful, or very clear and definite objects and ideas; for, if the conceptions we have of things be not very definite, clear, and distinct, the idea and the word are not likely to float down the stream of time together, they will be jostled and separated. Be very careful in spelling the Indian words; spell them in different ways, where our letters don't square exactly with their sounds. Take notice of their musical tones, and whether these tones get in, as essential parts, into their speech; and, above all, remember that a *word* is a *thing*, and that it may be examined as a *record*, or considered like a coin or medal, as well as if it had the stamp of a king or mint upon it.

I will write more if this vessel does not sail to-day. God bless you and yours, and believe me, in haste, your affectionate cousin.

J. McDONNELL.

XXIII.

Difficulties of Studying the Indian Tongues of the United States.
By Dr. ALEXANDER WOLCOTT, Jr.

Dr. Wolcott will be remembered by the early inhabitants of Chicago, when that place was still a military post and the site of an Indian agency, the latter of which trusts he filled. In 1820, the Pottowattomie tribe of Indians and their confederates—the Illinois—Chippewas, and Ottowas—possessed the whole surrounding regions, roving as lords of the prairies. These numerous and fierce hunter-tribes, who traded their peltries for fineries, had many horses, loved rum and fine clothes, and despised all restraints, came in to him, at his agency, as the mouthpiece of the President, to transact their affairs, and they often lingered for days and weeks around the place, which gave him a good opportunity of becoming familiar with their manners, customs, and history.

Dr. Wolcott was a man of education, of high morals, dignified manners, and noble sentiments, with decidedly saturnine feelings, and a keen perception of the ridiculous. Constitutionally averse to much or labored personal effort, his leisure hours, in this seclusion from society, were hours devoted to reading and social converse, and his attention was appropriately called by Gen. Cass to the "Inquiries," No. 21, above referred to. The reply which he at length communicated was written in so happy a vein, that I obtained permission to publish the substance of it, in 1824, in my *Travels in the Central Portions of the Mississippi Valley*, p. 381. It declares an important truth, which all must concur in, who have attempted the study of the Indian languages, for they are required to perform the prior labor of ascertaining and generalizing the principles of their accidence and concord. When I first came to St. Mary's, in 1822, and began the study of the Chippewa, I asked in vain the simple question how the plural was formed. It was formed, in truth, in twelve different ways, agreeably to the vowels of terminal syllables; but this could not be declared until quires of paper had been written over, the whole vocabulary explored, and days and nights devoted to it. My first interpreter could not tell a verb from a noun, and was incapable of translating the simplest sentence literally. Besides his ignorance, he was so great a liar that I never knew when to believe him. He sometimes told the Indians the reverse of what I said, and often told me the reverse of what they said.

XXIV.

Examination of the Elementary Structure of the Algonquin Language as it appears in the Chippewa Tongue. By HENRY R. SCHOOLCRAFT.

INTRODUCTORY NOTE.

SAULT STE. MARIE, May 31, 1823.

SIR: In order to answer your inquiries, I have improved my leisure hours, during the part of the summer following our arrival here (6th July last), and the entire winter and spring, in examining the words and forms of expression of the Chippewa, or (as the Indians pronounce it) Odjibwa, tongue. I have found, as I anticipated, my most efficient aid, in this inquiry, in Mr. Johnston, and the several members of his intelligent family; my public interpreter being too unprecise and profoundly ignorant of the rules of grammar to be of much use in the investigation. Mr. Johnston, as you are aware, perhaps, came from the north of Ireland, where his connections are highly respectable, during the first term of General Washington's administration. He brought letters from high sources to the Governor-General of Canada; but having, while at Montreal, fallen in with Don Andrew Tod, a countryman, who had the monopoly of the fur trade of Louisiana, in a spirit of enterprise and adventure, he threw himself into that, at the time, fascinating pursuit, and visited Michilimackinac. Circumstances determined him to fix his residence at St. Mary's, where he has resided, making frequent visits to Montreal and Great Britain, about thirty years. His children have been carefully instructed in the English language and literature, and the whole family are familiar with the Indian. Without such proficient aid, I should have labored against serious impediments at every step; and, with them, I have found the inquiry, in a philological point of view, involved in many, and some of them insuperable difficulties. The results I communicate to you, rather as an earnest of what may be hereafter done in this matter, than as completely fulfilling inquiries which it would require Horne Tooke himself, with the aid of the Bodleian library, to unravel.

With respect, &c.,

HENRY R. SCHOOLCRAFT.

His Excellency Gov. LEWIS CASS.

EXAMINATION OF THE ODJIBWA.

1, 2. *Simple Sounds.*—The language is one of easy enunciation. It has sixteen simple consonental and five vowel sounds. Of these, two are labials, *b* and *p*; five dentals, *d*, *t*, *s*, *z*, *j*, and *g* soft; two nasals, *m* and *n*; and four gutturals, *k*, *q*, *c*, and *g* hard. There is a peculiar nasal combination in *ng*, and a peculiar terminal sound of *g*, which may be represented by *gk*. Of the mixed dipthongal and consonental sounds, those most difficult to English organs are the sounds in *aiw* and *auw*.

3. *Letters not used.*—The language is wholly wanting in the sound of *th*. It drops the sound of *v* entirely, substituting *b*, in attempts to pronounce foreign words. The sound of *l* is sometimes heard in their necromantic chants; but, although it appears to have been known to the old Algonquin, it is supplied, in the Odjibwa of this day, exclusively by *n*. It also eschews the sounds of *f*, *r*, and *x*, leaving its simple consonental powers of utterance, as above denoted, at sixteen. In attempts to pronounce English words having the sound of *f*, they substitute *p*, as in the case of *v*. The sound of *r* is either dropped, or takes the sound of *au*. Of the letter *x* they make no use; the nearest approach I have succeeded in getting from them is *ek-is*, showing that it is essentially a foreign sound to them. The aspirate *h* begins very few words, not exceeding five in fifteen hundred, but it is a very frequent sound in terminals, always following the slender or Latin sound of *a*, but never its broad sound in *au*, or its peculiarly English sound as heard in the *a* of *may*, *pay*, *day*. The terminal syllable of the tribal name (Odjibwa), offers a good evidence of this rule, this syllable being never sounded by the natives either *wah* or *wau*, but always *wa*. These rules of utterance appear to be constant and imperative, and the natives have evidently a nice ear to discriminate sounds.

Rule of Euphony.—In the construction of words, it is required that a consonant should *precede* or *follow* a vowel. In dissyllables wherein two consonants are sounded in juxtaposition, it happens from the joining of two syllables, the first of which ends and the last begins with a consonant, as *muk-kuk,* a box, and *os-sin,* a stone; the utterance in these cases being confluent. But in longer compounds this juxtaposition is generally avoided by throwing in a vowel for the sake of euphony, as in the term *assinebwoin,* the *e* in which is a mere connective, and has no meaning by itself. Nor is it allowable for vowels to follow each other in syllabication, except in the restricted instances where the being or existence of a thing or person is affirmed, as in the vowel-words *i-e-e* and *i-e-a,* the animate and inanimate forms of this declaration. In these cases, there is a distinct accent on each vowel.

4. *Accent.*—The accent generally falls on full or broad vowels, and never on short vowels; such accented vowels are always significant, and if they are repeated in a compound word, the accents are also repeated, the only difference being that there are primary and secondary accents. Thus, in the long descriptive name for a horse, *Pa-bá-zhik-ó-ga-zhé,* which is compounded of a numeral term and two nouns, meaning, the animal with solid hoofs; there are three accents, the first of which is primary, while the others succeed each other with decreased intensity. By a table of words which I have constructed, and had carefully pronounced over by the natives, it is denoted that dissyllables are generally accented on the final syllable, trisyllables on the second, and words of four syllables on the second and fourth. But these indications may not be constant or universal, as it is perceived that the accents vary agreeably to the distribution of the full and significant vowels.

5. *Emphasis.*—Stress is laid on particular words in sentences to which the speaker designs to impart force, and the whole tone of the entire sentiment and passages is often adapted to convey particular impressions. This trait more frequently comes out in the private narrative of real or imaginary scenes, in which the narrator assumes the very voice and tone of the real or supposed actor. Generally, in their dealings and colloquial intercourse, there is a significant stress laid on the terms, *meenungaika,* certainly; *kaigait,* truly; *kaugaigo,* nothing at all; *tiau,* behold; *wohow,* who; *auwanain,*

were; and other familiar terms of inquiry, denial, or affirmation in daily use.

6. *Conjugation.*—The simplest form in which their verbs are heard, is in the third person singular of the indicative, as *he speaks, he says, he loves, he dances*, or in the first person present of the imperative. The want of a distinction between the pronouns *he* and *she*, is a defect which the language shares, I believe, with other very ancient and rude tongues. Conjugations are effected for persons, tenses, and number, very much as they are in other rude languages, particularly those of the transpositive class. The verb is often a single root, or syllable, as *saug*, love; but owing to the tendency of adding qualifying particles, their verbs are cluttered up with other meanings. The word *saug* is therefore never heard as an element by itself. In the first place, it takes before it the pronoun, and in the second place, the object of action; so that *nesaugeau*, I love him, or her, or a person, is one of the simplest of their colloquial phrases. And of this term, the *e*, being the fourth syllable, is mere verbiage, means nothing by itself, and is thrown in for euphony.

Tenses are formed by adding *gee* to the pronoun for the perfect, and *gah* for the future, and *gahgee* for the second future. These terms play the part, and supply the want of, auxiliary verbs. The imperative is made in *gah*, and the potential in *dau* where the second future is *daugee*. The subjunctive is made by prefixing the word *kishpin*, meaning if. The inflection *nuh*, asks a question, and as it can be put to all the forms of the conjugation, it establishes an interrogative mood. The particle *see*, negatives the verb, and thus all verbs can be conjugated positively and negatively.

To constitute the plural, the letter *g* is added to the conjugations; thus, *nesaugeaug* means, I love them. But this is an animate plural, and can only be added to words of the vital class. Besides, if the verb or noun to be made plural does not end in a vowel, but in a consonant, the *g* cannot be added without interposing a vowel. It results, therefore, that the vowel class of words have their plurals in *äg, eeg, ig, og*, or *ug*. But, if the class of words be non-vital and numerical, the plural is made in the letter *n*. But this letter cannot, as in the other form, be added, unless the word terminate in a vowel, when the regular

plurals are *än, een, in, on,* or *un.* This simple principle clears up
one cause of perplexity in the conjugations, and denotes a philosophical method, which divides the whole vocabulary into two
classes; while this provision *supersedes,* it answers the purpose of
gender. There is, in fact, no gender required by the conjugations,
it being sufficient to denote the *vitality* or *non-vitality* of the class.
Nothing can be clearer. This is one of the leading traits of the
grammar of the language, upon the observance of which the best
speakers pride themselves.

It does not, however, result that, because there is no gender
required in the conjugations, the idea of sexuality is unknown to
the nomenclature. Quite the contrary. The tenses for male and
female, in the chief orders of creation, are *iaba* and *nozha.* These
words prefixed to the proper names of animals, produce expressions of precisely the same meaning, and also the same inelegance;
as if we should say, male goose, female goose, male horse, and
female horse, male man and female man. The term for man
(*inini*) is masculine, and that for woman (*equa*) feminine in its
construction. It is only in the conjugations that the principle
of gender becomes lost in that of vitality.

7. *Active and passive voices.*—The distinction between these two
classes of verbs is made by the inflection *ego.* By adding this
form to the active verb, its action is reversed, and thrown back
on the nominative. Thus, the verb to carry is *nim bemön,* I
carry; *nim bemön-ego,* I am carried. *Adowawa* is the act of
thumping, as a log by the waves on the shore. *Adowawa-ego*
is a log that is thumped by the waves on shore. *Nesaugeah,* I
love; *Nesaugeigo,* I am loved. In the latter phrase, the personal
term *au* is dropped, and the long sound of *e* slips into *i,* which
converts the inflection into *igo* instead of *ego.*

8. *Participles.*—My impression is, that the Indians are in the
habit of using participles, often to the exclusion of other proper
forms of the verb. The vocabulary contains abundantly the indicative forms of the verb. To run, to rise, to see, to eat, to tie, to burn,
to strike, to sing, to cry, to dance, are the common terms of parlance; but as soon as these terms come to be connected with the
action of particular persons, this action appears to be spoken of
as if existing—both the past and future tenses being thrown away;
and the senses appear to be, I, you, he, or they; running, rising,

seeing, eating, tying, burning, striking, singing, crying, dancing. At least, I have not been able to convince myself that the action is not referred to as existing. When the participles should be used, they, on the contrary, employ the indicative forms, by which such sentences are made as, he run, he walk, for running, walking.

The general want of the substantive verb, in their colloquial phrases, constantly leads to imperfect forms of syntax. Thus, *nëbä* is the indicative, first person of the verb to sleep; but if the term, I am sleeping, be required, the phrase is *ne nëbä*, simply, I sleep. So, too, *tshägiz* is the first person indicative to burn; but the colloquial phrase, I am burned, or burning, is *nen tshägiz*—the verb remaining in the indicative, and not taking the participle form.

It is not common to address persons by their familiar names, as with us—as John, or James. The very contrary is the usage of Indian society, the object being to conceal all personal names, unless they be forced out. If it be required to express this sentence, namely: Adario has gone out (or temporarily departed), but will soon return; the equivalent is *Ogima, ke mahjaun, panema, ke takooshin.* This sentence literally retranslated is, Chief, he gone; by and by, he (will) return—the noun chief being put for the personal noun Adario. It will be perceived that the pronoun *ke* is repeated after the noun, making, chief, he gone. *Panema* is an adverb which is undeclinable under all circumstances, and *tahkooshin*, the future tense of the verb to arrive, or come (by land). The phraseology is perfectly loaded with local or other particulars, which constantly limit the action of verbs to places, persons, and things.

XXV.

A Vocabulary of the Odjibwa Algonquin Language. By H. R. Schoolcraft.

On referring to the manuscript of this vocabulary, it is found to fill a large folio volume, which puts it out of my power to insert it in this connection. It is hoped to bring it into the series of the Ethnological volumes, now in the process of being published at Philadelphia, under the auspices of Congress.

APPENDIX.

No. 2.

THE EXPEDITION TO ITASCA LAKE IN 1882.

29

SYNOPSIS.

1. INDIAN LANGUAGES.

I. II. Observations on the Grammatical Structure and Flexibility of the Odjibwa Substantive. By HENRY R. SCHOOLCRAFT.

III. Principles Governing the Use of the Odjibwa Noun-adjective. By HENRY R. SCHOOLCRAFT.

IV. Some Remarks respecting the Agglutinative Position and Properties of the Pronoun. By HENRY R. SCHOOLCRAFT.

2. NATURAL HISTORY.

V. Zoology.

    1. Limits of the Range of the Cervus Sylvestris in the Northwestern parts of the United States. By HENRY R. SCHOOLCRAFT.—*Northwest Journal.*

    2. Description of the Fringilia Vespertina, discovered by Mr. Schoolcraft in the Northwest. By WILLIAM COOPER.—*Annals of the New York Lyceum of Natural History.*

    8. A list of Shells collected by Mr. Schoolcraft during his Expedition to the Sources of the Mississippi in 1832. By WILLIAM COOPER.

VI. Botany.

    1. List of Species and Localities of Plants collected during the Exploratory Expeditions of Mr. Schoolcraft in 1831 and 1832. By DOUGLASS HOUGHTON, M. D., *Surgeon to said Expeditions.*

VII. Mineralogy and Geology.

    1. A Report on the Existence of Deposits of Copper in the Trap Rocks of Upper Michigan. By Dr. DOUGLASS HOUGHTON.

    2. Remarks on the Occurrence of Native Silver, and the Ores of Silver, in the Stratification of the Basins of Lakes Huron and Superior. By HENRY R. SCHOOLCRAFT.

    8. A General Summary of the Localities of Minerals observed in the Northwest. By HENRY R. SCHOOLCRAFT.

    4. Geological Outlines of the Valley of Takwymenon in the Basin of Lake Superior. By HENRY R. SCHOOLCRAFT.

    5. Suggestions respecting the Geological Epoch of the Deposit of Red Sandstone of St. Mary's Falls, Michigan. By HENRY R. SCHOOLCRAFT.

3. INDIAN TRIBES.

VIII. Condition and Disposition.
1. Official Report to the War Department, of an Expedition through Upper Michigan and Northern Wisconsin in 1831. By HENRY R. SCHOOLCRAFT.
2. Brief Notes of a Tour in 1831, from Galena, in Illinois, to Fort Winnebago, on the source of Fox River, Wisconsin. By HENRY R. SCHOOLCRAFT.
3. Official Report of the Expedition to Itasca Lake in 1832. By HENRY R. SCHOOLCRAFT.
4. Report of the Vaccination of the Indians in 1832, under the authority of an Act of Congress. By Dr. DOUGLASS HOUGHTON.

4. TOPOGRAPHY AND GEOGRAPHY.

IX. Astronomical and Barometrical Observations.
1. Table of Geographical Positions observed in 1836. By J. N. NICOLLET.

5. SCENERY.

X. Letters on the Scenery of Lake Superior. By MELANCTHON WOOLSEY. *Vide* Southern Literary Messenger, 1836.

APPENDIX.

---

1. INDIAN LANGUAGE.

I.

*Observations on the Grammatical Structure and Flexibility of the Odjibwa Substantive.**

INQUIRY 1.

Observations on the Ojibwai substantive. 1. The provision of the language for indicating gender—Its general and comprehensive character—The division of words into animate and inanimate classes. 2. Number—its recondite forms, arising from the terminal vowel in the word. 8. The grammatical forms which indicate possession, and enable the speaker to distinguish the objective person.

MOST of the researches which have been directed to the Indian languages, have resulted in elucidating the principles governing the use of the verb, which has been proved to be full and varied in its inflections. Either less attention has been paid to the other parts of speech, or results less suited to create high expectations of their flexibility and powers have been attained. The Indian verb has thus been made to stand out, as it were in bold relief, as a shield to defects in the substantive and its accessories, and as, in fact, compensating, by its multiform appendages of prefix and suffix—by its tensal, its pronominal, its substantive, its adjective, and its adverbial terminations, for barrenness and rigidity in all other parts of speech. Influenced by this reflection, I shall defer,

---

* Mr. Du Ponceau did me the honor, in 1884, to translate these two inquiries on the substantive in full, for the prize paper on the Algonquin, before the National Institute of France.

in the present inquiry, the remarks I intend offering on the verb, until I have considered the substantive, and its more important adjuncts.

Palpable objects, to which the idea of sense strongly attaches, and the actions or condition, which determine the relation of one object to another, are perhaps the first points to demand attention in the invention of languages. And they have certainly imprinted themselves very strongly, with all their materiality, and with all their local, and exclusive, and personal peculiarities upon the Indian. The noun and the verb not only thus constitute the principal elements of speech, as in all languages; but they continue to perform their first offices, with less direct aid from the auxiliary parts of speech, than would appear to be reconcilable with a clear expression of the circumstances of time and place, number and person, quality and quantity, action and repose, and the other accidents, on which their definite employment depends. But to enable the substantives and attributives to perform these complex offices, they are provided with inflections, and undergo changes and modifications, by which words and phrases become very concrete in their meaning, and are lengthened out to appear formidable to the eye. Hence the polysyllabic, and the descriptive character of the language, so composite in its aspect and in its forms.

To utter succinctly, and in as few words as possible, the prominent ideas resting upon the mind of the speaker, appear to have been the paramount object with the inventors of the language. Hence, concentration became a leading feature. And the pronoun, the adjective, the adverb, and the preposition, however they may be disjunctively employed in certain cases, are chiefly useful as furnishing materials to the speaker, to be worked up into the complicated texture of the verb and the substantive. Nothing, in fact, can be more unlike, than the language, viewed in its original, elementary state—in a vocabulary, for instance, of its primitive words, so far as such a vocabulary can now be formed, and the same language as heard under its oral, amalgamated form. Its transpositions may be likened to a picture, in which the copal, the carmine, and the white lead, are no longer recognized as distinct substances, but each of which has contributed its share towards the effect. It is the painter only who possesses the principle, by which one element has been curtailed, another augmented, and all, however seemingly discordant, made to coalesce.

Such a language may be expected to abound in derivatives and compounds; to afford rules for giving verbs substantive, and substantives verbal qualities; to concentrate the meaning of words upon a few syllables, or upon a single letter, or alphabetical sign; and to supply modes of contraction and augmentation, and, if I may so say, *short cuts*, and *by-paths* to meanings, which are equally novel and interesting. To arrive at its primitives, we must pursue an intricate thread, where analogy is often the only guide. We must divest words of those accumulated syllables, or particles, which, like the molecules of material matter, are clustered around the primitives. It is only after a process of this kind, that the *principle of combination*—that secret wire, which moves the whole machinery can be searched for, with a reasonable prospect of success. The labor of analysis is one of the most interesting and important, which the subject presents. And it is a labor which it will be expedient to keep constantly in view, until we have separately considered the several parts of speech, and the grammatical laws by which the language is held together; and thus established principles and provided materials wherewith we may the more successfully labor.

1. In a general survey of the language as it is spoken, and as it must be written, there is perhaps no feature which obtrudes itself so constantly to view, as the principle which separates all words, of whatever denomination, into animates and inanimates, as they are applied to objects in the animal, vegetable, or mineral kingdom. This principle has been grafted upon most words, and carries its distinctions throughout the syntax. It is the gender of the language; but a gender of so unbounded a scope, as to merge in it the common distinctions of a masculine and feminine, and to give a twofold character to the parts of speech. The concords which it requires, and the double inflections it provides, will be mentioned in their appropriate places. It will be sufficient here to observe, that animate nouns require animate verbs for their nominatives, animate adjectives to express their qualities, and animate demonstrative pronouns to mark the distinctions of person. Thus, if we say, "I see a man; I see a house," the termination of the verb must be changed. What was in the first instance *wâb imâ*, is altered

to *wâb indân.* *Wâb,* is here the infinitive, but the root of this verb is still more remote. If the question occur, "Is it a good man, or a good house," the adjective, which, in the inanimate form is *onishish-i,* is, in the animate *onishish-in'.* If the question be put, "Is it this man, or this house," the pronoun *this,* which is *mâ bum,* in the animate, is changed to *mâ ndun,* in the inanimate.

Nouns animate embrace the tribes of quadrupeds, birds, fishes, insects, reptiles, crustacæ, the sun, and moon, and stars, thunder, and lightning, for these are personified; and whatever either possesses animal life, or is endowed, by the peculiar opinions and superstitions of the Indians, with it. In the vegetable kingdom, their number is comparatively limited, being chiefly confined to trees, and those only while they are referred to, as whole bodies, and to the various species of fruits, and seeds, and esculents. It is at the option of the speaker to employ nouns, either as animates or inanimates: but it is a choice seldom resorted to, except in conformity with stated exceptions. These conventional exceptions are not numerous, and the more prominent of them, may be recited. The cause of the exceptions it is not always easy to perceive. It may, however, generally be traced to a particular respect paid to certain inanimate bodies, either from their real or fancied properties—the uses to which they are applied, or the ceremonies to which they are dedicated. A stone, which is the altar of sacrifice to their Manitoes; a bow, formerly so necessary in the chase; a feather, the honored sign of martial prowess; a kettle, so valuable in the household; a pipe, by which friendships are sealed and treaties ratified; a drum, used in their sacred and festive dances; a medal, the mask of authority; vermilion, the appropriate paint of the warrior; wampum, by which messages are conveyed, and covenants remembered. These are among the objects, in themselves inanimates, which require the application of animate verbs, pronouns, and adjectives, and are thereby transferred to the animate class.

It is to be remarked, however, that the names for animals, are only employed as animates, while the objects are referred to as whole and complete species. But the gender must be changed, when it becomes necessary to speak of separate numbers. Man, woman, father, mother, are separate nouns, so long as the individuals are meant; but hand, foot, head, eye, ear, tongue, are inanimates. Buck, is an animate noun, while his entire carcass is referred to, whether living or dead; but neck, back, heart, windpipe, take the inanimate form. In like manner, eagle, swan, dove, are distinguished as animates; but beak, wing, tail, are arranged with inanimates. So oak, pine, ash, are animate; branch, leaf, root, inanimates.

Reciprocal exceptions, however, exist to this rule—the reasons for which, as in the former instance, may generally be sought, either in peculiar opinions of the Indians, or in the peculiar qualities or uses of the objects. Thus the talons of the eagle, and the claws of the bear, and of other animals, which furnish ornaments for the neck, are invariably spoken of, under the animate form. The hoofs and horns of all quadrupeds, which are applied to various economical and mystic purposes; the castorum of the beaver, and the nails of man, are similarly situated. The vegetable creation also furnishes some exceptions of this nature; such are the names for the outer bark of all trees (except the birch), and the branches, the roots, and the resin of the spruce, and its congeners.

In a language, which considers all nature as separated into two classes of bodies, characterized by the presence or absence of life; neuter nouns will scarcely be looked for, although such may exist without my knowledge. Neuters are found amongst the verbs and the adjectives, but it is doubtful whether they render the nouns to which they are applied neuters, in the sense we attach to that term. The subject in all its bearings is interesting, and a full and minute description of it would probably elicit new light respecting some doubtful points in the language, and contribute something towards a curious collateral topic—the history of Indian opinions. I have stated the principle broadly, without filling up the subject of exceptions as fully as it is in my power, and without following its bearings upon points which will more properly come under discussion at other stages of the inquiry. A sufficient outline, it is believed, has been given, and having thus met, at the threshold, a principle deeply laid at the foundation of the language, and one which will be perpetually recurring, I shall proceed to enumerate some other prominent features of the substantive.

2. No language is perhaps so defective, as to be totally without

number. But there are, probably, few which furnish so many
modes of indicating it, as the Odjibwai. There are as many modes
of forming the plural, as there are vowel sounds, yet there is no
distinction between a limited and unlimited plural; although there
is, in the pronoun, an *inclusive* and an *exclusive* plural. Whether
we say *man* or *men*, *two men* or *twenty men*, the singular *inin'i*, and
the plural *inin'iwug*, remains the same. But if we say *we*, or *us*,
or *our men* (who are present), or *we*, or *us*, or *our Indians* (in general), the plural *we*, and *us*, and *our*—for they are rendered by the
same form—admit of a change to indicate whether the objective
person be *included* or *excluded*. This principle, of which full examples will be given under the appropriate head, forms a single
and anomalous instance of the use of particular plurals. And it
carries its distinctions, by means of the pronouns, separable and
inseparable, into the verbs and substantives, creating the necessity
of double conjugations and double declensions, in the plural forms
of the first person. Thus, the term for "Our Father," which, in
the inclusive form is *Kôsinân*, is, in the exclusive, *Nôsinân*.

The particular plural, which is thus, by the transforming power
of the language, carried from the pronoun into the texture of the
verb and substantive, is not limited to any fixed number of persons or objects, but arises from the operations of the verb. The
general plural is variously made. But the plurals making inflections take upon themselves an additional power or sign, by which
substantives are distinguished into animate and inanimate. Without this additional power, all nouns plural would end in the
vowels *a, e, i, o, u*. But to mark the gender, the letter *g* is added
to animates, and the letter *n* to inanimates, making the plurals
of the first class terminate in *âg, eeg, ig, ôg, ug*, and of the second
class in *ân, een, in, ôn, un*. Ten modes of forming the plural are
thus provided, five of which are animate, and five inanimate
plurals. A strong and clear line of distinction is thus drawn between the two classes of words; so unerring, indeed, in its application, that it is only necessary to inquire how the plural is formed,
to determine whether it belonged to one or the other class. The
distinctions which we have endeavored to convey will, perhaps,
be more clearly perceived, by adding examples of the use of each
of the plurals.

Animate Plural.

a. Odjibwâi,	a Chippewa.	Odjibwaig,	Chippewas.
e. Ojee,	a Fly.	Oj-eeg,	Flies.
i. Kosénan,	Our father, (in.)	Kosenân-ig,	Our fathers. (in.)
o. Ahmô,	a Bee.	Ahm-ôg,	Bees.
u. Ais,	a Schell.	Ais-ug,	Shells.

Inanimate Plural.

a. Ishkôdai,	Fire.	Ishkôdain,	Fires.
e. Wadôp,	Alder.	Wadôp-een,	Alders.
i. Adetaig,	Fruit.	Adetaig-in,	Fruits.
o. Nôdin,	Wind.	Nôdin-ôn,	Winds.
u. Meen,	Berry.	Meen-un,	Berries.

Where a noun terminates with a vowel in the singular, the addition of the *g*, or *n*, shows at once, both the plural and the gender. In other instances, as in *peenai*, a partridge—*seebi*, a river —it requires a consonant to precede the plural vowel, in conformity with a rule previously stated. Thus, *peenai*, is rendered *peenai-wug*—and *seebi*, *seebi-wun*. Where the noun singular terminates in the broad, instead of the long sound of, *a*, as in *ôgimâ*, a chief, *ishpatinâ*, a hill, the plural is *ogim-ag*, *ishpatinân*. But these are mere modifications of two of the above forms, and are by no means entitled to be considered as additional plurals.

Comparatively few substances are without number. The following may be enumerated:—

Missun',	Firewood.	Ussáimâ,	Tobacco.
Pinggwi,	Ashes.	Naigow,	Sand.
Méjim,	Food.	Ahwun,	Mist.
Kôn,	Snow.	Kimmiwun,	Rain.
Mishk'wi,	Blood.	Ossâkumig,	Moss.
Ukkukkuzhas,	Coals.	Unitshimin,	Peas.

Others may be found, and indeed, a few others are known. But it is less an object, in this lecture, to pursue exceptions into their minutest ramifications, than to sketch broad rules, applicable, if not to every word, to at least a majority of words in the language.

There is, however, one exception from the general use of number, so peculiar in itself, that not to point it out would be an unpardonable remissness in giving the outlines of a language, in which it is an object neither to extenuate faults nor to overrate

beauties. This exception consists in the want of number in the
third person of the declensions of animate nouns, and the conjugation of animate verbs. Not that such words are destitute of
number, in their simple forms, or when used under circumstances
requiring no change of these simple forms—no prefixes and no
inflections. But it will be seen, at a glance, how very limited
such an application of words must be, in a transpositive language.

Thus *mang* and *kâg* (loon and porcupine) take the plural inflection *wug*, becoming *mang wug* and *kag wug* (loons and porcu-
pines). So, in their pronominal declension :—

My loon	Ni mang	oom	
Thy loon	Ki mang	oom	
My porcupine	Ni gâg	oom	
Thy porcupine	Ki gag	oom	
My loons	Ni mang	oom	ug
Thy loons	Ki mang	oom	ug
My porcupines	Ni gâg	oom	ug
Thy porcupines	Ki gâg	oom	ug

But his loon, or loons (*o mang oom un*), his porcupine or por-
cupines (*o gâg oom un*), are without number. The rule applies
equally to the class of words in which the pronouns are inseparable. Thus, my father and thy father, *nôs* and *kôs*, become my
fathers and thy fathers, by the numerical inflection *ug*, forming
nôsug and *kôsug*. But *ôsun*, his father or fathers, is vague, and
does not indicate whether there be one father or twenty fathers.
The inflection *un*, merely denotes the *object*. The rule also applies
equally to sentences in which the noun is governed by or governs
the verb. Whether we say, "I saw a bear," *ningi wâbumâ muk-
wah*, or "a bear saw me," *mukwah ningi wâbumig*, the noun, itself,
undergoes no change, and its number is definite. But *ogi wâbumân muk-wun*, "he saw bear," is indefinite, although both the verb
and the noun have changed their endings. And if the narrator
does not subsequently determine the number, the hearer is either
left in doubt, or must resolve it by a question. In fine, the whole
acts of the third person are thus rendered questionable. This
want of precision, which would seem to be fraught with so much
confusion, appears to be obviated in practice, by the employment
of adjectives, by numerical inflections in the relative words of
the sentence, by the use of the indefinite article, *paizhik*, or by
demonstrative pronouns. Thus, *paizhik mukwun ogi wâbumân*,

conveys with certainty the information "he saw *a* bear." But in this sentence both the noun and the verb retain the objective inflections, as in the former instances. These inflections are not uniformly *un*, but sometimes *een*, as in *ogeen*, his mother, and sometimes *ôn*, as in *odakeek-ôn*, his kettle, in all which instances, however, the number is left indeterminate. It may hence be observed, and it is a remark which we shall presently have occasion to corroborate, that the plural inflection to inanimate nouns (which have no objective form), forms the objective inflection to animate nouns, which have no number in the third person.

8. This leads us to the consideration of the mode of forming possessives, the existence of which, when it shall have been indicated by full examples, will present to the mind of the inquirer, one of those tautologies in grammatical forms, which, without imparting additional precision, serve to clothe the language with accumulated verbiage. The strong tendency to combination and amalgamation, existing in the language, renders it difficult, in fact, to discuss the principles of it in that elementary form which could be wished. In the analysis of words and forms we are constantly led from the central point of discussion. To recur, however, from these collateral unravellings to the main thread of inquiry, at as short and frequent intervals as possible, and thus to preserve the chain of conclusions and proofs, is so important, that, without keeping the object distinctly in view, I should despair of conveying any clear impressions of those grammatical features which impart to the language its peculiar character.

It has been remarked that the distinctions of number are founded upon a modification of the five vowel sounds. Possessives are likewise founded upon the basis of the vowel sounds. There are five declensions of the noun to mark the possessive, ending in the possessive in *âm, eem, im, ôm, um, oom*. Where the nominative ends with a vowel, the possessive is made by adding the letter *m*, as in *maimai*, a woodcock, *ni maimaim*, my woodcock, &c. Where the nominative ends in a consonant, as in *ais*, a shell, the full possessive inflection is required, making *nin daisim*, my shell. In the latter form, the consonant *d* is interposed between the pronoun and noun, and sounded with the noun, in conformity with a general rule. Where the nominative ends in the broad in lieu of the long sound of *a*, as in *ogimâ*, a chief, the possessive is *âm*. The sound of *i*, in the third declension, is that of *i* in pin, and the sound of *u*, in the fifth declension, is that of *u* in bull. The latter will be uniformly represented by *oo*.

The possessive declensions run throughout both the animate and inanimate classes of nouns, with some exceptions in the latter, as knife, bowl, paddle, &c.

Inanimate nouns are thus declined.

Nominative.

Ishkôdai, Fire.

Possessive.

My,	Nin	Dishkod-aim.
Thy,	Ki	Dishkod-aim.
His,	O	Dishkod-aim.
Our,	Ki	Dishkod-aim-inân. (in.)
—	Ni	Dishkod-aim-inân. (ex.)
Your,	Ki	Dishkod-aim-iwâ.
Their,	O	Dishko-aim-iwâ.

Those words which form exceptions from this declension, take the separable pronouns before them as follows:—

Môkoman,	A Knife.
Ni môkoman,	My Knife.
Ki môkoman,	Thy Knife.
O môkoman,	His Knife, &c.

Animate substantives are declined precisely in the same manner as inanimate, except in the third person, which takes to the possessive inflections, *aim, eem, im, ôm, oom*, the objective particle *un*, denoting the compound inflection of this person, both in the singular and plural, *aimun, eemun, imun, ômun, oomun*, and the variation of the first vowel sound, *âmun*. Thus, to furnish an example of the second declension, *pizhiki*, a bison, changes its forms to *nim, bizhik-im*, my bison—*ke bizhik-im*, thy bison, *O bizhik-imun*, his bison, or bisons.

The cause of this double inflection in the third person, may be left for future inquiry. But we may add further examples in aid of it. We cannot simply say, "The chief has killed a bear," or, to reverse the object upon which the energy of the verb is exerted, "The bear has killed a chief." But, *ogimâ ogi nissân mukwun*, literally, "Chief he has has killed him bear," or, *mukwah ogi* *nissân ogimâ*n, "Bear he has killed him chief." Here the verb and the noun are both objective in *un*, which is sounded *ân*, where it comes after the broad sound of *a*, as in *missân*, objective of the verb to kill. If we confer the powers of the English possessive ('*s*), upon the inflections *aim*, *eem*, *im*, *ôm*, *oom*, and *âm*, respectively, and the meaning of *him*, and of course *he*, *her*, *his*, *hers*, *they*, *theirs* (as there is no declension of the pronoun, and no number to the third person), upon the objective particle *un*, we shall then translate the above expression, *o bizhik-eemum*, his bison's hisn. If we reject this meaning, as I think we should, the sentence would read, "His bison," him, a mere tautology.

It is true, it may be remarked, that the noun possessed, has a corresponding termination, or pronominal correspondence, with the pronoun possessor, also a final termination indicative of its being the *object* on which the verb exerts its influence—a mode of expression, which, so far as relates to the possessive, would be deemed superfluous, in modern languages; but may have some analogy in the Latin accusatives *am*, *um*, *em*.

It is a constant and unremitting aim in the Indian languages to distinguish the actor from the object, partly by prefixes, and partly by inseparable suffixes. That the termination *un*, is one of these inseparable particles, and that its office, while it confounds the number, is to designate the object, appears probable from the fact, that it retains its connection with the noun, whether the latter follow or precede the verb, or whatever its position in the sentence may be.

Thus we can, without any perplexity in the meaning say, *Waimittigôzhiwug ogi sagiân Pontiac-un*, "Frenchmen, they did love Pontiac him. Or to reverse it, *Pontiac-un Waimittigôzhiwug ogi sagiân*, " Pontiac, he did Frenchmen he loved." The termination *un*, in both instances, clearly determines the object beloved. So in the following instance, *Sagunoshug ogi sagiân Tecumseh-un*, "Englishmen, they did love Tecumseh," or *Tecumseh-un Sagunoshug oji sagiân*, "Tecumseh, he did Englishmen he loved."

In tracing the operation of this rule, through the doublings of the language, it is necessary to distinguish every modification of sound, whether it is accompanied or not accompanied by a modification of the sense. The particle *un*, which thus marks *the third person and persons*, is sometimes pronounced *wun*, and sometimes *yun*, as the harmony of the word to which it is suffixed may require. But not the slightest change is thereby made in its meaning.

Wâbojeeg ogi meegân-ân nâdowaisi-wun.
Wâbojeeg fought his enemies. L.* W. he did fight them, his enemy, or enemies.
O sâgi-ân inini-wun.
He, or she, loves a man. L. He, or she, loves him-man, or men.
Kigo-yun waindji pimmâdizziwâd.
They subsist on fish. L. Fish or fishes, they upon them, they live.
Ontwa o sagiân odi-yun.
Ontwa loves his dog. L. O. he loves him, his dog, or dogs.

In these sentences, the letters *w* and *y* are introduced before the inflection *un*, merely for euphony's sake, and to enable the speaker to utter the final vowel of the substantive, and the inflective vowel, without placing both under the accent. It is to be remarked in these examples, that the verb has a corresponding inflection with the noun, indicated by the final consonant *n*, as in *sagiâ-n*, objective of the verb *to love*. This is merely a modification of *un*, where it is requisite to employ it after broad *a* (*aw*), and it is applicable to nouns as well as verbs whenever they end in that sound. Thus, in the phrase, "He saw a chief," *O wâbumâ-n O gimâ-n*, both noun and verb terminate in *n*. It is immaterial to the sense, which precedes. And this leads to the conclusion, which we are in some measure compelled to state in anticipation of our remarks on the verb: That verbs must not only agree with their nominatives in number, person, and *gender* (we use the latter term for want of a more appropriate one), but also with their objectives. Hence, the objective sign *n* in the above examples. Sometimes this sign is removed from the ending of the verb, to make room for the plural of the nominative person, and is subjoined to the latter. Thus,

O sagiâ(wâ)n.
They love them (him or them).

In this phrase, the interposed syllable (*wâ*) is, apparently, the plural—it is a reflective plural—of *he*—the latter being indicated, as usual, by the sign *O*. It has been observed, above, that the deficiency in number, in the third person, is sometimes supplied "by numerical inflections in the relative words of the sentence," and this interposed particle (*wâ*) affords an instance in point.

* L. for *literally*.

The number of the nominative pronoun appears to be thus rendered precise, but the objective is still indefinite.

When two nouns are used without a verb in the sentence, or when two nouns compose the whole matter uttered, being in the third person, both have the full objective inflection. Thus,

> Os-(un).   Odi-(yun).
> His father's dog.   L. His father—his dog or dogs.

There are certain words, however, which will not admit the objective *un*, either in its simple or modified forms. These are rendered objective in *een*, or *ôn*.

> O wâbumâ-(n), ossin-(een).
> He sees the stone.   L. He sees him—stone or stones.
> O wâbumâ-(n) mittig o mish-(een).   L. He sees him, tree or trees.
> He sees an oak tree.
> O mittig wâb (een), gyai o bikwuk-(ôn).
> His bow and his arrows.   L. His bow him, and his arrows, him or them.
> Odyâ | wâ | wâ (n), akkik-(ôn).
> They possess a kettle.   L. They own them, kettle or kettles.

The syllable *wâ*, in the verb of the last example included between bars (instead of parentheses), is the reflective plural *they* pointed out in a preceding instance.

I shall conclude these remarks, with full examples of each pronominal declension.

a. First declension, forming the first and second persons in *aim*, and the third in *aimun.*

Nominative.

> Pinâi, a partridge.
> Pinâi-wug, partridges.

First and second person.

> My,   Nim Bin-aim.
> Thy,   Ki Bin-aim.
> Our,   Ki Bin-aim inân.   Inclusive plural.
> Our,   Ni Bin-aiminûn.   Exclusive plural.
> Your,   Ki Bin-aim wâ.

Third person.

> His,   O Bin-aim (un).
> Their, O Bin-aim iwâ (n).

e. Second declension forming the first and second persons in *eem*, and the third in *eemun.*

Nominative.

Ossin, a stone.
Ossineen, stones.

First and second persons.

My,   Nin Dossin-eem.
Thy,   Ki Dossin-eem.
Our,   Ki Dossin-eeminân. (in.)
Our,   Ni Dossin-eeminân. (ex.)
Your,   Ke Dossin-eemewâ.

Third person.

His,   O Dossin-eem(un).
Their,   O Dossin-eemewâ (n).

i. Third declension forming the first and second persons in *im*, and the third in *imun.*

Nominative.

Ais, a shell.
Aisug, shells.

First and second persons.

My,   Nin Dais-im.
Thy,   Ki Dais-im.
Our,   Ki Dais-iminân. (in.)
Our,   Ni Dais-iminân. (ex.)
Your,   Ki Dais-imiwâ.

Third person.

His,   O Dais-im (un).
Their,   O Dais-imewâ (n).

o. Fourth declension forming the first and second persons in *ôm*, and the third in *ômun.*

Nominative.

Monidô, a Spirit.
Monidôg, Spirits.

First and second persons.

My, Ni Monid-ôm.
Thy, Ki Monid-ôm.
Our, Ki Monid-ôminân. (in.)
Our, Ni Monid-ôminân. (ex.)
Your, Ki Monid-ômiwâ.

Third person.

His, O monid-ôm (un).
Their, O Monid-ômewâ (n).

u. (*oo*) Fifth declension forming the first and second persons in *oom*, and the third in *oomun.*

Nominative.

Môz, a Moose.
Môzôg, Moose.

First and second persons.

My, Ni Môz-oom.
Thy, Ki Môz-oom.
Our, Ki Môz-oominân. (in.)
Our, Ni Môz-oominân. (ex.)
Your, Ki Môz-oomiwu.

Third person.

His, O Môz oom (un).
Their, O Môz oomiwa (n).

aw. Additional declension, required when the noun ends in the broad, instead of the long sound of a, forming the possessive in *âm*, and the objective in *âmun.*

Nominative.

Ogimâ, a Chief.
Ogimâg, Chiefs.

First and second persons.

My, Ni Dôgim âm.
Thy, Ki Dôgim âm.
Our, Ki Dôgim âminân. (in.)
Our, Ni Dôgim âminân. (ex.)
Your, Ki Dôgim âmiwâ.

S. plural.

Aindâ-yân-in.	My homes.
Aindâ-yun-in.	Thy homes.
Aindâ-jin.	His homes.
Aindâ-yâng-in.	Our homes. (ex.)
Aindâ-yung-in.	Our homes. (in.)
Aindâ-yaig-in.	Your homes.
Aindâ-wâdjin.	Their homes.

By these examples, it is perceived that the final *d* in *aindâd* is not essential to its primitive meaning; and that the place of the pronoun is, in respect to this word, invariably a suffix. *Aindâd* means, truly, not home, but his home. The plural is formed by the inflection *in*, except in the third person, where the sound of *d* sinks in *j*.

INQUIRY 2.

Further remarks on the substantive—Local, diminutive, derogative, and tensal inflections—Mode in which the latter are employed to denote the disease of individuals, and to indicate the past and future seasons—Restricted or sexual terms—Conversion of the substantive into a verb, and the reciprocal character of the verb by which it is converted into a substantive—Derivative and compound substantives—Summary of the properties of this part of speech.

IN the view which has been taken of the substantive in the preceding Inquiry, it has been deemed proper to exclude several topics, which, from their peculiarities, it was believed could be more satisfactorily discussed in a separate form. Of this character are those modifications of the substantive by which locality, diminution, a defective quality, and the past tense are expressed; by which various adjective and adverbial significations are given; and, finally, the substantives themselves converted into verbs. Such are also the mode of indicating the masculine and feminine (both merged, as we have shown, in the animate class), and those words which are of a strictly *sexual* character, or are restricted in their *use* either to males or females. Not less interesting is the manner of forming derivatives, and of conferring upon the derivatives so formed a *personality*, distinguished as either animate or inanimate, at the option of the speaker.

Much of the flexibility of the substantive is derived from these properties, and they undoubtedly add much to the figurative character of the language. Some of them have been thought analogous to case, particularly that inflection of the noun which indicates the locality of the object. But if so, then there would be equally strong reasons for establishing an *adjective*, and an *adverbial*, as well as a *local* case, and a plurality of forms in each. But it is believed that no such necessity exists. There is no regular declension of these forms, and they are all used under limitations and restrictions incompatible with the true principles of case.

It is under this view of the subject, that the discussion of these forms has been transferred, together with the other accidents of the substantive just adverted to, and reserved as the subject-matter of a separate inquiry. And in now proceeding to express the conclusions at which we have arrived touching these points, it will be an object so to compress and arrange the materials before us, as to present within a small compass the leading facts and examples upon which each separate position depends.

1. That quality of the noun which, in the shape of an inflection, denotes the relative situation of the object, by the contiguous position of some accessory object, is expressed in the English language by the prepositions *in*, *into*, *at*, or *on*. In the Indian, they are denoted by an inflection. Thus, the phrase "In the box," is rendered in the Indian by one word, *mukukoong*. Of this word, *mukuk*, simply, is box. The termination, *oong*, denoting the locality, not of the box, but of the object sought after. The expression appears to be precise, although there is no definite article in the language.

The substantive takes this form, most commonly, after a question has been put, as *Anindi ni môkoman-ais?* "Where is my pen-knife?" *Mukukoong* (in the box), *addôpowin-ing* (on the table), are definite replies to this question. But the form is not restricted to this relation. *Chimân-ing n'guh pôz*, "I shall embark in the canoe;" *wakyigum n'yhu izhâ*, "I shall go into the house," are perfectly correct, though somewhat formal expressions, when the canoe or the house are present to the speaker's view.

The meaning of these inflections has been restricted to *in*, *into*, *at*, and *on*, but they are the more appropriate forms of expressing the first three senses, there being other modes besides these of expressing the preposition *on*. These modes consist in the use of prepositions, and will be explained under that head. The choice of the one or the other is, however, with the speaker. Generally, the inflection is employed when there is some circumstance or condition of the noun either concealed or not fully apparent. Thus, *Muzzinyigun-ing*, is the appropriate term for "In the book," and *may* also be used to signify "On the book." But if it is meant only to signify *on* the book, something visible being referred to, the preposition *ogidj* would be used, that word indicating with certainty *on*, and never *in*. *Wakyigun-ing* indicates with clearness "In the house;" but if it is necessary to say "On the house," and it be meant at the same time to exclude any reference to the interior, the expression would be changed to *ogidj wakyegun*.

It will be proper further to remark in this place, in the way of limitation, that there is also a separate preposition signifying *in*. It is *pinj*. But the use of this word does not, in all cases, supersede the necessity of inflecting the noun. Thus, the expression *pindigain*, is literally walk in, or enter. But if it is intended to say, "Walk in the house," the local, and not the simple form of house must be used; and the expression is, *Pindigain waky'igun-ing*, "Enter in the house," the verbal form which this preposition *pinj* puts on, having no allusion to the act of *walking*, but merely implying position.

The local inflection, which, in the above examples, is *ing* and *oong*, is further changed to *aing* and *eeng*, as the ear may direct—changes which are governed chiefly by the terminal vowel of the noun. Examples will best supply the rule, as well as the exceptions to it.

SIMPLE FORM.		LOCAL FORM.	
		a. First inflection in *aing*.	
Ishkodai	Fire	Ishkod-aing	In, &c. the fire.
Muskodai	Prairie	Muskod-aing	In, &c. the prairie.
Mukkuddai	Powder	Mukkud-aing	In, &c. the powder.
Pimmedai	Grease	Pimmid-aing	In, &c. the grease.

SIMPLE FORM.			LOCAL FORM.

e. Second inflection in *eeng.**

Seebi	River	Seeb-eeng	In, &c. the river.
Neebi	Water	Neeb-eeng	In, &c. the water.
Miskwi	Blood	Miskw-eeng	In, &c. the blood.
Unneeb	Elm	Unneeb-eeng	In, &c the elm.

i. Third inflection in *ing.*

Kôn	Snow	Kôn-ing	In, &c. the snow.
Min	Berry	Meen-ing	In, &c. the berry.
Chimân	Canoe	Chimân-ing	In, &c. the canoe.
Muzziny'egun	Book	Muzziny'egun-ing	In, &c. the book.

o. Fourth inflection in *oong.*

Azhibik	Rock	Azhibik-oong	In, &c. the rock.
Gizhig	Sky	Gizhig-oong	In, &c. the sky.
Kimmiwun	Rain	Kimmiwun-oong	In, &c. the rain.
Akkik	Kettle	Akkik-oong	In, &c. the kettle.

Throw it in the fire.
1. Puggidôn ishkod-aing.
  Go into the prairie.
2. Muskôdaing izhân.
  He is in the elm.
3. Unnib-eeng iâ.
  It is on the water.
4. Nib-eeng attai.
  Put it on the table.
5. Addôpôwin-ing attôn.
  Look in the book.
6. Enâbin muzziny'iguh-ing.
  You stand in the rain.
7. Kimmiwun-oong ki nibow.
  What have you in that box?
8. Waigonain aitsig mukuk-oong?
  Put it in the kettle.
9. Akkik-oong attôn, or Pôdawain.

My bow is not in the lodge; neither is it in the canoe, nor on the rock.

10. Kâwin *pindig* iâsi ni mittigwâb; kâwiuh gyai chimân-*ing*; kâwin gyai âzhibik-*oong*.

An attentive inspection of these examples will show that the local form pertains either to such nouns of the animate class as are in their nature inanimate, or at most possessed of vegetable

---

* The double vowel is here employed to indicate the long sound of *i*, as *i* in machine.

life. And here another conclusion presses upon us; that where these local terminations, in all their variety, are added to the names of animated beings, when such names are the nominatives of adjectives or adjective-nouns, these words are converted into terms of qualification, indicating *like, resembling, equal.* Thus, if we wish to say to a boy, "He is like a man," the expression is, *Inin-ing izzhinâgozzi;* or, if to a man, "He is like a bear," *Mukk-oong izzhinâgozzi;* or, to a bear, "He is like a horse, *Pabaizhiko-gâzh-ing izzhinâgozzi.* In all these expressions, the word *izzhi* is combined with the pronominal inflection *â* (or *nâ*) and the animate termination *gozzi.* And the inflection of the nominative is merely an adjective corresponding with *izzhi*—a term indicative of the general qualities of persons or animated beings. Where a comparison is instituted, or a resemblance pointed out, between inanimate instead of animate objects, the inflection *gozzi* is changed to *gwud,* rendering the expression, which was, in the animate form, *izzhinâ*gozzi, in the inanimate form *izzhinâ*gwud.

There is another variation of the local form of the noun, in addition to those above instanced, indicative of locality in a more general sense. It is formed by *ong* or *nong*—frequent terminations in geographical names. Thus, from *Ojibwai,* Chippewa, is formed *Ojibwai*nong, "Place of the Chippewas." From *Wamatti-gozhiwug,* Frenchmen, is formed *Wamittigozhi*nong, "Place of Frenchmen." From *Ishpatinâ,* Hill, *Ishpati*nong, "Place of the hill," &c. The termination *ing,* is also sometimes employed in this more general sense, as in the following names of places:—

Monomonikâning. In the place of wild rice.
Moninggwunikâning. In the place of sparrows.
Ongwashagooshing. In the place of the fallen tree, &c.

2. The diminutive forms of the noun are indicated by *ais, eas, ôs,* and *aus,* as the final vowel of the word may require. Thus, *Ojibwai,* a Chippewa, becomes *Ojibw-ais,* a little Chippewa: *Inin'i,* a man, *inin-ees,* a little man: *Amik,* a beaver, *amik-ôs,* a young beaver: *Ogimâ,* a chief, *ogim-âs,* a little chief, or a chief of little authority. Further examples may be added.

	SIMPLE FORM.	DIMINUTIVE FORM.
	—ais.	
A woman	Eekwâ	Eekwâz-ais.
A partridge	Pinâ	Pin-ais.
A woodcock	Mâimâi	Mâim-ais.
An island	Minnis	Minnis-ais.
A grape	Shômin	Shômin-ais.
A knife	Môkoman	Môkoman-ais.
	—ees.	
A stone	Ossin	Ossin-ees.
A river	Seebi	Seeb-ees.
A pigeon	Omimi	Omim-ees.
A bison	Pizhiki	Pizhik-ees.
A potato	Opin	Opin-ees.
A bird	Pinâisi	Pinâish-ees.
	—ôs.	
A moose	Môz	Môz-ôs.
An otter	Nigik	Nigik-ôs.
A reindeer	Addik	Addik-ôs.
An elk	Mushkôs	Mushkôs-ôs.
A hare	Wâbôs	Wâbôs-ôs.
A box	Mukuk	Mukuk-ôs.
	—aus.	
A bass	Ogâ	Og-âs.
A medal	Shôniâ	Shôni-âs.
A bowl	Onâgun	Onâg-âns.
A bed	Nibâgun	Nibâg-aûns.
A gun	Pâshkizzigun	Pâshkizzig-âns.
A house	Wakyigun	Wakyig-âns.

In the last four examples, the letter *n*, of the diminutive, retains its full sound.

The use of diminutives has a tendency to give conciseness to the language. As far as they can be employed they supersede the use of adjectives, or prevent the repetition of them. And they enable the speaker to give a turn to the expression, which is often very successfully employed in producing ridicule or contempt. When applied to the tribes of animals, or to inorganic objects, their meaning, however, is, very nearly, limited to an inferiority in size or age. Thus, in the above examples, *pizhik-ees*, signifies a calf; *omim-ees*, a young pigeon; and *ossin-ees*, a pebble, &c. But *inin-ees*, and *ogim-âs*, are connected with the idea of mental or conventional as well as bodily inferiority.

1. I saw a little chief, standing upon a small island, with an inferior medal about his neck.

Ogimâs n'gi wâbamâ nibowid minnisainsing onâbikowân shoniâsun.

2. Yamoyden threw at a young pigeon.

Ogi pukkitaiwun omimeesun Yamoyden.

8. A buffalo calf stood in a small stream.

Pizhikees ki nibowi sibeesing.

4. The little man fired at a young moose.

Ininees ogi pâshkizwân môzôsun.

5. Several diminutive-looking bass were lying in a small bowl, upon a small table.

Addôpowinaising attai onâgâns abbiwâd ogâsug.

Some of these sentences afford instances of the use, at the same time, of both the local and diminutive inflections. Thus, the word *minnisainsing*, signifies literally, "in the little island;" *seebees ing*, "in the little stream;" *addôpowinais ing*, "on the small table."

3. The preceding forms are not the only ones by which adjective qualities are conferred upon the substantive. The syllable *ish*, when added to a noun, indicates a bad or dreaded quality, or conveys the idea of imperfection or decay. The sound of this inflection is sometimes changed to *eesh*, *oosh*, or *aush*. Thus, *Chimân*, a canoe, becomes *Chimânish*, a bad canoe; *Ekwai*, a woman, *Ekwaiwish*, a bad woman; *nibi*, water, becomes *nibeesh*, turbid or strong water; *mittig*, a tree, becomes *mittigoosh*, a decayed tree; *akkik*, a kettle, *akkikoosh*, a worn-out kettle. By a further change, *wibid*, a tooth, becomes *wibidâsh*, a decayed or aching tooth, &c. Throughout these changes the final sound of *sh* is retained, so that this sound alone, at the end of a word, is indicative of a faulty quality.

In a language in which the expressions *bad-dog* and *faint-heart* are the superlative terms of reproach, and in which there are few words to indicate the modifications between positively good and positively bad, it must appear evident that adjective inflections of this kind must be convenient, and sometimes necessary modes of expression. They furnish a means of conveying censure and dislike, which, though often mild, is sometimes severe. Thus, if one person has had occasion to refuse the offered hand of another —for it must be borne in mind that the Indians are a hand-shaking people as well as the Europeans—the implacable party has it at his option, in referring to the circumstance, to use the adjective form of hand, not *onindj*, but *oninjeesh*, which would be deemed contemptuous in a high degree. So, also, instead of *odâwai winini*, a trader, or man who sells, the word may be changed to *odâwai wininiwish*, implying a bad or dishonest trader. It is seldom that a more pointed or positive mode of expressing personal disapprobation or dislike is required; for, generally speaking, more is implied by these modes than is actually expressed.

The following examples are drawn from the inorganic as well as organic creation, embracing the two classes of nouns, that the operation of these forms may be fully perceived.

	SIMPLE FORM.	ADJECTIVE FORM.
	—ish.	
A bowl	Onâgun	Onâgun-ish.
A house	Wakyigun	Wakyigun-ish.
A pipe	Opwâgun	Opwâgun-ish.
A boy	Kweewizais	Kweewizais-ish.
A man	Inini	Ininiw-ish.
	—eesh.	
Water	Neebi	Neeb-ish.
A stone	Ossin	Ossin-eesh.
A potato	Opin	Opin-eesh.
A fly	Ojee	Oj-eesh.
A bow	Mittigwâb	Mittigwâb-eesh
	—oosh.	
An otter	Neegik	Neegik-oosh.
A beaver	Ahmik	Ahmik-oosh.
A reindeer	Addik	Addik-oosh.
A kettle	Akkeek	Akkek-oosh.
An axe	Wagâkwut	Wagâkwut-oosh.
	—aush.	
A foot	Ozid	Ozid-âsh.
An arm	Onik	Onik-âsh.
An ear	Otowug	Otowug-âsh.
A hoof	Wunnussid	Wunnussid-âsh.
A rush mat	Appukwa	Appukw-âsh.

These forms cannot be said, strictly, to be without analogy in the English, in which the limited number of words terminating in *ish*, as saltish, blackish, furnish a correspondence in sound with the first adjective form.

It may subserve the purposes of generalization to add, as the result of the foregoing inquiries, that substantives have a diminutive form, made in *ais*, *ees*, *ôs*, or *ûs;* a derogative form, made in *ish*, *eesh*, *oosh*, or *ûsh;* and a local form, made in *aing*, *eeng*, *ing*, or *oong*. By a principle of accretion, the second or third may be added to the first form, and the third to the second.

EXAMPLE.

Serpent, s.	Kinai/bik.		
———— s. diminutive.	————ôns,	implying	Little serpent.
———— s. derogative.	————ish,	"	Bad serpent.
———— s. local.	————ing,	"	In (the) serpent.
———— s. dim. and der.	————ônsish,	"	Little bad serpent.
———— s. dim. and lo.	————ônsing,	"	In (the) little serpent.
———— s. dim. der. and lo.	————ônsishing,	"	In (the) little bad serpent.

4. More attention has, perhaps, been bestowed upon these points than their importance demanded; but, in giving anything like a comprehensive sketch of the substantive, they could not be omitted; and, if mentioned at all, it became necessary to pursue them through their various changes and limitations. Another reason has presented itself. In treating of an unwritten language, of which others are to judge chiefly from examples, it appeared desirable that the positions advanced should be accompanied by the data upon which they respectively rest—at least, by so much of the data employed as to enable philologists to appreciate the justice or detect the fallacy of our conclusions. To the few who take any interest in the subject at all, minuteness will not seem tedious, and the examples will be regarded with deep interest.

As much of our time as we have already devoted to these lesser points of inquiry, it will be necessary, at this place, to point out other inflections and modifications of the substantive, to clear it from obscurities, that we may go into the discussion of the other parts of speech unincumbered.

Of these remaining forms, none is more interesting than that which enables the speaker, by a simple inflection, to denote that the individual named has ceased to exist. This delicate mode of conveying melancholy intelligence, or alluding to the dead, is effected by placing the object in the past tense.

Aiekid-ôpun aieko Garrangula-bun.
So the deceased Garrangula spoke.

The syllable *bun*, in this sentence, added to the noun, and *ôpun* added to the verb, place both in the past tense. And, although the death of the Indian orator is not mentioned, that fact would be invariably inferred.

Names which do not terminate in a vowel sound, require a vowel prefixed to the tensal inflection, rendering it *ôbun* or *ebun*. Inanimate as well as animate nouns take these inflections.

PRESENT.	PAST FORM.
Tecumseh,	Tecumsi-bun.
Tammany,	Tamani-bun.
Skenandoah,	Skenandoa-bun.
Nôs (my father),	Nos-êbun.
Pontiac,	Pontiac-ibun.
Waub Ojeeg,	Waub Ojeeg-ibun.
Tarhe,	Tarhi-bun.
Mittig (a tree),	Mittig-ôbun.
Akkik (a kettle),	Akkik-obun.
Môz (a moose),	Môz-obon.

By prefixing the particle *Tah* to these words, and changing the inflection of the animate nouns to *iwi*, and the inanimates to *iwun*, they are rendered future. Thus, *Tah Pontiac-iwi; Tah Mittig-iwun,* &c.

The names for the seasons only come under the operation of these rules, when the year before the last, or the year after the next, is referred to. The last and the ensuing season are indicated as follows:—

	PRESENT.	LAST.	NEXT.
Spring,	Seegwun,	Seegwun-oong,	Segwung.
Summer,	Neebin,	Neebin-oong,	Neebing.
Autumn,	Tahgwâgi,	Tahgwâg-oong,	Tahgwàgig.
Winter,	Peebôn,	Peebôn-oong,	Peebông.

I spent last winter in hunting.
Ning'i nunda-wainjigai peebônoong.
I shall go to Detroit next spring.
Ninjah izhâ Wâwiâ'tunong seegwung.

5. *Sexual Nouns.*—The mode of indicating the masculine and feminine having been omitted in the preceding Inquiry, as not being essential to any concordance with the verb or adjective, is, nevertheless, connected with a striking peculiarity of the language—the exclusive use of certain words by one or the other sex. After having appeared to the founders of the language a distinction not necessary to be engrafted in the syntax, there are yet a limited number of words to which the idea of sex so strongly

attaches, that it would be deemed the height of impropriety in a female to use the masculine, and in a male to use the feminine expressions.

Of this nature are the words *Neeji* and *Nindongwai*, both signifying my friend, but the former is appropriated to males and the latter to females. A Chippewa cannot, therefore, say to a female, my friend; nor a Chippewa woman to a male, my friend. Such an interchange of the terms would imply arrogance or indelicacy. Nearly the whole of their interjections—and they are numerous —are also thus exclusively appropriated; and no greater breach of propriety in speech could be committed, than a woman's uttering the masculine exclamation of surprise, *Tyâ!* or a man's descending to the corresponding female interjection, *N'yâ!*

The word *Neenimoshai*, my cousin, on the contrary, can only be applied, like husband and wife, by a male to a female, or a female to a male. If a male wishes to express this relation of a male, the term is *Neetowis;* and the corresponding female term *Neendongwooshai.*

The terms for uncle and aunt are also of a twofold character, though not restricted like the preceding in their use. *Neemishomai*, is my uncle by the father's side; *Neezhishai*, my uncle by the mother's side. *Neezigwoos*, is my paternal aunt; *Neewishai*, my maternal aunt.

There are also exclusive words to designate elder brother and younger brother; but, what would not be expected after the foregoing examples, they are indiscriminately applied to younger brothers and sisters. *Neesgai*, is my elder brother, and *neemissai*, my elder sister. *Neeshemai*, my younger brother or younger sister, and may be applied to any brother or sister except the eldest.

The number of words to which the idea of sex is attached, in the usual acceptation, is limited. The following may be enumerated.

MASCULINE.		FEMININE.	
Inin'i,	A man.	Ekwai',	A woman.
Kwee'wizais,	A boy.	Ekwa'zais,	A girl.
Oskinahwai,	A young man.	Oskineegakwai,	A young woman.
Akiwaixi,	An old man.	Mindimô'ed,	An old woman.
Nôsai,	My father.	Nin Gah,	My mother.
Ningwisis,	My son.	Nin dânis,	My daughter.
Ni ningwun,	My son-in-law.	Nis sim,	My daughter-in-law.

MASCULINE.		FEMININE.	
Ni nûbaim,	My husband.	Nimindimôimish,	My wife.
Nimieshomiss,	My grandfather.	Nôkômiss,	My grandmother.
Ogimâ,	A chief.	Ogemâkwâ,	A chiefess.
Addik,	A reindeer.	Neetshâni,	A doe.
Annimoosh,	A dog.	Kiskisshâi,	A bitch.

The sex of the brute creation is most commonly denoted by prefixing the words *Iâbai*, male, and *Nôzhai*, female.

6. *Reciprocal Changes of the Noun.*—The pronominal particles with which verbs as well as substantives are generally encumbered, and the habit of using them in particular and restricted senses, leave but little occasion for the employment of either the present or past infinitive. Most verbs are transitives. A Chippewa does not say I love, without indicating, by an inflection of the verb, the object beloved: and thus the expression is constantly, I love him, or her, &c. Neither does the infinitive appear to be generally the ultimate form of the verb.

In changing their nouns into verbs, it will not, therefore, be expected that the change should uniformly result in the infinitive, for which there is so little use, but in such of the personal forms of the various moods as circumstances may require. Most commonly, the third person singular of the indicative, and the second person singular of the imperative, are the simplest aspects under which the verb appears; and hence these forms have been sometimes mistaken for, and reported as the present infinitive. There are some instances in which the infinitive is employed. Thus, although an Indian cannot say I love, thou lovest, &c., without employing the objective forms of the verb to love, yet he can say I laugh, I cry, &c.; expressions in which, the action being confined to the speaker himself, there is no transition demanded. And in all similar instances the present infinitive, with the proper pronoun prefixed, is employed.

There are several modes of transforming a substantive into a verb. The following examples will supply the rules, so far as known, which govern these changes:—

	INDICATIVE.	IMPERATIVE.
Chimân, a canoe.	Chimai, he paddles.	Chimain, paddle thou.
Pashkizzigun, a gun.	Pashkizzigai, he fires.	Pashkizzigain, fire thou.
Jeesidyigun, a broom.	Jeesidyigai, he sweeps.	Jeesidyigain, sweep thou.

31

	INDICATIVE.	IMPERATIVE.
Weedjeeagun, a helper.	Weedôkagai, he helps.	Weedjeei-wain, help thou.
Ojibwâi, a Chippewa.	Ojibwâmoo, he speaks Chippewa.	Ojibwâmoon, speak thou Chippewa.

Another class of nouns is converted into the first person, indicative, of a pseudo-declarative verb, in the following manner:—

Monido,	A spirit.	Ne Monidôw,	I (am) a spirit.
Wassaiâ,	Light.	Ne Wassaiâw,	I (am) light.
Ishkodai,	Fire.	Nin Dishkodaiw,	I (am) fire.
Weendigô,	A monster.	Ni Weendigôw,	I (am) a monster.
Addik,	A deer.	Nin Daddikoow,	I (am) a deer.
Wakyigun,	A house.	Ni Wakyiguniw,	I (am) a house.
Pinggwi,	Dust, ashes.	Nim Binggwiw,	I (am) dust, &c.

The word *am*, included in parenthesis, is not in the original, unless we may suppose the terminals *ow*, *aw*, *iw*, *oow*, to be derivatives from *Iaw*. These changes are reciprocated by the verb, which, as often as occasion requires, is made to put on a substantive form. The particle *win*, added to the indicative of the verb, converts it into a substantive. Thus—

Keegido,	He speaks.	Keegidowin,	Speech.
Pâshkizzigai,	He fires.	Pashkizzigaiwin,	Ammunition.
Agindasoo,	He counts.	Agindasoowin,	Numbers.
Wahyiâzhinggai,	He cheats.	Wahyiazhinggaiwin,	Fraud.
Minnikwâi,	He drinks.	Minnikwâiwin,	Drink.
Kubbâshi,	He encamps.	Kubbâishiwin,	An encampment.
Meegâzoo,	He fights.	Meegâzoowin,	A fight.
Ojeengai,	He kisses.	Ojeendiwin,	A kiss.
Annôki,	He works.	Annôkiwta,	Work.
Pâpi,	He laughs.	Pâpiwin,	Laughter.
Pimâdizzi,	He lives.	Pimâdoiziwin,	Life.
Onwâibi,	He rests.	Onwaibiwin,	Rest.
Annamiâ,	He prays.	Annamiâwin,	Prayer.
Nibâ,	He sleeps.	Nibâwin,	Sleep.
Odâwai,	He trades.	Odâwaiwin,	Trade.

Adjectives are likewise thus turned into substantives:—

Keezhaiwâdizzi,	He generous.	Keezhaiwâdizziwin,	Generosity.
Minwaindum,	He happy.	Minwaindumowin,	Happiness.
Keezhaizeâwizzi,	He industrious.	Keezhaizhâwizziwin,	Industry.
Kittimâgizzi,	He poor.	Kittimâgizziwin,	Poverty.
Aukkoossi,	He sick.	Aukkoossiwin,	Sickness.
Kittimishki,	He lazy.	Kittimishkiwin,	Laziness.
Nishkâdizzi,	He angry.	Nishkâdizziwin,	Anger.
Baikâdizzi,	She chaste.	Baikâdizziwin,	Chastity.

In order to place the substantives thus formed in the third person, corresponding with the indicative from which they were changed, it is necessary only to prefix the proper pronoun. Thus, *Ogeezhaiwâdizziwin*, his generosity, &c.

7. *Compound Substantives.*—The preceding examples have been given promiscuously from the various classes of words, primitive and derivative, simple and compound. Some of these words express but a single idea, as, *ôs*, father—*gah*, mother—*môz*, a moose —*kâg*, a porcupine—*mang*, a loon—and appear to be incapable of further division. All such words may be considered as primitives, although some of them may be contractions of dissyllabic words. There are also a number of dissyllables, and *possibly* some trisyllables, which, in the present state of our analytical knowledge of the language, may be deemed both simple and primitive. Such are *neebi*, water; *ossin*, a stone; *geezis*, the sun; *nodin*, wind. But it may be premised, as a principle which our investigations have rendered probable, that all polysyllabic words, all words of three syllables, *so far as examined,* and most words of two syllables, are compounds.

The application of a syntax, formed with a view to facilitate the rapid conveyance of ideas by consolidation, may, it is presumable, have early led to the coalescence of words, by which all the relations of object and action, time and person, were expressed. And in a language which is only spoken, and not written, the primitives would soon become obscured and lost in the multiform appendages of time and person, and the recondite connection of actor and object. And this process of amalgamation would be a progressive one. The terms that sufficed in the condition of the simplest state of nature, or in a given latitude, would vary with their varying habits, institutions, and migrations. The introduction of new objects and new ideas would require the invention of new words, or what is much more probable, existing terms would be modified or compounded to suit the occasion. No one who has paid much attention to the subject, can have escaped noticing a confirmation of this opinion, in the extreme readiness of our western Indians to bestow, on the instant, names, and appropriate names—to any new object presented to them. A readiness not attributable to their having at command a stock of generic polysyllables—for these it would be very awkward to wield—but, as appears more probable, to the powers of the syntax, which permits the resolution of new compounds from existing roots, and often concentrates, as remarked in another place, the entire sense of the parent words, upon a single syllable, and sometimes upon a single letter.

Thus it is evident that the Chippewas possessed names for a living tree, *mittig*, and a string, *aiáb*, before they named the bow *mittigwáb*—the latter being compounded under one of the simplest rules from the two former. It is further manifest that they had named earth *akki*, and (any solid, stony, or metallic mass) *ábik*, before they bestowed an appellation upon the kettle, *akkeek*, or *akkik*, the latter being derivatives from the former. In process of time these compounds became the bases of other compounds, and thus the language became loaded with double, and triple, and quadruple compounds, concrete in their meaning and formal in their utterance.

When the introduction of metals took place, it became necessary to distinguish the clay from the iron pot, and the iron from the copper kettle. The original compound, *akkeek*, retained its first meaning, admitting the adjective noun *piwábik* (itself a compound) iron, when applied to a vessel of that kind, *piwábik akkeek*, iron kettle. But a new combination took place to designate the copper kettle, *miskwákeek*, red metal kettle; and another expression to denote the brass kettle, *ozawábik akkeek*, yellow metal kettle. The former is made up from *miskówábik*, copper (literally *red-metal*—from *miskwá*, red, and *ábik*, the generic above mentioned), and *akkeek*, kettle. *Ozawábik*, brass, is from *ozawá*, yellow, and the generic *ábek*—the term *akkeek* being added in its separate form. It may, however, be used in its connected form of *wukkeek*, making the compound expression *ozawábik wukkeek*.

In naming the horse *paibáizhikógazhi, i. e.* the animal with solid hoofs, they have seized upon the feature which most strikingly distinguished the horse from the cleft-footed animals, which were the only species known to them at the period of the discovery. And the word itself affords an example, at once, both of their powers of concentration, and brief, yet accurate description, which it may be worth while to analyze. *Paizhik* is one, and is also used as the indefinite article—the only article the language possesses. This word is further used in an adjective sense, figuratively indicating, united, solid, undivided. And it acquires a plural signification by doubling, or repeating the first syllable, with a slight variation of the second. Thus, *Pai-baizhik* denotes not *one*, or *an*, but several; and when thus used in the context, renders the noun governed plural. *Oskuzh* is the nail, claw, or horny part of the foot of beasts, and supplies the first substantive member of the compound *gauzh*. The final vowel is from *ahwaisi*, a beast; and the marked *o*, an inseparable connective, the office of which is to make the two members coalesce, and harmonize. The expression thus formed becomes a substantive, specific in its application. It may be rendered plural like the primitive nouns, may be converted into a verb, has its diminutive, derogative, and local form, and, in short, is subject to all the modifications of other substantives.

Most of the modern nouns are of this complex character. And they appear to have been invented to designate objects, many of which were necessarily unknown to the Indians in the primitive ages of their existence. Others, like their names for a copper-kettle and a horse, above mentioned, can date their origin further back than the period of the discovery. Of this number of nascent words, are most of their names for those distilled or artificial liquors, for which they are indebted to Europeans. Their name for water, *neebi*, for the fat of animals, *weenin*, for oil or grease, *pimmidai*, for broth, *nâbôb*, and for blood, *miskwi*, belong to a very remote era, although all but the first appear to be compounds. Their names for the tinctures or extracts derived from the forest, and used as dyes, or medicines, or merely as agreeable drinks, are mostly founded upon the basis of the word *âbo*, a liquid, although this word is never used alone. Thus—

Shomin-âbo,	Wine,	From Shomin, a grape, âbo, a liquor.
Ishkodai-wâbo,	Spirits,	From Ishkôdâi, fire, &c.
Mishimin-âbo,	Cider,	From Mishimin, an apple, &c.
Tôtôsh-âbo,	Milk,	From Tôtôsh, the female breast, &c.
Sheew-âbo,	Vinegar,	From Sheewun, sour, &c.
Annibeesh-âbo,		From Annibeeshun, leaves, &c.
Ozhibiegun-aubo,		From Ozhibiêgai, he writes, &c.

In like manner their names for the various implements and utensils of civilized life, are based upon the word *Jeegun*, one of those primitives, which, although never disjunctively used, denotes, in its modified forms, the various senses implied by our words instrument, contrivance, machine, &c. And by prefixing to this generic a substantive, verb, or adjective, or parts of one or each, an entire new class of words is formed. In these combinations, the vowels *e* and *o* are sometimes used as connectives.

Keeahkeebô-jeegun,	A saw,	From Keeshkeezhun, v. a. to cut.
Seeseebô-jeegun,	A file,	From Seesee, to rub off, &c.
Wassakoonen-jeegun,	A candle,	From Wassakooda, bright, biskoona, flame, &c.
Beeseebô-jeegun,	A coffee-mill,	From Beesâ, fine grains, &c.
Minnikwâd-jeegun,	A drinking-vessel,	From Minnekwâi, he drinks, &c.
Tâshkeebôd-jeegun,	A saw-mill,	From Taushkâ, to split, &c.
Mudwâiabeed-jeegun,	A violin,	From Mudwâwâi, sound, âiâb, a string, &c.

Sometimes this termination is shortened into *gun*, as in the following instances :—

Onâ-gun,	A dish.
Tikkina-gun,	A cradle.
Neeba-gun,	A bed.
Puddukkyi-gun,	A fork.
Puggimmâ-gun,	A war-club.
Opwâ-gun,	A pipe.
Wassâitshie-gun,	A window.
Wakkyi-gun,	A house.
Pôdahwâ-gun,	A fire-place.
Sheema-gun,	A lance.

Another class of derivatives is formed from *wyân*, indicating, generally, an undressed skin. Thus—

Muk-wyân,	A bear skin,	From Mukwah, a bear, and wyaun, a skin.
Wazhusk-wyân,	A muskrat skin,	From Wazhusk, a muskrat, &c.
Wabôs-wyân,	A rabbit skin,	From Wabôs, a rabbit, &c.
Neegik-wyân,	An otter skin,	From Neegih, an otter, &c.
Ojeegi-wyân,	A fisher skin,	From Ojeeg, a fisher, &c.
Wabizhais-ewyân, a martin skin, from wabizhais, a martin, &c.		

Wâbiwyan, a blanket, and *bubbuggiwyan*, a shirt, are also formed from this root. As the termination *wyân*, is chiefly restricted to undressed skins, or peltries, that of *waigin* is, in like manner, generally applied to dressed skins or to cloths. Thus—

Monido-waigin,	Blue cloth, strouds,	From Monido, spirits, &c.
Misk-waigin,	Red cloth,	From Miskwâ, red, &c.
Nondâ-waigin,	Scarlet.	
Peezhiki-waigin,	A buffalo robe,	From Peezhiki, a buffalo, &c.
Addik-waigin,	A cariboo skin,	From Addik, a cariboo, &c.
Ozhauwushk-waigin,	Green cloth,	From Ozhâwushkwâ, green.

An interesting class of substantives is derived from the third person singular of the present indicative of the verb, by changing the vowel sound of the first syllable, and adding the letter *d* to that of the last, making the terminations in *aid, âd, eed, id, ood.* Thus, *Pimmoossâ,* he walks, becomes *pâmmoossâd,* a walker.

aid.

Munnissai,	He chops.	Mânissaid,	A chopper.
Oshibeigai,	He writes.	Wâshibeigaid,	A writer.
Nundowainjeegai,	He hunts.	Nûndowainjeegaid,	A hunter.

âd.

Neebâ,	He sleeps.	Nâbâd,	A sleeper.
Kwâbahwâ,	He fishes (with scoop net).	Kwyâbahwâd,	A fisher (with scoop net).
Puggidowâ,	He fishes (with seine).	Pâgidowâd,	A fisher (with seine).

eed.

Annokee,	He works.	Anokeed,	A worker.
Jeessakea,	He juggles.	Jossakeed,	A juggler.
Munnigobee,	He pulls bark.	Mainigobeed,	A bark puller.

id.

Neemi,	He dances.	Nâmid,	A dancer.
Weesinni,	He eats.	Wâssinid,	An eater.
Pimâdizzi,	He lives.	Paimaudizzid,	A living being.

ood.

Nugamoo,	He sings.	Naigumood,	A singer.
Keegido,	He speaks.	Kâgidood,	A speaker.
Keewonimoo,	He lies.	Kâwunimood,	A liar.

This class of words is rendered plural in *ig*—a termination, which, after *d* final in the singular, has a soft pronunciation, as if written *jig.* Thus, *Nâmid,* a dancer, *nâmidjig,* dancers.

The derogative form is given to these generic substantives by introducing *ish,* or simply *sh,* in place of the *d,* and changing the latter to *kid,* making the terminations in *ai, aishkid,* in *â, âshkid,* in *e, eeshkid,* in *i, ishkid,* and in *oo, ooshkid.* Thus, *naindowainjeeg-aid,* a hunter, is changed to *naindowainjeegaishkid,* a bad or unprofitable hunter. *Naibâd,* a sleeper, is changed to *naibâshkid,* a sluggard. *Jossakeed,* a juggler, to *jossakeeshkid,* a vicious juggler.

Wásinnid, an eater, to *wássinishkid,* a gormandizer. *Kágidood,* a speaker, *kágidooshkid,* a babbler. And in these cases the plural is added to the last educed form, making *kágidooshkidjig,* babblers, &c.

The word *nittá,* on the contrary, prefixed to those expressions, renders them complimentary. For instance, *nittá naigumood,* is a fine singer, *nittá kágidood,* a ready speaker, &c.

Flexible as the substantive has been shown to be, there are other forms of combination that have not been adverted to—forms, by which it is made to coalesce with the verb, the adjective, and the preposition, producing a numerous class of compound expressions. But it is deemed most proper to defer the discussion of these forms to their several appropriate heads.

Enough has been exhibited to demonstrate its prominent grammatical rules. It is not only apparent that the substantive possesses number and gender, but it also undergoes peculiar modifications to express locality and diminution, to denote adjective qualities and to indicate tense. It exhibits some curious traits connected with the mode of denoting the masculine and feminine. It is modified to express person and to distinguish, living from inanimate masses. It is rendered possessive by a peculiar inflection, and provides particles, under the shape either of prefixes or suffixes, separable or inseparable, by which the actor is distinguished from the object—and all this, without changing its proper substantive character, without putting on the aspect of a pseudo adjective, or a pseudo verb. Its changes to produce compounds are, however, its most interesting, its most characteristic trait. Syllable is heaped upon syllable, word upon word, and derivative upon derivative, until its vocabulary is crowded with long and pompous phrases, most formidable to the eye.

So completely transpositive do the words appear, that like chessmen on a board, their elementary syllables can be changed at the will of the player, to form new combinations to meet new contingencies, so long as they are changed in accordance with certain general principles and conventional rules; in the application of which, however, much depends upon the will or the skill of the player. What is most surprising, all these changes and combinations, all these qualifications of the object, and distinctions of the person, the time, and the place, do not supersede the use of adjectives, and pronouns, and verbs, and other parts of speech woven into the texture of the noun, in their elementary and conjunctive forms.

III.

Principles Governing the Use of the Odjibwa Noun-Adjective.

INQUIRY 3.

Observations on the adjective—Its distinction into two classes denoted by the presence or absence of vitality—Examples of the animates and inanimates—Mode of their conversion into substantives—How pronouns are applied to these derivatives, and the manner of forming compound terms from adjective bases to describe the various natural phenomena—The application of these principles in common conversation, and in the description of natural and artificial objects—Adjectives always preserve the distinction of number—Numerals—Arithmetical capacity of the language—The unit exists in duplicate.

1. IT has been remarked that the distinction of words into animates and inanimates, is a principle intimately interwoven throughout the structure of the language. It is, in fact, so deeply imprinted upon its grammatical forms, and is so perpetually recurring, that it may be looked upon, not only as forming a striking peculiarity of the language, but as constituting the fundamental principle of its structure, from which all other rules have derived their limits, and to which they have been made to conform. No class of words appears to have escaped its impress. Whatever concords other laws impose, they all agree, and are made subservient in the establishment of this.

It might appear to be a useless distinction in the adjective, when the substantive is thus marked; but it will be recollected that it is in the plural of the substantive only that the distinction is marked; and we shall presently have occasion to show that redundancy of forms is, to considerable extent, obviated in practice.

For the origin of the principle itself, we need look only to nature, which endows animate bodies with animate properties and qualities, and *vice versâ*. But it is due to the tribes who speak this language, to have invented one set of adjective symbols to express the ideas peculiarly appropriate to the former, and another set applicable exclusively to the latter; and to have given the words good and bad, black and white, great and small, handsome and ugly, such modifications as are practically competent to indicate the general nature of the objects referred to, whether provided with, or destitute of, the vital principle. And not only so, but, by the figurative use of these forms, to exalt inanimate masses into the class of living beings, or to strip the latter of its properties of life—a principle of much importance to their public speakers.

This distinction is shown in the following examples, in which it will be observed that the inflection *izzi* generally denotes the personal, and *au*, *un*, or *wud*, the impersonal forms.

	ADJ. INANIMATE.	ADJ. ANIMATE.
Bad,	Monaudud,	Monaudizzi.
Ugly,	Gushkoonaugwud,	Gushkoonaugoozzi.
Beautiful,	Bishegaindaugwud,	Bisheguindaugoozi.
Strong,	Söngun,	Söngizzi.
Soft,	Nökun,	Nökizzi.
Hard,	Mushkowau,	Mushkowizzi.
Smooth,	Shoiskwau,	Shoiskoozzi.
Black,	Mukkuddäwau,	Mukkuddäwizzi.
White,	Waubishkau,	Waubishkizzi.
Yellow,	Ozahwau,	Ozahwizzi.
Red,	Miskwau,	Miskwizzi.
Blue,	Ozhahwushkwau,	Ozhahwushkwizzi.
Sour,	Sheewun,	Sheewizzi.
Sweet,	Weeshköbun,	Weeshköbizzi.
Light,	Naugun,	Naungizzi.

It is not, however, in all cases, by mere modifications of the adjective that these distinctions are expressed. Words totally different in sound, and evidently derived from radically different roots, are, in some few instances, employed; as in the following examples:—

	ADJ. INANIMATE.	ADJ. ANIMATE.
Good,	Onisheshin,	Minno.
Bad,	Monaudud,	Mudjee.
Large,	Mitshau,	Mindiddo.
Small,	Pungee,	Uggaushe.
Old,	Geekau,	Gitizzi.

It may be remarked of these forms, that, although the impersonal will, in some instances, take the personal inflections, the rule is not reciprocated, and *minno*, and *mindiddo*, and *gilizzi*, and all words similarly situated, remain unchangeably animates. The word *pungee* is limited to the expression of quantity, and its correspondent, *uggaushi*, to size or quality. *Kishedä* (hot) is restricted to the heat of a fire; *keezhautä*, to the heat of the sun. There is still a third term to indicate the natural heat of the body; *kizzizoo*. *Mitshau* (large) is generally applied to countries, lakes, rivers, &c.; *mindiddo*, to the body; and *gitshee*, indiscriminately. *Onishishin*, and its correspondent, *onishishshä*, signify handsome or fair, as well as good. *Kwonaudy*, a. a., and *kwonaudyewun*, a. i., mean, strictly, handsome, and imply nothing further. *Minno* is the appropriate personal form for good. *Mudjee* and *monaudud* may reciprocally change genders, the first by the addition of *iee*, and the second by altering *ud* to *izzi*.

Distinctions of this kind are of considerable importance in a practical point of view, and their observance or neglect is noticed with scrupulous exactness by the Indians. The want of inanimate forms to such words as happy, sorrowful, brave, sick, &c., creates no confusion, as inanimate nouns cannot, strictly speaking, take upon themselves such qualities; and when they do—as they sometimes do—by one of those extravagant figures of speech which are used in their tales of transformations, the animate form answers all purposes; for in these tales the whole material creation may be clothed with animation. The rule, as exhibited in practice, is limited, with sufficiént accuracy, to the boundaries prescribed by nature.

To avoid a repetition of forms, were the noun and the adjective both to be employed in their usual relation, the latter is endowed with a pronominal or substantive inflection; and the use of the noun in its separate form is thus wholly superseded. Thus, *onishishin*, a. i., and *onishishsha*, a. a., become *wänishishing*, "That which is good or fair," and *wänishishid*, "He who is good or fair." The following examples will exhibit this rule under each of its forms:—

COMPOUND OR NOUN-ADJECTIVE ANIMATE.

Black,	Mukkuddawizzi,	Mäkuddäwizzid.
White,	Waubishkizzi,	Wyaubishkizzid.
Yellow,	Ozahwizzi,	Wäzauwizzid.
Red,	Miskwizzi,	Mäskoozzid.
Strong,	Söngizzi,	Swöngizzid.

Black,	Mukkuddāwau,	Mäkuddäwaug.
White,	Waubishkau,	Wyaubishkaug.
Yellow,	Ozahwau,	Wäzhauwaug.
Red,	Miskwau,	Maiskwaug.

The animate forms, in these examples, will be recognized as exhibiting a further extension of the rule, mentioned in the preceding Inquiry, by which substantives are formed from the indicative of the verb by a permutation of the vowels; and these forms are likewise rendered plural in the manner there mentioned. They also undergo changes to indicate the various persons.   For instance, *onishisha* is thus declined to mark the person :—

Wänishish-eyaun,	I (am) good or fair.
Wänishish-eyun,	Thou (art) good or fair.
Wänishish-id,	He (is) good or fair.
Wänishish-eyaung,	We (are) good or fair. (ex.)
Wänishish-eyung,	We (are) good or fair. (in.)
Wänishish-eyaig,	Ye (are) good or fair.
Wänishish-idjig,	They (are) good or fair.

The inanimate forms, being without person, are simply rendered plural by *in*, changing *maiskwaug* to *maiskwaug-in*, &c. &c. The verbal signification which these forms assume, as indicated in the words am, art, is, are, is to be sought in the permutative change of the first syllable.   Thus, *o* is changed to *wä, muk* to *mäk, waub* to *wy-aub, ozau* to *wäzau, misk* to *maisk*, &c.   The pronoun, as is usual in the double compounds, is formed wholly by the inflections *eyaun, eyun*, &c.

The strong tendency of the adjective to assume a personal or pronomico-substantive form, leads to the employment of many words in a particular or exclusive sense; and, in any future practical attempts with the language, it will be found greatly to facilitate its acquisition, if the adjectives are arranged in distinct classes, separated by this characteristic principle of their application. The examples we have given are chiefly those which may be considered strictly animate or inanimate, admit of double forms, and are of general use.   Many of the examples recorded in the original manuscripts employed in these inquiries, are of a more concrete character, and, at the same time, a more limited use. Thus, *shaugwewe* is a weak person; *nŏkaugumme*, a weak drink;

nŏkaugwud, a weak or soft piece of wood. *Sussägau* is fine, but can only be applied to personal appearance; *beesau,* indicates fine grains. *Keewushkwä* is giddy, and *keewushkwäbee,* giddy with drink—both being restricted to the third person. *Söngun* and *songizzi* are the personal and impersonal forms of strong, as given above, but *mushkowaugumme* is strong drink. In like manner, the two words for hard, as above, are restricted to solid substances. *Sunnuhgud* is hard (to endure). *Waindud* is easy (to perform). *Söngodää* is brave; *shaugedää,* cowardly; *keezhinghowizzi,* active; *kizheekau,* swift; *onaunegoozzi,* lively; *minwaindum,* happy; *gushkaindum,* sorrowful; but all these forms are confined to the third person of the indicative, singular. *Pibbigwun* is a rough or knotted substance; *pubbiggozzi,* a rough person. *Keenwau* is long or tall (any solid mass). *Kaynozid* is a tall person. *Tahkozid* a short person. *Wassayau* is light; *wassaubizzoo,* the light of the eye; *wasshauzhä,* the light of a star or any luminous body. *Keenau* is sharp; *keenaubikud,* a sharp knife or stone. *Keezhaubikeday* is hot metal, a hot stove, &c. *Keezhaugummeday* is hot water. *Uubudgeetön* is useful, a useful thing. *Wauweeug* is frivolous, anything frivolous in word or deed. *Tubbushish* appears to be a general term for low. *Ishpimming* is high in the air. *Ishpau* is applied to any high fixture, as a house, &c. *Ishpaubikau* is a high rock. *Taushkaubikau,* a split rock.

These combinations and limitations meet the inquirer at every step; they are the current phrases of the language; they present short, ready, and often beautiful modes of expression; and, as they shed light both upon the idiom and genius of the language, I shall not scruple to add further examples and illustrations. Ask a Chippewa the name for a rock, and he will answer *awzhebik.* The generic import of *awbik* has been explained. Ask him the name for red rock, and he will answer *miskwaubik;* for white rock, and he will answer *waubaubik;* for black rock, *mukkuddäwaubik;* for yellow rock, *ozahwaubik;* for green rock, *ozhahwushkwaubik;* for bright rock, *wassayaubik;* for smooth rock, *shoishkwaubik,* &c.—compounds in which the words red, white, black, yellow, &c., unite with *aubik.* Pursue this inquiry, and the following forms will be elicited:—

Impersonal.

Miskwaubik-ud,	It (is) a red rock.
Waububik-ud,	It (is) a white rock.
Mukkudäwaubik-ud,	It (is) a black rock.
Ozahwaubik-ud,	It (is) a yellow rock.
Wassayaubik-ud,	It (is) a bright rock.
Shoiskwaubik-ud,	It (is) a smooth rock.

Personal.

Miskwaubik-izzi,	He (is) a red rock.
Waubaubik-izzi,	He (is) a white rock.
Mukkuddawaubik-izzi,	He (is) a black rock.
Ozahwaubik-izzi,	He (is) a yellow rock.
Wassyaubik-izzi,	He (is) a bright rock.
Shoiskwaubik-izzi,	He (is) a smooth rock.

Add *bun* to these terms, and they are made to have passed away; prefix *tah* to them, and their future appearance is indicated. The word "is" in the translations, although marked with parentheses, is not deemed wholly gratuitous. There is, strictly speaking, an idea of existence given to these compounds, by the particle *au*, in *aubic*, which seems to be indirectly a derivative from that great and fundamental root of the language *Iau*. *Bik* is apparently the radix of the expression for "rock."

Let this mode of interrogation be continued, and extended to other adjectives, or the same adjectives applied to other objects, and results equally regular and numerous will be obtained. *Minnis*, we shall be told is an island; *miskominnis*, a red island; *mukkuddäminnis*, a black island; *waubeminnis*, a white island, &c. *Annokwut*, is a cloud; *miskwaunakwut*, a red cloud; *mukkuddawukwut*, a black cloud; *waubahnokwut*, a white cloud; *ozahwushkwahnakwut*, a blue cloud, &c. *Neebe* is the specific term for water; but is not generally used in combination with the adjective. The word *guma*, like *aubo*, appears to be a generic term for water, or potable liquids. Hence, the following terms:—

Gitshee,	Great.	Gitshiguma,	Great water.
Nokun,	Weak.	Nökauguma,	Weak drink.
Mushkowau,	Strong.	Mushkowauguma,	Strong drink.
Weeshkobun,	Sweet.	Weeshkobauguma,	Sweet drink.
Sheewun,	Sour.	Sheewauguma,	Sour drink.
Weesugun,	Bitter.	Weesugauguma,	Bitter drink.
Minno,	Good.	Minwauguma,	Good drink.
Monaudud,	Bad.	Mahnauguma,	Bad drink.

Miskwau,	Red.	Miskwauguma,	Red drink.
Ozahwau,	Yellow.	Ozahwauguma,	Yellow drink.
Weenun,	Dirty.	Weenauguma,	Dirty water.
Peenud,	Clean.	Peenauguma,	Clean water.

From *minno*, and from *monaudud*, good and bad, are derived the following terms: *Minnopogwud*, it tastes well; *minnopogoozzi*, he tastes well; *mawzhepogwud*, it tastes bad; *mawzhepogoozzi*, he tastes bad. *Minnomaugwud*, it smells good; *minnomaugoozzi*, he smells good; *mauzhemaugud*, it smells bad; *mauzhemaugoozzi*, he smells bad. The inflections *gwud*, and *izzi*, here employed, are clearly indicative, as in other combinations, of the words *it* and *him*.

Baimwa, is sound; *baimwäwa*, the passing sound; *minwäwa*, a pleasant sound; *maunwäwa*, a disagreeable sound; *mudwayaushkau*, the sound of waves dashing on the shore; *mudwayaunnemud*, the sound of winds; *mudwayaukooshkau*, the sound of falling trees; *mudwäkumigishin*, the sound of a person falling upon the earth; *mudwaysin*, the sound of any inanimate mass falling on the earth. These examples might be continued *ad infinitum*. Every modification of circumstances, almost every peculiarity of thought, is expressed by some modification of the orthography. Enough has been given to prove that the adjective combines itself with the substantive, the verb, and the pronoun, that the combinations thus produced are numerous, afford concentrated modes of conveying ideas, and oftentimes, happy turns of expression. Numerous and prevalent as these forms are, they do not, however, preclude the use of adjectives in their simple forms. The use of the one or the other appears to be generally at the option of the speaker. In most cases brevity or euphony dictates the choice. Usage results from these applications of the principles. There may be rules resting upon a broader basis; but if so, they do not appear to be very obvious. Perhaps the simple adjectives are often employed before verbs and nouns, in the first and second persons singular.

Ningee minno neebau-nabun,	I have slept well.
Ningee minno weesin,	I have eaten a good meal.
Ningee minno pimmoossay,	I have walked well, or a good distance.
Kägät minno geezhigud,	It (is) a very pleasant day.
Kwanaudy ningödahs,	I have a handsome garment.
Ke minno iau nuh,	Are you well?
Auneende ain deyun,	What ails you?

Keeshamonedo aupädush shäwainamik,	God prosper you.
Aupädush shäwaindaugoozzeyun,	Good luck attend you.
Aupädush nau kinwainzh pimmaudizziyun.	May you live long.
Onauneegoozzin,	Be (thou) cheerful.
Ne minwaindum waubumenaun,	I (am) glad to see you. ●
Kwanaudj kweeweezains,	A pretty boy.
Kägät söngeedää,	He (is) a brave man.
Kägät onishishsha,	She (is) handsome.
Gitshee kinözee,	He (is) very tall.
Uggausau bäwizzi,	She (is) slender.
Gitshee sussaigau,	He (is) fine dressed.
Bishegnaindaugoozziwug meegwunug,	They (are) beautiful feathers.
Ke daukoozzinuh,	Are you sick?
Monaudud muundun muskeekee,	This (is) bad medicine.
Monaudud aindauyun,	My place of dwelling (is) bad.
Aindauyaun mitshau,	My place of dwelling (is) large.
Ne mittigwaub onishishsha,	My bow (is) good.
Ne bikwukön monaududön,	But my arrows (are) bad.
Ne minwaindaun appaukoozzegun,	I love mild or mixed tobacco.
Kauweekau neezhikay ussämau ne suggus-wannausee,	But I never smoke pure tobacco.
Monaudud maishkowaugumig,	Strong drink (is) bad.
Keeguhgee budjeëgonaun,	It makes us foolish.
Gitshee Monedo neebe ogee özhetön,	The Great Spirit made water.
Ininewug dush ween ishködäwaubo ogee oz-hetönahwau,	But man made whiskey.

These expressions are put down promiscuously, embracing verbs and nouns as they presented themselves, and without any effort to support the opinion, which may or may not be correct, that the elementary forms of the adjectives are most commonly required before verbs and nouns in the first and second persons. The English expression is thrown into Indian in the most natural manner, and, of course, without always giving adjective for adjective or noun for noun. Thus, God is rendered, not *monedo*, but *Geezha monedo, merciful spirit.* Good luck is rendered by the compound phrase, *shäwaindaugoozzegun,* indicating in a very general sense, the influence of kindness or benevolence on *success in life.* *Söngedää* is, alone, *a brave man,* and the word *kägät* prefixed, is an adverb. In the expression "mild tobacco," the adjective is entirely dispensed with in the Indian, the sense being sufficiently rendered by the compound noun *appaukoozzegun,* which always means the Indian weed or smoking mixture. *Ussamau,* on the contrary, without the adjective, signifies pure tobacco. *Bikwukön,* signifies blunt or lumpy-headed arrows; *assowaun*, is the barbed arrow. *Kwonaudj kweeweezains* means, not simply "pretty boy," but *pretty little boy;* and there is no mode of using the word boy but in this diminutive form, the word itself being a derivative *kewewe coryugal*, with the regular diminutive in *ains*. *Onaunee-goozzin*, embraces the pronoun, verb, and adjective, *be thou cheerfull*. In the last phrase of the examples, "man" is rendered men (*inin-eewuy*) in the translation, as the term *man* cannot be employed in the general plural sense it conveys in this connection in the original. The word "whiskey" is rendered by the compound phrase, *ishködawaubo*, literally *fire-liquor*, a generic for all kinds of ardent spirits.

These aberrations from the literal terms will convey some conceptions of the difference of the two idioms, although, from the limited nature and object of the examples, they will not indicate the full extent of the difference. In giving anything like the spirit of the original, much greater deviations in the written forms must appear. And in fact, not only the structure of the language, but the mode and *order of thought* of the Indians is so essentially different, that any attempts to preserve the English idiom, to give letter for letter, and word for word, must go far to render the translation pure nonsense.

2. Varied as the adjective is in its changes, it has no comparative inflection. A Chippewa cannot say that one substance is hotter or colder than another, or of two or more substances unequally heated, that this or that is the hottest or coldest, without employing adverbs or accessory adjectives; and it is accordingly by adverbs and accessory adjectives that the degrees of comparison are expressed.

Pimmaudizziwin, is a very general substantive expression, indicating *the tenor of being or life*. *Izzhewäbizziwin*, is a term near akin to it, but more appropriately applied to the *acts, conduct, manner, or personal deportment of life*. Hence the expressions—

Nem bimmaudizziwin,	My tenor of life.
Ke bimmaudizziwin,	Thy tenor of life.
O pimmaudizziwin,	His tenor of life, &c.
Nin dizhewäbizziwin,	My personal deportment.
Ke dizhewäbizziwin,	Thy personal deportment.
O Izzhewäbizziwin,	His personal deportment, &c.

32

To form the positive degree of comparison from these terms, *minno*, good, and *mudjee*, bad, are introduced between the pronoun and verb, giving rise to some permutations of the vowels and consonants, which affect the sound only. Thus—

Ne minno pimmaudizziwin,	My good tenor of life.
Ke minno pimmaudizziwin,	Thy good tenor of life.
Minno pimmaudizziwin,	His good tenor of life.
Ne mudjee pimmaudizziwin.	My bad tenor of life.
Ke mudjee pimmaudizziwin,	Thy bad tenor of life.
Mudjee pimmaudizziwin,	His bad tenor of life.

To place these forms in the comparative degree, *nahwudj*, *more*, is prefixed to the adjective; and the superlative is denoted by *mahmowee*, an adverb or an adjective as it is variously applied, but the meaning of which is, in this connection, *most*. The degrees of comparison may be, therefore, set down as follows:—

Positive,	Kishedä.	Hot (restricted to the heat of a fire),
Comparative,	Nahwudj kishedä.	More hot,
Superlative,	Mahmowee kishedä.	Most hot.

Your manner of life is good,	Ke dizzhewäbizziwin onishishin.
Your manner of life is better,	Ke dizzhewäbizziwin nahwudj onishishin.
Your manner of life is best,	Ke dizzhewäbizziwin mahwowee onishishin.
His manner of life is best,	Odizzhewäbizziwin mahmowee onishishinine.
Little Turtle was brave,	Mikkenoköns söngedääbun.
Tecumseh was braver,	Tecumseh nahwudj söngedääbun.
Pontiac was bravest,	Pontiac mahmowee söngedääbun.

3. The adjective assumes a negative form when it is preceded by the adverb. Thus, the phrase *songedää*, he is brave, is changed to *kahween söngedääsee*, he is not brave.

POSITIVE.

Neebwaukah,	He is wise.
Kwonaudjewe,	She is handsome.
Oskineegee,	He is young.
Shaugweewee,	He is feeble.
Geekkau,	He is old.
Mushkowizzi,	He is strong

NEGATIVE.

Kahween neebwaukah-see,	He is not wise.
Kahween kwonaudjewee-see,	She is not handsome.
Kahween oskineegee-see,	He is not young.
Kahween Shaugweewee-see,	He is not feeble.
Kahween Geekkau-see,	He is not old.
Kahween Mushkowizzi-see,	He is not strong.

From this rule the indeclinable adjectives, by which is meant those adjectives which do not put on the personal and impersonal forms by inflection, but consist of radically different roots, form exceptions.

Are you sick?	Ke dahkoozzi nuh?
You are not sick!	Kahween ke dahkoozzi-see
I am happy,	Ne minwaindum.
I am unhappy,	Kahween ne minwainduz-see.
His manner of life is bad,	Mudjee izzhewabizzi.
His manner of life is not bad,	Kahween mudjee izzhewabizzi-see.
It is large,	Mitshau muggud.
It is not large,	Kahween mitshau-seenön.

In these examples, the declinable adjectives are rendered negative in *see;* the indeclinable, remain as simple adjuncts to the verbs; and the *latter* put on the negative form.

4. In the hints and remarks which have now been furnished respecting the Chippewa adjective, its powers and inflections have been shown to run parallel with those of the substantive, in its separation into animates and inanimates; in having the pronominal inflections; in taking an inflection for tense—a topic which, by the way, has been very cursorily passed over—and in the numerous modifications to form the compounds. This parallelism has also been intimated to hold good with respect to number—a subject deeply interesting in itself, as it has its analogy only in the ancient languages—and it was therefore deemed best to defer giving examples, till they could be introduced without abstracting the attention from other points of discussion.

Minno and *mudjee,* good and bad, being of the limited number of personal adjectives which modern usage permits being applied, although often improperly applied to inanimate objects, they, as well as a few other adjectives, form exceptions to the use of number. Whether we say "a good man" or "a bad man," "good men" or "bad men," the words *minno* and *mudjee* remain the same. But all the declinable and coalescing adjectives—adjectives which join on, and, as it were, *melt into* the body of the substantive—take the usual plural inflections, and are governed by the same rules in regard to their use, as the substantive; personal adjectives requiring personal plurals, &c.

ADJECTIVES ANIMATE.
Singular.

Onishishewe mishemin,	Good apple.
Kwonaudjewe eekwä,	Handsome woman.
Songedää inine,	Brave man.
Bishegaindaugoozzi peenasee,	Beautiful bird.
Ozahwizzi ahmo,	Yellow bee.

Plural.

Onishishewe-wug mishemin-ug,	Good apples.
Kwonaudjewe-wug eekwä-wug,	Handsome women.
Songedää-wug inine-wug,	Brave men.
Bishegaindaugoozzi-wug peenasee-wug,	Beautiful birds.
Ozahwizzi-wug ahm-ög,	Yellow bees.

ADJECTIVES INANIMATE.
Singular.

Onishishin mittig,	Good tree.
Kwonaudj tshemaun,	Handsome canoe.
Monaudud ishkoda,	Bad fire.
Weeshkobun aidetaig,	Sweet fruit.

Plural.

Onishishin-ön mittig-ön,	Good trees.
Kwonaudjewun-ön tshemaun-un,	Handsome canoes.
Monaudud-ön ishkod-än,	Bad fires.
Weeshkobun-ön aidetaig-in,	Sweet fruits.

Peculiar circumstances are supposed to exist in order to render the use of the adjective, in this connection with the noun, necessary and proper. But, in ordinary instances, as the narration of events, the noun would precede the adjective; and oftentimes, particularly where a second allusion to objects previously named became necessary, the compound expressions would be used. Thus, instead of saying "the yellow bee," *wazzahwizzid* would distinctly convey the idea of that insect, *had the species been before named.* Under similar circumstances, *kainwaukoozzid, agausheid, söngaunemud, mushkowaunemud,* would respectively signify, "a tall tree," "a small fly," "a strong wind," "a hard wind." And these terms would become plural in *jig,* which, as before mentioned, is a mere modification of *ig,* one of the five general animate plural inflections of the language.

Kägät wahwinaudj abbenöjeeug, is an expression indicating they are *very handsome children*. But *beeweezheewug monetösug* denotes *small insects*. *Minno neewugizzi*, is "good tempered," "he is good tempered." *Mawshininewugizzi*, is "bad tempered," both having their plural in *wug*. *Nin nuneenahwaindum*, "I am lonesome." *Nin nuneenahwaindaumin*, "we (excluding you) are lonesome." *Waweea*, is a term generally used to express the adjective sense of *round*. *Kwy*, is the scalp; *weewikwy*, his scalp. Hence, *weewukwon*, "hat," *wayweewukwonid*, "a wearer of the hat;" and its plural, *wayweewukwonidjig*, "wearers of the hats"—the usual term applied to Europeans, or white men generally. These examples go to prove that under every form in which the adjective can be traced, whether in its simplest or most compound state, it is susceptible of number.

The numerals of the language are converted into adverbs by the inflection *ing*, making one, *once*, &c. The unit exists in duplicate.

Päzhīk,	One, *general unit.*	} Aubeding,	Once.
Ingoot,	One, *numerical unit.*		
Neesh,	Two.	Neeshing,	Twice.
Niswee,	Three.	Nissing,	Thrice.
Neewin,	Four.	Neewing,	Four times.
Naunun,	Five.	Nauning,	Five times.
N'goodwaswä,	Six.	N'goodwautsking,	Six times.
Neeshwauswä,	Seven.	Neeshwautshing,	Seven times.
Shwauswe,	Eight.	Shwautshing,	Eight times.
Shonguaswe,	Nine.	Shongutshing,	Nine times.
Metauswe,	Ten.	Meetaushing,	Ten times.

These inflections can be carried as high as they can compute numbers. They count decimally. After reaching ten, they repeat, ten and one, ten and two, &c. to twenty. Twenty is a compound signifying two tens; thirty, three tens, &c.; a mode which is carried up to one hundred—*n'goodwak*. *Wak* then becomes the word of denomination, combining with the names of the digits until they reach a thousand, *meetauswauk*, literally *ten hundred*. Here a new compound term is introduced, made by prefixing twenty to the last denominator, *neeshtonnah duswak*, which doubles the last term, thirty triples it, forty quadruples it, &c. till the computation reaches to ten thousand, *n'goodwak dushing n'goodwak*, one hundred times one hundred. This is the probable extent of all certain computation. The term *gitshee* (great), prefixed to the last denomination, leaves the number indefinite.

There is no form of the numerals corresponding to second, third, fourth, &c. They can only further say, *nittum*, first, and *ishkwaudj*, last.

IV.

Some Remarks respecting the Agglutinative Position and Properties of the Pronoun.

INQUIRY 4.

Nature and principles of the pronoun—Its distinction into preformative and subformative classes—Personal pronouns—The distinction of an inclusive and exclusive form in the number of the first person plural—Modifications of the personal pronouns to imply existence, individuality, possession, ownership, position, and other accidents—Declension of pronouns to answer the purpose of the auxiliary verbs—Subformatives, how employed to mark the persons—Relative pronouns considered—Their application to the causative verbs—Demonstrative pronouns—Their separation into two classes, animates and inanimates—Example of their use.

PRONOUNS are buried, if we may so say, in the structure of the verb. In tracing them back to their primitive forms, through the almost infinite variety of modifications which they assume, in connection with the verb, substantive, and adjective, it will facilitate analysis to group them into preformative and subformative, which include the terms that have already been made use of—pronominal prefixes, and suffixes—and which admit of the further distinction of separable and inseparable. By separable, is intended those forms which have a meaning by themselves, and are thus distinguished from the inflective and subformative pronouns, and pronominal particles, significant only in connection with another word.

1. Of the first class, are the personal pronouns *Neen* (I), *Keen* (Thou), and *Ween* or *O* (He or She). They are declined, to form the plural persons, in the following manner:—

I,	Neen.	We,	Keen owind. (in.)
		We,	Neem owind. (ex.)
Thou,	Keen.	Ye,	Keen owau.
He or she,	Ween or O.	They,	Ween owau.

Here the plural persons are formed by a numerical inflection of the singular. The double plural of the first person, of which both the rule and examples have been incidentally given in the remarks on the substantive, is one of those peculiarities of the language which may, perhaps, serve to aid in a comparison of it with other dialects, kindred and foreign. As a mere conventional agreement for denoting whether the person addressed be included or excluded, it may be regarded as an advantage to the language. It enables the speaker, by the change of a single consonant, to make a full and clear discrimination, and relieves the narration from doubts and ambiguity, where doubts and ambiguity would otherwise often exist. On the other hand, by accumulating distinctions, it loads the memory with grammatical forms, and opens a door for improprieties of speech. We are not aware of any inconveniences in the use of a general plural; but, in the Indian, it would produce confusion. And it is, perhaps, to that cautious desire of personal discrimination, which is so apparent in the structure of the language, that we should look for the reason of the duplicate forms of this word. Once established, however, and both the distinction, and the necessity of a constant and strict attention to it, are very obvious and striking. How shall he address the Deity? If he say, "Our Father who art in heaven," the inclusive form of *our* makes the Almighty one of the suppliants, or family. If he use the exclusive form, it throws him out of the family, and may embrace every living being but the Deity. Yet, neither of these forms can be used very well in prayer, as they cannot be applied directly *to* the object addressed. It is only when speaking *of* the Deity, under the name of father, to other persons, that the inclusive and exclusive forms of the word *our* can be used. The dilemma may be obviated by the use of a compound descriptive phrase, *Wä ö se mig o yun*, signifying, "Thou, who art the father of all," or "universal father." In practice, however, the question is cut short by those persons who have embraced Christianity. It has seemed to them that, by the use of either of the foregoing terms, the Deity would be thrown into too remote a relation to them; and I have observed that in prayer they invariably address Him by the term used by children for the father of a family—that is, *nosa*, "my father."

The other personal pronouns undergo some peculiar changes

when employed as preformatives before nouns and verbs, which
it is important to remark. Thus *neen*, is sometimes rendered *ne*,
or *nin*, and sometimes *nim*. *Keen*, is rendered *ke*, or *kin*. In
compound words, the mere signs of the first and second pronouns, *N* and *K*, are employed. The use of *ween* is limited; and
the third person, singular and plural, is generally indicated by
the sign *O*.

The particle *suh*, added to the complete forms of the disjunctive
pronouns, imparts a verbal sense to them; and appears, in this
instance, to be a succedaneum for the substantive verb. Thus
Neen, I, becomes *neensuh*, it is I. *Keen*, thou, becomes *keensuh*,
it is thou; and *ween*, he or she, *weensuh*, it is he or she. This
particle may be also added to the plural forms.

Keenowind suh,	It is we. (in.)
Neenowind suh,	It is we. (ex.)
Keenowau suh,	It is ye, or you.
Weenowau suh,	It is they.

If the word *aittah*, be substituted for *suh*, a set of adverbial
phrases are formed:—

Neen aittah,	I only.	Neen aittah wind,	We, &c. (ex.)
		Keen aittah wind,	We, &c. (in.)
Keen aittah,	Thou only.	Keen aittah wau,	You, &c.
Ween aittah,	He or she only.	Ween aittah wau,	They, &c.

In like manner, *nittum*, first, and *ishkwaudj*, last, give rise to
the following arrangement of the pronoun:—

Neen nittum,	I first.
Keen nittum,	You or thou first.
Ween nittum,	He or she first.
Keen nittum ewind,	We first. (in.)
Neen nittum ewind,	We first. (ex.)
Keen nittum ewau,	Ye or you first.
Ween nittum ewau,	They first.

ISHKWAUDJ.

Neen ishkwaudj,	I last.
Keen ishkwaudj,	Thou last.
Ween ishkwaudj,	He or she last.
Keenowind ishkwaudj,	We last. (in.)
Neenowind ishkwaudj,	We last. (ex.)
Keenowau ishkwaudj,	Ye or you last.
Weenowau ishkwaudj,	They last.

The disjunctive forms of the pronoun are also sometimes preserved before verbs and adjectives.

NEEZHIKA. Alone, (an.)

Neen neezhika,	I alone.
Keen neezhika,	Thou alone.
Ween neezhika,	He or she alone.
Keenowind neezhika,	We alone. (in.)
Neenowind neezhika,	We alone. (ex.)
Keenowau neezhika,	Ye or you alone.
Weenowau neezhika,	They alone.

To give these expressions a verbal form, the substantive verb, with its pronominal modifications, must be superadded. For instance, *I am* alone, &c. is thus rendered:—

Neen neezhika nindyau,	I am alone + aumin.
Keen neezhika keedyau,	Thou art alone + aum.
Ween nezhika iyau,	He or she is alone, &c. + wug.

In the subjoined examples, the noun ow, body, is changed to a verb, by the permutation of the vowel, changing ow, to AUW; which last takes the letter *d* before it when the pronoun is prefixed:—

I am a man,	Neen nin dauw.
Thou art a man,	Keen ke dauw.
He is a man,	Ween ah weeh.
We are men, (in.)	Ke dauw we min.
We are men, (ex.)	Ne dauw we min.
Ye are men,	Ke dauw mim.
They are men,	Weenowau ah weeh wug.

In the translation of these expressions, "man" is used as synonymous with "person." If the specific term *inine* had been introduced, in the original, the meaning thereby conveyed would be, in this particular connection, "I am a man," with respect to *courage*, &c. in opposition to effeminacy. It would not be simply declarative of *corporeal existence*, but of existence in a *particular state or condition*.

In the following phrases, the modified forms, or the signs only, of the pronouns are used:—

N'debaindaun,	I own it.
Ke debaindaun,	Thou ownst it.
O debaindaun,	He or she owns it.

N'debaindaum-in,	We own it. (ex.)
Ke debaindaum-in,	Wé own it. (in.)
Ke debaindaun-ewau,	Ye own it.
O debaindaun-ewau,	They own it.

These examples are cited as exhibiting the manner in which
the *prefixed* and preformative pronouns are employed, both in
their full and contracted forms. To denote possession, nouns
specifying the things possessed are required; and, what would
not be anticipated had not full examples of this species of declension been given in another place, the purposes of distinction are
not affected by a simple change of the pronoun, as *I* to *mine*, &c.,
but by a subformative inflection of the *noun*, which is thus made
to have a reflective operation upon the pronoun speaker. It is
believed that sufficient examples of this rule, in all the modifications of inflection, have been given under the head of the substantive. But as the substantives employed to elicit these modifications were exclusively *specific* in their meaning, it may be
proper here, in further illustration of an important principle, to
present a generic substantive under their compound forms.

I have selected for this purpose one of the primitives. IE-AU',
is the abstract term for matter. It is in the animate form. Its
inanimate correspondent is IE-EE'. These are two important roots.
And they are found in combination, in a very great number of
derivative words. It will be sufficient here, to show their connection with the pronoun, in the production of a class of terms
in very general use.

Animate Forms.
Possessive.

SINGULAR.		PLURAL.	
Nin dyĕ aum,	Mine.	Nin dyĕ auminsun,	Ours. (ex.)
		Ke dyĕ auminauh,	Ours. (in.)
Ke dyĕ aum,	Thine.	Ke dyĕ aumewau,	Yours.

Objective.

O dyĕ aum-un,	His or Hers.	O dyĕ aumewaun,	Theirs.

Inanimate Forms.
Possessive.

SINGULAR.		PLURAL.	
Nin dyĕ eem,	Mine.	Nin dyĕ eeminaun,	Ours. (ex.)
		Ke dyĕ eeminaun,	Ours. (in.)
Ke dyĕ eem,	Thine.	Ke dyĕ eemewau,	Yours.

Objective.

O dyŝ eem.　　　　　His or Hers.　　　　O dyŝ eemewau,　　　Theirs. (pos. in.)

In these forms the noun is singular throughout. To render it plural, as well as the pronoun, the appropriate general plurals *ug* and *un*, or *ig* and *in*, must be superadded. But it must be borne in mind, in making these additions, "that the plural inflection to inanimate nouns (which have no objective case), forms the objective case to animate, which have no number in the third person." (p. 461.) The particle *un*, therefore, which is the appropriate plural for the inanimate nouns in these examples, is only the objective mark of the animate.

The plural of I, is *naun*, the plural of thou and he, *wau*. But as these inflections would not coalesce smoothly with the possessive inflections, the connective vowels *i* and *e* are prefixed, making the plural of I, *inaun*, and of theu, &c., *ewau*.

If we strike from these declensions the root IE, leaving its animate and inanimate forms AU and EE, and adding the plural of the noun, we shall then, taking the *animate* declension as an instance, have the following formula of the pronominal declensions:

Pronoun singular.	Place of the noun.	Possessive inflection.	Objective inflection to the noun singular.	Connective vowel.	Plural inflection of the pronoun.	Objective inflection, noun plural.	Plural of the noun.
Ne		aum		i	naun		ig
Ke		aum		e	wau		g
O		sum	un				
O		aum		e	waù	n	

To render this formula of general use, six variations (five in addition to the above) of the possessive inflection are required, corresponding to the six classes of substantives, whereby *aum* would be changed to *äm, eem, im, öm,* and *oom*, conformably to the examples heretofore given in treating of the substantive. The objective inflection would also be sometimes changed to *een*, and sometimes to *oan*.

Having thus indicated the mode of distinguishing the person, number, relation, and gender, or what is deemed its technical equivalent, the mutations words undergo, not to mark the distinctions of *sex*, but the presence or absence of *vitality*, I shall now advert to the inflections which the pronouns take for *tense*, or rather to form the auxiliary verbs, have, had, shall, will, may, &c.; a very curious and important principle, and one which clearly demonstrates that no part of speech has escaped the transforming genius of the language. Not only are the three great modifications of time accurately marked in the verbal form of the Chippewas, but, by the inflection of the pronoun, they are enabled to indicate some of the oblique tenses, and thereby to conjugate their verbs with accuracy and precision.

The particle *gee* added to the first, second, and third person singular, of the present tense, changes them to the perfect past, rendering I, thou, he, I. did, have, or had; thou didst, hast, or hadst; he or she did, have, or had. If *gah* be substituted for *gee*, the first future tense is formed, and the perfect past added to the first future, forms the conditional future. As the eye may prove an auxiliary in the comprehension of forms which are not familiar, the following tabular arrangement of them is presented.

First person, I.

Nin gee,	I did, have, had.
Nin gah,	I shall, will.
Nin gah gee,	I shall have, will have.

Second person, Thou.

Ke gee,	Thou didst, hast, hadst.
Ke gah,	Thou shalt, wilt.
Ke gah gee,	Thou shalt have, wilt have.

Third person, He or She.

O gee,	He or she did, have, had.
O gah,	He or she did, have, had.
O gah gee,	He or she shall have, will have.

The present and imperfect tense of the potential mood is formed by *dau*, and the perfect by *gee* suffixed, as in other instances.

First person, I.

Nin dau,	I may, can, &c.
Nin dau gee,	I may have, can have, &c.

Second person, Thou.

Ke dau,	Thou mayst, canst, &c.
Ke dau gee,	Thou mayst have, canst have, &c.

Third person, He or She.

O dau,	He or she may, can, &c.
O dau gee,	He or she may have, can have, &c.

In conjugating the verbs through the plural person, the singular terms for the pronoun remain, and they are rendered plural by a retrospective action of the pronominal inflections of the verb. In this manner the pronoun-verb auxiliary has a general application, and the necessity of double forms is avoided.

The preceding observations are confined to the formative or *prefixed* pronouns. The inseparable suffixed or subformative are as follows:—

Yaun,	My.
Yun,	Thy.
Id or d,	His or hers.
Yaung,	Our. (ex.)
Yung,	Our. (in.)
Yaig,	Your.
Waud,	Their.

These pronouns are exclusively employed as suffixes, and as suffixes to the descriptive compound substantives, adjectives, and verbs. Both the rule and examples have been stated under the head of the substantives, p. 463, and adjectives, p. 492. Their application to the verb will be shown as we proceed.

2. *Relative Pronouns.*—In a language which provides for the distinction of person by particles prefixed or suffixed to the verb, it will scarcely be expected that separate and independent relative pronouns should exist, or if such are to be found, their use, as separate parts of speech, must, it will have been anticipated, be quite limited; limited to simple interrogatory forms of expression, and not applicable to the indicative or declaratory. Such will be found to be the fact in the language under review; and it will be perceived from the subjoined examples, that in all instances requiring the relative pronoun *who*, other than the simple interrogatory forms, this relation is indicated by the inflections of the verb, or adjective, &c. Nor does there appear to be any declension of the separate pronoun corresponding to *whose* and *whom*.

The word *Ahwaynain*, may be said to be uniformly employed in the sense of *who*, under the limitations we have mentioned. For instance—

Who is there?	Ahwaynain e-mah ai-aud?
Who spoke?	Ahwaynain kau keegœdood?
Who told you?	Ahwaynain kau weendumoak?
Who are you?	Ahwaynain iau we yun?
Who sent you?	Ahwaynain waynönik?
Who is your father?	Ahwaynain kös?
Who did it?	Ahwaynain kau tödung?
Whose dog is it?	Ahwaynain way dyid?
Whose pipe is that?	Ahwaynain döpwaugunid en-eu?
Whose lodge is it?	Ahwaynain way weegewomid?
Whom do you seek?	Ahwaynain nain dau wau bumud?
Whom have you here?	Ahwaynain oh-amau ai auwaud?

Not the slightest variation is made in these phrases between who, whose, and whom.

Should we wish to change the interrogative, and to say he who is there, he who spoke, he who told you, &c., the separable personal pronoun *ween* (he) must be used in lieu of the relative; and the following forms will be elicited:—

Ween, kau unnönik,	He (who) sent you.
Ween, kau geedood,	He (who) spoke.
Ween, *ai*-aud e-mah,	He (who) is there.
Ween, kau weendumoak,	He (who) told you.
Ween, kau tödung,	He (who) did it, &c.

If we object that, in these forms, there is no longer the relative pronoun *who*, the sense being simply he sent you, he spoke, &c., it is replied that, if it be intended only to say he sent you, &c., and not he *who* sent you, &c., the following forms are used:—

Ke gee unnönig,	He (sent) you.
Ainnözhid,	He (sent) me.
Ainnönaud,	He (sent) him.
Iau e-mau,	He is there.
Ke geedo,	He (spoke).
Ke gee weendumaug,	He (told) you.
Ke to dum,	He did it.

We reply to this answer of the native speaker, that the particle *kau* prefixed to a verb, denotes the past tense; that in the former series of terms in which this particle appears, the verbs are in the perfect indicative, and in the latter, they are in the present indicative, marking the difference only between *sent* and *send*, *spoke* and *speak*, &c.; and that there is absolutely no relative pronoun in either series of terms. We further observe, that the personal pronoun *ween*, prefixed to the first set of terms, may be prefixed, with equal propriety, to the second set, and that its use or disuse is perfectly optional with the speaker, as he may wish to give additional energy or emphasis to the expression. To these positions, after reflection, discussion, and examination, we receive an assent, and thus the uncertainty is terminated.

We now wish to apply the principle thus elicited to verbs causative, and to other compound terms—to the adjective verbs, for instance—and to the other verbal compound expressions, in which the objective and the nominative persons are incorporated as a part of the verb, and are not prefixes to it. This may be shown in the causative verb—

TO MAKE HAPPY.

Mainwaindumäid,	He (who) makes *me* happy.
Mainwaindumäik,	He (who) makes *thee* happy.
Mainwaindumäaud,	He (who) makes *him* happy.
Mainwaindumäinung,	He (who) makes *us* happy. (in.)
Mainwaindumäyaug,	He (who) makes *us* happy. (ex.)
Mainwaindumäinnaig,	He (who) makes *ye* or *you* happy.
Mainwaindumäigowaud,	He (who) makes *them* happy.

And so the forms might be continued throughout all the objective persons—

Mainwaindum ä yun,	*Thou* (who) makest me happy, &c.

The basis of these compounds is *minno*, "good," and *aindum*, "the mind." Hence, *minwaindum*, "he happy." The adjective, in this connection, cannot be translated "good," but its effect upon the noun is to denote that state of the mind which is at rest with itself. The first change from this simple compound, is to give the adjective a verbal form; and this is effected by a permutation of the vowels of the first syllable—a rule of very extensive application—and by which, in the present instance, the phrase "he happy," is changed to "he makes happy," (*mainwaindum*.) The next step is to add the suffix personal pronouns, *id, ik, aud*, &c., rendering the expressions, "he makes *me* happy," &c. But, in adding these increments, the vowel *e* is thrown between the adjective-verb and the pronoun suffixed, making the expression, not *mainwaindum-yun*, but *mainwaindum ëyun*. Generally, the vowel *e*, in this situation, is a connective, or introduced merely for the sake of euphony. And those who maintain that it is here employed as a personal pronoun, and that the relative *who* is implied by the final inflection, overlook the inevitable inference, that if the marked *e* stands for *me* in the first phrase, it must stand for *thee* in the second, *he* in the third, *us* in the fourth, &c. As to the meaning and office of the final inflections *id*, *ik*, &c., whatever they may, in an involuted sense, *imply*, it is quite clear, by turning to the list of *suffixed personal pronouns*, and *animate plurals*, that they mark the persons, I, thou, he, &c., we, ye, they, &c.

Take, for example, *minwaindumëigowaud*, "he (who) makes them happy." Of this compound, *minwaindum*, as before shown, signifies "he makes happy." But as the verb is in the singular number, it implies that but *one person* is made happy; and the suffixed personal pronouns *singular*, mark the distinctions between *me*, *thee*, and *he*, or *him*.

Minwaindum-e-ig is the verb plural, and implies that several persons are made happy; and, in like manner, the suffixed personal pronouns *plural*, mark the distinctions between we, ye, they, &c.; for it is a rule of the language, that a strict concordance must exist between the number of the verb and the number of the pronoun. The termination of the verb consequently always indicates whether there be one or many objects to which its energy is directed. And as animate verbs can be applied only to animate objects, the numerical inflections of the verb are understood to mark the number of persons. But this number is indiscriminate, and leaves the sense vague until the pronominal suffixes are superadded. Those who, therefore, contend for the *sense* of the relative pronoun "who" being given in the last-mentioned phrase, and all phrases similarly formed by a succedaneum, contend for something like the following form of translation: "He makes them happy—him!" or "Him—he (meaning 'who') makes them happy."

The equivalent for *what*, is *waygonain*.

What do you want?	Waygonain wau iauyun?
What have you lost?	Waygonain kau wonetöyun?
What do you look for?	Waygonain nain dahwau bundahmun?
What is this?	Waygonain ewinain maundun?
What will you have?	Waygonain kad iauyun?
What detained you?	Waygonain kau oon dahme egöyun?
What are you making?	Waygonain wayzhetöyun?
What have you there?	Waygonain e-mau iauyun?

The use of this pronoun, like the preceding, appears to be confined to simple interrogative forms. The word *auneen*, which sometimes supplies its place, or is used for want of the pronoun *which*, is an adverb, and has considerable latitude of meaning. Most commonly, it may be considered as the equivalent for *how*, in what manner, or at what time.

What do you say?	Auneen akeedöyun?
What do you call this?	Auneen aizheneekaudahmun maundun?(i.)
What ails you?	Auneen aindeeyun?
What is your name?	Auneen aizheeksusoyun?
Which do you mean, this or that? (an.)	Auneen ah-ow ainud, woh-ow gämau ewaidde?
Which do you mean, this or that? (in.)	Auneen eh-eu ewaidumun oh-oo gämau ewaidde?
Which boy do you mean?	Auneen ah-ow-ainud?

By adding to this word the particle *de*, it is converted into an adverb of place, and may be rendered *where*.

Where do you dwell?	Auneende aindauyun?
Where is your son?	Auneende ke gwiss?
Where did you see him?	Auneende ke waubumud?
Where did you see it?	Auneende ke waubundumun?
Where are you going?	Auneende azhauyun?
Where did you come from?	Auneende ka oonjeebauyun?
Where is your pipe?	Auneende ke döpwaugun?
Where is your gun?	Auneende ke baushkizzigun?

By a still further modification, it is rendered an adverb of inquiry of the cause or motive.

Why do you do so?	Auneeshween eh eu todumun?
Why do you say so?	Auneeshween eh eu ekeedoyun?
Why are you angry?	Auneeshween nishkaudizzeyun?
Why will you depart?	Auneeshween wee matyauyun?
Why will you not depart?	Auneeshween matyauseewun?
Why have you come?	Auneeshween ke peëzhauyun?
Tell me why?	Weendumowishin auneeshween?
Wherefore is it so?	Auneeshween eh-eu izzhewaibuk? (in.)
Wherefore did you strike him?	Auneeshween ke pukketaywud?

3. Demonstrative pronouns are either animate or inanimate, and may be arranged as follows:—

33

ANIMATE.		INANIMATE.
Mau-bum (impersonal),	} This.	Maun-dun (inanimate proper).
Woh-ow (personal),		Oh-oo (inanimate conventional).
Ah-ow,	That.	Eh,eu.
Mau-mig,	These.	Mau-min.
Ig-eu (personal),	} Those.	In-eu (inanimate proper).
I-goo (impersonal),		O-noo (inanimate conventional).

These words are not always used merely to ascertain the object, but often, perhaps always, when the object is present to the sight, have a substantive meaning, and are used without the noun. It creates no uncertainty, if a man be standing at some distance to say, *Ah-ow;* or if a canoe be lying at some distance, to say, *Eh-eu;* the meaning is clearly, that *person,* or that *canoe,* whether the noun be added or not. Or, if there be two animate objects standing together, or two inanimate objects lying together, the words *maumig* (a.), or *maumin* (i.), if they be near, or *ig-eu,* (a.), or *in-eu* (i.), if they be distant, are equally expressive of the *materiality* of the objects, as well as their relative position. Under other circumstances the noun would be required, as where two animate objects of diverse character—a man and a horse for instance—were standing near each other; or a canoe and a package of goods were lying near each other—and, in fact, under all circumstances—the noun *may* be used after the demonstrative pronoun, without violating any rule of grammar, although not without the imputation, in many instances, of being over-formal and unnecessarily minute. What is deemed redundant, however, in oral use, and amongst a people who supply much by sight and gesticulation, becomes quite necessary in writing the language; and, in the following sentences, the substantive is properly employed after the pronoun:—

This dog is very lean,	Gitshee bukaukuddoozo woh-ow annemoosh.
These dogs are very lean,	Gitshee bukaukuddoozowug o-goo annemooshug.
Those dogs are fat,	Ig-eu annemooshug ween-in-oowug.
That dog is fat,	Ah-ow annemoosh ween-in-oo.
This is a handsome knife,	Gagait onishishin maundun mokomahn.
These are handsome knives,	Gagait wahwinaudj o-noo mokomahnun.
Those are bad knives,	Monaududön in-euwaidde mokomahnun.
Give me that spear,	Meezhishin eh-eu ahnitt.
Give me those spears,	Meezhishin in-eu unnewaidde ahnitteen.
That is a fine boy,	Gagait kwonaudj ah-ow kweewezains.
Those are fine boys,	Gagait wahwinaudj ig-euwaidde kweewezainsug.

This boy is larger than that, Nahwudj mindiddo woh-ow kweewezains ewaidde dush.

That is what I wanted, Meeh-eu wau iauyaumbaun.

This is the very thing I wanted, Mee-suh oh-oo wau iauyaumbaun.

In some of these expressions, the pronoun combines with an adjective, as in the compound words *ineuwaidde* and *igeuwaidde*, *those yonder* (in.), and *those yonder* (an.). Compounds which exhibit the full pronoun in coalescence with the adverb *ewaidde*, yonder.

2. NATURAL HISTORY.

V.

ZOOLOGY.

1. *Limits of the Range of the Cervus Sylvestris in the Northwestern parts of the United States.* By Henry R. Schoolcraft. (Northwest Journal.)

2. *Description of the Fringilia Vespertina, discovered by Mr. Schoolcraft in the Northwest.* By William Cooper. (Annals of the New York Lyceum of Natural History.)

3. Conchology.—*List of Shells collected by Mr. Schoolcraft, in the Western and Northwestern Territory.* By William Cooper.

HELIX.

1. Helix albolabris, *Say.* Near Lake Michigan.

2. Helix alternata, *Say.* Banks of the Wabash, near and above the Tippecanoe. Mr. Say remarks, that these two species, so common in the Atlantic States, were not met with in Major Long's second expedition, until their arrival in the secondary country at the eastern extremity of Lake Superior.

PLANORBIS.

3. Planorbis campanulatus, *Say.* Itasca (or La Biche) Lake, the source of the Mississippi.

4. Planorbis trivolvis, *Say.* Lake Michigan. These two

species were also observed by Mr. Say, as far east as the Falls of Niagara.

LYMNEUS.

5. LYMNEUS UMBROSUS, *Say*, Am. Con. iv. pl. xxxi. Fig. 1. Lake Winnipec, Upper Mississippi, and Rainy Lake.

6. LYMNEUS REFLEXUS, *Say*, l. c. pl. xxxi. Fig. 2. Rainy Lake, Seine River, and Lake Winnipec.

7. LYMNEUS STAGNALIS. Lake a la Crosse, Upper Mississippi.

PALUDINA.

8. PALUDINA PONDEROSA, *Say*. Wisconsin River.

9. PALUDINA VIVIPARA, *Say*, Am. Con. i. pl. x. The American specimens of this shell are more depressed than the European, but appear to be identical in species.

MELANIA.

10. MELANIA VIRGINICA, *Say*. Lake Michigan.

ANODONTA.

11. ANODONTA CATARACTA, *Say*. Chicago, Lake Michigan. This species, Mr. Lea remarks, has a great geographical extension.

12. ANODONTA CORPULENTA, *Nobis*. Shell thin and fragile, though less so than others of the genus; much inflated at the umbones, margins somewhat compressed; valves connate over the hinge in perfect specimens; surface dark brown, in old shells; in younger, of a pale dingy green, and without rays, in all I have examined; beaks slightly undulated at the tip. The color within is generally of a livid coppery hue, but sometimes, also, pure white.

Length of a middling sized specimen, four and a half inches, breadth, six and a quarter. It is often eighteen inches in circumference round the border of the valves, with a diameter through the umbones of three inches. Inhabits the Upper Mississippi, from Prairie du Chien to Lake Pepin.

This fine shell, much the largest I have seen of the genus, was first sent by Mr. Schoolcraft, to the Lyceum, several years ago. So far as I am able to discover, it is undescribed, and a distinct and remarkable species. It may be known by its length being greater in proportion to its breadth than in the other American species, by the subrhomboidal form of the posterior half, and generally, by the color of the nacre, though this is not to be relied on. It appears to belong to the genus SYMPHYNOTA of Mr. Lea.

ALASMODONTA.

13. ALASMODONTA COMPLANATA, *Barnes*. SYMPHYNOTA COMPLANATA, *Lea*. Shell Lake, River St. Croix, Upper Mississippi. Many species of shells found in this lake grow to an extraordinary size. Some of the present collected by Mr. Schoolcraft, measure nineteen inches in circumference.

14. ALASMODONTA RUGOSA, *Barnes*. St. Croix River, and Lake Vaseux, St. Mary's River.

15. ALASMODONTA MARGINATA, *Say*. Lake Vaseux, St. Mary's River; very large.

16. ALASMODONTA EDENTULA? *Say*. ANODON AREOLATUS? *Swainson*. Lake Vaseux. The specimens of this shell are too old and imperfect to be safely determined.

UNIO.

17. UNIO TUBERCULATUS, *Barnes*. Painted Rock, Upper Mississippi.

18. UNIO PUSTULOSUS, *Lea*. Upper Mississippi, Prairie du Chien, to Lake Pepin.

19. UNIO VERRUCOSUS, *Barnes, Lea*. St. Croix River of the Upper Mississippi.

20. UNIO PLICATUS, *Le Sueur, Say*. Prairie du Chien, and River St. Croix.

The specimens of U. PLICATUS sent from this locality by Mr. Schoolcraft have the nacre beautifully tinged with violet, near the posterior border of the shell, and are also much more ventricose than those found in more eastern localities, as Pittsburg, for example; at the same time, I believe them to be of the same species. Similar variations are observed in other species; the specimens from the south and west generally exhibiting a greater development.

21. UNIO TRIGONUS, *Lea*. From the same locality as the last, and like it unusually ventricose.

22. UNIO EBENUS, *Lea.* Upper Mississippi, between Prairie du Chien and Lake Pepin.

23. UNIO GIBBOSUS, *Barnes.* St. Croix River, Upper Mississippi.

24. UNIO RECTUS, *Lamarck.* U. PRÆLONGUS, *Barnes.* Upper Mississippi, from Prairie du Chien to Lake Pepin, and the River St. Croix. The specimens collected by Mr. Schoolcraft, vary much in the color of the nacre. Some have it entirely white, others rose purple, and others entirely of a very fine dark salmon color. This species inhabits the St. Lawrence as far east as Montreal.

25. UNIO SILIQUOIDEUS, *Barnes,* and U. INFLATUS, *Barnes.* Upper Mississippi, between Prairie du Chien and Lake Pepin. Large, ponderous, and the epidermis finely rayed.

26. UNIO COMPLANATUS, *Lea.* U. PURPUREUS, *Say.* Lake Vaseux, St. Mary's River. Lake Vaseux is an expansion of the River St. Mary, a tributary of the upper lakes. This shell does not appear to exist in any of the streams flowing into the Mississippi.

27. UNIO CRASSUS, *Say.* Upper Mississippi, Prairie du Chien.

28. UNIO RADIATUS, *Barnes.* Lake Vaseux. The specimen is old and imperfect, but I believe it to be the U. RADIATUS of our conchologists, which is common in Lake Champlain and also inhabits the St. Lawrence.

29. UNIO OCCIDENS, *Lea.* U. VENTRICOSUS, *Say,* Am. Con. U. VENTRICOSUS, *Barnes?* Wisconsin and St. Croix Rivers, and Shell Lake. Epidermis variously colored, and marked with numerous rays.

30. UNIO VENTRICOSUS, *Barnes.* Upper Mississippi, from Prairie du Chien to Lake Pepin and Shell Lake. The varieties of this, and the preceding pass insensibly into each other. Those from Shell Lake are of extraordinary size.

31. UNIO ALATUS, *Say.* SYMPHYNOTA ALATA, *Lea.* Upper Mississippi, and Shell Lake. Found also in Lake Champlain, by the late Mr. Barnes.

32. UNIO GRACILIS, *Barnes.* SYMPHYNOTA GRACILIS, *Lea.* Upper Mississippi, and Shell Lake. The specimens brought by Mr. Schoolcraft are larger and more beautiful than I have seen from any other locality.

VI.

BOTANY.

1. *A List of Species and Localities of Plants collected in the North-western Expeditions of Mr. Schoolcraft of* 1831 *and* 1832. By DOUGLASS HOUGHTON, M. D., Surgeon to the expeditions.

The localities of the following plants are transcribed from a catalogue kept during the progress of the expeditions, and embrace many plants common to our country, which were collected barely for the purpose of comparison. A more detailed account will be published at some future day.

Aster tenuifolius, Willdenow. Upper Mississippi.
" *sericea,* Nuttall. River de Corbeau, Missouri Ter.
" *lævis?* Willdenow. St. Croix River, Northwest Ter.
" *concolor,* Willdenow. Fox River, Northwest Ter.
" (*N. Spec.*). Sources of Yellow River, Northwest Ter.
Andropogon furcatus, Willdenow. Sources of Yellow River, Northwest Ter.
Alopecurus geniculatus, Linnæus. Sault Ste. Marie, M. T.
Aira flexuosa. Sault Ste. Marie, M. T.
Allium tricoccum, Aiton. Ontonagon River of Lake Superior.
" *cernuum,* Roth. River de Corbeau to the sources of the Miss.
" (*N. Spec.*). St. Louis River of Lake Superior.
Amorpha canescens, Nuttall. Upper Mississippi.
Artemisia canadensis, Mx. Lake Superior to the sources of the Miss.
" *sericea,* Nuttall. Keweena Point, Lake Superior.
" *gnaphaloides,* Nuttall. Fox River, Northwest Ter.
Arabis hirsuta, De Candolle. Upper Mississippi.
" *lyrata,* Linn. Lake Superior to the sources of the Miss.
Arundo canadensis, Mx. Lake Superior.
Arenaria lateriflora, Linn. Lake Superior to the sources of the Miss.
Alnus glauca, Mx. St. Croix River to the sources of the Miss.

Alliona albida, Walter. Yellow River, Northwest Ter.

Aronia sanguinea. Lake Superior to the sources of the Miss.

Alectoria jubata. Lake Superior to the sources of the Miss.

Aletris farinosa. Prairies of Michigan Ter.

Bidens beckii, Torrey. St. Croix River to the sources of the Miss.

Bunias maritima, Willdenow. Lake Michigan.

Baptisia cœrulea, Michaux. Fox River, Northwest Ter.

Blitum capitatum. Northwest Ter.

Betula papyracea, Willdenow. Lake Superior to the sources of
     the Miss.

   "   *glandulosa.* Savannah River, Northwest Ter.

Bartramia fontana. Lake Superior.

Bromus canadensis, Michaux. Upper Mississippi.

Batschia canescens. Plains of the Mississippi.

   "   "   Var. (or *N. Spec.*). Lake Superior.

Carex paucifolia. Sault Ste. Marie, Mich. Ter.

   "   *scirpoides*, Schkuhr. Sault Ste. Marie, Mich. Ter.

   "   *limosa*, Linn. Sault Ste. Marie, Mich. Ter.

   "   *curata*, Gmelin. Sault Ste. Marie, Mich. Ter.

   "   (apparently *N. Spec.* allied to *C. scabrata.*) Sources of the
       Miss.

   "   *washingtoniana*, Dewy. Lake Superior.

   "   *lacustris*, Willdenow. Lake Superior.

   "   *œdere*, Ehrhart. Leech Lake.

   "   *logopodioides*, Schkuhr. Savannah River, Northwest Ter.

   "   *rosea*, Var. Lake Superior.

   "   *festucacea*, Schkuhr. St. Louis River of Lake Superior.

Cyperus mariscoides, Elliott. Upper Mississippi.

   "   *alterniflorus*, Schwinitż. River St. Clair, Mich. Ter.

Cnicus pitcheri, Torrey. Lakes Michigan and Superior.

Coreopsis palmata, Nuttall. Prairies of the Upper Mississippi.

Cardamine pratensis. Lake Superior to the sources of the Miss.

Calamagrostis coarctata, Torrey. Lake Winnipec.

Cetraria icelandica. Lakes Superior and Michigan.

Corydalis aurea, Willdenow. Cass Lake, Upper Mississippi.

   "   *glauca*, Persoon. Lake Superior.

Cynoglossum amplexicaule, Michaux. Sault Ste. Marie.

Cassia chamæcrista. Upper Mississippi.

Corylus americana, Walter. Lake Superior to the sources of the Miss.

" *rostrata*, Willdenow. Lake Superior to the sources of the Miss.

Cistus canadensis, Willdenow. Lake Superior to the sources of the Miss.

Cornus circinata, L'Heritier. Lake Superior to the sources of the Miss.

Cypripedium acaule, Aiton. Lake Superior to the sources of the Miss.

Cymbidium pulchellum, Swartz. Lake Superior to the sources of the Miss.

Corallorhiza multiflora, Torrey. Lake Superior.

Convallaria borealis, Willdenow. Lake Superior to the sources of the Mississippi.

" *trifolia*, Linn. Lake Superior.

Cenchrus echinatus, Linn. Upper Mississippi.

Cerastium viscosum, Linn. Lake Superior.

" *oblongifolium*, Torrey. Michigan Ter.

Campanula acuminata, Michaux. St. Louis River of Lake Superior.

Chrysosplenium oppositifolium. Lake Superior to the Mississippi.

Cinna arundinacea, Willdenow. Upper Mississippi.

Drosera linearis, Hooker. Lake Superior.

" *rotundifolia.* Lake Superior to the sources of the Miss.

" *americana*, Muhlenberg. Lake Superior to the sources of the Miss.

Dracocephalum virginicum, Willdenow. Red Cedar River, Northwest Territory.

Delphinium virescens, Nuttall. Upper Mississippi.

Danthonia spicata, Willdenow. Mauvais River of Lake Superior.

Dirca palustris, Willdenow. Ontonagon River of Lake Superior.

Equisetum limosum, Torrey. Lake Superior.

" *palustre*, Willdenow. Lake Superior.

" *variegatum*, Smith. Lake Michigan.

Erigeron integrifolium, Bigelow. Falls of Peckagama, Upper Miss.

" *purpureum*, Willdenow. Falls of Peckagama, Upper Miss.

" (*N. Spec.*). Sources of St. Croix River, Northwest Ter.

Erigeron heterophyllum, Var. or (*N. Spec.*). Sources of St. Croix
    River, Northwest Ter.
Eryngium aquaticum, Jussieu. Galena, Ill.
Euphorbia corollata, Willdenow. Red Cedar River.
Eriophorum virginicum, Linn. Lake Superior.
    "   *alpinum*, Linn. Lake Superior.
    "   *polystachyon*, Linn. Lake Superior.
Empetrum nigrum, Michaux. Lake Superior.
Erysimum chiranthoides, Linn. Lake Superior.
Eriocaulon pellucidum, Michaux. Lake Superior.
Euchroma coccinea, Willdenow. Lake Superior to the Mississippi.
Elymus striatus, Willdenow. St. Croix River, Northwest Ter.
    "   *virginicus*, Linn. St. Croix River, Northwest Ter.
Festuca nutans, Willdenow. Lake Winnipec.
Glycera fluitans, Brown. Savannah River, Northwest Ter.
Gyrophora papulosa. Lake Superior.
Gentiana crinita, Willdenow. Lake Michigan.
Geranium carolinianum. Lake Superior to the Mississippi.
Galium lanceolatum, Torrey. Red Cedar River to the Mississippi.
Gerardia pedicularis. Fox River, Northwest Ter.
    "   *maritima*, Rafinesque. Lake Michigan.
Galeopsis tetrahit, Var. Falls of St. Mary, Mich. Ter.
Gnaphalium plantaginium, Var. Sources of the Mississippi.
Goodyera pubescens, Willdenow. Lake Superior.
Hippophæ canadensis, Willdenow. Lake Superior.
    "   *argentea*, Pursh. Lake Superior.
Hedeoma glabra, Persoon. Lake Michigan to the sources of the
    Miss.
Hydropeltis purpurea, Michaux. Northwest Ter.
Hippuris vulgaris. Yellow River to sources of the Mississippi.
Hudsonia tomentosa, Nuttall. Lake Superior.
Hypericum canadense. Lake Superior.
    "   *prolificum*, Willdenow. Lake Michigan.
Hieracium fasciculatum, Pursh. Pukwàewa Lake, Northwest Ter.
Hierochloa borealis, Roemer & Schultes. Lake Superior.
Holcus lanatus. Savannah River, Northwest Ter.
Houstonia longifolia, Willdenow. St. Louis River of Lake Supe-
    rior.
Heuchera americana, Linn. St. Louis River of Lake Superior.

Hypnum crista-castrensis. Sources of the Mississippi.

Hordeum jubatum. Upper Red Cedar Lake.

Helianthus decapetalis. Northwest Ter.

" *gracilis,* Torrey. Upper Lake St. Croix, Northwest Ter.

Hyssopus anisatus, Nuttall. Upper Mississippi.

" *scrophularifolius,* Willdenow. Upper Mississippi.

Inula villosa, Nuttall. Upper Mississippi.

Ilex canadensis, Michaux. Lake Superior.

Juncus nodosus. St. Mary's River.

" *polycephalus,* Michaux. Lake Superior.

Kœleria nitida, Nuttall. Lake Winnipec.

Lycopodium dendroideum, Michaux. Lake Superior to the sources of the Mississippi.

" *annotinum,* Willdenow. Lake Superior to the sources of the Mississippi.

Lonicera hirsuta, Eaton. Lake Superior to the sources of the Miss.

" *sempervirens,* Aiton. Lake Superior.

Lechea minor. Upper Mississippi.

Linhea borealis, Willdenow. Lake Superior to the sources of the Miss.

Lathyrus palustris. Lake Superior.

" *decaphyllus,* Pursh. Leech Lake.

" *maritimus,* Bigelow. Lake Superior.

Lobelia kalmii, Linnæus. Lake Superior.

" *claytoniana,* Michaux. Upper Mississippi.

" *puberula?* Michaux. Yellow River, Northwest Ter.

Liatris scariosa, Willdenow. Upper Mississippi.

" *cylindrica,* Michaux. Upper Mississippi.

Lysimachia revoluta, Nuttall. Lake Superior.

" *thyrsifolia,* Michaux. Lake Superior.

Ledum latifolium, Aiton. Lake Superior to the sources of the Miss.

Myrica gale, Willdenow. Lake Superior.

Malva (N. Spec.). Upper Mississippi.

Monarda punctata, Linnæus. Upper Mississippi.

" *oblongata,* Aiton. Upper Mississippi.

Microstylis ophioglossoides, Willdenow. Lac la Biche [Itasca].

Myriophyllum spicatum. Lake Superior.

Mitella cordifolia, Lamarck. Lake Superior.

Menyanthes trifoliata. Lake Superior to the sources of the Miss.

Myosotis arvensis, Sibthorp. St. Clair River, Mich. Ter.

Nelumbium luteum, Willdenow. Upper Mississippi.

Œnothera biennis, Var. Bois Brulé River of Lake Superior.

    "   *serrulata*, Nuttall. Upper Mississippi.

Psoralea argophylla, Pursh. Falls of St. Anthony.

Primula farinosa, Var. *Americana*, Torrey. Lakes Huron and
        Superior.

    "   *mistassinica*, Michaux. Keweena Point, Lake Superior.

Pinguicula (*N. Spec.*). Presque Isle, Lake Superior.

Parnassia americana, Muhlenberg. Lake Michigan.

Pedicularis gladiata, Michaux. Fox River.

Pinus nigra, Lambert. Lake Superior.

    " *banksiana*, Lambert. Lake Superior.

Populus tremuloides, Michaux. Northwest Ter.

    " *lævigata*, Willdenow. Upper Mississippi.

Prunus depressa, Pursh. Lakes Superior and Michigan.

Petalostemon violaceum, Willdenow. Upper Mississippi.

    "   *candidum*, Willdenow. Upper Mississippi.

Potentilla tridentata, Aiton. Lake Superior.

    "   *fruticosa*, Linnæus. Lakes Superior and Michigan.

Pyrola uniflora, Mauvais River of Lake Superior.

Polygonum amphibium, Linnæus. St. Croix River.

    "   *cilinode*, Michaux. Lake Superior.

    "   *articulatum*, Linnæus. Lake Superior.

    "   *coccinium*, Willdenow. St. Croix River.

Polygala polygama, Walter. Northwest Ter.

Phlox aristata, Michaux. Upper Mississippi.

Poa canadensis. Upper Mississippi.

Pentstemon gracile, Nuttall. Upper Red Cedar Lake.

    "   *grandiflorum*, Nuttall. Falls of St. Anthony.

Physalis lanceolata, Var. (or *N. Spec.*). Lac la Biche [Itasca].

Quercus coccinea, Wangenheim. Upper Red Cedar Lake.

    "   *obtusiloba*, Michaux. Upper Mississippi.

Ranunculus filiformis, Michaux. Falls of St. Mary, Mich. Ter.

    "   *pusillus*, Pursh. Mich. Ter.

    "   *prostratus*, Lamarck. Lake Superior to the Mississippi.

    "   *lacustris*, Beck & Tracy. Upper Mississippi.

Rudbeckia hirta, Linnæus.   Upper Mississippi and Michigan Ter.
"   *digitata*, Aiton.   Upper Mississippi.
Rubus parviflorus, Nuttall.   Lake Superior to the sources of the Miss.
"   *hispidus*, Linnæus.   Lake Superior.
"   *saxatilis*, Var. *canadensis*, Michaux.   Lake Superior.
Rosa gemella, Willdenow.   Lake Superior.
"   *rubifolia*, Brown.   Michigan Ter.
Ribes albinervum, Michaux.   Sources of the St. Croix River.
Saururus cernuus, Linnæus.   Upper Mississippi.
Streptopus roseus, Michaux.   Lake Superior.
Sisymbrium brachycarpum, Richardson.   Lake Superior.
"   *chiranthoides*, Linnæus.   Lake Superior.
Swertia deflexa, Smith.   Bois Brulé River of Lake Superior.
Silphium terebinthinaceum, Elliott.   Michigan Territory to the Miss.
"   *gummiferum*.   Fox River to the Mississippi.
Stachys aspera, Var. Michaux.   Lake Superior.
Sterocaulon paschale.   Lake Superior.
Struthiopteris pennsylvanica, Willdenow.   Lake Superior.
Scirpus frigetur?   Lake of the Isles, Northwest Ter.
"   *palustris*, Linnæus.   Lake Superior to the Mississippi.
Salix prinoides, Pursh.   Mauvais River of Lake Superior.
"   *longifolia*, Muhlenberg.   Upper Mississippi.
Spiræa opulifolia, Var. *tomentella*, De Candolle.   Lake Superior.
Sorbus americana, Willdenow.   Lake Huron to the head of Lake Superior.
Smilax rotundifolia, Linnæus.   Lake Superior to the Mississippi.
Silene antirrhina, Linnæus.   Lac la Biche.
Saxifraga virginiensis, Michaux.   Lake Superior.
Scutellaria ambigua, Nuttall.   Upper Mississippi.
Solidago virgaurea, Var. *alpina*.   Lake Superior.
Stipa juncea, Nuttall.   Usawa R.
Symphora racemosa, Michaux.   Source of the Miss. R.
Senecio balsamitæ, Var.   Falls of Peckagama, Upper Miss.
Sagittaria heterophylla, Pursh.   Upper Miss.
Tanacetum huronensis, Nuttall.   Lakes Michigan and Superior.
Tussilago palmata, Willdenow.   Lake Michigan.
Tofeldia pubens, Michaux.   Lake Superior.
Triglochin maritimum, Linnæus.   Lake Superior.

Thalyctrum corymellum, De Candolle.   St. Louis River.

Triticum repens, Linnæus.   Leech Lake.

Troximon virginicum, Pursh.   Lake Winnipec.

Talinum teretifolium, Pursh.   St. Croix River.

Tradescantia virginica.   Upper Mississippi.

Utricularia cornuta, Michaux.   Lake Superior.

   "   *purpurea*, Walter.   Lac Chetac, N. W. Ter.

Uraspermum canadense, Lake Superior to the Miss.

Viola lanceolata, Linnæus.   Sault Ste. Marie.

   "   *pedata*, Var. (or *N. Spec.*).   Lac la Biche, sources of the Miss.

Viburnum oxycoccus, Pursh.   Lake Superior.

   "   *lentago*.   Lake Superior.

Vernonia novoboracensis, Willdenow.   Upper Miss.   .

Verbena bracteosa, Michaux.   Upper Miss.

   "   *stricta*, Ventenat.   Upper Miss.

Zapania nodiflora, Michaux.   Galena, Illinois.

Zigadenus chloranthus, Richardson.   Sandy shores of Lake Michigan.

Zizania aquatica, Pursh.   Illinois to the sources of the Miss.

                              VII.

                   MINERALOGY AND GEOLOGY.

1. *A Report on the Existence of Deposits of Copper in the Geological Basin of Lake Superior.* By Dr. D. HOUGHTON.

                              FREDONIA, N. Y., November 14, 1881.

   SIR: In fulfilment of the duties assigned to me in the late expedition into the Indian country, under the direction of H. R. Schoolcraft, Esq., Indian Agent, I would beg leave to transmit to you the following observations relative to the existence of copper in the country bordering on the southern shore of Lake Superior.

   It is without doubt true that this subject has long been viewed with an interest far beyond its actual merit. Each mass of native copper which this country has produced, however insulated, or however it may have been separated from its original position, appears to have been considered a sure indication of the existence of that metal in beds; and hence we occasionally see, upon maps of that section of our country, particular portions marked as containing "copper mines," where no copper now exists. But, while it is certain that a combination of circumstances has served to mislead the public mind with regard to the geological situation and existing quantity of that metal, it is no less certain that a greater quantity of insulated native copper has been discovered upon the borders of Lake Superior, than in any other equal portion of North America.

Among the masses of native copper which have engaged the attention of travellers in this section of country, one, which from its great size was early noticed, is situated on the Ontonagon River, a stream which empties its waters into the southern part of Lake Superior, 331 miles above the Falls of the Ste. Marie. The Ontonagon River is, with some difficulty, navigable by batteaux 36 miles, at which place, by the union of two smaller streams—one from an easterly and the other from a westerly direction—the main stream is formed. The mass of copper is situated on the western fork, at a distance of six or eight miles from the junction.

The face of the country through the upper half of the distance from Lake Superior is uneven, and the irregularity is given it by hills of marly clay, which occasionally rise quite abruptly to the height of one or two hundred feet. No rock was observed *in situ*, except in one place, where, for a distance, the red sandstone was observed, forming the bed of the river.

The mass of copper lies, partly covered by water, directly at the foot of a clay hill, from which, together with numerous boulders of the primitive rocks, it has undoubtedly been washed by the action of the water of the river. Although it is completely insulated, there is much to interest in its examination. Its largest surface measures three and a half by four feet, and this, which is of malleable copper, is kept bright by the action of the water, and has the usual appearance of that metal when worn. To one surface is attached a small quantity of rock, singularly bound together by threads of copper, which pass through it in all directions. This rock, although many of its distinctive characters are lost, is evidently a dark colored serpentine, with small interspersed masses of milky quartz.

The mass of copper is so situated as to afford but little that would enable us to judge of its original geological position. In examining the eastern fork of the river, I discovered small water-worn masses of trap-rock, in which were specks of imbedded carbonate of copper and copper black; and with them were occasionally associated minute specks of serpentine, in some respects resembling that which is attached to the large mass of copper; and facts would lead us to infer that the trap formation which appears on Lake Superior east of the Ontonagon River, crosses this section of country at or near the source of that river, and at length forms one of the spurs of the Porcupine Mountains.

Several smaller masses of insulated native copper have been discovered on the borders of Lake Superior, but that upon Ontonagon River is the only one which is now known to remain.

At as early a period as before the American Revolution, an English mining company directed their operations to the country bordering on Lake Superior, and Ontonagon River was one point to which their attention was immediately directed. Traces of a shaft, sunk in the clay hill, near a mass of copper, are still visible—a memento of ignorance and folly.

Operations were also commenced on the southern shore of Lake Superior, near the mouth of a small stream, which, from that circumstance, is called Miners' River. Parts of the names of the miners, carved upon the sandstone rock at the mouth of the river, are still visible. What circumstance led to the selection of this spot does not now appear. No mineral traces are at this day perceptible, except occasional discolorations of the sandstone rock by what is apparently a mixture of the carbonates of iron and copper; and this is only to be observed where water, holding in solution an extremely minute portion of these salts, has trickled slowly over those rocks.

It does not, in fact, appear that the red sandstone, which constitutes the principal rock formation of the southern shore of Lake Superior, is in any instance metalliferous in any considerable degree. If this be true, it would require but little reflection to convince one of the inexpediency of conducting mining operations at either of the points selected for that purpose; and it is beyond a doubt true, that the company did not receive the least inducement to continue their labors.

In addition to these masses of native copper, an ore of that metal has long been known to the lake traders as the green rock, in which the characteristic substances are the green and blue carbonates of copper, accompanied by copper black. It is situated upon Keweena Point, 280 miles above the falls of the Ste. Marie. The ore is embraced by what is apparently a recently formed crag; and, although it is of a kind and so situated as to make an imposing appearance, there is little certainty of its existence in large quantities in this formation. The ore forms a thin covering to the pebbles of which the body of the rock is composed, and is rarely observed in masses separate from it. The crag is composed of angular fragments of trap-rock, and the formation is occasionally traversed by broad and continuous belts of calc. spar, here and there tinged with copper. Although the ore was not observed in any considerable quantity, except at one point, it apparently exists in minute specks through a greater part of the crag formation, which extends several miles, forming the shore of the lake.

This examination of the crag threw new interest upon the trap formation, which had been first observed to take the place of the sandstone at the bottom of a deep bay, called Montreal Bay, on the easterly side of Keweena Point. The trap-rock continues for a few miles, when the crag before noticed appears to lie directly upon it, and to form the extremity of the point; the crag, in turn, disappears, and the trap-rock is continued for a distance of six or eight miles upon the westerly side of the point, when the sandstone again reappears.

The trap-rock is of a compact granular texture, occasionally running into the amygdaloid and toadstone varieties, and is rich in imbedded minerals, such as amethystine quartz, smoky quartz, carnelian, chalcedony, agate, &c., together with several of the ores of copper. Traces of copper ore in the trap-rock were first noticed on the easterly side of Keweena Point, and near the commencement of the trap formation. This ore, which is an impure copper black, was observed in a vein of variable thickness, but not in any part exceeding two and a half inches. It is sufficiently compact and hard to receive a firm polish, but it is rather disposed to break into small irregular masses. A specimen furnished, upon analysis, 47.5 per cent. of pure copper.

34

On the western side of Keweena Point, the same ore appears
under different circumstances, being disseminated through the
body of the trap-rock, in grains varying in size from a pin's head
to a pea. Although many of these grains are wholly copper black,
they are occasionally only depositions of the mineral upon specks
of carnelian, chalcedony, or agate, or are more frequently composed, in part, of what is apparently an imperfect steatite. The
ore is so connected with, and so much resembles in color the
rock, of which it may be said to be a constituent part, that they
might easily, during a hasty examination, be confounded. A
random specimen of the rock furnished, upon analysis, 3.2 per
cent. of pure copper. The rock continues combined with that
mineral for nearly the space of three miles. Extremely thin veins
of copper black were observed to traverse this same rock; and
in enlargements of these were discovered several masses of amorphous native copper. The latter mineral appeared in two forms
—the one consisting of compact and malleable masses, varying
from four to ten ounces each; and the other, of specks and fasciculi
of pure copper, binding together confused masses of copper green,
and partially disintegrated trap-rock; the latter was of several
pounds' weight. Each variety was closely embraced by the rock,
although the action of the water upon the rock had occasionally
exposed to view points of the metal. In addition to the accompanying copper green, which was in a disintegrated state, small
specks of the oxide of copper were associated in most of the native
specimens.

Circumstances would not permit an examination of any portion
of the trap formation, except that bordering directly upon the lake.
But facts would lead us to infer that that formation extends from
one side of Keweena Point to the other, and that a range of thickly
wooded hills, which traverses the point, is based upon, if not formed
of that rock. An Indian information, which, particularly upon such
a subject, must be adopted with caution, would sanction the opinion
that the prominent constituents are the same wherever the rock is
observed.

After having duly considered the facts which are presented, I
would not hesitate to offer, as an opinion, that the trap-rock formation was the original source of the masses of copper which have
been observed in the country bordering on Lake Superior; and

that, at the present day, examinations for the ores of copper could not be made in that country with hopes of success, except in the trap-rock itself; which rock is not certainly known to exist upon any place upon Lake Superior, other than Keweena Point.

If this opinion be a correct one, the cause of failure of the mining company in this region is rendered plain. Having considered each insulated mass of pure metal as a true indication of the existence of a bed in the vicinity, operations were directed to wrong points; when, having failed to realize their anticipations, the project was abandoned without further actual investigation. We would be induced to infer that no attempts were made to learn the original source of the metal which was discovered, and thus, while the attention was drawn to insulated masses, the ores, ordinary in appearance, but more important *in sitû*, were neglected; and perhaps, from the close analogy in appearance to the rock with which they were associated, no distinction was observed.

What quantity of ore the trap-rock of Keweena Point may be capable of producing, can only be determined by minute and laborious examination. The indications which were presented by a hasty investigation are here embodied, and with deference submitted to your consideration.

I have the honor to be,

Sir, your obedient servant,

DOUGLASS HOUGHTON.

Hon. LEWIS CASS, *Secretary of War.*

2. *Remarks on the Occurrence of Native Silver and Ores of Silver in the Stratification of the Basins of Lakes Huron and Superior.* By HENRY R. SCHOOLCRAFT.

Traces of this metal which have been found in the drift and boulder stratum of both Lakes Huron and Superior, indicate the existence of the metal in place. During my residence at St. Mary's, two specimens of its occurrence were brought to my notice. The first of these consisted of points of native silver in a moderately large mass of native copper, found in 1828, near the entrance of the *Nama* or Sturgeon River into Keweena Lake, of the large peninsula of that name, in Lake Superior. Like the majority of such masses of the region, it had no adhering portion of rock or vein stone, from which a judgment might be formed of its original position.

I had, the prior year, set up my mineralogical cabinet in my office, and stated to the Indians, who roved over large tracts, my solicitude to collect specimens of the mineral productions of the country of every description, and, indeed, of its zoology, always acknowledging their comity, in bringing me specimens in any department of natural history, by some small present; and I found this to be a means of extending my inquiries.

Subsequently, I received a boulder specimen from the shores of Lake Huron, containing veins of native silver. Part of the metal had been detached. I submitted these specimens to the Lyceum of Natural History at New York, in 1825. The following remarks are taken from their annals.

Mineralogical and Chemical Characters.—By examining this mineral, it will be perceived to possess the color, lustre, malleability, and other obvious characters of native silver. It is so soft as to be easily cut by the knife; and in a state of purity which permits it to spread under the hammer. These characters serve to distinguish it from antimonial silver, which is not *malleable;* from native antimony which tarnishes on exposure, &c. The metal occurs in thin, massive veins in the rock. These veins sometimes intersect, but never cross each other. It is also disseminated in small particles through the stone, or spread in flattened masses over its surface. Some of these masses were detached by the discoverer, but have been preserved, and are presented to the Lyceum with the more solid and undisturbed portions.

By submitting a small portion of the metal to the action of nitric acid, I obtained an imperfect solution. On repeating the experiment, and adding a little sulphuric acid, the action was more brisk, and a clear and apparently perfect solution effected. By standing, however, a pulpy, white precipitate appeared at the bottom of the glass. This was collected and submitted to the action of the blowpipe, on a basis of charcoal. The result gave a number of minute, metallic globules, possessing greater lustre, malleability, and ductility, than the original mass. I repeated the latter experiment, adding to the nitro-sulphuric solution muriate of soda. A more perfect precipitation of the white powder was effected; but the results with the blowpipe remained the same.

Geognostic Position.—It is a rolled mass. An opinion of the specific character of the rock may be dubious, from the smallness of the specimen. It appears to have been detached from a stratum of gneiss, and is essentially composed of quartz. The blackish color of some parts of this latter mineral would, at first glance, lead us to attribute this color to the presence of hornblende; but, on closer examination, it will be perceived to be owing to a dark-colored steatite, which, in certain parts of the rock, is well developed, soft, and easily cut. A little calcspar is intermingled with the steatite.

Locality.—I am indebted to the politeness of Lieut. Lewis S. Johnston, of the British Indian Department, at Malden (U. C.), for the opportunity of adding this specimen to the mineralogical cabinet of the Lyceum. This gentleman, as he informed me, obtained it from an Indian, who picked it up on the southeastern shores of Lake Huron, near Point aux Barques, in Michigan Territory. That part of Lake Huron was cursorily examined by me, in the year 1820, in the course of the expedition conducted by Gov. Cass, through the upper lakes, &c. I consider it remarkable, even in a region abounding in rolled rocks, for the great number and variety of granite, gneiss, hornblende, and trap boulders, scattered along the shores of the lake. The water here is generally shallow and dangerous to approach in vessels; these boulder stones sometimes extending and presenting themselves above water for a mile or more from land. But we could not satisfy ourselves by an examination necessarily partial, that either of the primitive species mentioned, existed there in any other condition than as rolled masses, or displacements of rock strata, contiguous, perhaps, but not observed. Dr. Bigsby has informed me, that he observed the gneiss *in situ*, on the northwestern shores of this lake. The nearest rock in place, and that which in fact constitutes the abraded and caverned promontory of Point aux Barques, is gray sandstone.

The occurrence of this metal in the copper-bearing and other metalliferous rocks of this region, may be confidently affirmed.[*]

---

[*] At the date of this publication, it is known that this metal occurs, both as a constituent of the mass copper in Lake Superior, and is also developed in veins in the stratification.

8. *A General Summary of the Localities of Minerals observed in the Northwest in* 1831 *and* 1832. By HENRY R. SCHOOLCRAFT.

CLASS I. *Bodies not metallic, containing an acid.*

1. CALCAREOUS SPAR. Keweena Point, Lake Superior. Imbedded in small globular masses, in the trap-rock; also forming veins in the same formation. Some of the masses break into rhombic forms, and possess a certain but not perfect degree of transparency; others are opaque, or discolored by the green carbonate of copper. Also in the trap-rock between Fond du Lac and Old Grand Portage, Lake Superior, in perfect, transparent rhombs, exhibiting the property of double refraction. Also, at the lead mines, in Iowa County, in the marly clay formation, often exhibiting imperfect prisms, variously truncated.

2. CALCAREOUS TUFA. Mouth of the River Brulé, of Lake Superior. In small, friable, broken masses, in the diluvial soil. Also, in the gorge below the Falls of St. Anthony. In detached, vesicular masses, amidst debris.

3. COMPACT CARBONATE OF LIME. In the calcareous cliffs of horizontal formation, commencing at the Falls of St. Anthony. Carboniferous.

4. SEPTARIA. In the reddish clay soil, between Montreal River and Lapointe, Lake Superior.

5. GYPSUM. In the sandstone rock at the Point of Grand Sable West, Lake Superior. In orbicular masses, firmly imbedded. Not abundant. Granular, also imperfectly foliated.

6. CARBONATE OF MAGNESIA. Serpentine rock, at Presque Isle, Lake Superior. Compact.

7. HYDRATE OF MAGNESIA? With the preceding.

CLASS II. *Earthy compounds, amorphous or crystalline.*

8. COMMON QUARTZ. Huron Islands, Lake Superior; also the adjoining coast. In very large veins or beds. White, opaque.

9. GRANULAR QUARTZ. Falls of Peckagama, Upper Mississippi. *In situ.*

10. SMOKY QUARTZ. In the trap-rock, Keweena Point, Lake Superior, crystallized. In connection with amethystine quartz.

11. AMETHYST: With the preceding. Also, at the Pic Bay, and at Gargontwa, north shore of Lake Superior, in the trap-rock, in perfect crystals, of various intensity of color.

12. CHALCEDONY. Keweena Point, Lake Superior. In globular or orbicular masses, in amygdaloid rock. Often, in detached masses along the shores.

13. CARNELIAN. With the preceding.

14. HORNSTONE. In detached masses, very hard, on the shores of Lake Superior. Also, at Dodgeville, Iowa County, Mich. Ter., in fragments or nodular masses in the clay soil.

15. JASPER. In the preceding locality. Common and striped, exceedingly difficult of being acted on by the wheel. Not observed *in situ.*

16. AGATE. Imbedded in the trap-rocks of Lake Superior, and also detached, forming a constituent of its detritus. Variously colored. Often made up of alternate layers of chalcedony, carnelian, and cacholong. Sometimes zoned, or in fortification points. Specimens not taken from the rock are not capable of being scratched by quartz or flint, and are incapable of being acted on by the file; consequently, *harder* than any of the described species.

17. CYANITE. Specimens of this mineral, in flat, six-sided prisms, imbedded in a dark primitive rock, were brought out from Lac du Flambeau outlet, where the rock is described as existing *in situ.* The locality has not been visited, but there are facts brought to light, within the last two or three years, to justify the extension of the primitive to that section of country.

18. PITCHSTONE. A detached mass of this mineral, very black and lava-like, was picked up in the region of Lake Superior, where the volcanic mineral, trachyte, is common among the rolled masses. Neither of these substances have been observed *in situ.*

19. MICA. Huron Islands, Lake Superior. In granite.

20. SCHORL. Common. Outlet of Lac du Flambeau. Also, in a detached mass of primitive rock at Green Bay.

21. FELDSPAR. Porcupine mountains, Lake Superior.

22. BASALT. Amorphous. Granite Point, Lake Superior.

23. STILBITE. Amygdaloid rock, Keweena Point, Lake Superior.

24. ZEOLITE. Mealy. With the preceding.

25. ZEOLITE. Radiated. Lake Superior. This mineral consists of fibres, so delicate and firmly united as to appear almost compact, radiating from a centre. Some of the masses produced by this radiation measure 2.5 inches in diameter. They are of a uniform, pale, yellowish red. This mineral has not been traced *in situ*, being found in detached masses of rock, and sometimes as water-worn portions of radii. Its true position would seem to be the trap-rock.

26. ASBESTUS. Presque Isle, Lake Superior. In the serpentine formation.

27. HORNBLENDE. Very abundant as a constituent of the primitive rocks on the Upper Mississippi, and in the basin of Lake Superior. Often in distinct crystals.

28. DIALLAGE, GREEN. Lake Superior. In detached masses, connected with primitive boulders. *Harder* than the species.

29. SERPENTINE, COMMON. Presque Isle, Lake Superior.

30. SERPENTINE, PRECIOUS. With the preceding. Color, a light pistachio green, and takes a fine polish. Exists in veins in the common variety.

31. PSEUDOMORPHOUS SERPENTINE. With the preceding. This beautiful green mineral constitutes a portion of the veins of the precious serpentine. Its crystalline impressions are very distinct.

32. ARGILLITE. River St. Louis, northwest of Lake Superior. Nearly vertical in its position.

CLASS III. *Combustibles.*

33. PEAT. Marine sand formation composing the shore of Lake Superior, between White-fish Point and Grand Marrais. Also, on the island of Michilimackinac.

CLASS IV. *Ores and Metals.*

34. NATIVE COPPER. West side of Keweena Point, Lake Superior. Imbedded in a vein with carbonate of copper, and copper black, in the trap-rock.

35. COPPER BLACK. With the preceding.

36. CARBONATE OF COPPER, GREEN. With the preceding.

These two minerals (35 and 36) characterize the trap-rock of the peninsula of Keweena, Lake Superior, from Montreal Bay, extending to and around its extremity, west, to Sand-hill Bay.

The entire area may be estimated to comprise a rocky, serrated coast of about seventy-five miles in length, and not to exceed seven or eight miles in width. The principal veins are at a point called Roche Verd, and along the coast which we refer to as the Black Rocks. At the latter, native copper is one of the constituents of the vein.

Green and blue carbonate of copper was also observed in limited quantity, in small rounded masses, at one of the lead diggings near Mineral Point, Iowa County.

37. CHROMATE OF IRON. Presque Isle, Lake Superior.

38. SULPHURET OF LEAD. Lead mines of Iowa County, Michigan Territory.

39. EARTHY CARBONATE OF LEAD. Brigham's mine, Iowa County, Mich. Ter. Also, in small masses, of a yellowish white, dirty color, and great comparative weight, at several of the lead mines (diggings) in the more westerly and southern parts of the county.

4. *Geological Outline of the Taquimenon Valley of Lake Superior.* By HENRY R. SCHOOLCRAFT.

The River Takquimenon originates on a plateau between the northern shores of Lake Michigan and the southeastern coast of Lake Superior. At a central point on this plateau, there lies a lake of moderate size, which, in the translated Indian phrase, is called Heartsblood Lake. A little to the west of this lake, and, perhaps, connected with it, originates the head stream of the North Manistic River of Lake Michigan, running southwest. Towards the northeast the Takwymenon takes its way, winding through level grassy plains, till it reaches the rim of the geological basin that circumscribes Lake Superior. The height of this point is conjectural. It is probably one hundred and fifty feet above the level of the lake.

To comprehend the geography of the region, it is necessary to advert to the fact that the sandstone formation, which appears in the picturesque form of the Pictured Rocks, is last seen in its range eastward at La Pointe des Grande Sable, where its surface is of a compact structure and dull red color. Between this locality and the bold cape of Point Iroquois, at the head of St. Mary's River, there intervenes an extensive formation of gravel, boulders, and sand. The length of this line of coast is about ninety miles, its breadth to the basinic rim, perhaps thirty. It is covered with small pines, spruce, birch, and poplar, with frequent sphagnous tracts and ponds; the lake shore, where the sands are continually accumulated, being higher than the interior portions. It has, from early days, been a favorite resort for beaver, from which it is called by the natives, Namikong, meaning, excellent place of beavers.

This tract of the Namikong is primarily due to diluvial formations, with a comparatively recent hem of lake action, consisting of sands and pebbles pushed up by the waves of Lake Superior. Through this tract, from the plateaux, four small rivers make their way to the lake. They are, in their order, from west to east, the river of Grand Mauvais, the Twin River, the Shelldrake, and the Tacquimenon, which enters the lake fifteen miles from Point Iroquois.

Of these streams, the Tacquimenon carries the largest body of water into the lake. It is already a stream of seventy feet wide, and three feet deep, when it reaches the rim of sandstone rocks referred to. Over these, it is plunged, at a single perpendicular leap, forty feet, falling like a curtain. It drops into a vast concavity in the sand rock, where the water is of unfathomable depth, black and still. I had reached this point in a canoe manned by Indians. They had urged their way up a very rapid brawling bed for six miles above the lower falls, and when we reached this still, deep, and dark basin, they said that care was required to keep from under the suction of the falling sheet.

The lower falls of the stream are probably twelve or fourteen feet. They are broken into several fan-shaped cascades, and present a picturesque appearance—an idea which has also impressed the Chippewas, for they refer to it as a favorite locality of fairies. Hence their name for it. Immediately below these falls the river winds about, making a peninsula, which is covered with deciduous trees and a fertile soil. The amount of water power at this point is such as must command attention whenever the country justifies settlement.

5. *Suggestions respecting the Geological Epoch of the Deposit of Sandstone Rock at St. Mary's Falls.* By HENRY R. SCHOOLCRAFT.

Lake Superior presents to the eye the singular spectacle of a body of pure translucent water, five hundred miles in length from east to west, and one hundred and·eighty or two hundred miles wide. This vast mass of water is thought to have an extreme depth—I know not on what principles—of nine hundred feet deep. It lies at an elevation of six hundred feet above the Atlantic ocean, at high water.

From this depth there has been protruded from its bottom two species of formations, which were thus elevated by volcanic forces, namely, the trap and the granitical series. Cones and high mural cliffs, with large rents, make this basis one of great inequalities. To fill up these, the sedimentary rocks, by a natural law of gravitation, let fall the dissolved and suspended matter which constitutes the horizontal strata, such as the neutral and deep-colored sandstones. This process also gives origin to grauwackes and the grauwacke slates and the argillites. But these horizontal deposits do not all retain their horizontality. They were tilted up by other volcanic forces, after the deposition and hardening of the sandstones, as we see them at the north foot of the Porcupine Mountains and along the rugged valley of the St. Louis River.

This secondary upheaval or series of upheavals, is conceived to furnish proof of epochs. Strata of the same mineral constitution and system of formation which are upheaved, are clearly of posterior age to the horizontal. Some of these strata of the secondary epoch have only had their horizontality disturbed, while others are quite vertical. Yet, the disturbances of an epoch are only relative, and it remains true that any disturbance, however slight, in the fundamental series, throws the epoch beyond the newer fletz and tertiary formations.

Some theory of this kind is necessary in scrutinizing the position of the St. Mary's sandstone, which is manifestly of the palaozoic era. It has felt the impulse of disturbance, although it appears to be little. Evidences of this are most perceptible in the British Channel, on the north side of the Island of St. Joseph.

This channel, and, indeed, the entire course of the river up to Lake Superior, is the line of juxtaposition between the rocks of elder and the secondary epoch. At the extreme foot of Sugar Island occurs the remains of a stratum of the sandstone era, consisting of white quartz filled with coarse red jasper pebbles. I observed remains of this stratum of remarkable rock, which have been broken off and swept away in the basin of Lake Huron, deposited in boulder masses on its southern shores.

The sandstone of St. Mary's is, structurally, brittle, fissile, and worthless, as a building material. Its substructure is complicated and made up of thin layers exactly deposited, as if from watery suspension, but deposited without disturbance. These sub-layers of construction, are sometimes cut off by parallel lines at right angles, or by new series of layers diagonally formed, or in echelon.

3. INDIAN TRIBES.

VIII.

CONDITION AND DISPOSITION.

1. *Official Report of an Expedition through Upper Michigan and Northern Wisconsin in 1831.*

SAULT STE. MARIE, Sept. 21, 1831.

SIR: In compliance with instructions to endeavor to terminate the hostilities between the Chippewas and Sioux, I proceeded into the Chippewa country with thirteen men in two canoes, having the necessary provisions and presents for the Indians, an interpreter, a physician to attend the sick, and a person in charge of the provisions and other public property. The commanding officer of Fort Brady furnished me with an escort of ten soldiers, under the command of a lieutenant; and I took with me a few Chippewas, in a canoe provided with oars, to convey a part of the provisions. A flag was procured for each canoe. I joined the expedition at the head of the portage, at this place, on the 25th of June; and, after visiting the Chippewa villages in the belt of country between Lake Superior and the Mississippi, in latitudes 44° to 46°, returned on the 4th of September, having been absent seventy-two days, and travelled a line of country estimated to be two thousand three hundred and eight miles. I have now the honor to report to you the route pursued, the means employed to accomplish the object, and such further measures as appear to me to be necessary to give effect to what has been done, and to insure a lasting peace between the two tribes.

Reasons existed for not extending the visit to the Chippewa bands on the extreme Upper Mississippi, on Red Lake, and Red River, and the River De Corbeau. After entering Lake Superior, and traversing its southern shores to Point Chegoimegon, and the adjacent cluster of islands, I ascended the Mauvaise River to a portage of 8¼ miles into the Kaginogumac, or Long Water Lake. This lake is about eight miles long, and of very irregular width. Thence, by a portage of 280 yards, into Turtle Lake; thence, by a portage of 1,075 yards, into Clary's Lake, so called; thence, by a portage of 425 yards, into Lake Polyganum; and thence, by a portage of 1,050 yards, into the Namakagon River, a branch of the River St. Croix of the Upper Mississippi. The distance from Lake Superior to this spot is, by estimation, 124 miles.

We descended the Namakagon to the Pukwaewa, a rice lake, and a Chippewa village of eight permanent lodges, containing a population of 53 persons, under a local chief called Odabossa. We found here gardens of corn, potatoes, and pumpkins, in a very neat state of cultivation. The low state of the water, and the consequent difficulty of the navigation, induced me to leave the provisions and stores at this place, in charge of Mr. Woolsey, with directions to proceed (with part of the men, and the aid of the Indians) to *Lac Courtorielle*, or Ottowa Lake, and there await my arrival. I then descended the Namakagon in a light canoe, to its discharge into the St. Croix, and down the latter to Yellow River, the site of a trading-post and an Indian village, where I had, by runners, appointed a council. In this trip I was accompanied by Mr. Johnson, sub-agent, acting as interpreter, and by Dr. Houghton, adjunct professor of the Rensselaer school. We reached Yellow River on the 1st of August, and found the Indians assembled. After terminating the business of the council (of which I shall presently mention the results), I reascended the St. Croix and the Namakagon, to the portage which intervenes between the latter and Lac Courtorielle. The first of the series of carrying-places is about three miles in length, and terminates at the Lake of the Isles (*Lac des Isles*); after crossing which, a portage of 750 yards leads to *Lac du Gres*. This lake has a navigable outlet into Ottowa Lake, where I rejoined the advanced party (including Lieutenant Clary's detachment) on the 5th of August.

Ottowa Lake is a considerable expanse of water, being about twelve miles long, with irregular but elevated shores. A popu-lous Chippewa village and a trading-post are located at its outlet, and a numerous Indian population subsists in the vicinity. It is situated in a district of country which abounds in rice lakes, has a proportion of prairie or burnt land, caused by the ravages of fire, and, in addition to the small fur-bearing animals, has several of the deer species. It occupies, geographically, a central situa-tion, being intermediate, and commanding the communications between the St. Croix and Chippewa Rivers, and between Lake Superior and the Upper Mississippi. It is on the great slope of land descending towards the latter, enjoys a climate of compara-tive mildness, and yields, with few and short intervals of ex-treme want, the means of subsistence to a population which is still essentially erratic. These remarks apply, with some modi-fications, to the entire range of country (within the latitudes men-tioned) situated west and south of the high lands circumscribing the waters of Lake Superior. The outlet of this Lake (Ottowa) is a fork of Chippewa River, called Ottowa River.

I had intended to proceed from this lake, either by following down the Ottowa branch to its junction with the main Chippewa, and then ascending the latter into Lac du Flambeau, or by de-scending the Ottowa branch only to its junction with the north-west fork, called the Ochasowa River; and, ascending the latter to a portage of sixty *pauses*, into the Chippewa River. By the latter route time and distance would have been saved, and I should, in either way, have been enabled to proceed from Lac du Flambeau to Green Bay by an easy communication into the Upper Ouisconsin, and from the latter into the Menomonie River, or by Plover Portage into Wolf River. This was the route I had de-signed to go on quitting Lake Superior; but, on consulting my Indian maps, and obtaining at Ottowa Lake the best and most recent information of the distance and the actual state of the water, I found neither of the foregoing routes practicable, without extending my time so far as to exhaust my supplies. I was finally determined to relinquish the Lac du Flambeau route, by learning that the Indians of that place had dispersed, and by knowing that a considerable delay would be caused by reassembling them.

The homeward route by the Mississippi was now the most eligible, particularly as it would carry me through a portion of country occupied by the Chippewas, in a state of hostility with the Sioux, and across the disputed line at the mill. Two routes, to arrive at the Mississippi, were before me—either to follow down the outlet of Ottowa Lake to its junction with the Chippewa, and descend the latter to its mouth, or to quit the Ottowa Lake branch at an intermediate point, and, after ascending a small and very serpentine tributary, to cross a portage of 6,000 yards into Lake Chetac. I pursued the latter route.

Lake Chetac is a sheet of water about six miles in length, and it has several islands, on one of which is a small Chippewa village and a trading-post. This lake is the main source of Red Cedar River (called sometimes the *Folle Avoine*), a branch of the Chippewa River. It receives a brook at its head from the direction of the portage, which admits empty canoes to be conveyed down it two *pauses*, but is then obstructed with logs. It is connected by a shallow outlet with Weegwos Lake, a small expanse which we crossed with paddles in twenty-five minutes. The passage from the latter is so shallow that a portage of 1,295 yards is made into Balsam of Fir or *Sapin* Lake. The baggage is carried this distance, but the canoes are brought through the stream. Sapin Lake is also small; we were thirty minutes in crossing it. Below this point, the river again expands into a beautiful sheet of water, called Red Cedar Lake, which we were an hour in passing; and afterward into *Bois François*, or Rice Lake. At the latter place, at the distance of perhaps sixty miles from its head, I found the last fixed village of Chippewas on this stream, although the hunting camps, and other signs of temporary occupation, were more numerous below than on any other part of the stream. This may be attributed to the abundance of the Virginia deer in that vicinity, many of which we saw, and of the elk and moose, whose tracks were fresh and numerous in the sands of the shore. Wild rice is found in all the lakes. Game, of every species common

to the latitude, is plentiful. The prairie country extends itself
into the vicinity of Rice Lake; and for more than a day's march
before reaching the mouth of the river, the whole face of the
country puts on a sylvan character, as beautiful to the eye as it
is fertile in soil, and spontaneously productive of the means of
subsistence. A country more valuable to a population having
the habits of our northwestern Indians could hardly be conceived
of; and it is therefore cause of less surprise that its possession
should have been so long an object of contention between the
Chippewas and Sioux.

About sixty miles below Rice Lake commences a series of
rapids, which extend, with short intervals, 24 miles. The remainder of the distance, to the junction of this stream with the
Chippewa, consists of deep and strong water. The junction itself
is characterized by commanding and elevated grounds, and a
noble expanse of waters. And the Chippewa River, from this
spot to its entrance into the Mississippi, has a depth and volume,
and a prominence of scenery, which mark it to be inferior to
none, and superior to most of the larger tributaries of the Upper
Mississippi. Before its junction, it is separated into several
mouths, from the principal of which the observer can look into
Lake Pepin. Steamboats could probably ascend to the falls.

The whole distance travelled, from the shores of Lake Superior
to the mouth of the Chippewa, is, by estimation, 643 miles, of
which 138 should be deducted for the trip to Yellow River leaving the direct practicable route 505 miles. The length of the
Mauvaise to the portage is 104; of the Namakagon, from the
portage, 161; of the Red Cedar, 170; of the Chippewa, from the
entrance of the latter, 40. Our means of estimating distances
was by time, corrected by reference to the rapidity of water and
strength of wind, compared with our known velocity of travelling
in calm weather on the lakes. These estimates were made and
put down every evening, and considerable confidence is felt in
them. The courses were accurately kept by a canoe compass.
I illustrate my report of this part of the route by a map protracted by Dr. Houghton. On this map, our places of encampment, the sites and population of the principal Indian villages,
the trading-posts, and the boundary lines between the Sioux and

Chippewa, are indicated. And I refer you to it for several details which are omitted in this report.

The present state of the controversy between the Sioux and the Chippewas will be best inferred from the facts that follow. In stating them, I have deemed it essential to preserve the order of my conferences with the Indians, and to confine myself, almost wholly, to results.

Along the borders of Lake Superior, comparatively little alarm was felt from the hostile relation with the Sioux. But I found them well informed of the state of the difficulties, and the result of the several war-parties that had been sent out the last year. A system of information and advice is constantly kept up by runners; and there is no movement meditated on the Sioux borders, which is not known and canvassed by the lake bands.

They sent warriors to the scene of conflict last year, in consequence of the murder committed by the Sioux on the St. Croix. Their sufferings from hunger during the winter, and the existence of disease at Torch Lake (*Lac du Flambeau*), and some other places, together with the entire failure of the rice crop, had produced effects, which were depicted by them and by the traders in striking colors. They made these sufferings the basis of frequent and urgent requests for provisions. This theme was strenuously dwelt upon. Whatever other gifts they asked for, they never omitted the gift of food. They made it their first, their second, and their third request.

At Chegoimegon, on Lake Superior (or *La Pointe*, emphatically so called), I held my first and stated council with the Indians. This is the ancient seat of the Chippewa power in this quarter. It is a central and commanding point, with respect to the country lying north, and west, and south of it. It appears to be the focus from which, as radii from a centre, the ancient population emigrated; and the interior bands consequently look back to it with something of the feelings of parental relation. News from the frontiers flies back to it with a celerity which is peculiar to the Indian mode of express. I found here, as I had expected, the fullest and most recent information from the lines. Mozojeed, the principal man at Ottowa Lake, had recently visited them for the purpose of consultation; but returned on the alarm of an attack upon his village.

The Indians listened with attention to the message transmitted to them from the President, and to the statements with which it was enforced. Pezhickee, the venerable and respected chief of the place, was their speaker in reply. He lamented the war, and admitted the folly of keeping it up; but it was carried on by the Chippewas in self-defence, and by volunteer parties of young men, acting without the sanction of the old chiefs. He thought the same remark due to the elder Sioux chiefs, who probably did not sanction the crossing of the lines, but could not restrain their young men. He lived, he said, in an isolated situation, did not mingle in the interior broils, and did not deem himself responsible for acts done out of his own village, and certainly not for the acts of the villages of Torch Lake, Ottowa Lake, and the St. Croix. He had uniformly advised his people to sit still and remain at peace, and he believed that none of his young men had joined the war-parties of last year. The Government, he said, should have his hearty co-operation in restoring peace. He referred to the sub-agency established here in 1826, spoke of its benefits, and wished to know why the agent had been withdrawn, and whether he would be instructed to return? In the course of his reply, he said that formerly, when the Indians lived under the British government, they were usually told what to do, and in very distinct terms; but they were now at a loss. From what had been said and done at the treaty of Fond du Lac, he expected the care and protection of the American government, and that they would advance towards, instead of (as in the case of the sub-agency) withdrawing from them. He was rather at a loss for our views respecting the Chippewas, and he wished much for my advice in their affairs.

I thought it requisite to make a distinct reply to this point. I told him that when they lived under the British government, they were justified in shaping their course according to the advice they received; but that, on the transfer of the country, their allegiance was transferred with it. And when our Government hoisted its flag at Mackinac (1796), it expected from the Indians living within our boundaries the respect due to it; and it acknowledged, at the same time, the reciprocal obligations of care and protection. That it always aimed to fulfil these obligations, of which facts within his own knowledge and memory would afford ample proofs. I referred him to the several efforts the Government had made to establish a lasting peace between the Chippewas and Sioux; for which purpose the President had sent one of his principal men (alluding to Gov. Cass), in 1820, who had visited their most extreme northwestern villages, and induced themselves and the Sioux to smoke the pipe of peace together at St. Peter's. . In accordance with these views, and acting on the information then acquired, the President had established an agency for their tribe at Sault Ste. Marie, in 1822. That, in 1825, he had assembled at Prairie du Chien all the tribes who were at variance on the Upper Mississippi, and persuaded them to make peace, and, as one of the best means of insuring its permanency, had fixed the boundaries of their lands. Seeing that the Chippewas and Sioux still continued an harassing and useless contest, he had sent me to remind them of this peace and these boundaries, which, I added, you, Perikee, yourself agreed to, and signed, in my presence. I come to bring you back to the terms of this treaty. Are not these proofs of his care and attention? Are not these clear indications of his views respecting the Chippewas? The chief was evidently affected by this recital. The truth appeared to strike him forcibly; and he said, in a short reply, that he was now *advised;* that he would hereafter feel himself to be advised, &c. He made some remarks on the establishment of a mission school, &c., which, being irrelevant, are omitted. He presented a pipe, with an ornamented stem, as a token of his friendship, and his desire of peace.

I requested him to furnish messengers to take belts of wampum and tobacco, with three separate messages, viz: to Yellow River, to Ottowa Lake, and to Lac du Flambeau, or Torch Lake; and also, as the water was low, to aid me in the ascent of the Mauvaise River, and to supply guides for each of the military canoes, as the soldiers would here leave their barge, and were unacquainted with the difficulties of the ascent. He accordingly sent his oldest son (Che-che-gwy-ung) and another person, with the messages, by a direct trail, leading into the St. Croix country. He also furnished several young Chippewas to aid us on the Mauvaise, and to carry baggage on the long portage into the first intermediate lake west of that stream.

After the distribution of presents, I left Chegoimegon on the 18th of July. The first party of Indians met at the Namakagon, belonging to a Chippewa village called Pukwaewa, having, as its geographical centre and trading-post, Ottowa Lake. As I had directed part of the expedition to precede me there, during my journey to Yellow River, I requested these Indians to meet me at Ottowa Lake, and assist in conveying the stores and provisions to that place—a service which they cheerfully performed. On ascending the lower part of the Namakagon, I learned that my messenger from Lake Superior had passed, and, on reaching Yellow River, I found the Indians assembled and waiting. They were encamped on an elevated ridge, called Pekogunagun, or the Hip Bone, and fired ·a salute from its summit. Several of the neighboring Indians came in after my arrival. Others, with their chiefs, were hourly expected. I did not deem it necessary for all to come in, but proceeded to lay before them the objects of my visit, and to solicit their co-operation in an attempt to make a permanent peace with the Sioux, whose borders we then were near. Kabamappa, the principal chief, not being a speaker, responded to my statements and recommendations through another person (Sha-ne-wa-gwun-ai-be). He said that the Sioux were of bad faith; that they never refused to smoke the pipe of peace with them, and they never failed to violate the promise of peace thus solemnly made. He referred to an attack they made last year on a band of Chippewas and half-breeds, and the murder of four persons. Perpetual vigilance was required to meet these inroads. Yet he could assert, fearlessly, that no Chippewa warparty from the St. Croix had crossed the Sioux line for years; that the murder he had mentioned was committed within the Chippewa lines; and although it was said, at the treaty of Prairie du Chien, that the first aggressor of territorial rights should be punished, neither punishment was inflicted by the Government, nor had any atonement or apology thus far been made for this act by the Sioux. He said his influence had been exerted in favor of peace; that he had uniformly advised both chiefs and warriors to this effect; and he stood ready now to do whatever it was reasonable he should do on the subject.

I told him it was not a question of recrimination that was before us. It was not even necessary to go into the inquiry of who had spilt the first blood since the treaty of Prairie du Chien. The treaty had been violated. The lines had been crossed. Murders had been committed by the Chippewas and by the Sioux. These murders had reached the ears of the President, and he was resolved to put a stop to them. I did not doubt but that the advice of the old chiefs, on each side, had been pacific. I did not doubt but that his course had been *particularly* so. But rash young men, of each party, had raised the war-club; and when they could not go openly, they went secretly. A stop must be put to this course, and it was necessary the first movement should be made *somewhere*. It was proper it should be made here, and be made at this time. Nothing could be lost by it; much might be gained; and if a negotiation was opened with the Sioux chiefs while I remained, I would second it by sending an explanatory message to the chiefs and to their agent. I recommended that Kabamappa and Shakoba, the war-chief of Snake River, should send jointly wampum and tobacco to the Petite Corbeau and to Wabisha, the leading Sioux chiefs on the Mississippi, inviting them to renew the league of friendship, and protesting their own sincerity in the offer. I concluded by presenting him with a flag, tobacco, wampum, and ribbons, to be used in the negotiation. After a consultation, he said he would not only send the messages, but, as he now had the protection of a flag, he would himself go with the chief Shakoba to the Petite Corbeau's village. I accompanied these renewed offers of peace with explanatory messages, in my own name, to Petite Corbeau and to Wabisha, and a letter to Mr. Taliaferro, the Indian agent at St. Peter's, informing him of these steps, and soliciting his co-operation. A copy of this letter is hereunto annexed. I closed the council by the distribution of presents; after which the Indians called my attention to the conduct of their trader, &c.

Information was given me immediately after my arrival at Yellow River, that Neenaba, a popular war-leader from the Red Cedar fork of Chippewa River, had very recently danced the war-dance with thirty men at Rice Lake of Yellow River, and that his object was to enlist the young men of that place in a war-party against the Sioux. I also learned that my message for Ottowa Lake had been promptly transmitted through Neenaba, whom I was now anxious to see. I lost not an hour in reascending the St. Croix and the Namakagon. I purchased two additional canoes of the Indians, and distributed my men in them, to lighten the draught of water, and facilitate the ascent; and, by pushing early and late, we reached Ottowa Lake on the fifth day in the morning. Neenaba had, however, delivered his message, and departed. I was received in a very friendly and welcome manner, by Mozojeed, of the band of Ottowa Lake; Wabezhais, of the Red Devil's band of the South Pukwaewa; and Odabossa, of the Upper Namakagon. After passing the usual formalities, I prepared to meet them in council the same day, and communicate to them the objects of my mission.

In the course of the conference at this place, I obtained the particulars of a dispute which had arisen between the Chippewas of this quarter, which now added to their alarm, as they feared the latter would act in coincidence with their ancient enemies, the Sioux. The reports of this disturbance had reached me at the Sault, and they continued, with some variations, until my arrival here. The following are the material facts in relation to this new cause of disquietude: In the summer of 1827, Okunzhewug, an old woman, the wife of Kishkemun, the principal chief of Torch Lake, a man superannuated and blind, attended the treaty of Butte des Morts, bearing her husband's medal. She was treated with the respect due to the character she represented, and ample presents were directed to be given to her; among other things, a handsome hat. The latter article had been requested of her by a young Menomonie, and refused. It is thought a general feeling of jealousy was excited by her good reception. A number of the Menomonies went on her return route as far as the Clover Portage, where she was last seen. Having never returned to her village, the Chippewas attributed her death to the Menomonies. Her husband died soon after; but she had numerous and influential relatives to avenge her real or supposed murder. This is the account delivered by the Chippewas, and it is corroborated by reports from the traders of that section of the country. Her singular disappearance and secret death at the Clover Portage, is undisputed; and whether caused or not by any agency of the Menomonies, the belief of such agency, and that of the most direct kind, is fixed in the minds of the Chippewas, and has furnished the basis of their subsequent acts in relation to the Menomonie hunting-parties who have visited the lower part of Chippewa

River. Two women belonging to one of these parties were killed by a Chippewa war-party traversing that part of the country the ensuing year. The act was disclaimed by them as not being intentional, and it was declared they supposed the women to be Sioux. On a close inquiry, however, I found the persons who committed this act were relatives of Okunzewug, which renders it probable that the murder was intentionally perpetrated. This act further widened the breach between the two hitherto fraternal tribes; and the Chippewas of this quarter began to regard the Menomonie hunting-parties, who entered the mouth of the Chippewa River, as intruders on their lands. Among a people whose means of verbal information is speedy, and whose natural sense of right and wrong is acute, the more than usual friendship and apparent alliance which have taken place between the Menomonies and Sioux, in the contest between the Sacs and Foxes, and the murder by them jointly of the Fox chief White Skin and his companions at a smoking council, in 1830, have operated to increase the feeling of distrust; so much so, that it was openly reported at Chegoimegon, at Yellow River, and Ottowa Lake, that the Menomonies had formed a league with the Sioux against the Chippewas also, and they were fearful of an attack from them. A circumstance that had given point to this fear, and made it a subject of absorbing interest, when I arrived at Ottowa Lake, was the recent murder of a Menomonie chief by a Chippewa of that quarter, and the demand of satisfaction which had been made (it was sometimes said) by the Indian agent at Prairie du Chien, and sometimes by the commanding officer, with a threat to march troops into the country. This demand, I afterward learned from the Indians at Rice Lake, and from a conversation with General Street, the agent at Prairie du Chien, had not been made, either by himself or by the commanding officer; and the report had probably arisen from a conversation held by a subaltern officer in command of a wood or timber-party near the mouth of the Chippewa River, with some Chippewas who were casually met. Its effects, however, were to alarm them, and to lead them to desire a reconciliation with the Menomonies. I requested them to lose no time in sending tobacco to the Menomonies, and adjusting this difference. Mozojeed observed that the murder of the Menomonie had been committed by a person *non compos,* and he deplored the folly of it, and disclaimed all agency in it for himself and his band. The murderer, I believe, belonged to his band; he desired a reconciliation. He also said the measures adopted at Yellow River, to bring about a firm peace with the Sioux, had his fullest approbation, and that nothing on his part should be wanting to promote a result in every view so wise and so advantageous to the Indians. In this sentiment, Wabezhais and Odabossa, who made distinct speeches, also concurred. They confirmed their words by pipes, and all the assembly made an audible assent. I invested Mozojeed with a flag and a medal, that he might exert the influence he has acquired among the Indians beneficially for them and for us, and that his hands might thus be officially strengthened to accomplish the work of pacification. I then distributed presents to the chiefs, warriors, women, and children, in the order of their being seated, and immediately embarked, leaving them under a lively and enlivened sense of the good-will and friendship of the American government, on this first official visit to them, and with a sincere disposition, so far as could be judged, to act in obedience to its expressed and known wishes.

The Indians at Torch Lake being dispersed, and my message to them not having been delivered, from this uncertainty of their location, I should have found reasons for not proceeding in that direction, independent of the actual and known difficulties of the route at that time. I was still apprehensive that my appearance had not wholly disconcerted the war-party of Neenaba, and lost no time in proceeding to his village on the Red Cedar fork. We found the village at Lake Chetac, which in 1824 was 217 strong, almost totally deserted, and the trading-house burnt. Scattering Indians were found along the river. The mutual fear of interruption was such that Mr. B. Cadotte, Sen., the trader at Ottowa Lake, thought it advisable to follow in our train for the purpose of collecting his credits at Rice Lake.

While at breakfast on the banks of Sapin Lake, a returning war-party entered the opposite side of it; they were evidently surprised, and they stopped. After reconnoitring us, they were encouraged to advance, at first warily, and afterward with confidence. There were eight canoes, with two men in each; each man had a gun, war-club, knife, and ammunition-bag: there was nothing else except the apparatus for managing the canoe. They were all young men, and belonged to the vicinity of Ottowa Lake. Their unexpected appearance at this place gave me the first information that the war-party at Neenaba had been broken up. They reported that some of their number had been near the mill, and that they had discovered signs of the Sioux being out, in the moose having been driven up, &c. In a short conference, I recited to them the purpose of the council at Ottowa Lake, and referred them to their chiefs for particulars, enjoining their acquiescence in the proposed measures.

I found at Rice Lake a band of Chippewas, most of them young men, having a prompt and martial air, encamped in a very compact form, and prepared at a moment's notice, for action. They saluted our advance with a smartness and precision of firing that would have done honor to drilled troops. Neenaba was absent on a hunting-party; but one of the elder men pointed out a suitable place for my encampment, as I intended here to put new bottoms to my bark canoes. He arrived in the evening, and visited my camp with forty-two men. This visit was one of ceremony merely; as it was late, I deferred anything further until the following day. I remained at this place part of the 7th, the 8th, and until 3 o'clock on the 9th of August. And the following facts present the result of several conferences with this distinguished young man, whose influence is entirely of his own creation, and whose endowments, personal and mental, had not been misrepresented by the Indians on my route, who uniformly spoke of him in favorable terms. He is located at the most advanced point towards the Sioux borders, and, although not in the line of ancient chiefs, upon him rests essentially the conduct of affairs in this quarter. I therefore deemed it important to acquire his confidence and secure his influence, and held frequent conversations with him. His manner was frank and bold, equally free from servility and repulsiveness. I drew his attention to several subjects. I asked him whether the saw-mill on the lower part of the Red Cedar, was located on Chippewa lands? He said, Yes. Whether it was built with the consent of the Chippewas? He said, No; it had been built, as it were, by stealth. I asked him if anything had been subsequently given them in acknowledgment of their right to the soil? He said, No; that the only ac-

knowledgment was their getting tobacco to smoke when they
visited the mill; that the Sioux claimed it to be on their side of
the line, but the Chippewas contended that their line ran to a certain bluff and brook below the mill. I asked him to draw a map
of the lower part of Chippewa River, with all its branches, showing the exact lines as fixed by the treaty at Prairie du Chien, and
as understood by them. I requested him to state the facts respecting the murder of the Menomonie, and the causes that led to
it; and whether he, or any of his band, received any message
from the agent or commanding officer at Prairie du Chien, de-
manding the surrender of the murderer? To the latter inquiry
he answered promptly, No. He gave in his actual population af
142; but it is evident that a very considerable additional population, particularly men, resort there for the purpose of hunting a
part of the year.

The day after my arrival, I prepared for and summoned the
Indians to a council, with the usual formalities. I opened it by
announcing the objects of my visit. Neenaba and his followers
listened to the terms of the message, the means I had adopted to
enforce it, and, finally, to the request of co-operation on the part
of himself and band, with strict attention. He confined his reply
to an expression of thanks, allusions to the peculiarity of his
situation on an exposed frontier, and general sentiments of
friendship. He appeared to be mentally embarrassed by my
request to drop the war-club, on the successful use of which he
had relied for his popularity, and whatever of real power he possessed. He often referred to his young men, over whom he
claimed no superiority, and who appeared to be ardently attached
to him. I urged the principal topic upon his attention, presenting it in several lights. I finally conferred on him, personally,
a medal and flag, and directed the presents intended for his band
to be laid, in gross, before him.

After a pause, Neenaba got up, and spoke to the question, connecting it with obvious considerations, of which mutual rights,
personal safety, and the obligation to protect the women and
children, formed the basis. The latter duty was not a slight one.
Last year, the Sioux had killed a chief on the opposite shore of
the lake, and, at the same time, decoyed two children, who were
in a canoe, among the rice, and killed and beheaded them. He

said, in allusion to the medal and flag, that these marks of honor were not necessary to secure his attention to any requests made by the American government. And after resuming his seat awhile (during which he overheard some remarks not pleasing to him, from an Indian on the opposite side of the ring), he finally got up and declined receiving them until they were eventually pressed upon him by the young warriors. Everything appeared to proceed with great harmony, and the presents were quickly distributed by one of his men. It was not, however, until the next day, when my canoes were already put in the water, that he came with his entire party, to make his final reply, and to present the peace-pipe. He had thrown the flag over one arm, and held the war-club perpendicularly in the other hand. He said that, although he accepted the one, he did not drop the other; he held fast to both. When he looked at the one, he should revert to the counsels with which it had been given, and he should aim to act upon those counsels; but he also deemed it necessary to hold fast the war-club; it was, however, with a determination to use it in defence, and not in attack. He had reflected upon the advice sent to the Chippewas by the President, and particularly that part of it which counselled them to sit still upon their lands; but while they sat still, they also wished to be certain that their enemies would sit still. And the pipe he was now about to offer, he offered with a request that it might be sent to the President, asking him to use his power to prevent the Sioux from crossing the lines. The pipe was then lit, handed round, the ashes knocked out, and a formal presentation of it made. This ceremony being ended, I shook hands with them, and immediately embarked.

On the second day afterward, I reached the saw-mill, the subject of such frequent allusion, and landed there at 7 o'clock in the morning. I found a Mr. Wallace in charge, who was employed, with ten men, in building a new dam on a brook of the Red Cedar, the freshet of last spring having carried away the former one. I inquired of him where the line between the Sioux and Chippewas crossed. He replied that the line crossed above the mill, he did not precisely know the place; adding, however, in the course of conversation, that he believed the land in this vicinity originally belonged to the Chippewas. He said it was seven years since any Sioux had visited the mill; and that the latter was owned by persons at Prairie du Chien.

The rapids of the Red Cedar River extend (according to the estimates contained in my notes) about twenty-four miles. They commence a few miles below the junction of Meadow River, and terminate about two miles below the mills. This extension of falling water, referred to in the treaty as a fixed point, has led to the existing uncertainty. The country itself is of a highly valuable character for its soil, its game, its wild rice, and its wood. We found the butternut among those species which are locally included under the name of *Bois franc*, by the traders. The land can, hereafter, be easily brought into cultivation, as it is inter-spersed with prairie; and its fine mill privileges will add to its value. Indeed, one mile square is intrinsically worth one hundred miles square of Chippewa country, in some other places.

The present saw-mills (there are two), are situated 65 miles from the banks of the Mississippi. They are owned exclusively by private citizens, and employed for their sole benefit. The boards are formed into rafts; and these rafts are afterward attached together, and floated down the Mississippi to St. Louis, where they command a good price. The business is understood to be a profitable one. For the privilege, no equivalent has been paid either to the Indians or to the United States. The first mill was built several years ago, and before the conclusion of the treaty of Prairie du Chien, fixing boundaries to the lands. A permit was given for building, either verbal or written, as I have been informed, by a former commanding officer at Prairie du Chien. I make these statements in reference to a letter I have received from the Department since my return, but which is dated June 27th, containing a complaint of one of the owners of the mill, that the Chippewas had threatened to burn it, and requesting me to take the necessary precautionary measures. I heard nothing of such a threat, but believe that the respect which the Chippewas have professed, through me, for the American government, and the influence of my visit among them, will prevent a resort to any measures of violence; and that they will wait the peaceable adjustment of the line on the rapids. I will add that, *wherever* that line may be determined, in a reasonable probability, to fall, the mill itself cannot be supplied with logs for any length of time, if *it is now so supplied*, without cutting them on Chippewa lands, and rafting them down the Red Cedar. Many of the logs heretofore sawed at this mill, have been rafted *up stream*, to the mill. And I understood from the person in charge of it, that he was now anxious to ascertain new sites for chopping; that his expectations were directed up the stream, but that his actual knowledge of the country, in that direction, did not embrace a circumference of more than five miles.

The line between the Chippewa and Sioux, as drawn on the MS. map of Neenaba, strikes the rapids on Red Cedar River at a brook and bluff a short distance below the mill. It proceeds thence, across the point of land between that branch of the main Chippewa, to an island in the latter; and thence, up stream, to the mouth of Clearwater River, as called for by the treaty, and from this point to the bluffs of the Mississippi Valley (where it corners on Winnebago land), on Black River, and not to the "*mouth*" of Black River, as erroneously inserted in the 5th article of the treaty; the Chippewas never having advanced any claims to the lands at the mouth of Black River. This map, being drawn by a Chippewa of sense, influence, and respectability, an exact copy of it is herewith forwarded for the use of the Department, as embracing the opinions of the Chippewas on this point. The lines and geographical marks were drawn on paper by Neenaba himself, and the names translated and written down by Mr. Johnston.

It is obvious that the adjustment of this line must precede a permanent peace on this part of the frontiers. The number of Chippewas particularly interested in it is, from my notes, 2,102; to which, 911 may be added for certain bands on Lake Superior. It embraces 27 villages, and the most influential civil and war chiefs of the region. The population is enterprising and warlike. They have the means of subsistence in *comparative* abundance. They are increasing in numbers. They command a ready access to the Mississippi by water, and a ready return from it by land. Habits of association have taught them to look upon this stream as the theatre of war. Their young men are carried into it as the natural and almost only means of distinction. And it is in coincidence with all observation to say that they are now, as they were in the days of Captain Carver, the terror of the east bank of this river, between the St. Croix and Chippewa Rivers. No other tribe has now, or has had, within the memory of man, a village or permanent possession on this part of the shore. It is landed on in fear. It is often passed by other nations by stealth, and at night. Such is not an exaggerated picture. And with a knowledge of their geographical advantages, and numbers, and distribution, on the tributary streams, slight causes, it may be imagined, will often excite the young and thoughtless portion of them to raise the war-club, to chant the war-song, and follow the war-path.

To remove these causes, to teach them the folly of such a contest, to remind them of the treaty stipulations and promises solemnly made to the Government, and to the Sioux, and to induce them to renew those promises, and to act on fixed principles of political faith, were the primary objects committed to me; and they were certainly objects of exalted attainment, according as well with the character of the Government as with the spirit and moral and intellectual tone of the age. To these objects I have faithfully, as I believe, devoted the means at my command. And the Chippewas cannot, hereafter, err on the subject of their hostilities with the Sioux, without knowing that the error is disapproved by the American government, and that a continuance in it will be visited upon them in measures of severity.

Without indulging the expectation that my influence on the tour will have the effect to put an end to the spirit of predatory warfare, it may be asserted that this spirit has been checked and allayed; and that a state of feeling and reflection has been produced by it, which cannot fail to be beneficial to our relations with them, and to their relations with each other. The messages sent to the Sioux chiefs, may be anticipated to have resulted in restoring a perfect peace during the present fall and ensuing winter, and will thus leave to each party the undisturbed chase of their lands. The meditated blow of Steenaba was turned aside, and his war-party arrested and dispersed at the moment it was ready to proceed. Every argument was used to show them the folly and the insecurity of a continuance of the war. And the whole tenor and effect of my visit has been to inform and reform these remote bands. It has destroyed the charm of their seclusion. It has taught them that their conduct is under the supervision of the American government; that they depend on its care and protection; that no other government has power to regulate trade and send traders among them; finally, that an adherence to foreign counsels, and to anti-pacific maxims, can be visited upon them in measures of coercion. That their country, hitherto deemed nearly inaccessible, can be penetrated and traversed by men and troops, with baggage and provisions, even in midsummer, when the waters are lowest; and that, in proportion as they comply with political maxims, as benevolent as they are just, will they live at peace with their enemies, and have the means of subsistence for an increased population among themselves. The conduct of the traders in this quarter, and the influence they have exerted, both moral and political, cannot here be entered upon, and must be left to some other occasion, together with statistical details and other branches of information not arising from particular instructions.

It may be said that the Indians upon the St. Croix and Chippewa Rivers, and their numerous branches, have been drawn into a close intercourse with Government. But it will be obvious that a perseverance in the system of official advice and restraints, is essential to give permanence to the effects already produced, and to secure a firm and lasting peace between them and the Sioux. To this end, the settlement of the line upon the Red Cedar Fork is an object which claims the attention of the Department; and would justify, in my opinion, the calling together the parties interested, at some convenient spot near the junction of the Red Cedar River with the Chippewa. Indeed, the handsome elevation, and the commanding geographical advantages of this spot, render it one which, I think, might be advantageously occupied as a military post. Such an occupancy would have the effect to keep the parties at peace; and the point of land, on which the work is proposed to be erected, might be purchased from the Sioux, together with such part of the disputed lands near the mills as might be deemed necessary to quiet the title of the Chippewas. By acquiring this portion of country for the purposes of military occupancy, the United States would be justified in punishing any murders committed upon it; and I am fully convinced that no measure which could, at this time, be adopted, would so certainly conduce to a permanent peace between the tribes. I

therefore beg leave, through you, to submit these subjects to the
consideration of the honorable the Secretary of War, with every
distrust in my own powers of observation, and with a very full
confidence in his.

    I have the honor to be, sir,

        Very respectfully, your obedient servant,

             H. R. SCHOOLCRAFT.

To ELBERT HERRING, ESQ., *Com. Ind. Affairs.*

2. *Brief Notes of a Tour in* 1831, *from Galena, in Illinois, to Fort
Winnebago, on the source of Fox River, Wisconsin.* By HENRY
R. SCHOOLCRAFT.

Time admonishes me of my promise to furnish you some
account of my journey from Galena to Fort Winnebago. But I
confess, that time has taken away none of those features which
make me regard it as a task. Other objects have occupied so
much of my thoughts, that the subject has lost some of its vivid-
ness, and I shall be obliged to confine myself more exclusively
to my notes than I had intended. This will be particularly true
in speaking of geological facts. Geographical features impress
themselves strongly on the mind. The shape of a mountain is
not easily forgotten, and its relation to contiguous waters and
woods is recollected after the lapse of many years. The succession of plains, streams, and settlements is likewise retained in the
memory, while the peculiar plains, the soils overlaying them, and
all the variety of their mineral and organic contents, require to
be perpetuated by specimens and by notes, which impose neither
a slight nor a momentary labor.

Limited sketches of this kind are, furthermore, liable to be mis-
conceived. Prominent external objects can only be brought to
mind, and these often reveal but an imperfect notion of the pervading character of strata, and still less knowledge of their mineral contents. Haste takes away many opportunities of observation; and scanty or inconvenient means of transporting hand
specimens, often deprive us of the requisite data. Indeed, I
should be loath to describe the few facts I am about to communicate, had you not personally visited and examined the great
carboniferous and sandstone formation on the Mississippi and

Wisconsin, and thus got the knowledge of their features. The parallelism which is apparent in these rocks, by the pinnacles which have been left standing on high—the wasting effects of time in scooping out valleys and filling up declivities—and the dark and castle-looking character of the cherty limestone bluffs, as viewed from the water, while the shadows of evening are deepening around, are suited to make vivid impressions. And these broken and denuded cliffs offer the most favorable points for making geological observations. There are no places inland where the streams have cut so deep. On gaining the height of land, the strata are found to be covered with so heavy a deposit of soil, that it is difficult to glean much that can be relied on respecting the interior structure.

The angle formed by the junction of the Wisconsin with the Mississippi, is a sombre line of weather-beaten rocks. Gliding along the current, at the base of these rocks, the idea of a "hill country," of no very productive character, is naturally impressed upon the observer. And this impression came down, probably, from the days of Marquette, who was the first European, that we read of, who descended the Wisconsin, and thus became the true discoverer of the Mississippi. The fact that it yielded lead ore, bits of which were occasionally brought in by the natives, was in accordance with this opinion; and aided, it may be supposed, in keeping out of view the real character of the country. I know not how else to account for the light which has suddenly burst upon us from this bank of the Mississippi, and which has at once proved it to be as valuable for the purposes of agriculture as for those of mining, and as sylvan in its appearance as if it were not fringed, as it were, with rocks, and lying at a great elevation above the water. This elevation is so considerable as to permit a lively descent in the streams, forming numerous mill-seats. The surface of the country is not, however, broken, but may be compared to the heavy and lazy-rolling waves of the sea after a tempest. These wave-like plains are often destitute of trees, except a few scattering ones, but present to the eye an almost boundless field of native herbage. Groves of oak sometimes diversify those native meadows, or cover the ridges which bound them. Very rarely does any rock appear above the surface. The highest elevations, the Platte Mounds, and the Blue Mound,

36

are covered with soil and with trees. Numerous brooks of limpid
water traverse the plains, and find their way into either the Wisconsin, Rock River, or the Mississippi. The common deer is still
in possession of its favorite haunts; and the traveller is very often
startled by flocks of the prairie-hen rising up in his path. The
surface soil is a rich black alluvion; it yields abundant crops of
corn, and, so far as they have been tried, all the cereal gramina.
I have never, either in the West or out of the West, seen a richer
soil, or more stately fields of corn and oats, than upon one of the
plateaux of the Blue Mound.

Such is the country which appears to be richer in ores of lead
than any other mineral district in the world—which yielded forty
millions of pounds in seven years—produced a single lump of
ore of two thousand cubic feet—and appears adequate to supply
almost any amount of this article that the demands of commerce
require.

The River of Galena rises in the mineral plains of Iowa county,
in that part of the Northwestern Territory which is attached, for
the purposes of temporary government, to Michigan. It is made
up of clear and permanent springs, and has a descent which affords a very valuable water-power. This has been particularly
remarked at the curve called Mill-seat Bend. No change in its
general course, which is southwest, is, I believe, apparent after it
enters the northwest angle of the State of Illinois. The town of
Galena, the capital of the mining country, occupies a somewhat
precipitous semicircular bend, on the right (or north) bank of the
river, six or seven miles from its entrance into the Mississippi.
Backwater, from the latter, gives the stream itself the appearance, as it bears the name, of a "river," and admits steamboat
navigation thus far. It is a rapid brook immediately above the
town, and of no further value for the purpose of navigation.
Lead is brought in from the smelting furnaces, on heavy ox-
teams, capable of carrying several tons at a load. I do not know
that water *has been*, or that it *cannot* be made subservient in the
transportation of this article from the mines. The streams themselves are numerous and permanent, although they are small, and
it would require the aid of so many of these, on any projected
route, that it is to be feared the supply of water would be inadequate. To remedy this deficiency, the Wisconsin itself might be

relied on. Could the waters of this river be conducted in a canal along its valley from the portage to the bend at Arena, they might, from this point, be deflected in a direct line to Galena. This route would cut the mine district centrally, and afford the upper tributaries of the Pekatolika and Fever Rivers as feeders. Such a communication would open the way to a northern market, and merchandise might be supplied by the way of Green Bay, when the low state of water in the Mississippi prevents the ascent of boats. It would, at all times, obviate the tedious voyage, which goods ordered from the Atlantic cities have to perform through the straits of Florida and Gulf of Mexico. A railroad could be laid upon this route with equal, perhaps superior advantages. These things may seem too much like making arrangements for the next generation. But we cannot fix bounds to the efforts of our spreading population, and spirit of enterprise. Nor, after what we have seen in the way of internal improvement, in our own day and generation, should we deem anything too hard to be accomplished.

I set out from Galena in a light wagon, drawn by two horses, about ten o'clock in the morning (August 17th), accompanied by Mr. B. It had rained the night and morning of the day previous, which rendered the streets and roads quite muddy. A marly soil, easily penetrated by rain, was, however, as susceptible to the influence of the sun, and, in a much shorter period than would be imagined, the surface became dry. Although a heavy and continued shower had thoroughly drenched the ground, and covered it with superfluous water, but very little effects of it were to be seen at this time. We ascended into the open plain country, which appears in every direction around the town, and directed our course to Gratiot's Grove. In this distance, which, on our programme of the route, was put down at fifteen miles, a lively idea of the formation and character of the country is given. The eye is feasted with the boundlessness of its range. Grass and flowers spread before and beside the traveller, and, on looking back, they fill up the vista behind him. He soon finds himself in the midst of a sylvan scene. Groves fringe the tops of the most distant elevations, and clusters of trees—more rarely, open forests—are occasionally presented. The trees appear to be almost exclusively of the species of white oak and rough-bark hickory.

Among the flowers, the plant called rosin-weed attracts attention by its gigantic stature, and it is accompanied, as certainly as substance by shadow, by the wild indigo, two plants which were afterwards detected, of less luxuriant growth, on Fox River. The roads are in their natural condition; they are excellent, except for a few yards where streams are crossed. At such places there is a plunge into soft, black muck, and it requires all the powers of a horse harnessed to a wagon to emerge from the stream.

On reaching Gratiot's Grove, I handed letters of introduction to Mr. H. and B. Gratiot. These gentlemen appear to be extensively engaged in smelting. They conducted me to see the ore prepared for smelting in the log furnace; and also the preparation of such parts of it for the ash furnace as do not undergo complete fusion in the first process. The ash furnace is a very simple kind of air furnace, with a grate so arranged as to throw a reverberating flame upon the hearth where the prepared ore is laid. It is built against a declivity, and charged, by throwing the materials to be operated upon, down the flue. A silicious flux is used; and the scoria is tapped and suffered to flow out, from the side of the furnace, before drawing off the melted lead. The latter is received in an excavation made in the earth, from which it is ladled out into iron moulds. The whole process is conducted in the open air, with sometimes a slight shed. The lead ore is piled in cribs of logs, which are roofed. Hammers, ladles, a kind of tongs, and some other iron tools are required. The simplicity of the process, the absence of external show in buildings, and the direct and ready application of the means to the end, are remarkable, as pleasing characteristics about the smelting establishment.

The ore used is the common sulphuret, with a foliated, glittering and cubical fracture. It occurs with scarcely any adhering gangue. Cubical masses of it are found, at some of the diggings, which are studded over with minute crystals of calcareous spar. These crystals, when examined, have the form of the dog-tooth spar. This broad, square-shaped, and square-broken mineral, is taken from *east and west leads*, is most easy to smelt, and yields the greatest per centum of lead. It is estimated to produce fifty per cent. from the log furnace, and about sixteen more when treated with a flux in the ash furnace.

Miners classify their ore from its position in the mine. Ore from *east and west leads*, is raised from clay diggings, although these diggings may be pursued under the first stratum of rock. Ore from *north and south leads*, is termed "sheet minerals," and is usually taken from rock diggings. The vein or sheet stands perpendicularly in the fissure, and is usually struck in sinking from six to ten feet. The sheet varies in thickness from six or eight inches, in the broadest part, to not more than one. The great mass found at "Irish diggings" was of this kind.

I observed, among the piles of ore at Gratiot's, the combination of zinc with lead ore, which is denominated *dry bone*. It is cast by as unproductive. Mr. B. Gratiot also showed me pieces of the common ore which had undergone desulphuration in the log furnace. Its natural splendor is increased by this process, so as to have the appearance of highly burnished steel. He also presented me some uniform masses of lead, recrystallized from a metallic state, under the hearth of the ash furnace. The tendency to rectangular structure in these delicate and fragile masses is very remarkable. Crystallization appears to have taken place under circumstances which opposed the production of a complete and perfect cube or parallelogram, although there are innumerable rectangles of each geometric form.

In the drive from Gratiot's to Willow Springs, we saw a succession of the same objects that had formed the prominent features of the landscape from Galena. The platte mounds, which had appeared on our left all the morning, continued visible until we entered the grove that embraces the site of the springs. Little mounds of red earth frequently appeared above the grass, to testify to the labors of miners along this part of the route. In taking a hasty survey of some of the numerous excavations of Irish diggings, I observed among the rubbish small flat masses of a yellowish white amorphous mineral substance of great weight. I have not had time to submit it to any tests. It appears too heavy and compact for the earthy yellow oxide of lead. I should not be disappointed to find it an oxide of zinc. No rock stratum protrudes from the ground in this part of the country. The consolidated masses, thrown up from the diggings, appear to be silicated limestone, often friable, and not crystalline. Galena is found in open fissures in this rock.

We reached the springs in the dusk of the evening, and found good accommodations at Ray's. Distance from Galena thirty miles.

The rain fell copiously during the night, and on the morning (18th) gave no signs of a speedy cessation. Those who travel ought often, however, to call to mind the remark of Xenophon, that "pleasure is the result of toil," and not permit slight impediments to arrest them, particularly when they have definite points to make. We set forward in a moderate rain, but in less than an hour had the pleasure to perceive signs of its mitigating, and before nine o'clock it was quite clear. We stopped a short time at Bracken's furnace. Mr. Bracken gave me specimens of organic remains, in the condition of earthy calcareous carbonates, procured on a neighboring ridge. He described the locality as being plentiful in casts and impressions such as he exhibited, which appeared to have been removed from the surface of a shelly limestone. At Rock-Branch diggings, I found masses of calcareous spar thrown from the pits. The surface appears to have been much explored for lead in this vicinity. I stopped to examine Vanmater's lead. It had been a productive one, and affords a fair example of what are called east and west leads. I observed a compass standing on the line of the lead, and asked Mr. V. whether much reliance was to be placed upon the certainty of striking the lead by the aid of this instrument. He said that it was much relied on. That the course of the leads was definite. The present one varied from a due east and west line but nine minutes, and the lead had been followed without much difficulty. The position of the ore was about forty feet below the surface. Of this depth about thirty-six feet consisted of the surface rock and its earthy covering. A vein of marly clay, enveloping the ore, was then penetrated. A series of pits had been sunk on the course of it, and the earth and ore in the interstices removed, and drawn to the surface by a windlass and bucket. Besides the ore, masses of iron pyrites had been thrown out, connected with galena. In stooping to detach some pieces from one of these masses, I placed my feet on the verge of an abandoned pit, around which weeds and bushes had grown. My face was, however, averted from the danger; but, on beholding it, I was made sensible that the least deviation from a proper balance would have pitched me into it. It was forty feet deep. The danger I had just escaped fell to the lot of Mr. B.'s dog, who, probably deceived by the growth of bushes, fell in. Whether killed or not, it was impossible to tell, and we were obliged to leave the poor animal, under a promise of Mr. V., that he would cause a windlass to be removed to the pit, to ascertain his fate.

At eleven o'clock we reached Mineral Point, the seat of justice of Iowa county. I delivered an introductory letter to Mr. Ansley, who had made a discovery of copper ore in the vicinity, and through his politeness, visited the locality. The discovery was made in sinking pits in search of lead ore. Small pieces of green carbonate of copper were found on striking the rock, which is apparently silico-calcareous, and of a very friable structure. From one of the excavations, detached masses of the sulphuret, blue and green mingled, were raised. These masses are enveloped with ochery clay.

In riding out on horseback to see this locality, I passed over the ridge of land which first received the appellation of "Mineral Point." No digging was observed in process, but the heaps of red marly clay, the vigorous growth of shrubbery around them, and the number of open or partially filled pits, remain to attest the labor which was formerly devoted in the search for lead. And this search is said to have been amply rewarded. The track of discovery is conspicuously marked by these excavations, which often extend, in a direct line, on the cardinal points, as far as the eye can reach. Everywhere the marly clay formation appears to have been relied on for the ore, and much of it certainly appears to be *in situ* in it. It bears no traces of attrition; and its occurrence in regular leads forbids the supposition of its being an oceanic arrangement of mineral detritus. At Vanmater's, the metalliferous clay marl is overlaid by a grayish sedimentary limestone. Different is the geological situation of what is denominated *gravel ore*, of which I noticed piles, on the route from Gratiot's. This bears evident marks of attrition, and appears to have been uniformly taken from diluvial earth.

On returning to the village from this excursion, I found Mr. B. ready to proceed, and we lost no time in making the next point in our proposed route. A drive of five miles brought us to the residence of Colonel Dodge, whose zeal and enterprise in opening this portion of our western country for settlement, give him claims to be looked up to as a public benefactor. I here met the superintendent of the mines (Captain Legate), and after spending some time in conversation on the resources and prospects of the country, and partaking of the hospitalities politely offered by Colonel D. and his intelligent family, we pursued our way. The village of Dodgeville lies at the distance of four miles. Soon after passing through it some part of our tackle gave way, in crossing a gully, and I improved the opportunity of the delay to visit the adjacent diggings, which are extensive. The ore is found as at other mines, in regular leads, and not scattered about promiscuously in the red marl. Masses of brown oxide of iron were more common here than I had noticed them elsewhere. Among the rubbish of the diggings, fragments of hornstone occur. They appear to be, most commonly, portions of nodules, which exhibit, on being fractured, various discolorings.

Night overtook us before we entered Porter's Grove, which is also the seat of mining and smelting operations. We are indebted to the hospitality of Mr. M., of whom my companion was an acquaintance, for opening his door to us, at an advanced hour of the evening. Distance from Willow Springs, twenty-five miles.

There is no repose for a traveller. We retired to rest at a late hour, and rose at an early one. The morning (19th) was hazy, and we set forward while the dew was heavy on the grass. Our route still lay through a prairie country. The growth of native grass, bent down with dew, nearly covered the road, so that our horses' legs were continually bathed. The rising sun was a very cheerful sight, but as our road lay up a long ascent, we soon felt its wilting effects. Nine miles of such driving, with not a single grove to shelter us, brought us to Mr. Brigham's, at the foot of the Blue Mound, being the last house in the direction to Fort Winnebago. The distance from Galena is sixty-four miles, and this area embraces the present field of mining operations. In rapidly passing over it, mines, furnaces, dwelling-houses, mining villages, inclosed fields, upland prairies (an almost continued prairie), groves, springs, and brooks, have formed the prominent features of the landscape. The impulse to the settlement of the country was first given by its mineral wealth; and it brought here, as it were by magic, an enterprising and active population.

It is evident that a far greater amount of labor was a few years ago engaged in mining operations; but the intrinsic value of the lands has operated to detain the present population, which may be considered as permanent. The lands are beautifully disposed, well watered, well drained by natural streams, and easily brought into cultivation. Crops have everywhere repaid the labors of the farmer; and, thus far, the agricultural produce of the country has borne a fair price. The country appears to afford every facility for raising cattle, horses, and hogs. Mining, the cardinal interest heretofore, has not ceased in the degree that might be inferred from the depression of the lead market; and it will be pursued, with increased activity, whenever the purposes of commerce call for it. In the present situation of the country, there appear to be two objects essential to the lasting welfare of the settlements: first, a title to their lands from Congress; second, a northern market for the products of their mines and farms. To these, a *third* requisite may be considered auxiliary, namely, the establishment of the seat of territorial government at some point west of Lake Michigan, where its powers may be more readily exercised, and the reciprocal obligations of governor and people more vividly felt.

Mr. Brigham, in whom I was happy to recognize an esteemed friend, conducted us over his valuable plantation. He gave me a mass of a white, heavy metallic substance, taken as an accompanying mineral, from a lead of Galena, which he has recently discovered in a cave. Without instituting any examination of it but such as its external characters disclose, it may be deemed a native carbonate of lead. The mass from which it was broken weighed ninety or one hundred pounds. And its occurrence, at the lead, was not alone.

From the Blue Mound to Fort Winnebago is an estimated distance of fifty-six miles. The country is, however, entirely in a state of nature. The trace is rather obscure; but, with a knowledge of the general geography and face of the country, there is no difficulty in proceeding with a light wagon, or even a loaded team, as the Indian practice of firing the prairies every fall has relieved the surface from underbrush and fallen timber. After driving a few miles, we encountered two Winnebagoes on horseback, the forward rider having a white man in tie behind him.

The latter informed us that his name was H., that he had come
out to Twelve-mile Creek, for the purpose of locating himself
there, and was in pursuit of a hired man, who had gone off, with
some articles of his property, the night previous. With this relation, and a *boshu** for the natives, with whom we had no means
of conversing, we continued our way, without further incident, to
Duck Creek, a distance of ten miles. We here struck the path,
which is one of the boundary lines, in the recent purchase from
the Winnebagoes. It is a deeply marked horse path, cutting
quite through the prairie sod, and so much used by the natives as
to prevent grass from growing on it; in this respect, it is as well-
defined a landmark as "blazed tree," or "saddle." The sur-
veyor appointed to run out the lines, had placed mile-posts on the
route, but the Winnebagoes, with a prejudice against the practice
which is natural, pulled up many, and defaced others. When
we had gone ten miles further, we began to see the glittering of
water through the trees, and we soon found ourselves on the
margin of a clear lake. I heard no name for this handsome
sheet of water. It is one of the four lakes, which are connected
with each other by a stream, and have their outlet into Rock
River, through a tributary called the Guskihaw. We drove
through the margin of it, where the shores were sandy, and innumerable small unio shells were driven up. Most of these
small pieces appeared to be helices. Standing tent-poles, and
other remains of Indian encampments, appeared at this place. A
rock stratum, dark and weather-beaten, apparently sandstone,
jutted out into the lake. A little further, we passed to the left of
an abandoned village. By casting our eyes across the lake, we
observed the new position which had been selected and occupied
by the Winnebagoes. We often assign wrong motives, when we
undertake to reason for the Indian race; but in the present instance, we may presume that their removal was influenced by
too near a position to the boundary path.

We drove to the second brook, beyond the lake, and encamped.
Comfort in an encampment depends very much upon getting a

---

* This term is in use by the Algic or Algonquin tribes, particularly by the
Chippewas. The Winnebagoes, who have no equivalent for it, are generally acquainted with it, although I am not aware that they have, to any extent, adopted
it. It has been supposed to be derived from the French *bon jour*.

good fire. In this we totally failed last night, owing to our having but a small piece of spunk, which ignited and burned out without inflaming our kindling materials. The atmosphere was damp, but not sufficiently cooled to quiet the ever-busy mosquito. Mr. B. deemed it a hardship that he could not boil the kettle, so as to have the addition of tea to our cold repast. I reminded him that there was a bright moon, and that it did not rain; and that, for myself, I had fared so decidedly worse, on former occasions, that I was quite contented with the light of the moon and a dry blanket. By raising up and putting a fork under the wagon-tongue, and spreading our tent-cloth over it, I found the means of insulating ourselves from the insect hordes, but it was not until I had pitched my mosquito net within it that we found repose.

On awaking in the morning (20th), we found H., who had passed us the day before in company with the Winnebagoes, lying under the wagon. He had returned from pursuing the fugitive, and had overtaken us, after twelve o'clock at night. He complained of being cold. We admitted him into the wagon, and drove on to reach his camp at Twelve-mile Creek. In crossing what he denominated Seven-mile Prairie, I observed on our right a prominent wall of rock, surmounted with image-stones. The rock itself consisted of sandstone. Elongated water-worn masses of stone had been set up, so as to resemble, at a distance, the figures of men. The illusion had been strengthened by some rude paints. This had been the serious or the sportive work of Indians. It is not to be inferred, hence, that the Winnebagoes are idolaters. But there is a strong tendency to idolatry in the minds of the North American Indians. They do not bow before a carved image, shaped like Dagon or Juggernaut; but they rely upon their guardian spirits, or personal manitos, for aid in exigencies, and impute to the skins of animals, which are preserved with religious care, the power of gods. Their medicine institution is also a gross and bold system of semi-deification connected with magic, witchcraft, and necromancy. Their jossakeeds are impostors and jugglers of the grossest stamp. Their wabenos address Satan directly for power; and their metais, who appear to be least idolatrous, rely more upon the invisible agency of spirits and magic influence, than upon the physical properties of the medicines they exhibit.

On reaching Twelve-mile Creek, we found a yoke of steers of H., in a pen, which had been tied there two days and nights without water. He evinced, however, an obliging disposition, and, after refreshing ourselves and our horses, we left him to complete the labors of a "local habitation." The intermediate route to Fort Winnebago afforded few objects of either physical or mental interest. The upland soil, which had become decidedly thinner and more arenaceous, after reaching the Lake, appears to increase in sterility on approaching the Wisconsin. And the occurrence of *lost rocks* (primitive boulders), as Mr. B. happily termed them, which are first observed after passing the Blue Mound, becomes more frequent in this portion of the country, denoting our approach to the borders of the northwestern primitive formation. This formation, we have now reason to conclude, extends in an angle, so far south as to embrace a part of Fox River, above Apukwa Lake.

Anticipated difficulties always appear magnified. This we verified in crossing Duck Creek, near its entrance into the Wisconsin. We found the adjoining bog nearly dry, and drove through the stream without the water entering into the body of the wagon. It here commenced raining. Having but four miles to make, and that a level prairie, we pushed on. But the rain increased, and poured down steadily and incessantly till near sunset. In the midst of this rain-storm we reached the fort, about one o'clock, and crossed over to the elevated ground occupied by the Indian Department, where my sojourn, while awaiting the expedition, was rendered as comfortable as the cordial greeting and kind attention of Mr. Kinzie, the agent, and his intelligent family, could make it.

A recapitulation of the distances from Galena makes the route as follows, viz: Gratiot's Grove, fifteen miles; Willow Springs, fifteen; Mineral Point, seven; Dodgeville, nine; Porter's Grove, nine; Blue Mound, nine; Duck Creek, ten; Lake, ten; Twelve-mile Creek, twenty-four; Crossing of Duck Creek, eight; and Fort Winnebago, four; total, one hundred and twenty miles.

H. R. S.

To GEORGE P. MORRIS, ESQ., New York.

3. *Official Report of the Exploratory Expedition to the Actual Source of the Mississippi River in* 1832.

OFFICE OF THE INDIAN AGENCY OF SAULT STE. MARIE, }
 Sept. 1, 1832.

SIR: I had the honor to inform you, on the 15th ultimo, of my return from the sources of the Mississippi, and that I should communicate the details of my observations to you as soon as they could be prepared.

On reaching the remotest point visited heretofore by official authority, I found that the waters on that summit were favorable to my tracing this river to its utmost sources. This point having been left undetermined by prior expeditions, I determined to avail myself of the occasion to take Indian guides, with light canoes, and, after encamping my heavy force, to make the ascent. It was represented to be practicable in five days. I accomplished it, by great diligence, in three. The distance is 158 miles above Cass Lake. There are many sharp rapids, which made the trial severe. The river expands into numerous lakes.

After passing about forty miles north of Red Cedar Lake, during which we ascended a summit, I entered a fine large lake, which, to avoid repetitions in our geographical names, I called Queen Anne's Lake. From this point the ascent of the Mississippi was due south; and it was finally found to have its origin in a handsome lake, of some seven miles in extent, on the height of land to which I gave the name of Itasca.

This lake lies in latitude 47° 13' 25''. It lies at an altitude of 1,575 feet, by the barometer, above the Gulf of Mexico. It affords me satisfaction to say, that, by this discovery, the geographical point of the origin of this river is definitely fixed. Materials for maps and plans of the entire route have been carefully collected by Lieut. James Allen, of the U. S. Army, who accompanied me, with a small detachment of infantry, as high as Cass Lake; and, having encamped them at that point, with my extra men, he proceeded with me to Itasca Lake. The distance which is thus added to the Mississippi, agreeably to him, is 164 miles, making its entire length, by the most authentic estimates, to be 3,200 miles. In this distance there are numerous and arduous rapids, in which the total amount of ascent to be overcome is 178 feet.

Councils were held with the Indians at Fond du Lac, at Sandy Lake, Cass Lake, at the mouth of the Great De Corbeau River, &c.

In returning, I visited the military bands at Leech Lake; passing from thence to its source, and descending the whole length of the Crow-wing River, and thence to St. Anthony's Falls, I assembled the Sioux at the agency of St. Peter's, and at the Little Crow's village. The Chippewas of the St. Croix and Broule Rivers were particularly visited. Many thousands of the Chippewa and Sioux nations were seen and counselled with, including their most distinguished chiefs and warriors. Everywhere they disclaimed a connection with Black Hawk and his schemes. I left the Mississippi, about forty miles above the point where, in a few days, the Sauk chief was finally captured and his forces overthrown; and, reaching the waters of Lake Superior, at the mouth of the Brule, returned from that point to the agency at Sault de Ste. Marie.

The flag of the Union has secured respect from the tribes at every point; and I feel confident in declaring the Chippewas and Sioux, as tribes, unconnected with the Black Hawk movement.

I am, sir, very respectfully,
　　　　　Your obedient servant,
　　　　　　HENRY R. SCHOOLCRAFT,
　　　　　　　　　　　U. S. Ind. Agent.

C. HERRING, ESQ., Commissioner of Indian ⎫
　Affairs.　　　　　　　　　　　　　　　⎬

IV.

VACCINATION OF THE INDIANS.

4. Report of the number and position of the Indians vaccinated on the Exploratory Expedition to the Sources of the Mississippi, conducted by Mr. Schoolcraft, in 1832. By Dr. DOUGLASS HOUGHTON.

SAULT STE. MARIE, Sept. 21, 1832.

SIR: In conformity with your instructions, I take the earliest opportunity to lay before you such facts as I have collected, touching the vaccination of the Chippewa Indians, during the progress of the late expedition into their country: and also "of the prevalence, from time to time, of the smallpox" among them.

The accompanying table will serve to illustrate the "ages, sex, tribe, and local situation" of those Indians who have been vaccinated by me. With the view of illustrating more fully their local situation, I have arranged those bands residing upon the shores of Lake Superior; those residing in the Folle Avoine country (or that section of country lying between the highlands southwest from Lake Superior, and the Mississippi River); and those residing near the sources of the Mississippi River, separately.

Nearly all the Indians noticed in this table were vaccinated at their respective villages; yet I did not fail to vaccinate those whom we chanced to meet in their hunting or other excursions.

I have embraced, with the Indians of the frontier bands, those half-breeds, who, in consequence of having adopted more or less the habits of the Indian, may be identified with him.

But little difficulty has occurred in convincing the Indians of the efficacy of vaccination; and the universal dread in which they hold the appearance of the smallpox among them, rendered it an easy task to overcome their prejudices, whatever they chanced to be. The efficacy of the vaccine disease is well appreciated, even by the most interior of the Chippewa Indians; and so universal is this information, that only one instance occurred where the Indian had never heard of the disease.

In nearly every instance the opportunity which was presented for vaccination, was embraced with cheerfulness and apparent gratitude; at the same time manifesting great anxiety that, for the safety of the whole, each one of the band should undergo the operation. When objections were made to vaccination, they were not usually made because the Indian doubted the protective power of the disease, but because he supposed (never having seen its progress), that the remedy must nearly equal the disease which it was intended to counteract.

Our situation, while travelling, did not allow me sufficient time to test the result of the vaccination in most instances; but an occasional return to bands where the operation had been performed, enabled me, in those bands, either to note the progress of the disease, or to judge from the cicatrices marking the original situation of the pustules, the cases in which the disease had proved successful.

CHIPPEWA INDIANS. BANDS.	MALES. Under 10.	10 to 20.	20 to 40.	40 to 60.	60 to 80.	Over 80.	FEMALES. Under 10.	10 to 20.	20 to 40.	40 to 60.	60 to 80.	Over 80.	Males.	Females.	Total.
LAKE SUPERIOR															
Sault Ste. Marie	98	22	19	8	2	1	75	28	21	10	8	1	146	188	288
Grand Island	17	9	7	2	..	..	12	6	7	..	..	..	85	24	59
Keweena Bay	28	11	10	6	1	..	20	12	17	6	2	1	51	57	108
Ontonagon River	7	8	10	8	..	..	18	6	12	6	1	..	28	87	05
La Pointe	87	82	40	6	2	1	88	26	28	12	2	..	118	108	224
Fond du Lac	50	21	45	10	2	..	41	18	85	18	6	2	128	115	248
FOLLE AVOINE COUNTRY															
Lac du Flambeau	6	2	6	1	1	..	2	8	4	2	2	..	16	15	29
Ottowa Lake	11	4	8	1	..	..	10	7	8	2	..	..	24	22	46
Yellow River	11	2	6	1	..	..	11	8	6	2	1	..	20	28	48
Nama Kowagun of St. Croix River	4	1	2	1	..	..	4	..	8	2	..	..	8	0	17
Snake River	14	8	7	4	1	1	25	8	12	1	1	2	80	42	72
SOURCES OF THE MISSISSIPPI RIVER															
Sandy Lake	75	21	47	10	2	1	80	19	48	28	6	2	155	184	889
Lake Winnipeg	4	4	10	8	..	1	1	1	1	2	..	..	21	5	26
Cass, or Upper Red Cedar Lake	18	5	11	6	..	..	18	8	8	5	1	1	41	80	77
Leech Lake	76	48	78	16	4	1	66	41	61	25	2	1	218	226	480
Lake Superior	227	108	181	85	7	2	199	98	120	46	14	5	505	477	982
Folle Avoine Country	46	12	29	8	2	1	52	12	82	9	4	..	98	109	207
Sources of the Mississippi	178	78	141	85	6	2	201	64	118	55	9	4	480	451	881
Total	446	188	801	78	15	5	452	160	270	110	27	9	1088	1087	2070

About one-fourth of the whole number were vaccinated directly from the pustules of patients laboring under the disease; while the remaining three-fourths were vaccinated from crusts, or from virus which had been several days on hand. I did not pass by a single opportunity for securing the crusts and virus from the arms of healthy patients; and to avoid, as far as possible, the chance of giving rise to a disease of a spurious kind, I invariably made use of those crusts and that virus, for the purposes of vaccination, which had been most recently obtained. To secure, as far as possible, against the chances of escaping the vaccine disease, I invariably vaccinated in each arm.

Of the whole number of Indians vaccinated, I have either watched the progress of the disease, or examined the cicatrices of about seven hundred. An average of one in three of those vaccinated from crusts has failed, while of those vaccinated directly from the arm of a person laboring under the disease, not more than one in twenty has failed to take effect—when the disease did not make its appearance after vaccination, I have invariably, as the cases came under my examination, revaccinated until a favorable result has been obtained.

Of the different bands of Indians vaccinated, a large proportion of the following have, as an actual examination has shown, undergone thoroughly the effects of the disease; viz: Sault Ste. Marie, Keweena Bay, La Pointe, and Cass Lake, being seven hundred and fifty-one in number; while of the remaining thirteen hundred and seventy-eight, of other bands, I think it may safely be calculated that more than three-fourths have passed effectually under the influence of the vaccine disease: and as directions to revaccinate all those in whom the disease failed, together with instructions as to time and manner of vaccination, were given to the chiefs of the different bands, it is more than probable that, where the bands remained together a sufficient length of time, the operation of revaccination has been performed by themselves.

Upon our return to Lake Superior, I had reason to suspect, on examining several cicatrices, that two of the crusts furnished by the surgeon-general, in consequence of a partial decomposition, gave rise to a spurious disease, and these suspicions were confirmed when revaccinating with genuine vaccine matter, when the true disease was communicated. Nearly all those Indians

37

vaccinated with those two crusts, have been vaccinated, and passed regularly though the vaccine discase.

The answers to my repeated inquiries respecting the introduction, progress, and fatality of the smallpox, would lead me to infer that the disease has made its appearance at least five times, among the bands of Chippewa Indians noticed in the accompanying table of vaccination.

The smallpox appears to have been wholly unknown to the Chippewas of Lake Superior until about 1750; when a war-party, of more than one hundred young men, from the bands resident near the head of the lake, having visited Montreal for the purpose of assisting the French in their then existing troubles with the English, became infected with the disease, and but few of the party survived to reach their homes. It does not appear, although they made a precipitate retreat to their own country, that the disease was at this time communicated to any others of the tribe.

About the year 1770, the disease appeared a second time among the Chippewas, but, unlike that which preceded it, it was communicated to the more northern bands.

The circumstances connected with its introduction are related nearly as follows:—

Some time in the fall of 1767 or 8, a trader who had ascended the Mississippi, and established himself near Leech Lake, was robbed of his goods by the Indians residing at that lake; and, in consequence of his exertions in defending his property, he died soon after.

These facts became known to the directors of the Fur Company, at Mackinac; and, each successive year after, requests were sent to the Leech Lake Indians, that they should visit Mackinac, and make reparation for the goods they had taken, by a payment of furs, at the same time threatening punishment in case of a refusal. In the spring of 1770, the Indians saw fit to comply with this request; and a deputation from the band visited Mackinac, with a quantity of furs, which they considered an equivalent for the goods which had been taken. The deputation was received with politeness by the directors of the Company, and the difficulties readily adjusted. When this was effected, a cask of liquor and a flag closely rolled were presented to the Indians as a token of friendship. They were at the same time strictly enjoined neither to break the seal of the cask nor to unroll the flag, until they had reached the heart of their own country. This they promised to observe; but while returning, and after having travelled many days, the chief of the deputation made a feast for the Indians of the band at Fond du Lac, Lake Superior, upon which occasion he unsealed the cask and unrolled the flag for the gratification of his guests. The Indians drank of the liquor, and remained in a state of inebriation during several days. The rioting was over, and they were fast recovering from its effects, when several of the party were seized with violent pain. This was attributed to the liquor they had drunk; but the pain increasing, they were induced to drink deeper of the poisonous drug, and in this inebriated state several of the party died, before the real cause was suspected. Other like cases occurred; and it was not long before one of the war-party who had visited Montreal in 1750, and who had narrowly escaped with his life, recognized the disease as the same which had attacked their party at that time. It proved to be so; and of those Indians then at Fond du Lac, about three hundred in number, nearly the whole were swept off by it. Nor did it stop here; for numbers of those at Fond du Lac, at the time the disease made its appearance, took refuge among the neighboring bands; and although it did not extend easterly on Lake Superior, it is believed that not a single band of Chippewas north or west from Fond du Lac escaped its ravages. Of a large band then resident at Cass Lake, near the source of the Mississippi River, only one person, a child, escaped. The others having been attacked by the disease, died before any opportunity for dispersing was offered. The Indians at this day are firmly of the opinion that the smallpox was at this time communicated through the articles presented to their brethren by the agent of the Fur Company at Mackinac; and that it was done for the purpose of punishing them more severely for their offences.

The most western bands of Chippewas relate a singular allegory of the introduction of the smallpox into their country by a war-party, returning from the plains of the Missouri, as nearly as information will enable me to judge, in the year 1784. It does not appear that, at this time, the disease extended to the bands east of Fond du Lac; but it is represented to have been extremely fatal to those bands north and west from there.

In 1802 or 8, the smallpox made its appearance among the Indians residing at the Sault Ste. Marie, but did not extend to the bands west from that place. The disease was introduced by a voyager, in the employ of the Northwest Fur Company, who had just returned from Montreal; and although all communication with him was prohibited, an Indian imprudently having made him a visit, was infected with and transmitted the disease to others of the band. When once communicated, it raged with great violence, and of a large band scarcely one of those then at the village survived, and the unburied bones still remain, marking the situation they occupied. From this band the infection was communicated to a band residing upon St. Joseph's Island, and many died of it; but the surgeon of the military post then there, succeeded, by judicious and early measures, in checking it before the infection became general.

In 1824, the smallpox again made its appearance among the Indians at the Sault Ste. Marie. It was communicated by a voyager to the Indians upon Drummond's Island, Lake Huron; and through them several families at Sault Ste. Marie became infected. Of those belonging to the latter place, more than twenty in number, only two escaped. The disease is represented to have been extremely fatal to the Indians at Drummond's Island.

Since 1824, the smallpox is not known to have appeared among the Indians at the Sault Ste. Marie, nor among the Chippewas north or west from that place. But the Indians of these bands still tremble at the bare name of a disease which (next to the compounds of alcohol) has been one of the greatest scourges that has ever overtaken them since their first communication with the whites. The disease, when once communicated to a band of Indians, rages with a violence wholly unknown to the civilized man. The Indian, guided by present feeling, adopts a course of treatment (if indeed it deserves that appellation) which not unfrequently arms the disease with new power. An attack is but a warning to the poor and helpless patient to prepare for death, which will almost assuredly soon follow. His situation under these circumstances is truly deplorable; for while in a state that even, with proper advice, he would of himself recover, he adds fresh fuel to the flame which is already consuming him, under the delusive hope of gaining relief. The intoxicating

draught (when it is within his reach) is not among the last reme-
dies to which he resorts, to produce a lethargy from which he is
never to recover. Were the friends of the sick man, even under
these circumstances, enabled to attend him, his sufferings might
be, at least, somewhat mitigated; but they too are, perhaps, in a
similar situation, and themselves without even a single person to
minister to their wants. Death comes to the poor invalid, and,
perhaps, even as a welcome guest, to rid him of his suffering.

By a comparison of the number of Indians vaccinated upon
the borders of Lake Superior with the actual population, it will be
seen that the proportion who have passed through the vaccine
disease is so great as to secure them against any general prevalence of the smallpox; and perhaps it is sufficient to prevent the
introduction of the disease to the bands beyond, through this
channel. But in the Folle Avoine country it is not so. Of the
large bands of Indians residing in that section of country, only a
small fraction have been vaccinated; while of other bands, not a
single person has passed through the disease.

Their local situation undoubtedly renders it of the first importance that the benefits of vaccination should be extended to them.
Their situation may be said to render them a connecting link
between the southern and northwestern bands of Chippewas;
and while on the south they are liable to receive the virus of the
smallpox from the whites and Indians, the passage of the disease
through them to their more northern brethren would only be
prevented by their remaining, at that time, completely separated.
Every motive of humanity towards the suffering Indian, would
lead to extend to him this protection against a disease he holds in
constant dread, and of which he knows, by sad experience, the
fatal effects. The protection he will prize highly, and will give
in return the only boon a destitute man is capable of giving; the
deep-felt gratitude of an overflowing heart.

I have the honour to be,
Very respectfully, sir,
Your obedient servant,
DOUGLASS HOUGHTON.

HENRY R. SCHOOLCRAFT, ESQ.,
U. S. Ind. Agt., Sault de Ste. Marie.

4. TOPOGRAPHY AND GEOGRAPHY.

IX.

ASTRONOMICAL AND BAROMETRICAL OBSERVATIONS.

1. *A Table of Geographical Positions on the Mississippi River at Low Water, observed in* 1836.* By J. N. Nicollet.

PLACES OF OBSERVATION.	ESTIMATED DISTANCES BY WATER.		Altitudes above the Gulf of Mexico.†	North latitudes.			WEST OF GREENWICH.						Authorities, &c.
	From place to place.	From the Gulf of Mexico.					Longitudes in time.			Longitudes in arc.			
	Miles.	Miles.	Feet.	°	′	″	h.	m.	s.	°	′	″	
Mouths of the Mississippi—													
Northeast pass { The old Balize of the French and pilot-house	...	...	...	29	7	15.3	5	56	18.44	89	4	86.6	Captain A. Talcott.
Light-house at the	...	...	...	29	8	32.8	5	56	5.52	89	1	22.9	do.
South pass—light-house at the entrance	...	...	...	28	59	42.8	5	56	29.40	89	7	27.1	do.
The new Balize and pilot-house on the east bayou	...	...	...	28	59	49.5	5	57	16.88	89	18	58.2	do.
Southwest pass { The new light-house, completed January, 1840	...	...	...	28	58	50	5	57	25.80	89	21	27	do.
New Orleans Cathedral, and level of its front pavement	104	104	10.5	29	57	28	5	59	56	39	59	4	Albert Stein, C. E.
NOTE.—Level of the Mississippi above the Gulf of Mexico, 0.5 foot. Greatest depth of the Mississippi at low water, 118 feet. Range between high and low water, 18 feet.	...	...	...										
Red River, north end of the island, opposite the mouth	236	840	76	31	2	25	6	45		91	41	15	Nicollet.
Natchez, light-house	66	406	86	31	33	37	6	5	53.5	91	28	22.5	do.
general level of the city	...	...	264										
NOTE.—Range between high and low water, in 1835, 52 feet.													
Yazoo River, the mouth	128	584	...	32	28	00	6	8	58	90	59	80	Ferrer.

Locality			Height in feet†	Latitude °	′	″	Long. in time h	m	s	Longitude °	′	″	Authority
White River, Montgomery's Landing, one mile above the mouth	220	754	202	33	57	20	6	1	47	90	26	45	Nicollet.
New Madrid, Missouri	861	1,115		36	34	30	5	57	49	89	27	15	Ferrer.
Ohio River, north side of the mouth	101	1,216	324	37	00	25	5	56	10	89	2	30	Ferrer's longitude.
Cape Girardeau	41	1,257		37	18	39	5	57	8	89	17	00	Long's 1st expedition.
St. Genevieve, Catholic church, and level of its pavement	73	1,380	872	37	59	47	6	0	44.7	90	11	10	Nicollet.
St. Louis, garden of the Cathedral	60	1,390	882	38	37	28	6	1	2.6	90	15	39	do.
Illinois River, the mouth	36	1,426		38	58	12							Long's 1st expedition.
Moingonan River (Des Moines River), a small island at the mouth	168	1,594	444	40	21	43	6	6	10	91	32	30	Nicollet.
Montrose, or old Fort Des Moines, the mouth of the creek	15	1,609	470	40	30	34	6	6	4	91	31	00	do.
Flint River, the mouth, above Burlington	30	1,639	486	40	52	56				91	21	30	do.
Maskudeng, the middle mouth of the slough	39	1,678	505	41	14	47	6	5	26	91			do.
Rock Island, a quarter of a mile above Davenport's residence	44	1,722	528	41	31	50				90	29	00	do.
Head of the Upper Rapids, below Port Biron and Parkhurst	15	1,737	554	41	36	8	6	1	56	91	9	19.5	do.
Prairie du Chien (Kipi-saging), American Fur Company's house	195	1,932	612	43	8	6	6	4	37.3	91			do.
Summit of bluff on the eastern side of Prairie du Chien			1,010										do.
Cap-à-l'ail, the summit—height above the Mississippi, 355 feet	82	1,964	1,013										
Upper Iowa River, island at the mouth	14	1,978		43	29	26	6	4	40	91	10	00	do.
Hokah River (Root River), the mouth	28	2,001		43	47	00	6	4	46	91	11	30	do.
Prairie à la Crosse River, the mouth	8	2,004		43	49	00	6	4	56	91	14	00	do.
Sappah River, or Black River, opposite the old mouth	31	2,035	688	43	57	14	6	5	36	91	24	00	do.
Top of mountain on right bank, opposite the old mouth			1,214										do.
Dividing ridge between Sappah River and Prairie à la Crosse River, 6 miles east of Mississippi			1,103										do.

* Com. Doc. No. 237.

† The numbers in this column refer to the surface of the water in the Mississippi at the point mentioned, except when otherwise specially expressed.

TABLE OF GEOGRAPHICAL POSITIONS—Continued.
Mississippi River at Low Water.

PLACES OF OBSERVATION.	ESTIMATED DISTANCES BY WATER. From place to place.	From the Gulf of Mexico.	Altitudes above the Gulf of Mexico.	North latitudes.			WEST OF GREENWICH. Longitudes in time.			Longitudes in arc.			Authorities, &c.
	Miles.	*Miles.*	*Feet.*	°	′	″	*h.*	*m.*	*s.*	°	′	″	
Mountain Island, or *Montagne qui trempe à l'Eau* of the French	7	2,042	...	44	1	7	6	6	2	91	30	80	Nicollet.
Miniskah River, or White-water River	27	2,069	...	44	12	86	6	7	25	91	51	15	do.
Wazi-oju River, or Pinewood River (*Rivière aux Embarras* of the French)	1	2,070	...	44	13	20	6	7	22	91	50	80	do.
At Roque's, two and a half miles below Chippeway River	14	2,084	...	44	23	24	6	8	00	92	00	00	do.
Clear Water River, the mouth, northwest corner of Lake Pepin	...	...	...	44	36	20	6	9	40	92	25	00	do.
Reminicha (*Montagne la Grange* of the French), upper end of Lake Pepin	31	2,115	714	44	38	30	6	10	4	92	31	00	do.
Top of Reminicha	...	...	1,086				6	10		92			do.
La Hontan River, the mouth (Cannon River of the Americans, Canoe River of the French)	8	2,118	...	44	34	00	6	10	8	92	32	00	do.
St. Croix River, the mouth	32	2,150	729	44	45	80	6	11	00	92	45	00	do.
Upland on the banks of the Mississippi and Lake St. Croix	...	...	866										do.
St. Peter's, the mouth	42	2,192	744	41	52	46	6	12	19.6	93	4	54	do.
General level of the plateau on which Fort Snelling and the Indian agency stand	...	...	850										do.
Pilot Knob, the top	...	...	1,008										do.
Falls of St. Anthony, United States Cottage	8	2,200	856	44	58	40	6	12	42	93	10	80	do.
Ishkode-wabo River, or Rum River, the mouth	19	2,219	...	45	15	00							do.
Karishon River (Sioux), or Undeg River (Chippewas), Crow River of the Americans	10	2,229	...	45	16	00							do.

Place	No.	Feet		Lat. °	′	″	Long. h	m	s	Long. °	′	″	Authority
St. ... River, Wicha-niwa River of the Sioux	9	2,238	...	45	20	80	6	15	50	93	57	80	Nicollet
Migadiwin Creek, or War Creek, the mouth	18	2,256	...	45	18	14	6	16	80	94	7	80	do.
...nik River, or Clear-Water River, the mouth	24	2,280	...	45	24	25	6	16	48	94	12	00	do.
Round Island, at the lower end of Osakis Rapids	...	...	...	45	35	00	6	16	48	94	12	00	do.
Osakis River, the mouth	22	2,302	...	45	85	35	6	16	58	94	14	80	do.
...lb River, the mouth	8	2,305	...	45	87	00	6	17	14	94	18	80	do.
Pekushino River, the mouth	18	2,328	...	45	46	50	6	17	28	94	22	00	do.
Waboci River, or Swan River, a half mile above the mouth	18	2,841	1,098	45	54	80	6	17	4	94	16	00	do.
...os River, or Elk River, the mouth	10	2,851	...	46	4	00	6	17	15	94	18	45	do.
Nokay's River, the mouth	18	2,369	...	46	10	80	6	17		94	22		do.
Kagi-wigwan River, the mouth (Aile de Corbeau River of the French, ... River of the Americans) o.	12	2,881	1,180	46	16	50	6	17	81	94		45	do.
Nagadjika River, pp site the mouth	18	2,899	1,176	46	26	00	6			98	...		do.
Pine River, the mouth	80	2,429	...	46	35	80	6	13	80	98	22	80	do.
...w River, the mouth	65	2,494	1,253	46	40	10	6	12	88	93	9	80	do.
Sandy Lake River, the mouth	32	2,526	1,290	46	47	43	6	13	86	98	9	00	do.
w...n River, the mouth	88	2,564	1,340	46	00	00	6	13	47	98	26	45	do.
Kabikons, or Little Falls, the head of the falls	63	2,627	...	47	14	10	6	14	10	98	32	80	do.
...n River, or ... River, the mouth	21	2,648	...	47	11	4	6			98			do.
Eagle ... rush (Marais aux Nids d'Aigle of the French)	16	2,664	1,356	47	18	10	6	14	86	98	89	00	do.
...eh Lake River, the mouth	11	2,675	...	47	14	00	6	14	52	98	43	00	do.
Lake ..., the old trading-house on a tongue of land near the ...nce of the Mississippi	80	2,755	1,402	47	25	23	6	18	16	94	34	00	do.
Pemidji ..., or Lake Travers, the entrance of the Mississippi	45	2,800	1,456	47	28	46	6	19	22	94	50	80	do.
...den Lake, Schoolcraft's Island	90	2,800	1,575	47	13	35	6	20	8	95	2	00	do.
...st ...es of the Mississippi, at the summit of the Hauteurs de Terre, or dividing ridge, ...een the Mippi and Red River of the North	6	2,896	1,680										

TABLE OF GEOGRAPHICAL POSITIONS—Continued.
Regions of the Sources of the Mississippi.

PLACES OF OBSERVATION.	Altitudes above the Gulf of Mexico.	North latitudes.	WEST OF GREENWICH.		Authorities, &c.
			Longitudes in time.	Longitudes in arc.	
	Feet.	° ′ ″	h. m. s.	° ′ ″	
Gayashk River, or Little Gull River, the mouth . . .	1,131	46 18 50	6 17 44	94 26 00	Nicollet.
Gayashk Lake, or Little Gull Lake, end of Long Point .	1,152	46 24 28	6 17 30	94 22 30	do.
Kadicomeg Lake, or White-Fish Lake, the entrance of Pine River	1,192	46 40 25	6 16 10	94 2 30	do.
Lake Chanché, southwest end .	...	46 46 85	...	...	do.
Lake Eccleston, northwest end	...	46 57 00	...	...	do.
Leech Lake, Otter-tail Point .	1,880	47 11 40	6 17 20	94 20 00	do.
Leech Lake, the bay opposite Otter-tail Point . . .	...	47 7 22	6 17 28	94 22 00	do.
Kabekonang River, the junction of the upper fork, near the next-mentioned portage .	1,406	47 16 00	...	...	do.
Portage from Kabekonang River to La Place River, near the west end	1,540	47 15 00	...	...	do.
Assawa Lake, below the south end	1,532	47 12 10	6 19 40	94 55 00	do.
Highest ridge on the portage between Assawa Lake and Itasca Lake	1,695	...	...	...	do.
Cleared pine camp, on Leech Lake River	...	47 18 00	6 16 00	94 00 00	do.

5. SCENERY.

X.

(a) *Scenery of Lake Superior.* By HENRY R. SCHOOLCRAFT.

Few portions of America can vie in scenic attractions with this interior sea. Its size alone gives it all the elements of grandeur; but these have been heightened by the mountain masses which nature has piled along its shores. In some places, these masses consist of vast walls, of coarse gray, or drab-colored sandstone, placed horizontally, until they have attained many hundred feet in height above the water. The action of such an immense liquid area, forced against these crumbling walls by tempests, has caused wide and deep arches to be worn into the solid structure, at their base, into which the billows roll, with a noise resembling low-pealing thunder. By this means, large areas of the impending mass are at length undermined and precipitated into the lake, leaving the split and rent parts, from which they have separated, standing like huge misshapen turrets and battlements. Such is the varied coast, called the Pictured Rocks.

At other points of the coast, volcanic forces have operated, lifting up these level strata into positions nearly vertical, and leaving them to stand, like the leaves of a vast open book. At the same time, the volcanic rocks sent up from below, have risen in high mountains, with ancient gaping craters. Such is the condition of the disturbed stratification at the Porcupine Mountains.

The basin and bed of this lake act like a vast geological mortar, in which the masses of broken and fallen stones are whirled about and ground down, till all the softer ones, such as the sandstones, are brought into the state of pure yellow sand. This sand is driven ashore by the waves, where it is shoved up in long wreaths, and dried by the sun. The winds now take it up, and spread it inland, or pile it immediately along the coast, where it presents itself in mountain masses. Such are the great sand dunes of the Grande Sables.

There are yet other theatres of action for this sublime mass of inland waters, where the lake has manifested, perhaps, still more strongly, its abrasive powers. The whole force of its waters, under the impulse of a northwest tempest, is directed against prominent portions of the shore, which consist of black and hard volcanic rocks. Solid as these are, the waves have found an entrance in veins of spar, or minerals of softer texture, and have thus been led on their devastating course inland, tearing up large fields of amygdaloid, or other rock; or, left portions of them standing in rugged knobs, or promontories. Such are the east and west coasts of the great peninsula of Keweena, which have recently become the theatre of mining operations.

When the visitor to these remote and boundless waters comes to see this wide and varied scene of complicated geological disturbances and scenic magnificence, he is absorbed in wonder and astonishment. The eye, once introduced to this panorama of waters, is never done looking and admiring. Scene after scene, cliff after cliff, island after island, and vista after vista are presented. One day's scenes of the traveller are but the prelude to another; and when weeks, and even months, have been spent in picturesque rambles along its shores, he has only to ascend some of its streams, and go inland a few miles, to find falls, and cascades, and cataracts of the most beautiful or magnificent character. Go where he will, there is something to attract him. Beneath his feet are pebbles of agates; the water is of the most crystalline purity. The sky is filled, at sunset with the most gorgeous piles of clouds. The air itself is of the purest and most inspiring kind. To visit such a scene is to draw health from its purest sources, and while the eye revels in intellectual delights, the soul is filled with the liveliest symbols of God, and the most striking evidences of his creative power.

(b) *Letters of Mr. M. Woolsey, Southern Literary Messenger*, 1836. Oneöta, p. 322.

These spirited and graphic letters are unavoidably excluded. The evidence they bear to the purity of principle, justness of taste, and excellence of character of a young man, now no more, ought to preserve his name from oblivion. He accompanied me in 1831, as a volunteer, in a leisure moment, an admirer of nature, seeking health.

INDEX.

38

THE END.

CHIPPEWA INDIANS. BANDS.	MALES.						FEMALES.						Males.	Females.	Total.
	Under 10.	10 to 20.	20 to 40.	40 to 60.	60 to 80.	Over 80.	Under 10.	10 to 20.	20 to 40.	40 to 60.	60 to 80.	Over 80.			
LAKE SUPERIOR															
Sault Ste. Marie	93	22	19	8	2	1	75	28	21	10	8	1	145	188	288
Grand Island	17	9	7	2			12	5	7				85	24	59
Keweena Bay	23	11	10	6	1		20	12	17	5	2	1	51	57	108
Ontonagon River	7	8	10	8	2	1	18	5	12	6	1		28	87	65
La Pointe	87	32	40	6	2		88	25	28	12	2		118	106	224
Fond du Lac	50	21	45	10	2		41	18	35	18	6	2	128	115	248
FOLLE AVOINE COUNTRY															
Lac du Flambeau	6	2	6	1	1		2	8	4	2			16	15	29
Ottowa Lake	11	4	8	1			10	7	8	2			24	22	46
Yellow River	11	2	6	1			11	8	6	2	1		20	28	48
Nama Kowagun of St. Croix River	4	1	2	1			4		8				8	9	17
Snake River	14	8	7	4	1		25	8	12	1	1		80	42	72
SOURCES OF THE MISSISSIPPI RIVER															
Sandy Lake	75	21	47	10	2	1	86	19	48	28	6	2	155	184	889
Lake Winnipeg	4	4	10	8			1	1	1	2			21	5	26
Cass, or Upper Red Cedar Lake	18	5	11	6			18	8	8	5	1	1	41	86	77
Leech Lake	76	48	78	16	4	1	96	41	61	25	2	1	218	226	489
Lake Superior	227	108	181	35	7	2	199	93	120	46	14	5	505	477	982
Folle Avoine Country	46	12	29	8	2	1	52	12	32	9	4		98	109	207
Sources of the Mississippi	178	78	141	35	6	2	201	64	118	55	9	4	430	451	881
Total	446	188	301	78	15	5	452	169	270	110	27	9	1088	1087	2070

4. TOPOGRAPHY AND GEOGRAPHY.

IX.

ASTRONOMICAL AND BAROMETRICAL OBSERVATIONS.

1. *A Table of Geographical Positions on the Mississippi River at Low Water, observed in 1836.* By J. N. Nicollet.*

PLACES OF OBSERVATION.	ESTIMATED DISTANCES BY WATER. From place to place. (Miles.)	From the Gulf of Mexico. (Miles.)	Altitudes above the Gulf of Mexico.† (Feet.)	North latitudes. ° ' "	WEST OF GREENWICH. Longitudes in time. h. m. s.	Longitudes in arc. ° ' "	Authorities, &c.
Mouths of the Mississippi—							
Northeast pass { The old Balize of the French and pilot-house .	...	...	...	29 7 15.3	5 56 18.44	89 4 86.6	Captain A. Talcott.
Light-house at the entrance .	...	...	...	29 8 32.8	5 56 5.52	89 1 22.9	do.
South pass—light-house at the entrance .	...	...	...	28 59 42.3	5 56 29.40	89 7 27.1	do.
Southwest pass { The new Balize and pilot-house on the east bayon	...	...	...	28 59 49.5	5 57 16.88	89 18 58.2	do.
The new light-house, completed January, 1840	...	...	...	28 58 50	5 57 25.80	89 21 27	do.
New Orleans Cathedral, and level of its front pavement	104	104	10.5	29 57 23	5 59 56	89 59 4	
NOTE.—Level of the Mississippi above the Gulf of Mexico, 0.5 foot. Greatest depth of the Mississippi at low water, 118 feet. Range between high and low water, 18 feet.	...	...	...				Albert Stein, C. E.
Red River, north end of the island, opposite the mouth	286	840	76	31 2 25	6 45	91 41 15	Nicollet.
Natchez, light-house	66	406	80	31 33 37	6 5 53.5	91 28 22.5	do.
general level of the city	...	...	264				
NOTE.—Range between high and low water, in 1835, 52 feet.	...	...	...				
Yazoo River, the mouth .	128	584	...	32 28 00	6 8 58	00 50 80	Ferrer.

Locality	Intervening dist. (miles)	Total dist. (miles)	Feet	Lat. °	′	″	Long. h	m	s	Long. °	′	″	Authority
White River, Montgomery's Landing, one mile above the mouth	220	754	202	33	57	20	6	1	47	90	26	45	Nicollet.
New Madrid, Missouri	361	1,115		36	34	30	5	57	49	89	27	15	Ferrer.
Ohio River, north side of the mouth	101	1,216	824	37	00	25	5	56	10	89	2	30	Ferrer's longitude.
Cape Girardeau	41	1,257		37	18	39	5	57	8	89	17	00	Long's 1st expedition.
St. Genevieve, Catholic church, and level of its pavement	73	1,330	872	37	59	47	6	0	44.7	90	11	10	Nicollet.
St. Louis, garden of the Cathedral	60	1,390	882	38	37	28	6	1	2.6	90	15	39	do.
Illinois River, the mouth	36	1,426		38	58	12							Long's 1st expedition.
Moingonan River (Des Moines River), a small island at the mouth	168	1,594	444	40	21	43	6	6	10	91	32	30	Nicollet.
Montrose, or old Fort Des Moines, the mouth of the creek	15	1,609	470	40	30	34	6	6	4	91	31	00	do.
Flint River, the mouth, above Burlington	30	1,639	486	40	52	56							do.
Maskudeng, the middle mouth of the slough	39	1,678	505	41	14	47	6	5	26	91	21	30	do.
Rock Island, a quarter of a mile above Davenport	44	1,722	528	41	31	50							do.
H— of the Upper Rapids, below Port Biron and Parkhurst	15	1,737	554	41	36	8	6	1	56	90	29	00	do.
Prairie du Chien (Kipi-saging), American Fur Company's house	195	1,932	642	43	3	6	6	4	37.3	91	9	19.5	do.
Summit of bluff on the eastern side of Prairie du Chien			1,010										
Cap-à-l'ail, the summit—height above the Mississippi, 855 feet	82	1,964	1,018	43	29	26	6	4	40	91	10	00	do.
Upper Iowa River, island at the mouth	14	1,978		43	47	00	6	4	46	91	11	30	do.
Hokah River (Root River), the mouth	23	2,001		43	49	00	6	4	56	91	14	00	do.
Prairie à la Crosse River, the mouth	3	2,004	688	43	57	14	6	5	36	91	24	00	do.
Sappah River, or Black River, opposite the old mouth	31	2,035	688										do.
Top of mountain on right bank, opposite the old mouth			1,214										
Dividing ridge between Sappah River and Prairie à la Crosse River, 6 miles east of Mississippi			1,108										do.

* Com. Doc. No. 237.

† The numbers in this column refer to the surface of the water in the Mississippi at the point mentioned, except when otherwise specially expressed.

TABLE OF GEOGRAPHICAL POSITIONS—CONTINUED.
MISSISSIPPI RIVER AT LOW WATER.

PLACES OF OBSERVATION.	Estimated distances by water. From place to place. (Miles.)	From the Gulf of Mexico. (Miles.)	Altitudes above the Gulf of Mexico. (Feet.)	North latitudes. (° ′ ″)	West of Greenwich. Longitudes in time. (h. m. s.)	Longitudes in arc. (° ′ ″)	Authorities, &c.
Mountain Island, or *Montagne qui trempe à l'Eau* of the French	7	2,042	...	44 1 7	6 6 2	91 30 30	Nicollet.
Miniskah River, or White-water River	27	2,069	...	44 12 36	6 7 25	91 51 15	do.
Waxi-oju River, or Pinewood River (*Rivière aux Embarras* of the French)	1	2,070	...	44 13 20	6 7 22	91 50 30	do.
At Roque's, two and a half miles below Chippeway River	14	2,084	...	44 28 24	6 8 00	92 00 00	do.
Clear Water River, the mouth, northwest corner of Lake P in	...	...	...	44 36 20	6 9 40	92 25 00	do.
Reminicha (*la Grange* of the French), upper end of Lake Pepin	31	2,115	714	44 38 30	6 10 4	92 31 00	do.
Top of Reminicha	...	...	1,036	...	...	...	do.
La Hontan River, the mouth (Cannon River of the Americans, Canoe River of the French)	8	2,118	...	44 34 00	6 10 8	92 32 00	do.
St. Croix River, the mouth	82	2,150	729	44 45 30	6 11 00	92 45 00	do.
Upland on the banks of the Mississippi and Lake St. Croix	...	...	866	...	...	...	do.
St. Peter's, the mouth	42	2,192	744	44 52 46	6 12 19.6	93 4 54	do.
General level of the plateau on which Fort Snelling and the Indian agency stand	...	...	850	...	...	...	do.
Pilot Knob, the top	...	...	1,006	...	...	...	do.
Falls of St. Anthony, United States Cottage	8	2,200	856	44 58 40	6 12 42	93 10 80	do.
Ishkode-wabo River, or Rum River, the mouth	19	2,219	...	45 15 00	...	...	do.
Karishon River (Sioux), or Undeg River (Chippewas), Crow River of the Americans	10	2,229	...	45 16 00	...	...	do.

Locality		Elev.		Latitude			Long. (time)			Longitude			Nicollet.
St. Francis River, Wicha-niwa River of the Sioux	9	2,238	...	45	20	80	6	15	50	93	57	80	Nicollet.
Migadiwin Creek, or War Creek, the mouth	18	2,256	...	45	18	14	6	16	80	94	7	80	do.
Kawakomik River, or Clear-Water River, the mouth	24	2,280	...	45	24	25	6	16	48	94	12	00	do.
Round Island, at the lower end of Osakis Rapids	...	...	...	45	85	00	6	16	48	94	12	00	do.
Osakis River, the mouth	22	2,302	...	45	85	85	6	16	68	94	14	80	do.
Watab River, the mouth	8	2,305	...	45	87	85	6	17	14	94	18	80	do.
Pakushino River, the mouth	18	2,323	...	45	46	50	6	17	28	94	22	00	do.
Waboxi River, or Swan River, a half mile above the mouth	18	2,341	1,098	45	54	80	6	17	4	94	16	00	do.
Omoshkos River, or Elk River, the mouth	10	2,851	...	46	4	00	6	17	15	94	18	45	do.
Nokay's River, the mouth	18	2,369	...	46	10	80	6	17	81	94	22	45	do.
Kagi-wigwan River, the mouth (*Aile de Corbeau River* of the French, Crow-Wing River of the Americans)	12	2,381	1,130	46	16	50	6	17	81	94	22	45	do.
Nagadjita River, opposite the mouth	18	2,399	...	46	26	00	6	13	80	93	22	80	do.
Pine River, the mouth	30	2,429	1,176	48	85	00	6	12	88	93	9	80	do.
Willow River, the mouth	65	2,494	...	46	40	80	6	12	86	93	9	00	do.
Sandy Lake River, the mouth	32	2,526	1,253	46	47	10	6	13	47	93	26	45	do.
Swan River, the mouth	38	2,564	1,290	47	00	48	6	14	10	93	32	80	do.
Kabikons, or Little Falls, the head of the falls	63	2,627	1,840	47	14	50	6	14	86	93	39	00	do.
Wanomon River, or Vermilion River, the mouth	21	2,648	...	47	11	4	6	14	52	93	48	00	do.
Eagle Nest savannah (*Marais aux Nids d'Aigle* of the French)	16	2,664	...	47	18	10	6	18	16	94	34	00	do.
Leech Lake River, the mouth	11	2,675	1,356	47	14	00	6	18	89	98	89	48	do.
Lake Cass, the old trading-house on a tongue of land near the entrance of the Mississippi	30	2,755	1,402	47	25	28	6	18	16	94	84	00	do.
Pemidji Lake, or Lake Travers, the entrance of the Mississippi	45	2,800	1,456	47	28	46	6	19	22	94	50	80	do.
Itasca Lake, Schoolcraft's Island	90	2,890	1,575	47	18	85	6	20	8	95	2	00	do.
Utmost sources of the Mississippi, at the summit of the Hauteurs de Terre, or dividing ridge, between the Mississippi and Red River of the North	6	2,896	1,680										do.

INDEX.

THE END.